Lecture Notes in Civil Engineering

Volume 187

Lecture Notes in Civil Engineering (LNCE) publishes the latest developments in Civil Engineering—quickly, informally and in top quality. Though original research reported in proceedings and post-proceedings represents the core of LNCE, edited volumes of exceptionally high quality and interest may also be considered for publication. Volumes published in LNCE embrace all aspects and subfields of, as well as new challenges in, Civil Engineering. Topics in the series include:

- Construction and Structural Mechanics
- Building Materials
- Concrete, Steel and Timber Structures
- Geotechnical Engineering
- Earthquake Engineering
- Coastal Engineering
- Ocean and Offshore Engineering; Ships and Floating Structures
- Hydraulics, Hydrology and Water Resources Engineering
- Environmental Engineering and Sustainability
- Structural Health and Monitoring
- Surveying and Geographical Information Systems
- Indoor Environments
- Transportation and Traffic
- Risk Analysis
- Safety and Security

To submit a proposal or request further information, please contact the appropriate Springer Editor:

- Pierpaolo Riva at pierpaolo.riva@springer.com (Europe and Americas);
- Swati Meherishi at swati.meherishi@springer.com (Asia - except China, and Australia, New Zealand);
- Wayne Hu at wayne.hu@springer.com (China).

All books in the series now indexed by Scopus and EI Compendex database!

More information about this series at https://link.springer.com/bookseries/15087

T. G. Sitharam · Sreevalsa Kolathayar · Ravi Jakka
Editors

Earthquake Geotechnics

Select Proceedings of 7th ICRAGEE 2021

Springer

Editors
T. G. Sitharam
Indian Institute of Technology Guwahati
Guwahati, Assam, India

Sreevalsa Kolathayar
Department of Civil Engineering
National Institute of Technology
Surathkal, Karnataka, India

Ravi Jakka
Department of Earthquake Engineering
Indian Institute of Technology Roorkee
Roorkee, Uttarakhand, India

ISSN 2366-2557 ISSN 2366-2565 (electronic)
Lecture Notes in Civil Engineering
ISBN 978-981-16-5668-2 ISBN 978-981-16-5669-9 (eBook)
https://doi.org/10.1007/978-981-16-5669-9

This Springer imprint is published by the registered company Springer Nature Singapore Pte Ltd.
The registered company address is: 152 Beach Road, #21-01/04 Gateway East, Singapore 189721, Singapore

Preface

Earthquake geotechnics deals with seismic problems in geotechnical engineering and dynamic behavior of soils under earthquake vibrations. This volume presents select papers presented at the 7th International Conference on Recent Advances in Geotechnical Earthquake Engineering and Soil Dynamics. The papers discuss advances in the fields of soil dynamics and geotechnical earthquake engineering. Some of the themes include ground response analysis and local site effect, seismic slope stability and landslides, application of AI in geotechnical earthquake engineering, etc. A strong emphasis is placed on connecting academic research and field practice, with many examples, case studies, best practices, and discussions.

We thank all the staff of Springer for their full support and cooperation at all the stages of the publication of this book. We thank and acknowledge the service of authors and reviewers for their valuable time and efforts. We do hope that this book will be beneficial to students, researchers, and professionals working in the field of geotechnical earthquake engineering. The comments and suggestions from the readers and users of this book are most welcome.

Guwahati, India — T. G. Sitharam
Surathkal, India — Sreevalsa Kolathayar
Roorkee, India — Ravi Jakka

Contents

About the Editors

Prof. Dr. T. G. Sitharam is the Chairman of 7th ICRAGEE, and President of Indian Society for Earthquake Technology. Presently, he is the Director of Indian Institute of Technology, Guwahati, Assam since July 2019. He is a Senior Professor at Department of Civil Engineering, Indian Institute of Science, Bengaluru. He is also the Director of Central Institute of Technology, Kokrajhar, Assam (additional charge) since May 2021. He holds Masters in Civil engineering from Indian Institute of Science and Ph.D. from University of Waterloo, Canada and worked for his postdoctoral at University of Texas, USA. He has served as a visiting or honorary/visiting professor at several overseas universities. He has guided 40 Ph.D., more than 35 Masters Students, 20 postdoctoral students and several thousand industry professionals and teachers through continuing education workshops. Over the last 35 years, he has carried out research and development in the area of geotechnical and infrastructure engineering, seismic microzonation & soil dynamics and geotechnical earthquake engineering leading to about 500 technical papers, 20 books and five patents. He has two start-up companies and many technological innovations to his credit. Prof Sitharam has recently delivered the IGS annual lecture in IGC 2020 and recipient of several National and International awards. In 2020, He was listed in the world's top 2% of scientists for the most-cited scientists (with more than 6700 citations with an H-index of 46) in various disciplines by Stanford University in 2020. He is the Fellow of American Society of Civil Engineering, Fellow of the Institution of Civil Engineers (UK), Diplomate of Geotechnical Engineering from Academy of Geoprofessionals, ASCE, USA, and holds many other fellowships to his credit.

Dr. Sreevalsa Kolathayar pursued his M.Tech. from Indian Institute of Technology (IIT) Kanpur, Ph.D. from Indian Institute of Science (IISc) and served as International Research Staff at UPC BarcelonaTech Spain. He is presently Assistant Professor in the Department of Civil Engineering, National Institute of Technology (NIT), Karnataka. Dr. Kolathayar has authored six books and over 75 research articles. He is Associate Editor of two international journals. His broad research areas are geotechnical earthquake engineering, geosynthetics & geonaturals, and water geotechnics. He is currently the Secretary of the Indian chapter of International Association for

Coastal Reservoir Research (IACRR), and Executive Committee Member of Indian Society of Earthquake Technology. In 2017, The New Indian Express honored Dr. Kolathayar with 40 under 40—South India's Most Inspiring Young Teachers Award. He is the recipient of ISET DK Paul Research Award from Indian Society of Earthquake Technology, IIT Roorkee. He received "IEI Young Engineers Award" by The Institution of Engineers (India), in recognition of his contributions in the field of Civil Engineering. He was recently featured in Geostrata Magazine by American Society of Civil Engineers (ASCE). Dr. Sreevalsa is the Organizing Secretary of 7th International Conference on Recent Advances in Geotechnical Earthquake Engineering.

Prof. Ravi Jakka is working as Associate Professor in the Department of Earthquake Engineering, Indian Institute of Technology, Roorkee. He is also currently serving as Secretary, Indian Society of Earthquake Technology (ISET). He has graduated in Civil Engineering from Andhra University Engineering College in the year 2001. He has obtained masters and doctorate degrees from IIT Delhi in the years 2003 and 2007 respectively. His areas of interest are Dynamic Site Characterization, Soil Liquefaction, Seismic Slope Stability of Dams, Landslides, Foundations & Seismic Hazard Assessment. He has published over 100 articles in reputed international journals and conferences. He has supervised over 35 Masters Dissertations and six Ph.D. thesis, while he is currently guiding 10 Ph.D. Thesis. He has received prestigious DAAD and National Doctoral fellowships. He has obtained University Gold Medal from Andhra University. He also received 'Young Geotechnical Engineer Best Paper Award' from Indian Geotechnical Society. He was instrumental in the development of Earthquake Early Warning System for northern India, a prestigious national project. He is also the Organizing Secretary to 7th International Conference on Recent Advances in Geotechnical Earthquake Engineering. Recently, he took the responsibilities of the *International Journal of Geotechnical Earthquake Engineering* as an editor.

Control of Liquefaction Potential by Geosynthetic Reinforcements—A Study

Rajiv Chauhan and Satyendra Mittal

Abstract One of the major reasons for collapse of structures during earthquakes is the loss of strength due to generation of excess pore pressure and subsequent liquefaction of saturated sandy soils. The tests carried out on shake table studies indicate that geogrid and geocell can control the liquefaction behavior of sand when same is reinforced with these materials. The tests carried out on saturated sand at relative density of 35% show that average excess pore pressure falls from 3.31 to 2.41 kN/m^2 and 2.2 kN/m^2 after inclusion of 4 and 5 geogrid layers, respectively, at 0.1 g acceleration. For geocells, the excess pore pressure reduced to 2 kN/m^2 at same density and acceleration. With use of geocells, the liquefaction resistance increased by 31.6 at 35% density and 0.3 g acceleration. At 65% density, the liquefaction resistance increased by 26 and 35% with 4 and 5 layers of geogrid and by 41% with use of geocell, respectively. Effect of surcharge was also studied on reinforced sand to simulate actual field conditions. For surcharge of 5.94 kN/m^2 at 50% density, the average liquefaction resistance of saturated sand increased by 28 and 37% with 4, 5 layers of geogrid and by 43% for geocells, respectively, at 0.3 g acceleration.

Keywords Liquefaction · Geogrid layers · Geocell · Surcharge pressure · Shake table

1 Introduction

The enormous damage potential associated with liquefaction has been cause reported in many geotechnical studies of earthquake—induced liquefaction. Efforts have been made for developing procedures to evaluate the potential of liquefaction during earthquakes, thereby either limiting or eliminating damage during liquefaction. The phenomenon and factors of the liquefaction have been studied and developed by

R. Chauhan (✉)
Inder Kumar Gujral Punjab Technical University, Kapurthala, Punjab, India
e-mail: dr.rajivchauhan@ptu.ac.in

S. Mittal
Indian Institute of Technology Roorkee, , Roorkee, Uttrakhand, India

T. G. Sitharam et al. (eds.), *Earthquake Geotechnics*, Lecture Notes in Civil Engineering 187, https://doi.org/10.1007/978-981-16-5669-9_1

Florin and Ivanov [8], Seed and Lee [21], Seed and Idriss [22-23], Prakash and Gupta [20], Finn et al. [6, 7], Casagrande [4], DeAlba et al. [5], Gupta [10], Mittal [17], Singh [25] and several others. When sand is reinforced with short fibers and given cyclic loading, the presence of fiber leads to an increase in the number of cycles required to cause liquefaction under undrained loading conditions [14, 15, 18]. Soil reinforcement has also been considered as anti-liquefaction measures [1, 12, 13, 19, 27]. The liquefaction potential of flyash during earthquake was studied by Boominathan and Hari [3] and Sing et al. [26] which was reduced with use of fibers and stone columns, respectively. Gang et al. [9] modeled geosynthetic-encased column for liquefaction mitigation. Granular piles for liquefaction prevention have also been extensively used worldwide because of high permeability of granular material and shortening drainage path to dissipate excess pore pressure quickly during earthquakes. The liquefaction studies with geogrid as planar reinforcement and geocell as three-dimensional reinforcement in shake table are scanty. In the present paper, the liquefaction behavior of sand reinforced with HDPE geogrid layers and geocells has been presented. To understand the effect of reinforcement and sand density on the liquefaction behavior of the reinforced sand, a series of shake table tests were carried out on saturated samples with different accelerations, surcharge loads 2.0 material used in study.

2 Material Used in Study

2.1 Sand and Its Properties

The sand (classification as SP) collected locally from the bed of River Solani was used in experimental studies, as it contains fines on which liquefaction effect could be seen clearly. The sand was cleaned and air dried before use. The sand with specific gravity of 2.65 had less than 5% particles passing through IS: 75 micron sieve and showed no cohesion. From particle size distribution curve (Fig. 1), coefficient of uniformity was determined as 1.8 and D_{50} as 0.27.

2.2 Reinforcement

Geogrid

The uniaxial geogrid made from high-strength polyester yarns with black PVC coating having carbon black as 2%, and ultimate tensile strength of 88.7 kN/m was used as reinforcing material in the study (Fig. 2a). It was a stiff grid with rectangular openings of size 220 × 17 mm and unit weight of 6.5 N/mm^2. The geogrid sheets of size 800 × 400 mm were placed in shake table at vertical spacings of 100 mm

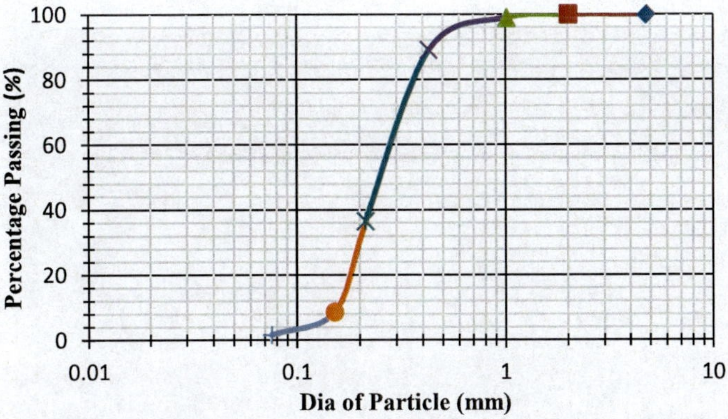

Fig. 1 Particle size distribution curve

c/c (in case of 4 geogrid layers) and 83 mm c/c for 5 geogrid layers, respectively, in 500 mm depth of sample. Its properties are summarized in Table 1.

Geocell

A typical geocell used in the study is shown in Fig. 2b. Geocells were having plan size of 350 × 350 mm with diamond pattern and height as 350 mm. Geocell had three-dimensional forms having interconnected cells. The geocells were filled with aggregates and placed at bottom of test tank.

2.3 Test Tank

The test bin is a water tight tank 1060 mm long, 600 mm wide, and 600 mm high as shown in Fig. 3.

The sides of the tank consist of rigid mild steel frame with 5-mm-thick steel panels. This tank is mounted on a horizontal shake table. The platform with wheels rests on four knife edges being rigidly fixed on two pairs of rails anchored to the foundation. This is driven in horizontal direction by a 3 H.P. A.C. motor through crank mechanism, for changing rotary motion into translatory motion. The amplitude of motion can be changed through two eccentric shafts. By changing the relative position of two shafts, the amplitude can be fixed as desired. The hand brake assembly is used for stopping the shake table. The pore pressure transducers were tried in experimental work for reading rise in pore water pressure, but due to overburden and sensitivity of the transducer, the observations given by transducers were not satisfactory. Therefore, the pore pressure measurement was performed with the help of glass tube piezometer of 5 mm diameter, attached to the tank through rubber tubes at heights of 80, 180, and 260 mm (B, M, T). These tubes were attached to steel pipes placed in the tank through side of wall of tank. At mouth of each steel pipe, a porous stone was wrapped,

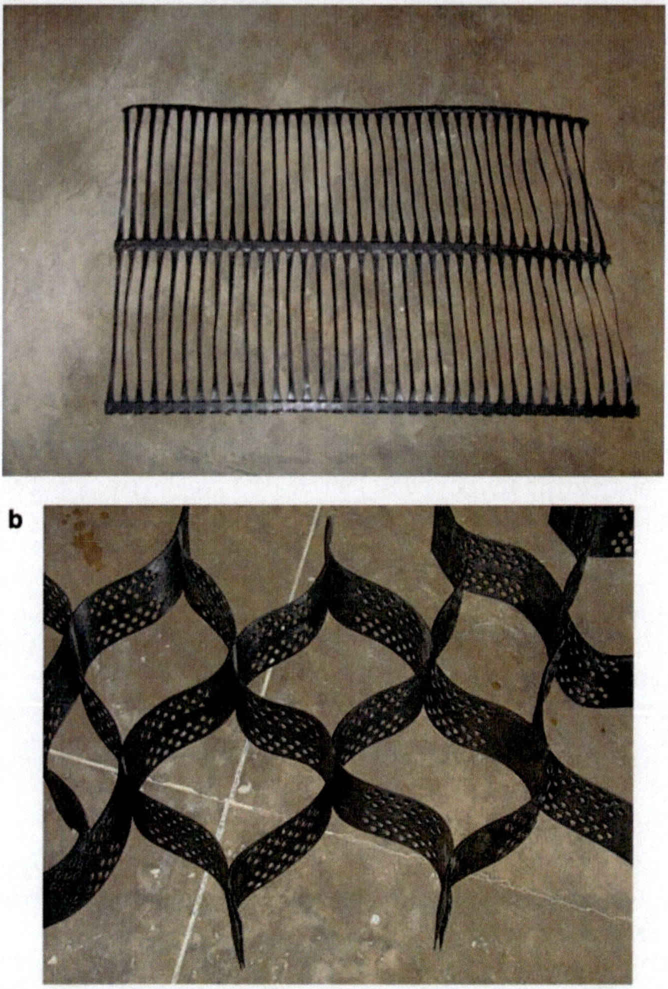

Fig. 2 **a** Geogrid, **b** geocell

Table 1 Properties of geogrid

Properties of geogrid	Value
Polymer	High-density polyethylene
Carbon black	2%
Ultimate strength (T_{ult})	88.7 kN/m
Unit weight	6.5 N/m^2
Typical strain at T_{ult}	11.50%

Fig. 3 View of shake table, surcharge weights, and piezometers

so as to allow rise in pore water pressure through this stone, and finally, it could be measured through piezometer. This technique was also adopted by Singh et al. [16], Maheshwari and Patel [26]. The shake table could produce steady state vibrations. One-dimensional harmonic excitation can be generated. The maximum amplitude of horizontal acceleration which could be generated in the shake table was up to 0.3 g. The shake table could be shaked at 4, 8, and 12 Hz frequency.

2.4 Surcharge Loads

To simulate the behavior of a sample at some depth below ground level, tests were conducted using surcharge loads also. The surcharge applied was through precast concrete blocks. Each block was weighing approximately 180 kg. These were placed on a 10-mm-thick steel plate placed on the soil sample for applying the load on to the soil sample uniformly. These concrete blocks were loaded and unloaded for test with the help of a chain pulley block [17]. The blocks were rigidly connected to each other through 4 Nos. anchor bolts and two channels, so that their position was not disturbed under the action of dynamic loads.

2.5 Preparation of Sample

Firstly, 150 L of clean water was filled in the tank. Air bubbles in the piezometer tube were removed. Porous stone wrapped in filter paper was tied at mouth of steel pipe attached with piezometer through a rubber tube, so that no sand particles pass and choke the tube. The porous stone before placing in steel tube was kept in boiled water for 5–10 min to remove air present, if any, in the pores of porous stone. After that, 480 kg of air dried sand was allowed to fall in tank to obtain 500 mm height of sample with the help of a funnel. After all the sand had settled in tank, the water overlying the sand deposit was removed by using a sponge. This water removed from top of sand layer was weighed for density calculation.

2.6 Initial Relative Density of Sample

Tests were performed keeping the relative density (D_r) of sand as 35, 50, and 65%, respectively. The density in the sample was maintained as per guidelines of ASTM D 5311–92(Reapproved [2]) and Ishihara [11]: Firstly 150 L of water was filled in the tank, which was sufficient to submerge all three piezometer tubes. The relative density (D_r) of sand is defined by:

$$D_r = (e_{max} - e)/(e_{max} - e_{min}) \tag{1}$$

where e_{max} is the maximum void ratio, e_{min} is the minimum void ratio, and e is the desired void ratio at a particular D_r. The void ratio (e) corresponding to $D_r = 35\%$ was calculated using:

$$e = e_{max} - D_r(e_{max} - e_{min}) \tag{2}$$

The values of e_{max} and e_{min} were 0.86 and 0.48, respectively, for the test sand used in the study. Knowing the value of void ratio (e) from above, dry unit weight of sand (γ_d) was determined by the following equation:

$$\gamma_d = G.\gamma_w/(1 + e) \tag{3}$$

For the soil used in this study, $G = 2.65$ and $\gamma_w = 9.81$ kN/m^3. The value of dry unit weight (γ_d) for D_r as 35, 50, and 65% were 15.34 kN/m^3, 15.86 kN/m^3, and 16.43 kN/m^3, respectively. Taking constant height of sand sample in the tank, i.e., 500 mm, the volume (V) occupied by the sand in the tank was determined using the plan dimensions of tank, i.e., 1060 × 600 mm. The dry weight of sand (W_d) was determined by the equation:

$$W_d = (\gamma_d \cdot V) \tag{4}$$

It was found that weight of dry sand required for $D_r = 35$, 50, and 65% were 484 kg, 500 kg, and 517 kg, respectively.

2.7 Shaking Duration

Seed and Idriss [24] gave empirical correlation of magnitude of earthquake to its duration as follows.

Magnitude	6	6.5	7	7.5	8	8.5
Duration (sec)	5	8	11.5	15	21	26

From the above table, we can say that duration and magnitude of earthquake relation can be used for dynamic studies related to soil liquefaction. This criterion for shaking duration was adopted in the study.

2.8 Test Procedure

The tests had been performed at an acceleration values of 0.1–0.3 g and at a frequency of 4 Hz. Before imparting the shaking, the values of pore water pressures rise in all the three piezometer were recorded. For an earthquake of 7.5 magnitude, the duration is 15 s as per Seed and Idriss [24], and therefore, same time duration for shaking was adopted in the present study. The rise in pore water pressures were recorded after shaking, till these got constant. This procedure was adopted in studying the liquefaction behavior of sand. The tests were conducted for cases given in Table 2.

3 Observations and Results

The tests results for different cases (Table 2) are given in Table 3, Figs. 5, 6, and 7. The pore water pressures were recorded at three locations points in the tank. The rise of pore water for virgin soil shows that with increase in acceleration, there is increase in excess pore pressure with maximum at lowest level piezometer also on account of higher overburden pressure (Fig. 5). This behavior is similar to studies conducted by Gupta [10].

Excess pore pressure ratio r_u has been used for liquefaction criteria, which is ratio of excess pore pressure to effective overburden pressure. Effective overburden pressure for bottom, middle, and top, i.e., σ_y' (B), σ_y' (M), σ_y' (T) was calculated by:

Table 2 Testing program

Soil type	D_r (%)	Acceleration (g)	Surcharge	Remarks
Virgin	35	0.1 0.2 0.3	Nil	Unreinforced soil
Virgin	50	0.1 0.2 0.3	Nil	Unreinforced soil
Virgin	65	0.1 0.2 0.3	Nil	Unreinforced soil
Soil with 4 and 5 layers of geogrid each @100 mm and 83 mm c/c, respectively	35	0.1 0.2 0.3	Nil	Reinforced soil
-Do-	50	0.1 0.2 0.3	Nil	Reinforced soil
-Do-	65	0.1 0.2 0.3	Nil	Reinforced soil
Soil with geocells	35	0.1 0.2 0.3	Nil	Reinforced soil
-Do-	50	0.1 0.2 0.3	Nil	Reinforced soil
-Do-	65	0.1 0.2 0.3	Nil	Reinforced soil
Soil with 4, 5 layers of geogrid and geocell	35	0.1 0.2 0.3	5.94 kN/m^2	Reinforced soil with surcharge
Soil with 4, 5 layers of geogrid and geocell	50	0.1 0.2 0.3	5.94 kN/m^2	Reinforced soil with surcharge
Soil with 4, 5 layers of geogrid and geocell	65	0.1 0.2 0.3	5.94 kN/m^2	Reinforced soil with surcharge

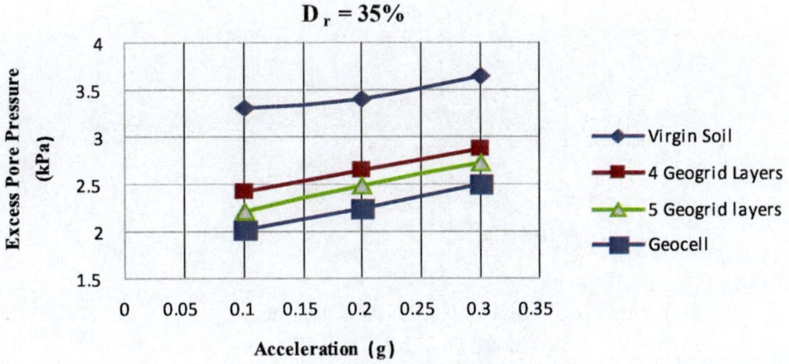

(a) For Dr = 35%

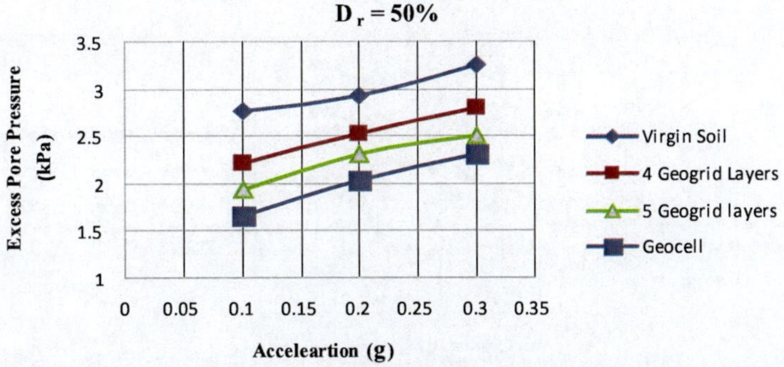

(b) For Dr = 50%

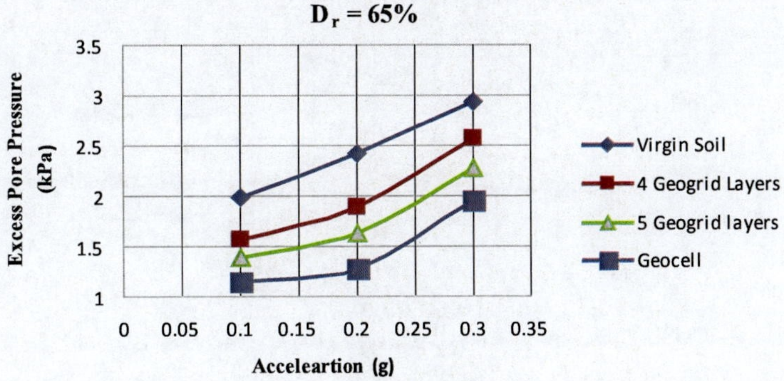

(c) For D = 65%

Fig. 4 Excess pore pressure versus acceleration

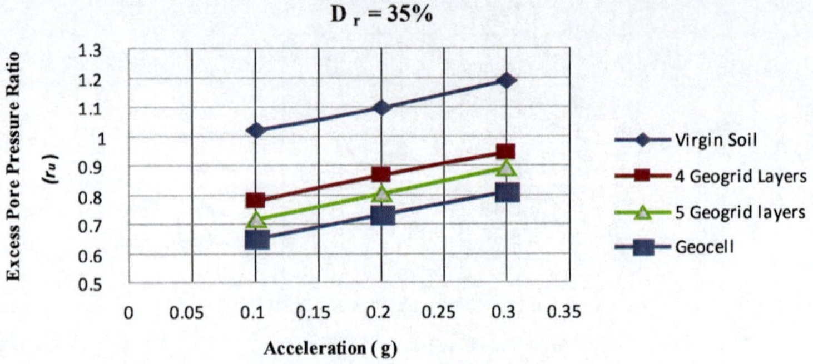

(a) For D_r = 35%

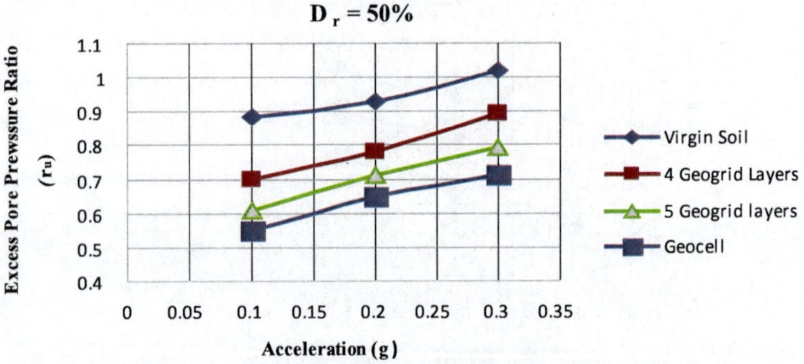

(b) For D_r = 50 %

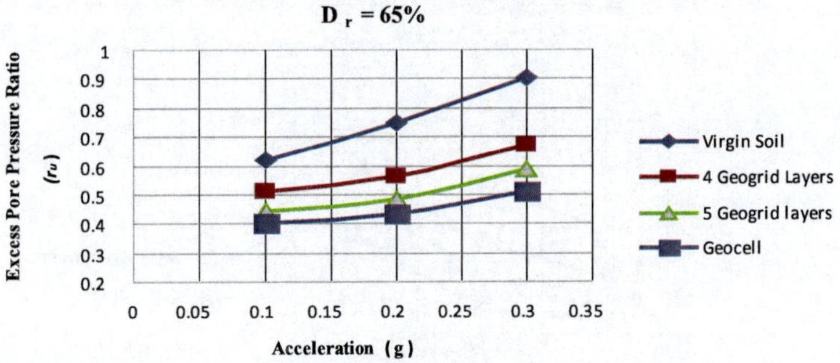

(c) For D_r = 65%

Fig. 5 Excess pore pressure ratio versus acceleration

$$(G - 1) \cdot \gamma_w \cdot h/(1 + \mathrm{e}) \tag{5}$$

where values of h were 420, 320, and 240 mm, respectively for B, M, T locations. Further e was the void ratio for D_r considered, γ_w unit weight of water. If this value is equal to or greater than 1.0, liquefaction is likely to occur, and if it is below 1.0,

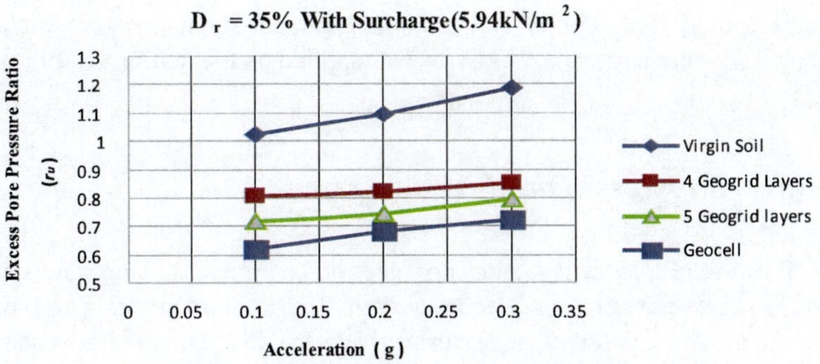

(a) For D_r =35% with Surcharge Load

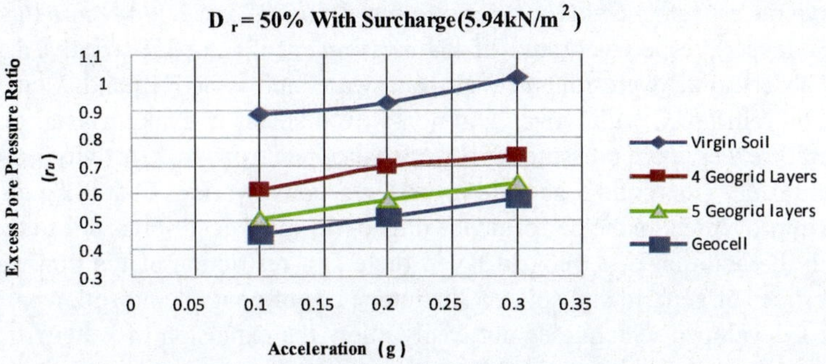

(b) For D_r = 50% with Surcharge Load

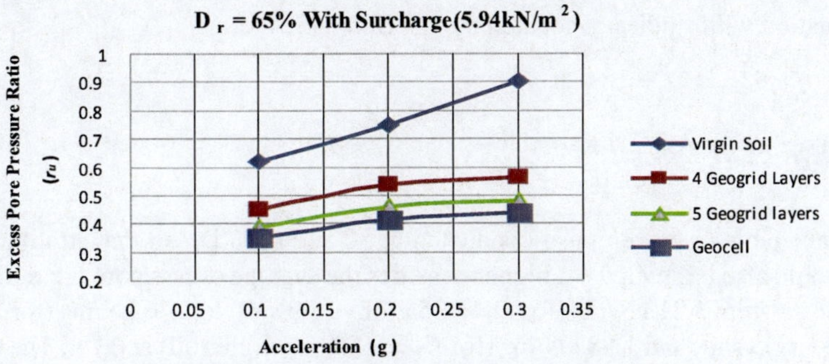

(c) For Dr = 65% with Surcharge Load

Fig. 6 Excess pore pressure ratio for surcharge load versus acceleration

then there will not be any liquefaction From Fig. 6, it is clear that average excess pore pressure ratio r_u varies from 1.10–1.18 for virgin soil at 0.1 g to 0.3 g acceleration with different densities, which means soil liquefies. Tests were conducted with and without application of surcharge. The surcharge loads were not placed in case of virgin soil as it showed signs of liquefaction. Different surcharge were applied on the sample. The surcharge of 5.94 kN/m^2 was applied on the reinforced soil sample.

3.1 Effect on Excess Pore Pressure Rise

When soil was reinforced with 4 and 5 nos. geogrid layers and shaking was imparted, the average excess pore pressure decreases from 3.31 kN/m^2(for virgin soil) to 2.41 and 2.21 kN/m^2(for reinforced soil), respectively for 35% D_r,at 0.1 g as shown in Fig. 5. The pore pressures were further reduced to 2.0 kN/m^2 after inclusion of geocell at 35% D_r and 0.1 g. It was, observed that at 0.3 g acceleration value the excess pore pressure reduces from 3.64 kN/m^2 (for virgin soil) to 2.87 and 2.72 kN/m^2 (for 4 and 5 geogrid layers), respectively, and 2.5 kN/m^2 (for geocells) for 35% relative density. Similar observations were found by Krishnaswami and Issac [12] and Altun et al. [1] for soil reinforced with woven and non-woven sheets in cyclic triaxial testing. The average excess pore pressure ratio r_u also dropped from 1.02 (for virgin soil) to 0.778 and 0.717 (for 4 and 5 geogrid layers), respectively, (Fig. 6) at 0.1 g on 35% $\mathrm{D_r}$. The improvement in the performance of geogrid-reinforced soil is due to the fact that with the addition of a geogrid layer, there is a restriction of the flow path at the interstices of geogrid and soil. At the interstices of geogrid and soil, a confined zone gets developed which does not easily allow the expulsion of water from the soil. The inclusion of the geogrid sheet makes the sand a composite material whose strength and stiffness is higher than that of sand alone. In case of geocell, the average excess pore pressure ratio $\mathrm{r_u}$ reduced to 0.65 at 0.1 g at same density, and this is in confirmation with studies conducted by and Singh [25].

3.2 Effect of Density

The shake table tests were also conducted at 50 and 65% D_r, so that stratification effect could also be studied. At higher densities the average excess pore pressure had come down from 3.31 kN/m^2 (for $D_r = 35\%$ of virgin soil) to 2.76 kN/m^2 (for $D_r =$ 50% of virgin soil) and 1.98 kN/m^2 (for $D_r = 65\%$ of virgin soil) at 0.1 g. The same trend of results were obtained by Mittal [17]. This decrease might be due to tension in sand particles and rearrangement of soil particles. At 0.3 g for 50% D_r,the average excess pore pressure ratio r_u reduced from 1.02 (for virgin soil) to 0.893 and 0.79 (for sand reinforced with 4 and 5 nos. geogrid layers, respectively) (Fig. 6). In case of geocell, the average excess pore pressure ratio r_u decreased to 0.71. Similarly, for 65% D_r, the average excess pore pressure ratio r_u came down from 0.902 to 0.669 and 0.585 (for 4 and 5 geogrid layers), respectively. With use of geocell, the average excess pore pressure ratio r_u came down from 0.902 (for virgin soil) to 0.51.

3.3 *Effect of Surcharge*

To study the effect of surcharge and behavior of soil particles during liquefaction at deeper layers, some tests had been conducted by applying surcharge also. Therefore, a surcharge load of 5.94 kN/m^2 was applied on sand sample. The average excess pore pressure ratio r_u was dropped from 1.185 (for virgin soil) to 0.85 and 0.794 (for sand reinforced with 4 and 5 geogrid layers, respectively) at 0.3 g for 35% D_r (Fig. 7), under surcharge condition. In case of geocell, the average excess pore pressure ratio r_u also came down to 0.72. It shows that the liquefaction behavior of sand during earthquakes can be effectively controlled by reinforcing the sands for overburden pressure also.

3.4 *Effect on Settlement and Pore Pressure Build Up Time*

Settlement behavior of sand with and without reinforcement was also studied. For virgin soil, 18.8 mm of settlement reduces to 15.2 and 13.6 mm with inclusion of 4 and 5 nos. geogrid layers, respectively, at 35% D_r for 0.1 g (Table 3). In case of geocell, the settlement recorded was 14.1 mm for 35% D_r and 0.1 g. This trend was further observed at higher acceleration values also. Excess pore pressure build up time, i.e., time to reach maximum pore pressure was also studied. Average excess pore pressures build time increases from 20.6 s (for virgin soil) to 50 s (for reinforced sand) at 0.1 g. Similarly, excess pore pressure dissipation time also increased from 80 s (for virgin soil) to 119 s (for reinforced sand) at 35% relative density for 0.1 g.

4 Conclusions

To understand behavior of sand with geogrid layers and geocell, shake table tests were conducted on saturated virgin and reinforced sand specimens at 35, 50, and 65% relative densities. These tests were done at accelerations varying from 0.1 g to 0.3 g at a frequency of 4 Hz. The conclusions drawn from this study are as follows:

1. When soil was reinforced with 4 nos. and 5 nos. geogrid layers and shaking was imparted, the rise in excess pore water pressure decreases from 3.3 kN/m^2 (for virgin soil) to 2.41 kN/m^2 and 2.21 kN/m^2 (for reinforced soil), respectively, at 0.1 g for 35% D_r. With inclusion of geocell, the excess pore water pressure decreased to 2.01 kN/m^2 at same density and acceleration value. This trend was further observed at higher acceleration values.
2. The liquefaction resistance of sand increases due to addition of geogrid layers. For 0.1 g acceleration at 35% D_r, the liquefaction resistance increased by about 23 and 29% with 4 and 5 nos. geogrid layers, respectively. Further by increasing

acceleration to 0.3 g, the liquefaction resistance further increased to 20 and 24% for same density.

3. In case of geocell, the liquefaction potential is reduced to 36% at 0.1 g for 35% D_r. On higher acceleration value, i.e., at 0.3 g, the liquefaction resistance increased to 31.6% at same density.
4. At higher density, i.e., on 50 and 65% D_r, the soil liquefaction resistance was of the order of 12–25% (with 4 nos. geogrid layers) and 22–35% (with 5 nos. geogrid layers), respectively, at 0.3 g. With inclusion of geocell, the liquefaction resistance was 30–43% at 50% and 65% D_r, respectively.
5. With the application of surcharge, the average liquefaction resistance increased to 28.0 and 32.9% with 4 and 5 layers of geogrid, respectively, for 0.3 g and 35% D_r. In case of geocell, the liquefaction resistance was 39.2% at same density and acceleration value. This shows that it works in good arrangement under overburden conditions also.
6. Similarly, at 50% D_r, the average liquefaction resistance factor was increased by 28 and 37% (with 4 and 5 layers of geogrid), respectively, at 0.3 g for surcharge conditions. When geocell was used, the liquefaction resistance increases to 43% at same density and 0.3 g.
7. Settlement and pore pressure build up time for reinforced sand showed significant improvement. Thus, it can be concluded that, severity of liquefaction of sand can be controlled appreciably by use of geogrid or geocell reinforcement. Hence, prior to construction of new embankment for railways, highways, etc., in earthquake prone zone, its damages on account of liquefaction may be stopped.

Acknowledgements The research is supported by a fellowship to first author from MHRD, Govt. of India. This support is gratefully acknowledged.

References

1. Altun S, Goktepe BA, Lav AM (2008) Liquefaction resistance of sand reinforced with Geosynthetics. Geosynth Int 15(5):322–332
2. ASTM D5311–92 (2004) Standard test method for load controlled cyclic triaxial strength of soil 1–10
3. Boominathan A, Hari S (2002) Liquefaction strength of fly ash reinforced with randomly distributed fibers. J Soil Dynam Earthquake Engg 22(9–12):1027–1033
4. Casagrande A (1976) Liquefaction and cyclic deformation of sand-a critical review. In: Harvard soil mechanics series. vol 88. Harvard University, Cambridge
5. DeAlba P, Seed HB, Chan CK (1976) Sand liquefaction in large-scale simple shear tests. J Geotech Eng Division ASCE 102(9):909–927
6. Finn WDL, Bransby PL, Pickering DJ (1970) Effect of strain history on liquefaction of sands. J Soil Mech Found Division ASCE 96(6):1917–1934
7. Finn WDL, Bransby PL, Pickering DJ (1976) Seismic pore water pressure generation and dissipation. In: Symposium on soil liquefaction, ASCE National Convention, Philadelphia, pp 169–198

8. Florin VA, Ivanov PA (1961) Liquefaction of saturated sandy soils. In: Proceedings of 15th conference on soil mechanics and foundation engineering, Paris, pp 107–111
9. Geng L, Tang L, Cong SY, Ling XZ, Lu J (2016) Three-dimensional analysis of geosynthetic-encased granular columns for liquefaction mitigation'. Geosynth Int 24:45–59
10. Gupta MK (1977) Liquefaction of sands during earthquakes. Ph.D. Thesis, University of Roorkee, Roorkee, India
11. Ishihara K (1996) In: Soil behavior in earthquake geotechnics. 1st edn. Oxford, Clarendon Press, pp 350
12. Krishnaswamy NR, Isaac NT (1994) Liquefaction potential of reinforced sand. J Geotextiles Geomembranes 13(1):23–41
13. Li J, Ding DW (2002) Nonlinear elastic behaviour of fiber-reinforced soil under cyclic loading. J Soil Dynam Earthquake Eng 22(9–12):977–983
14. Maher MH, Woods RD (1990) Dynamic response of sands reinforced with randomly distributed fibers. J Geotech Eng ASCE 116(7):1116–1131
15. Maher MH, Ho YC (1993) Behavior of fiber-reinforced cemented sand under static and cyclic loads. J Geotech Testing 16(3):330–338
16. Maheshwari BK, Patel AK (2010) Effects of non-plastic silts on liquefaction potential of solani sand. J Geotech Geol Engg 28(5):559–566
17. Mittal S (1988) Vibration table studies for prediction of Liquefaction—a critical study. M.Tech. Dissertation, University of Roorkee, Roorkee, India
18. Noorany I, Uzdavines, M.: Dynamic behavior of saturated sand reinforced with geosynthetic fabrics. In: Proceedings of 89th conference on geosynthetics. vol 2. San Diego, California, pp 385–396
19. Perlea VG, Koester J, Prakash S (1999) How liquefiable are cohesive soils? In: proceedings of conference on Earthquake Geotechnical engineering. Balkema, Rotterdam, The Netherlands, pp 611–18
20. Prakash S, Gupta MK (1970) Liquefaction and settlement characteristics of loose sand under vibrations. In: Proceedings of conference on dynamic waves in civil engineering, Swansea (U.K.), pp 323–338
21. Seed HB, Lee KL (1966) Liquefaction of saturated sands during cyclic loading. J Soil Mech Found Division ASCE 92(6):105–134
22. Seed HB, Idriss IM (1967) Analysis of soil liquefaction, Niigata earthquake. J Soil Mech Found Division ASCE 93(3):83–108
23. Seed HB, Idriss IM (1971) Simplified procedure for evaluating soil liquefaction potential. J Soil Mech Found Division ASCE 107(9):1249–1274
24. Seed HB, Idriss IM (1982) Ground motion and soil liquefaction during earthquakes, monograph series. Berkeley, CA, Earthquake Engineering Research Institute
25. Singh HP (2009) Liquefaction studies of composite materials. Ph.D. Thesis, Indian Institute of Technology, Roorkee, India
26. Singh HP, Maheshwari BK, Saran S, Paul DK (2010) Improvement in liquefaction resistance of pond ash using stone-sand columns. J Geotech Eng 4:23–30
27. Unnikrishnan N, Rajagopal K, Krishnaswamy NR (2002) Behaviour of reinforced clay under monotonic and cyclic loading. J Geotextiles Geomembranes 20(2):117–133
28. Vercueil D, Billet P, Cordary D (2016) Study of the liquefaction resistance of a saturated sand reinforced with geosynthetics. J Soil Dynam Earthquake Eng 16(7–8):417–425
29. Author F (2016) Article title Journal 2(5):99–110
30. Author F, Author S (2016) Title of a proceedings paper. In: Editor F, Editor S (eds) Conference 2016, LNCS, vol 9999. Springer, Heidelberg, pp 1–13
31. Author F, Author S, Author T (1999) Book title, 2nd edn. Publisher, Location
32. Author F (2010) Contribution title. In: 9th International proceedings on proceedings. Publisher, Location, pp 1–2
33. LNCS Homepage. http://www.springer.com/lncs. Last Accessed 21 Nov 2016

Analytical Solution for the Action of Seismic Active Earth Pressures of Unsaturated Backfills Behind Inclined Walls

S. Rajesh and R. Ganesh

Abstract Numerous studies are available to compute the seismic active earth pressures on inclined walls supporting fully dry or saturated soil backfills. However, in practice, unsaturated compacted soils or naturally formed soils are usually found behind earth retention structures. In this study, an analytical solution has been established for determining the magnitude and the location of the line of action of the total seismic active earth thrust, as well as its distribution against inclined walls with unsaturated backfills considering the influence of soil types, groundwater table positions, wall inclinations, and surcharge pressures on the ground surface. The analysis was performed based on the assumption of the planar failure surface within the framework of limit-equilibrium approach. The closed-form effective stress function coupled with Mohr–Coulomb failure criterion was adopted to evaluate the shear strength of unsaturated soils while the effect of quasi-static seismic inertial forces was included from the pseudostatic approach. With the help of the proposed method, the influence of different parameters on the magnitude and the location of the line of action of the total seismic active earth thrust has been studied. The present results showed considerable nonlinearity in the distributions of seismic active earth pressures of unsaturated soils.

Keywords Seismic active pressure · Unsaturated structures · Inclined walls

Notations

P_a	Seismic active earth thrust
p_a	Seismic active earth pressure
h	Point of application of P_a from the wall toe

S. Rajesh (✉)
Indian Institute of Technology, Kanpur, Uttar Pradesh 208016, India
e-mail: hsrajesh@iitk.ac.in

R. Ganesh
Indian Institute of Science, Bengaluru, Karnataka 560012, India

T. G. Sitharam et al. (eds.), *Earthquake Geotechnics*, Lecture Notes in Civil Engineering 187, https://doi.org/10.1007/978-981-16-5669-9_2

k_h	Horizontal seismic acceleration coefficient
k_v	Vertical seismic acceleration coefficient
g	Gravitational acceleration
μ	Inclination angle of the backfill-wall interface
q	Vertical surcharge pressure
z	Depth below the ground surface
H	Height of wall
L	Depth of water table
δ	Friction angle at the backfill-wall interface
c_a	Adhesion
c'	Effective cohesion
ϕ'	Effective friction angle
σ'	Effective stress
σ	Total stress
σ^{s}	Suction stress
τ	Shear strength of backfill
u_{a}	Pore air pressure
α and n	Empirical fitting parameters
γ_w	Unit weight of water
γ	Saturated unit weight of backfill
C'	Effective cohesive force
U_a	Force due to pore air pressure
$F_{\sigma^{\mathrm{s}}}$	Force due to suction stress
S_r	Degree of saturation at the residual state
S_e	Effective degree of saturation
S	Degree of saturation
R	Reactive force
C_a	Effective adhesive force
q_L	Inertial force on account of surcharge pressure
W	Self-weight of the failure wedge
β	Angle of failure surface with vertical
$\boldsymbol{\beta}$	Critical failure angle
K_a	Seismic active earth thrust coefficient
f_1, f_2 and f_3	Dimensionless functions

1 Introduction

The determination of earth pressure, exerted by the backfills under various wall movements, is often required to be precise for the safe stability assessment of any earth retaining structures. Over the last few decades, a number of studies concerning the determination of earth pressure against rigid walls have been reported in the literature by different researchers using different approaches, including, but not limited to, limit equilibrium method, upper bound analysis considering rigid blocks and method of stress characteristics [1–7]. Most of these analyses performed so far in the context of earth pressure assume that the backfill behind the walls is completely dry and/or saturated [8, 9]. However, in reality, unsaturated compacted soils or naturally formed soils are usually found behind these structures. Although significant contribution of negative pore pressure or matric suction in unsaturated backfills toward the stability of retaining wall system is expected, in the current practice of design and stability assessment, the increase in shear strength given by the matric suction component is discarded. However, for the realistic prediction on the earth pressure of unsaturated backfills and the stability of earth retaining structures, the increase in shear strength of unsaturated soils with respect to matric suction needs to be accounted in the analysis.

However, in contrast to dry/saturated soil, the evaluation of the shear strength of unsaturated soils is likely challenging owing to the changes occurred in the matric suction above the water table by several environmental factors. In this regard, Fredlund et al. [10] proposed an extended Mohr–Coulomb (EMC) failure criterion for the prediction of the shear strength of unsaturated soils. Many attempts have earlier been made to incorporate the influence of matric suction in the analysis of lateral earth pressure behind walls with the help of this EMC failure criterion [11–13]. However, it is now realized that the EMC failure criterion is ineffective in describing the increase in shear strength of unsaturated soils, as it does not account for the nonlinearity of the unsaturated soil shear strength with respect to matric suction [14]. Notwithstanding, several recent developments in the field have shown that the shear strength of unsaturated soils can be estimated reasonably using the Mohr–Coulomb failure criterion in combination with the effective stress function for unsaturated soils [14–16]. With the implementation of Mohr–Coulomb failure criterion in combination with the effective stress function for unsaturated soils, some attempts have been made to consider the influence of matric suction on the lateral earth pressure of unsaturated backfills behind walls under different flow conditions [17–20].

Until recently, all the studies aforementioned were concerned in examining the impact of unsaturated flow on the lateral earth pressure of unsaturated soils [17–20]. However, it seems no study is presently available for the determination of active pressure of unsaturated soils under seismic condition. As the occurrence of earthquake leads to detrimental effect on the stability of earth retaining structures, the impact of seismic forces on the modification of the earth pressure exerted by the unsaturated backfills needs to be understood. Accordingly, the aim of the present is to develop a simplified analytical solution within the framework of limit-equilibrium approach for determining the magnitude and the location of the line of action of the total seismic active earth thrust, as well as its distribution against inclined walls with unsaturated backfills. The analysis was performed considering a planar failure wedge and pseudostatic approach for the inclusion of seismic forces to avoid further complexity. The influence of different factors, including, soil types, groundwater table positions, wall inclinations, and surcharge pressures on the ground surface has been elucidated.

2 Details of the Problem Considered

The present work is concerned with the development of simplified analytical solutions for the determination of the total seismic active earth thrust (P_a) and the location of its action point (h) of an unsaturated backfill behind an inclined wall, as illustrated in Fig. 1. In accordance with the pseudostatic approach, the quasi-static seismic inertia forces generated by the uniform horizontal and vertical accelerations (i.e., $k_h g$ and $k_v g$) were considered, where k_h and k_v refer to the horizontal and vertical acceleration coefficients and g is the acceleration due to gravity.

The inclination angle of the backfill-wall interface IJ with the vertical is μ. A horizontal top surface of the unsaturated backfill is subjected to a vertical surcharge pressure q. Height of the wall is denoted as H. The ground water table ($W.T$) is assumed to be located at a depth L measured below the toe of the retaining wall. As the wall is considered to be rough, the line of action of the P_a need to be inclined at an angle of δ with normal to the backfill-wall interface, where δ is the friction angle at the backfill-wall interface. Further, the interface between backfill and wall is considered to have an adhesion component c_a. The magnitude of c_a is being taken equal to $c'(\tan\delta/\tan\phi')$ where c' and ϕ' represent effective value of cohesion and friction angle of an unsaturated backfill [19]. The unit weight of the unsaturated backfill is denoted as γ.

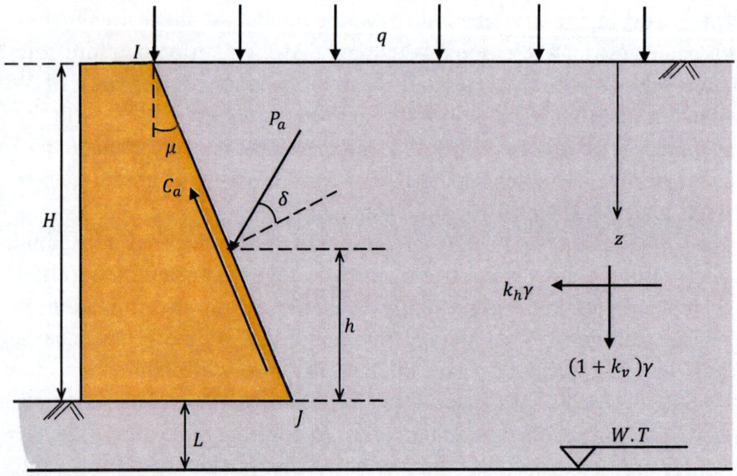

Fig. 1 Schematic diagram of the problem considered

3 Outline of the Proposed Analysis

3.1 Suction Stress and Shear Strength of Unsaturated Soils

According to Lu and Likos [14, 15], the effective stress function applicable for unsaturated soils under the framework of suction stress characteristic curve is expressed as follows

$$\sigma' = \sigma - u_{\mathrm{a}} - \sigma^{\mathrm{s}} \tag{1}$$

where σ' is the effective stress, σ is the total stress; σ^{s} represents the suction stress; u_{a} is the pore air pressure and the value of which is generally taken equal to zero relative to the atmospheric pressure [12, 17, 20].

Considering the van Genuchten's [21] soil water characteristic curve (SWCC) model, the expressions for suction stress σ^{s} and the degree of saturation S as a function of matric suction $(u_a - u_w)$ is as follows [14, 15]

$$\sigma^{\mathrm{s}} = -(u_a - u_w)\left[1 + \{\alpha(u_a - u_w)\}^n\right]^{(1/n-1)} \tag{2a}$$

$$S_e = \left[1 + \{\alpha(u_a - u_w)\}^n\right]^{(1/n-1)} \tag{2b}$$

where u_w is the pore water pressure, $S_e\ (= (S - S_r)/(1 - S_r))$ refers to the effective (normalized) degree of saturation; S is the degree of saturation; S_r is the residual degree of saturation; α and n are the empirical fitting parameters [21].

Because there is no alteration in flow conditions, the variation of matric suction along the depth z can be considered as hydrostatic [14, 17, 22] and therefore,

$$(u_a - u_w) = \gamma_w(H + L - z) \tag{3}$$

where γ_w is the unit weight of water.

Combining Eqs. (2a–2b) and (3), it can be shown that

$$\sigma^{\mathrm{s}} = -\gamma_w(H + L - z)\left[1 + \{\alpha\gamma_w(H + L - z)\}^n\right]^{(1/n-1)} \tag{4a}$$

$$S_e = \left[1 + \{\alpha\gamma_w(H + L - z)\}^n\right]^{(1/n-1)} \tag{4b}$$

Since $(u_a - u_w)$ varies linearly with depth below the ground surface, both the suction stress and unit weight γ of backfill also varies with depth. However, the influence of unit weight variation due to changes in the degree of saturation S is not explored in the present work [12–14, 17, 20]. Adopting Mohr–Coulomb failure criterion in combination with effective stress function (i.e., Eq. (1)), the shear strength of unsaturated soils can be evaluated as [23].

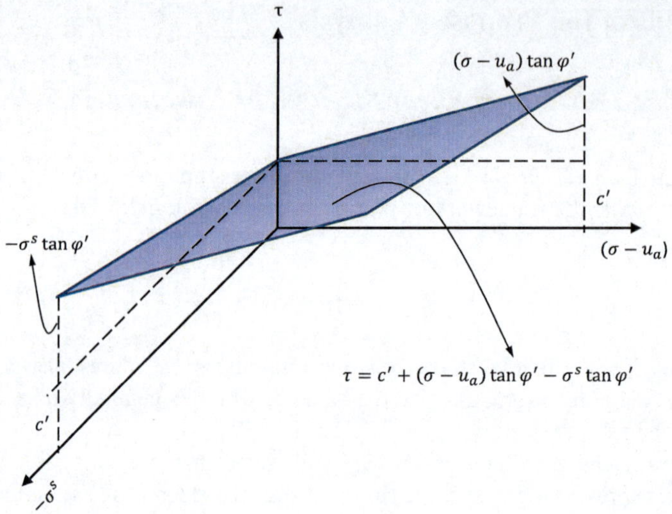

Fig. 2 Schematic diagram of the shear strength envelope for an unsaturated soil

$$\tau = c' + (\sigma - u_{\mathrm{a}})\tan\phi' - \sigma^{\mathrm{s}}\tan\phi' \tag{5}$$

where $(\sigma - u_{\mathrm{a}})$ is the net stress, c' and ϕ' represent effective value of cohesion and friction angle of an unsaturated backfill. Figure 2 illustrates the shear strength envelope for an unsaturated soil.

3.2 *Analytical Formulation*

Figure 3 shows the geometry of the failure wedge IJK considered in the present analysis under seismic condition including details of external and inertial forces.

1. Along the failure surface JK, the following forces can be identified:
- The effective cohesive force (C'),
- The force due to pore air pressure (U_a),
- The force due to suction stress (F_{σ^s}),
- The reactive force (R).
2. Along the backfill-wall interface IJ, the forces identified includes
- The effective adhesive force (C_a)
- The total seismic active earth thrust (P_a)
3. Along the top surface IK and within the failure wedge, quasi-static seismic inertial forces will be generated under the action of earthquake due to surcharge load ($q_L = q \times L_{IK}$) placed on the ground surface and the self-weight of failure wedge (W), where L_{IK} is the length of the top surface IK.

With reference to Fig. 3, the weight of wedge is given by

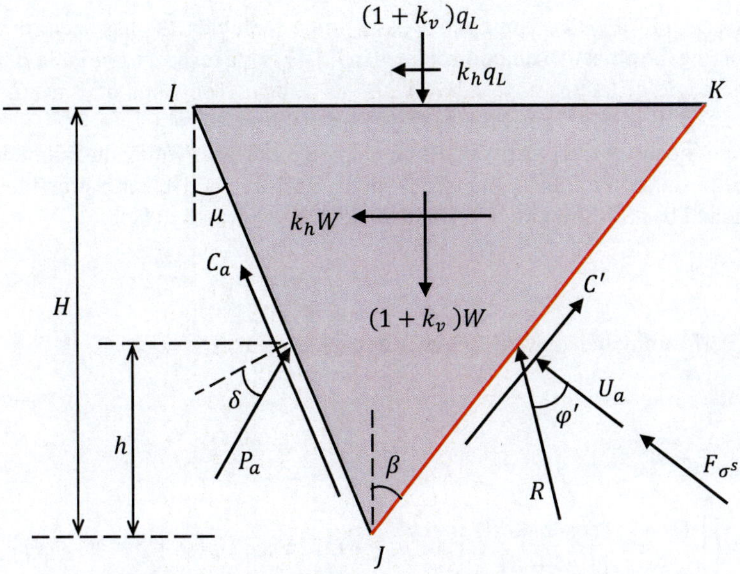

Fig. 3 Geometry of the failure wedge considered in the present analysis under seismic condition including details of external and inertial forces

$$W = 0.5\gamma H^2 \sin(\mu + \beta)/(\cos\mu \cos\beta) \tag{6}$$

And the value of F_{σ^s} is computed as follows

$$F_{\sigma^s} = \sec\beta \int_0^H \sigma^s dz \tag{7}$$

Considering force equilibrium conditions along the horizontal and vertical directions, the total seismic active earth thrust due to an unsaturated backfill can be obtained as follows

$$P_a = \left\{ \begin{array}{c} (\mathrm{W} + q_L)\begin{bmatrix} (1+k_v)\cos(\varphi' + \beta) \\ +k_h \sin(\varphi' + \beta) \end{bmatrix} \\ +F_{\sigma^s}\sin\varphi' \\ -C'\cos\varphi' \\ -C_a\cos(\varphi' + \beta + \mu) \end{array} \right\} / \sin(\mu + \delta + \beta + \varphi') \tag{8}$$

The results of the present study has been obtained in terms of seismic active earth thrust coefficient K_a, which is defined as

$$K_a = 2P_a/(\gamma H^2) \tag{9}$$

The critical value of seismic active earth thrust coefficient K_a has been sought by maximizing the objective function defined in Eq. (9) with respect to variable β, where β is the angle of failure surface JK with the vertical. The value of β maximizing the K_a, will be denoted hereafter as $\boldsymbol{\beta}$, is the critical failure angle.

Assuming the development of the internal failure surfaces within the failure wedge parallel to the failure plane, the distribution of seismic active earth pressure of an unsaturated backfill supported by an inclined wall is given by [19]

$$p_a = \cos\mu(\partial P_a/\partial z)|_{\beta=\boldsymbol{\beta}} \tag{10}$$

On performing the above differentiation, we obtain

$$p_a = (\gamma z + q) f_1 - c' f_2 + \sigma^s f_3 \tag{11}$$

where

$$f_1 = \left\{\begin{bmatrix} (1+k_v)\cos(\varphi' + \boldsymbol{\beta}) \\ +k_h \sin(\varphi' + \boldsymbol{\beta}) \end{bmatrix} \sin(\mu + \beta)\right\} / \left\{\cos\boldsymbol{\beta}\sin(\mu + \delta + \boldsymbol{\beta} + \varphi')\right\} \tag{12}$$

$$f_2 = \left\{\begin{matrix} \cos(\varphi' + \boldsymbol{\beta} + \mu)(\tan\delta/\tan\varphi') \\ +\cos\varphi'\cos\mu/\cos\boldsymbol{\beta} \end{matrix}\right\} / \left\{\sin(\mu + \delta + \boldsymbol{\beta} + \varphi')\right\} \tag{13}$$

$$f_3 = \left\{\sin\varphi'\cos\mu\right\} / \left\{\cos\boldsymbol{\beta}\sin(\mu + \delta + \boldsymbol{\beta} + \varphi')\right\} \tag{14}$$

Comparing simultaneously the moment generated by p_a and P_a about the toe of the retaining wall (I), the location of the line of the action of P_a is given by

$$h = \sec\mu\left\{\int_0^H p_a(H - z)dz\right\} / P_a \tag{15}$$

4 Results and Discussions

4.1 Seismic Active Earth Thrust Coefficient

Using the proposed formulation, solutions have been obtained for two different unsaturated backfill types (i.e., sand and clay) using the parameters listed in Table 1. The following values were assumed to be constant in the present study:$\gamma = 20$ KN/m^3, $\delta/\varphi' = 2/3$ and $k_v/k_h = 0.5$. Figures 4a–f show the variation of seismic active earth thrust coefficient K_a with k_h for different values of μ, L/H, $q/\gamma H$ and H for the cases of unsaturated sand and clay backfills. From Fig 4a–f, it can be seen that the values of the seismic active earth thrust coefficient K_a computed using the

Table 1 Typical unsaturated backfill types and their parameters [14, 17, 20]

Soil types	φ'	$c'/\gamma H$	n	$\alpha(\text{KPa}^{-1})$
Sand	30°	0	5	0.1
Clay	20°	0.05	2	0.005

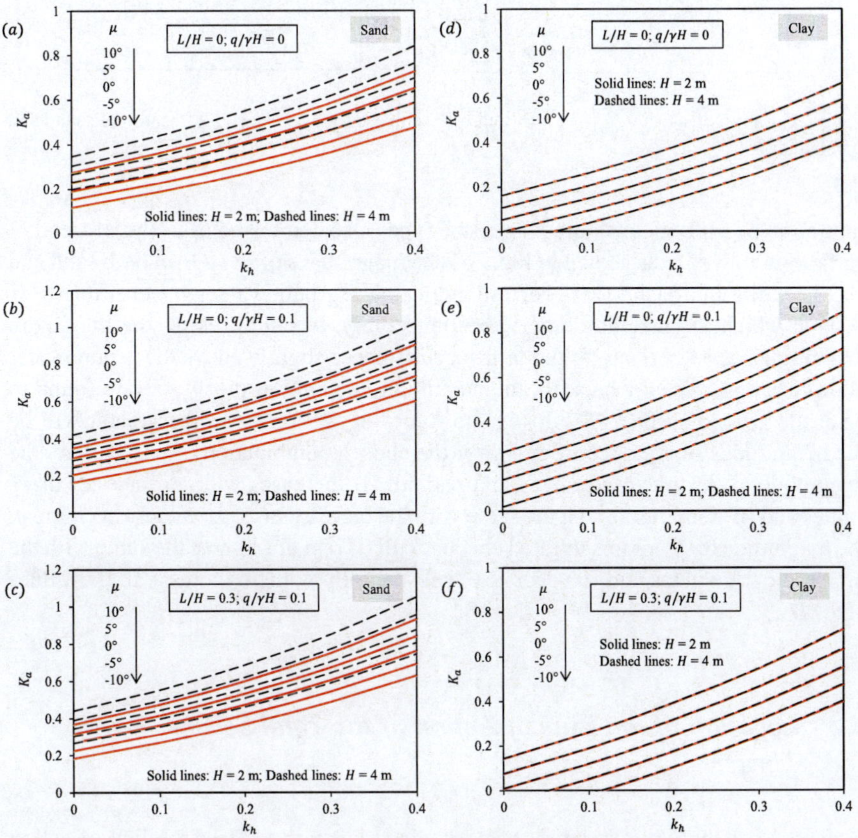

Fig. 4 Variation of seismic active earth thrust coefficient K_a with k_h for different values of, L/H, $q/\gamma H$ and H for (a–c) unsaturated sand backfill; (d–e) unsaturated clay backfill

present study for both unsaturated sand and clay backfills increase continuously with an increase in the values of k_h, $q/\gamma H$ and μ. For better understanding the results obtained in this study, the variations of (1) normalized suction stress $\sigma^s/\gamma H$ with normalized depth z/H and (2) normalized degree of saturation S_e with normalized depth z/H for both backfill types have been presented in Fig. 5a–b. As shown in Fig. 5a, the variation of $\sigma^s/\gamma H$ with z/H for two different H values is almost linear and identical for clay backfill; however, for sand, a nonlinear and distinct variation of $\sigma^s/\gamma H$ with z/H is observed for two different H values. The effective degree of

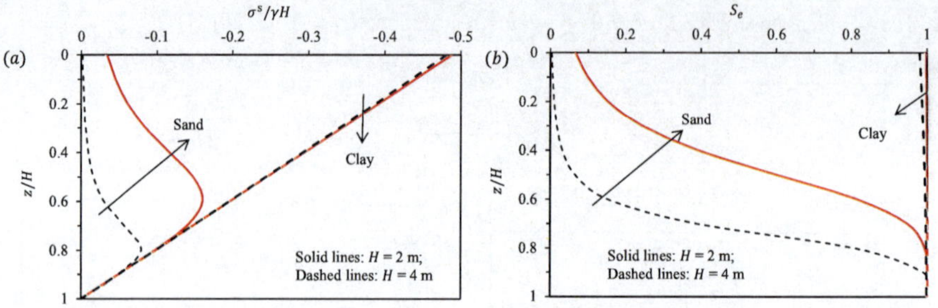

Fig. 5 Variation of **a** $\sigma^s/\gamma H$ with z/H and **b** S_e with z/H for $L/H = 0$ for unsaturated sand and clay backfills

saturation S_e for both unsaturated backfill types also demonstrates a similar trend of variations with z/H as shown in Fig. 5b. The magnitudes of $\sigma^s/\gamma H$ in both sand and clay backfills approach to zero corresponding to the groundwater table position (z/H = 1) at which the magnitude of S_e becomes unity. In sand backfill, the maximum magnitudes of $\sigma^s/\gamma H$ are found to lie at a distances slightly above the groundwater table; whereas, for clay backfill, the maximum magnitudes of $\sigma^s/\gamma H$ are found to be at the ground surface. The lower the values of suction stress the greater will be the magnitudes of K_a. Accordingly, for the chosen combination of parameters, the magnitude of K_a of unsaturated sand backfill (i) increases with an increase of H values and (ii) remains almost the same with the increase of L/H values. In contrast, the magnitude of K_a of unsaturated clay backfill (i) remains almost the same with the increase of H values and (ii) decreases substantially with an increase in the values of L/H.

4.2 *Location of the Line of Action of the Total Seismic Earth Thrust*

Figures 6a–d show the variation of normalized location h/H of the line of action of the total seismic earth thrust with k_h for different values of μ, $q/\gamma H$ and H with $L/H = 0$ for two different unsaturated backfill types. It can be observed from Fig. 6a–d that the values of h/H obtained corresponding to unsaturated sand vary significantly with the variation in the height of walls; whereas, these values obtained corresponding to unsaturated clay backfill remain almost the same. With an increase of μ, $q/\gamma H$ and k_h, the values of h/H seems to increases. The increase of h/H with an increase of μ and k_h is only marginal for sand backfills. Further, in the case of large walls with sand backfill, the changes in the values of μ seem to have negligible impact on the values of h/H. Since the cases analyzed includes nonzero surcharge pressure along the ground surface, it is of no surprise to have values of h/H larger than 1/3. What surprising is the value of h/H for the unsaturated clay backfill tends

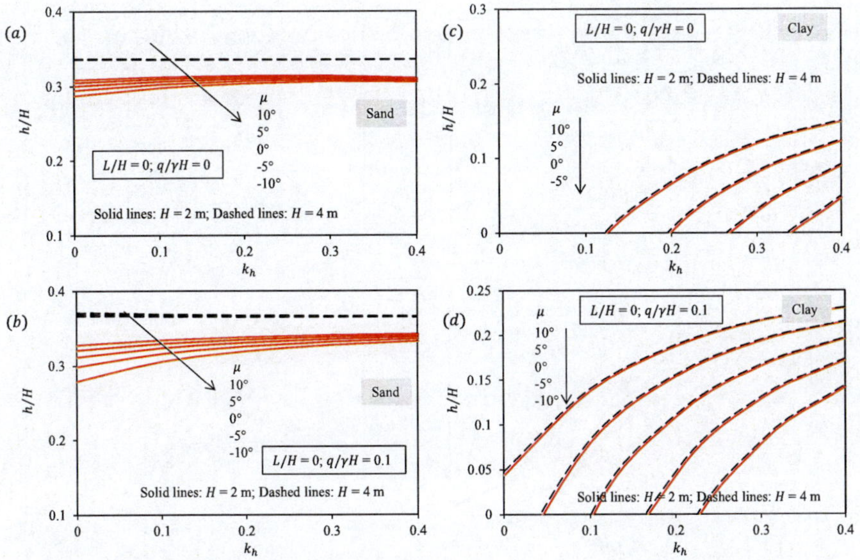

Fig. 6 Variation of normalized location h/H of the line of action of the total seismic earth thrust with k_h for different values of μ, $q/\gamma H$ and H with $L/H = 0$ for (a–b) unsaturated sand backfill and (c–d) unsaturated clay backfill

to become even negative for certain combination of parameters. This is attributed to the development of the negative pressure along the backfill-wall interface near the crest region of the walls. As the soil backfills are weak in resisting tension, one would need to eliminate the contribution of negative pressure for the realistic estimation of active earth thrust.

To compute the depth up to which the active earth pressure along the interface between backfill and wall is negative, the following equation is developed by equating the active earth pressure to zero:

$$z = \left\{ \left(c' f_2 - \sigma^{s} f_3 \right) / f_1 - q \right\} / \gamma \tag{16}$$

Note that Eq. (16) is implicit and may be solved efficiently using iterative methods. The procedure to eliminate the contribution of the negative earth pressure in computing the net earth thrust coefficient is presented elsewhere [20].

4.3 *Distribution of Seismic Active Earth Pressure*

Figure 7a–d show the distribution of normalized active earth pressure $p_a/\gamma H$ along the normalized depth z/H for different values of μ, k_h and H for unsaturated sand and clay backfills with $q/\gamma H = 0.1$ and $L/H = 0$. For the chosen combination of

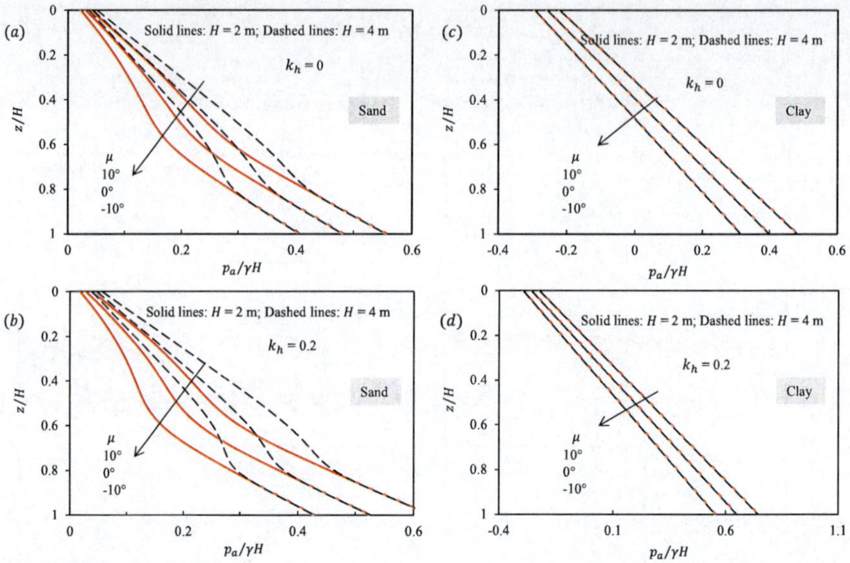

Fig. 7 Distribution of normalized active earth pressure $p_a/\gamma H$ along the normalized depth z/H with $q/\gamma H = 0.1$ and $L/H = 0$ for (a–b) unsaturated sand backfill and (c–d) unsaturated clay backfill

parameters, it can be observed from Fig. 6a–b that the values of $p_a/\gamma H$ of unsaturated sand backfills against inclined walls vary nonlinearly with depth below the ground surface. This is primarily attributed to the nonlinear suction stress variation associated with the unsaturated sand backfills (See Fig. 5a). On the other hand, the distributions of the seismic active earth pressures obtained for unsaturated clay backfills are found to be almost linear owing to the similar characteristics exhibited by its suction stress profiles. The value of $p_a/\gamma H$ at any depth is found to increases with an increase in the values of μ and k_h for both backfill types. Further, as expected, up to some depth below the crest of wall, negative earth pressure exists. The existence of such negative earth pressure will have the tendency to develop a tension crack, and therefore, this effect must be accounted in the realistic estimation of active earth pressure with unsaturated backfills.

4.4 Validation of Present Solutions

Table 2 presents the comparison of K_a obtained in this study with those reported in the literature for dry sand with $k_h = 0$, $q/\gamma H = 0$ and $\mu = 0$. The present values of K_a are found to be slightly lower than those reported by Soubra and Macuh [2] and Sahoo and Ganesh [6], especially for higher values of δ/φ'. This difference is possibly attributed to the usage of linear failure surface in the present analysis.

Table 2 Comparison of K_a obtained in this study with those reported in the literature for dry sand with $k_h = 0$, $q/\gamma H = 0$ and $\mu = 0$

Reference	$\varphi' = 20°$ and $\delta/\varphi' =$					$\varphi' = 30°$ and $\delta/\varphi' =$			
	0	0.5	1			0	0.5	1	
Present study	0.490	0.447	0.427			0.333	0.301	0.297	
Soubra and Macuh [2]	0.490	0.449	0.436			0.333	0.303	0.304	
Sahoo and Ganesh [6]	0.490	0.448	0.434			0.333	0.303	0.302	

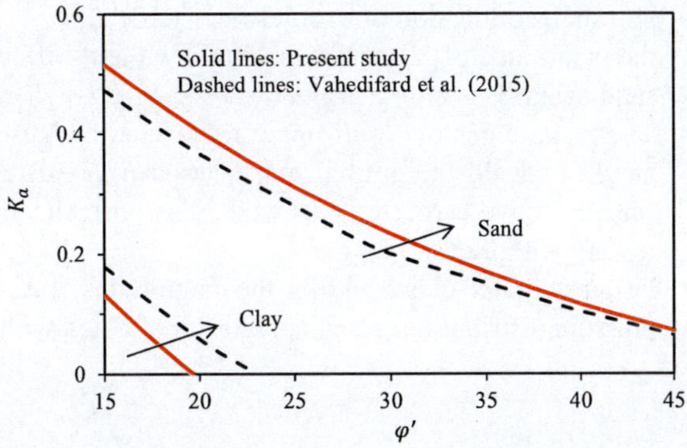

Fig. 8 Comparison of K_a obtained in this study with that reported by Vahedifard et al. [18] with $\delta/\varphi' = 0$, $k_h = 0$, $q/\gamma H = 0$, $L/H = 0$ and $\mu = -10°$

For unsaturated backfills, Fig. 8 presents the comparison of K_a values obtained in this study with that reported by Vahedifard et al. [18] for $\delta/\varphi' = 0$, $k_h = 0$, $q/\gamma H = 0$, $L/H = 0$ and $\mu = -10°$. It can be seen that the present solutions compare reasonably well with the results of Vahedifard et al. [18]. The difference between the two solutions is primarily due to the usage of different assumptions. Vahedifard et al. [18] analysis was also based on a limit equilibrium approach but used a rotational log-spiral slip mechanism. They computed the values of K_a for different unsaturated backfills by solely utilizing the moment equilibrium condition, assuming that the point of application of the active earth thrust is located at 1/3H from the toe of the retaining wall. However, in the present analysis, horizontal and vertical force equilibrium conditions are employed to determine the magnitude of active earth thrust while the moment equilibrium condition is used to obtain the location of its action point on the retaining wall.

5 Conclusions

In the present work, the impact of seismic forces on the active earth pressure of unsaturated backfills has been examined. The analysis was performed within the framework of limit equilibrium approach and the analytical solution for computing the magnitude and the location of the line of action of the total seismic active earth thrust, as well as its distribution against inclined walls with unsaturated backfills has been established. Possible influence of factors such as, soil types, groundwater table positions, wall inclinations, and surcharge pressures on the ground surface has been considered while developing the solution. Based on the present study results, the following conclusions can be drawn.

- For both unsaturated sand and clay backfills, the magnitude of the seismic active earth thrust coefficient K_a is found to increase continuously with an increase in the values of seismic acceleration coefficient k_h, normalized surcharge pressure $q/\gamma H$ and wall inclination angle μ.
- Normalized location of groundwater table L/H is found to significantly affect the magnitude of K_a of clay backfills while its effect on the magnitude of K_a of sand backfills is only marginal.
- The magnitudes of K_a of unsaturated sand backfills are found to vary with the height of wall; the normalized location h/H of the line of action of the total seismic active earth thrust as well as its normalized distributions are also found to vary with the wall height.
- For unsaturated clay backfills, the magnitudes of K_a, h/H and normalized active pressure distributions remain almost unaffected with the variation of wall height.

References

1. Peng J, Zhu Y, Zhou Y (2018) Derivation of Shukla's generalized expression of seismic passive earth pressure on retaining walls with cohesive-frictional backfill by the inclined slice element method. Soil Dyn Earthq Engrg 114:225–228
2. Soubra AH, Macuh B (2002) Active and passive earth pressure coefficients by a kinematical approach. Proc Inst Civil Eng Geotech Engrg 155(2):119–131
3. Kumar J (2001) Seismic passive earth pressure coefficients for sands. Can Geotech J 38(4):876–881
4. Mylonakis G, Klokinas P, Papantonopoulus C (2007) An alternative to the Mononobe-Okabe equations for seismic earth pressure. Soil Dyn Earthq Engrg 27(10):957–969
5. Ganesh R, Sahoo JP (2017) Seismic passive resistance of cohesive-frictional soil medium: kinematic limit analysis. Int J Geomech 17(8):04017029
6. Sahoo JP, Ganesh R (2018) Kinematic limit analysis approach for seismic active earth thrust coefficients of cohesive-frictional backfill. Int J Geomech 18(1):04017123
7. Kumar J, Chitikela S (2002) Seismic passive earth pressure coefficients using the method of characteristics. Can Geotech J 39(2):463–471
8. Barros PLA (2006) A Coulomb-type solution for active earth thrust with seepage. Géotechnique 56(3):159–164
9. Soubra AH, Kastner R, Benmansour A (1999) Passive earth pressures in the presence of hydraulic gradients. Géotechnique 49(3):319–330
10. Fredlund DG, Morgenstern NR, Widger RA (1978) The shear strength of unsaturated soils. Can Geotech J 15(3):313–321
11. Liang WB, Zhao JH, Li Y, Zhang CG, Wang S (2012) Unified solution of Coulomb's active earth pressure for unsaturated soils without crack. Appl Mech Mater 170–173:755–761. https://doi.org/10.4028/www.scientific.net/AMM.170-173.755

12. Pufahl DE, Fredlund DG, Rahardjo H (1983) Lateral earth pressures in expansive clay soils. Can Geotech J 20(2):228–241. https://doi.org/10.1139/t83-027
13. Zhang C, Zhao J, Zhang Q, Xu F (2010) Unified solutions for unsaturated soil shear strength and active earth pressure. In: GeoShanghai 2010: experimental and applied modeling of unsaturated soils. Geotechnical Special Publication, vol 202
14. Lu N, Likos WJ (2004) Unsaturated soil mechanics. Wiley, New York
15. Lu N, Likos WJ (2006) Suction stress characteristic curve for unsaturated soil. J Geotech Geoenviron Engrg 132(2):131–142
16. Lu N, Godt JW, Wu DT (2010) A closed-form equation for effective stress in unsaturated soil. Water Resour Res 46(5):W05515
17. Vahedifard F, Leshchinsky BA, Mortezaei K, Lu N (2015) Active earth pressures for unsaturated retaining structures. J Geotech Geoenviron Engrg 141(11):04015048
18. Li ZW, Yang XL (2018) Active earth pressure for soils with tension cracks under steady unsaturated flow conditions. Can Geotech J. https://doi.org/10.1139/cgj-2017-0713
19. Ganesh R, Rajesh S (2019) Analytical solution to estimate the point of application of resultant passive earth thrust against unsaturated retaining structures. Geomech Geoeng 1–8
20. Sahoo JP, Ganesh R (2018) Active earth pressure on retaining walls with unsaturated soil backfill. In: Bouassida M, Meguid M (eds) Ground improvement and earth structures. GeoMEast 2017. Sustainable Civil Infrastructures. Springer, Cham
21. van Genuchten MT (1980) A closed-form equation for predicting the hydraulic conductivity of unsaturated soils. Soil Sci Soc Am J 44(5):892–898
22. Yeh TC (1989) One-dimensional steady state infiltration in heterogeneous soils. Water Resour Res 25(10):2149–2158
23. Wang L, Hu W, Sun DA, Li L (2019) 3D stability of unsaturated soil slopes with tension cracks under steady infiltrations. Int J Numer Anal Methods Geomech 43(6):1184–1206

Shear Wave Velocity-Based Liquefaction Susceptibility of Soil Using Extreme Learning Machine (ELM) with Strength Pareto Evolutionary Algorithm (SPEA 2)

Ranajeet Mohanty, Sarat Kumar Das, and Madhumita Mohanty

Abstract In the present study, the multi-objective optimization problems observed in the prediction of shear wave velocity (V_s)-based liquefaction susceptibility of soil are solved in the framework of multi-objective feature selection (MOFS) algorithms. The learning algorithm, extreme learning machine, ELM and a multi-objective evolutionary algorithm (MOEA) algorithm, strength Pareto evolutionary algorithm (SPEA 2) are unified to form the MOFS algorithm. The proposed MOFS model is equally proficient in predicting the liquefied and non-liquefied cases for a highly unbalanced database of V_s with the ratio of liquefaction (L) to non-liquefaction (NL) cases being 287:124 ($L/NL = 2.31$). The additional merit of the present study is the identification of important input features; cyclic stress ratio (CSR), V_s, and moment magnitude of the earthquake (M_w). The representation of the results as Pareto front facilitates in the decision-making process. ELM + SPEA 2 is also found to be more efficient in equal prediction of majority and minority class instances; thus, it outperforms ELM + non-dominated sorting genetic algorithm (NSGA-II) as a classifier model. It was also observed that for training to testing ratios of 0.70:0.30 and 0.75:0.25, ELM + SPEA 2 has better generalization capacity than ELM + NSGA-II.

Keywords Extreme learning machine (ELM) · Liquefaction · Multi-objective feature selection (MOFS) · Shear wave velocity (V_s) · Strength Pareto evolutionary algorithm (SPEA 2)

1 Introduction

Evaluation of liquefaction susceptibility of soil based on shear wave velocity (V_s) test is preferable to standard penetration test and cone penetration test. It is more advantageous as: (i) It measures a fundamental soil property; (ii) it can correlate

R. Mohanty
National Institute of Technology, Rourkela, India

S. K. Das (✉) · M. Mohanty
Indian Institute of Technology (ISM), Dhanbad, India
e-mail: saratdas@iitism.ac.in

T. G. Sitharam et al. (eds.), *Earthquake Geotechnics*, Lecture Notes in Civil Engineering 187, https://doi.org/10.1007/978-981-16-5669-9_3

more directly with relative density, which has a strong effect on the cyclic behavior of saturated soil [5]; and (iii) it requires only minor corrections for fines content [6].

Various artificial intelligence (AI) methods have proved to be better than statistical methods [3, 10–14]. However, the problems they encounter are: (i) Complexity increases with the increase in the number of input parameters and (ii) occurrence of error in classifier model built from an unbalanced dataset for binary classification. Both these problems can be solved simultaneously by feature selection (FS) algorithm. These mutually conflicting objectives are attained by multi-objective evolutionary algorithms (MOEA). They can simultaneously reduce all the objective functions and give a Pareto front, which can find a trade-off solution between different objectives and have been applied to FS [2, 8]. The V_s database [6] considered here is highly imbalanced with the ratio of liquefaction (*L*) to non-liquefaction (*NL*) cases as 2.31. Thus, prediction of liquefaction susceptibility is considered as a binary classification problem, which faces the multi-objective optimization problems of decrease of the number of features and decrease of error. It is well-known that extreme learning machine (ELM) is more efficient as compared to artificial neural network (ANN) [1, 4, 19]. The proposed method implements multi-objective feature selection (MOFS) in a framework of ELM, a learning algorithm, and strength Pareto evolutionary algorithm (SPEA 2), an MOEA algorithm.

In an unbalanced database, the majority of the class instances are of one type, and the rest are minority class instances. A classifier model built from unbalanced data poses a major threat in machine learning [16]. Thus, an appropriate performance metric becomes necessary. ELM + non-dominated sorting genetic algorithm (NSGA-II) had been used to evaluate the liquefaction susceptibility of soil using MOFS [9]. SPEA 2 is a recently developed MOEA [18] and is more or equally efficient as NSGA-II. The performance capability of ELM + SPEA 2 is studied from optimal Pareto sets obtained on the basis of error rates. Also, the efficiency of ELM + SPEA 2 is compared with ELM + NSGA-II [9].

This paper deals with the classification problem of liquefaction susceptibility, where 1 and 0 indicate *L* and *NL*, respectively. The determination of the true performance of the proposed model can be reflected by certain basic measures [16] which are as follows:

TP = Total number of correctly classified liquefaction instances
TN = Total number of correctly classified non-liquefaction instances.
FN = Total number of misclassified liquefaction instances.
FP = Total number of misclassified non-liquefaction instances.

The overall accuracy of the model can be defined as [16]:

$$Overall\ accuracy = \frac{TP + TN}{TP + FP + FN + TN} \tag{1}$$

The accuracy of the class instance related to liquefaction of soil can be defined as:

$$A_L = \frac{TP}{TP + FN} \tag{2}$$

The accuracy of the class instance related to non-liquefaction can be defined as:

$$A_{NL} = \frac{TN}{TN + FP} \tag{3}$$

For the present study, the performance assessment of the classifier model is done based on error rates as:

$$E_L = 1 - A_L \tag{4}$$

$$E_{NL} = 1 - A_{NL} \tag{5}$$

For E_L and E_{NL}, the values 0 and 1 indicate a 100% and 0% accurate classifier model, respectively. For a good classifier model, the values of E_L and E_{NL} are equal.

Another performance metric, G_{mean} [7], is useful where the performance capability of each class carries importance and has a high value [15, 17]. It can be defined as:

$$G_{mean} = \sqrt{A_L \times A_{NL}} \tag{6}$$

It can efficiently classify both the majority and minority classes of data. G_{mean} in terms of error rate has been used here to assess the performance of the classifier model and is defined as:

$$G_{mean(error)} = 1 - G_{mean} \tag{7}$$

$G_{mean(error)}$ used for binary classification varies between 0 and 1 representing 100% and 0% accurate classifier model, respectively.

2 Methodology

2.1 Multi-objective Feature Selection (MOFS)

Extreme Learning Machine (ELM)

A weighted ELM algorithm is implemented to nullify the effect of imbalanced dataset [19]. In this algorithm, a weighting scheme is automatically produced from the class information. It can be defined as:

$$W_{ii} = \frac{1}{\#(t_i)} \tag{8}$$

where $\#(t_i)$ = number of samples of the corresponding class t_i, $i = 1,\ldots, m$.

Strength Pareto Evolutionary Algorithm (SPEA 2)

SPEA 2 locates and maintains a set of optimal Pareto solutions [18]. First, it calculates the raw fitness of each individuals as the sum of the strength values of the individuals that dominate the given individual, where strength is the number of individuals that a given individual dominates. Then, the density of the individual is calculated as $1/(\sigma^k + 2)$, where σ^k is the Euclidean distance of the objective values between a given individual and its nearest neighbors, and k is the square root of the size of the population and archive combined. After the filling of the archive population, truncation of the individuals is done by removing those population members which has small σ^k values. Then, by binary tournament selection process, the parents are selected for crossover and mutation operations [18].

Feature selection by ELM + SPEA 2

The features are presented using binary chromosomes: 0 specifying non-selection and 1 specifying selection of a particular feature. Uniform crossover and bitwise mutation are implemented. The major aims defined in the SPEA 2 algorithms are the minimization of:

1. The number of selected input features
2. E_L for training
3. E_{NL} for training
4. E_L for testing and
5. E_{NL} for testing.

3 Database and Preprocessing

The liquefaction ($L = 1$) and non-liquefaction ($NL = 0$) cases are indicated by a liquefaction index (LI). The V_s database [6] has total 411 case histories of liquefaction consisting of 287 liquefiable and 124 non-liquefiable results. Thus, it has a high-class imbalance as L/NL is 2.31. The details of the database are given in Table 1.

The characteristics of ELM algorithm include: transfer function (tan-sigmoid); range of normalization of input parameters (−1 to 1); number of hidden neurons (250); and the trade of constant, C (2^6) [9]. The properties of SPEA 2 algorithm are: population size (1000); number of generations (25); crossover probability (0.95); mutation probability (0.1); and mutation rate (0.1) [9]. The influence of data division on performance capability is studied by using the ratios of training and testing as: 0.70:0.30, 0.75:0.25, and 0.80:0.20.

Table 1 Statistical parameters of the shear wave velocity dataset (L:NL = 287:124)

Input	Symbol	Maximum	Minimum	Average	Standard deviation
Moment magnitude of the earthquake	M_w	9.00	5.90	7.11	0.53
Critical depth (m)	D	18.50	1.10	5.34	2.53
Depth of water table (m)	d_w	7.00	0.40	1.96	1.16
Total stress (kPa)	Σ	331.68	17.40	89.93	45.15
Effective stress (kPa)	σ'	176.08	11.03	56.34	25.64
Peak horizontal ground acceleration in terms of g	a_{max}	0.76	0.02	0.33	0.16
Non-linear shear mass participation factor	r_d	1.00	0.51	0.85	0.11
Uniform cyclic stress ratio	CSR	0.73	0.02	0.28	0.15
Corrected normalized shear wave velocity (m/s)	V_{s1}	362.90	81.70	166.97	40.28

4 Results and Discussion

4.1 ELM + SPEA 2 (Train:Test = 0.70:0.30)

The Pareto front in Fig. 1 shows the relation of error rate with the number of features. For the training to testing ratio of 0.70:0.30, the most optimal point is at five number of

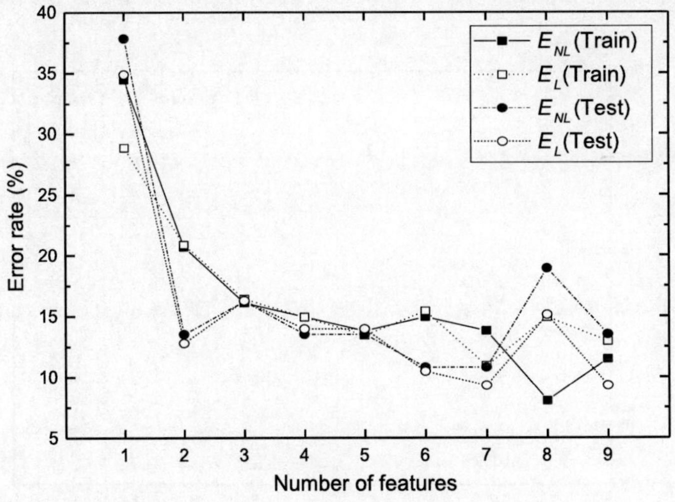

Fig. 1 Pareto optimal solutions for ELM + SPEA 2 algorithm (Train:Test = 0.70:0.30)

Table 2 Details of the optimal data points of ELM + SPEA 2 algorithm (Train:Test = 0.70:0.30)

Selected features									Train		Test	
									E_{NL} (%)	E_L(%)	E_{NL} (%)	E_L(%)
–	–	–	–	–	–	–	*CSR*	–	34.48	28.86	37.84	34.88
–	–	–	–	–	–	–	*CSR*	V_{s1}	20.69	20.90	13.51	12.79
M_w	–	–	–	–	–	–	*CSR*	V_{s1}	16.09	16.42	16.22	16.28
M_w	–	d_w	–	–	–	–	*CSR*	V_{s1}	14.94	14.93	13.51	13.95
M_w	–	–	σ	–	–	r_d	*CSR*	V_{s1}	13.79	13.43	13.51	13.95
M_w	–	d_w	σ	–	–	r_d	*CSR*	V_{s1}	14.94	15.42	10.81	10.47
M_w	d	d_w	–	σ'	–	r_d	*CSR*	V_{s1}	13.79	10.95	10.81	9.30
M_w	d	d_w	σ	σ'	–	r_d	*CSR*	V_{s1}	8.05	14.93	18.92	15.12
M_w	d	d_w	σ	σ'	a_{max}	r_d	*CSR*	V_{s1}	11.49	12.94	13.51	9.30

features (M_w, σ, r_d, *CSR*, and V_{s1}), and the most desired classifier model corresponds to it.

The details of the optimal data points are given in Table 2. The desired model has E_{NL} and E_L for training phase: 13.79 and 13.43% and testing phase: 13.51 and 13.95%, respectively, at five number of features.

4.2 ELM + SPEA 2 (Train:Test = 0.75:0.25)

The Pareto front for train:test ratio of 0.75:0.25 is illustrated in Fig. 2. It shows the

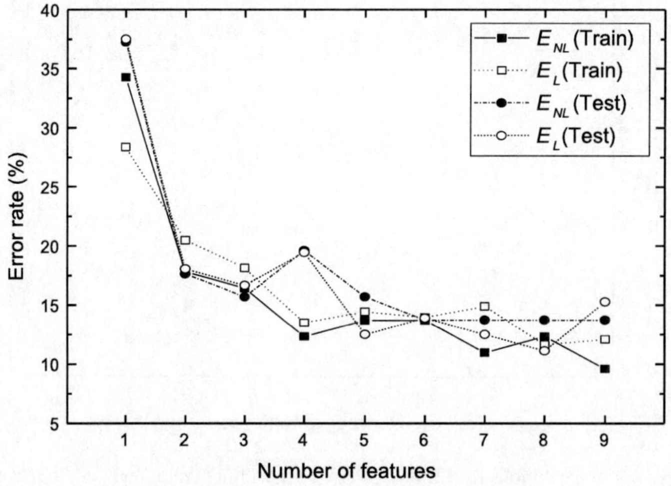

Fig. 2 Pareto optimal solutions for ELM + SPEA 2 algorithm (Train:Test = 0.75:0.25)

Table 3 Details of the optimal data points of ELM + SPEA 2 algorithm (Train:Test = 0.75:0.25)

Selected features									Train		Test	
									E_{NL} (%)	E_L(%)	E_{NL} (%)	E_L(%)
–	–	–	–	–	–	–	CSR	–	34.25	28.37	37.25	37.50
–	–	–	–	–	–	–	CSR	V_{s1}	17.81	20.47	17.65	18.06
–	–	d_w	–	–	–	–	CSR	V_{s1}	16.44	18.14	15.69	16.67
M_w	–	d_w	–	–	–	–	CSR	V_{s1}	12.33	13.49	19.61	19.44
M_w	–	–	σ	–	–	r_d	CSR	V_{s1}	13.70	14.42	15.69	12.50
M_w	–	d_w	–	σ'	–	r_d	CSR	V_{s1}	13.70	13.95	13.73	13.89
M_w	–	–	σ	σ'	$a_{\max}$	r_d	CSR	V_{s1}	10.96	14.88	13.73	12.50
M_w	d	d_w	σ	σ'	–	r_d	CSR	V_{s1}	12.33	11.63	13.73	11.11
M_w	d	d_w	σ	σ'	$a_{\max}$	r_d	CSR	V_{s1}	9.59	12.09	13.73	15.28

most optimal point at 6 number of features (M_w, d_w, σ', r_d, CSR, and V_{s1}), which corresponds to the most desired model.

Thus, from Table 3, it is observed that the most desired classifier model has E_{NL} and E_L for training phase: 13.70% and 13.95% and testing phase: 13.73% and 13.89%, respectively.

4.3 ELM + SPEA2 (Train:Test = 0.80:0.20)

Considering the train:test ratio of 0.80:0.20, the Pareto front obtained by using FS algorithm is presented in Fig. 3. The optimal points are found to appear at 2, 4, and 6 number of features. In the training phase, the difference of E_{NL} and E_L for two features is: 1.14%; four features: 0.56% and six features: 1.62%. For the testing phase, the difference for two features is: 1.35%; four features: 0.88% and six features: 1.59%. Therefore, the performance of the classifier is better for four features as the difference of error rates is least for it in both the cases of training as well as testing. Thus, the classifier model for four features is considered as the best.

From the detailed data points given in Table 4, the respective values of E_{NL} and E_L for training phase: 13.79% and 14.35% and testing phase: 16.67% and 15.79% are obtained for the most desired classifier model.

4.4 Comparison of ELM + SPEA 2 and ELM + NSGA-II

In both the models, ELM + NSGA-II [9] and ELM + SPEA 2, the same learning algorithm (ELM) and V_s database are used. The preprocessing stage of the MOEA algorithms (NSGA-II and SPEA 2) is also identical. The two models have been

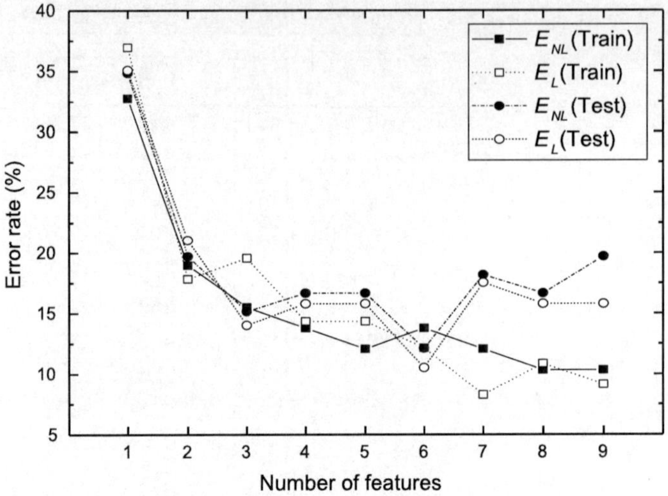

Fig. 3 Pareto optimal solutions for ELM + SPEA 2 algorithm (Train:Test = 0.80:0.20)

Table 4 Details of the optimal data points of ELM + SPEA 2 algorithm (Train:Test = 0.80:0.20)

Selected features									Train		Test	
									E_{NL} (%)	E_L(%)	E_{NL} (%)	E_L(%)
–	–	–	–	–	a_{max}	–	–	–	32.76	36.96	34.85	35.09
–	–	–	–	–	–	–	CSR	V_{s1}	18.97	17.83	19.70	21.05
–	–	d_w	–	–	–	–	CSR	V_{s1}	15.52	19.57	15.15	14.04
M_w	–	–	–	–	–	r_d	CSR	V_{s1}	13.79	14.35	16.67	15.79
M_w	–	d_w	–	–	–	r_d	CSR	V_{s1}	12.07	14.35	16.67	15.79
M_w	d	d_w	–	–	–	r_d	CSR	V_{s1}	13.79	12.17	12.12	10.53
M_w	–	d_w	–	σ'	a_{max}	r_d	CSR	V_{s1}	12.07	8.26	18.18	17.54
M_w	–	d_w	σ	σ'	a_{max}	r_d	CSR	V_{s1}	10.34	10.87	16.67	15.79
M_w	d	d_w	σ	σ'	a_{max}	r_d	CSR	V_{s1}	10.34	9.13	19.70	15.79

compared on the basis of their E_{NL} and E_L in Table 5. Corresponding to train:test ratio of (0.70:0.30), (0.75:0.25), and (0.80:0.20), the difference of E_{NL} and E_L in training phase for ELM + SPEA 2 model is: [(0.36%), (0.25%), and (0.56%)] and ELM + NSGA-II: [(0.85%), (0.25%), and (1.44%)]. Similarly, in the case of testing, the above said difference for ELM + SPEA 2 model is: [(0.44%), (0.16%), and (0.88%)] and ELM + NSGA-II: [(0.44%), (1.23%), and (1.11%)]. It can be clearly seen that the values obtained for ELM + SPEA 2 are less than or comparable with those of ELM + NSGA-II. Thus, ELM + SPEA 2 is the better classifier model as it is equally efficient in predicting majority (liquefaction) and minority (non-liquefaction) class instances.

Table 5 Comparison of ELM + NSGA-II and ELM + SPEA 2 based on E_{NL} and E_L

AI model (Train:Test)	Number of features	Training		Testing		Difference of E_{NL} and E_L(%)	
		E_{NL} (%)	E_L (%)	E_{NL} (%)	E_L (%)	Training	Testing
ELM + NSGA-II (0.70:0.30)	5	13.79	12.94	13.51	13.95	0.85	0.44
ELM + SPEA 2 (0.70:0.30)	5	13.79	13.43	13.51	13.95	0.36	0.44
ELM + NSGA-II (0.75:0.25)	8	13.70	13.95	13.73	12.50	0.25	1.23
ELM + SPEA 2 (0.75:0.25)	6	13.70	13.95	13.73	13.89	0.25	0.16
ELM + NSGA-II (0.80:0.20)	4	15.52	16.96	15.15	14.04	1.44	1.11
ELM + SPEA 2 (0.80:0.20)	4	13.79	14.35	16.67	15.79	0.56	0.88

On the basis of $G_{mean(error)}$, the detailed comparison of the two models is given in Table 6. The difference in values of $G_{mean(error)}$ in training and testing phase for ELM + SPEA 2 model is: [(0.12%), (0.02%), and (2.16%)] and ELM + NSGA-II: [(0.36%), (0.71%), and (1.64%)] in case of train:test ratio of (0.70:0.30), (0.75:0.25), and (0.80:0.20), respectively. ELM + SPEA 2 has better generalization capacity for train:test ratio of (0.70:0.30) and (0.75:0.25) as it has lesser difference between the $G_{mean(error)}$ due to training and testing.

Table 6 Comparison of ELM + NSGA-II and ELM + SPEA 2 based on $G_{mean(error)}$

AI model	Train:Test	Number of features	$G_{mean(error)}$ (%)		Difference in $G_{mean(error)}$ (%) (for training and testing)
			Training	Testing	
ELM + NSGA-II	0.70:0.30	5	13.37	13.73	0.36
ELM + SPEA 2		5	13.61	13.73	0.12
ELM + NSGA-II	0.75:0.25	8	13.83	13.12	0.71
ELM + SPEA 2		6	13.83	13.81	0.02
ELM + NSGA-II	0.80:0.20	4	16.24	14.60	1.64
ELM + SPEA 2		4	14.07	16.23	2.16

Table 7 Various V_s-based AI models to predict liquefaction susceptibility

References	*L:NL* (*L/NL*)	AI model	E_{NL} (%)	E_L(%)
Muduli and Das (2015)	88:98 (0.90)	MGGP	15.00	9.00
Hoang and Bui (2016)	107:78 (1.37)	KFDA-LSSVM	10	14
Present study	287:124 (2.31)	ELM + SPEA 2 (Train:Test = 0.70:0.30)	13.79(Train)	13.43(Train)
			13.51(Test)	13.95(Test)
		(Train:Test = 0.75:0.25)	13.70(Train)	13.95(Train)
			13.73(Test)	13.89(Test)
		(Train:Test = 0.80:0.20)	13.79(Train)	14.35(Train)
			16.67(Test)	15.79(Test)

4.5 *Comparison of ELM + SPEA 2 and Other V_s-based AI Models*

While predicting liquefaction susceptibility by multi-gene genetic programming (MGGP) based on a V_s-based dataset with *L:NL* ratio of 88:98 ($L/NL = 0.90$), the E_{NL} and E_L were obtained as 15% and 9%, respectively [12]. Similarly, when the combination of the kernel Fisher discriminant analysis and least squares support vector machine (KFDA-LSSVM) was used with a V_s-based dataset with *L:NL* ratio of 107:78 ($L/NL = 1.37$), the E_{NL} and E_L were obtained as 10 and 14%, respectively [3].

The V_s dataset implemented for the present study is much more imbalanced than the MGGP and KFDA-LSSVM models, since its *L/NL* ratio is 2.31 with the number of majority instances (liquefaction cases) more than double of the number of minority instances (non-liquefaction cases). It can be seen that values of E_{NL} and E_L of the present study for training and testing phases are comparable with those of the above said studies. The above AI models have been compared with the present study in Table 7.

5 Conclusion

The present study accomplished the development of ELM + SPEA 2 models for the classification problem of liquefaction susceptibility of soil considering a highly imbalanced dataset of V_s with *L: NL* (*L/NL*) as 287:124 (2.31). *CSR*, V_{s1}, and M_w are identified as the most important input parameters. The outcome is uninfluenced by the biased nature of the dataset. The proposed model can be used in new field conditions as it has good generalization capacity. For biased database, more accurate prediction models can be developed based on this study. ELM + SPEA 2 is more efficient in equal prediction of majority and minority class instances; thus, it can serve as a better prediction model than ELM + NSGA-II. It is also observed that for train:test ratio of (0.70:0.30) and (0.75:0.25), ELM + SPEA 2 has better generalization capacity than ELM + NSGA-II.

References

1. Bartlett PL (1998) The sample complexity of pattern classification with neural networks; the size of the weights is more important than the size of network. IEEE Trans Inf Theory 44(2):525–536
2. He J, Bi Y, Ding L, Li Z, Wang S (2016) Unsupervised feature selection based on decision graph. Neural Comput Applic 1–13
3. Hoang N-D, Bui DT (2016) Predicting earthquake-induced soil liquefaction based on a hybridization of kernel Fisher discriminant analysis and a least squares support vector machine: a multi-dataset study. Bull Eng Geol Env 77(1):191–204
4. Huang GB, Zhu QY, Siew CK (2006) Extreme learning machine: theory and applications. J Neurocomput 70:489–501
5. Idriss IM, Boulanger RW (2008) Semi-empirical procedures for evaluating liquefaction potential during earthquakes. Soil Dyn Earthq Eng 26:115–130
6. Kayen R, Moss RES, Thompson EM, Seed RB, Cetin KO, Kiureghian AD, Tanaka Y, Tokimatsu K (2013) Shear-wave velocity-based probabilistic and deterministic assessment of seismic soil liquefaction potential. J Geotech Geoenviron Eng 139:407–419
7. Kubat M, Matwin S (1997) Addressing the curse of imbalanced training sets: one-sided selection. In: Proceedings 14th international conference machine learning, Nashville, USA, pp 179–186
8. Li Z, Lu W, Sun Z, Xing W (2016) A parallel feature selection method study for text classification. Neural Comput Appl 1–12
9. Mohanty R, Mahamaya M, Das SK, Mohanty M (2019) Liquefaction susceptibility of soil using multi objective feature selection. In: Proceedings of the 7th international conference on earthquake geotechnical engineering, Rome, Italy. Proceedings in Earth and geosciences. vol 4. CRC press, pp 3981–3988
10. Muduli PK, Das SK (2013) CPT-based seismic liquefaction potential evaluation using multi-gene genetic programming approach. Indian Geotechnical Journal 44(1):86–93
11. Muduli PK, Das SK (2013) Evaluation of liquefaction potential of soil based on standard penetration test using multi-gene genetic programming model. Acta Geophys 62(3):529–543
12. Muduli PK, Das SK (2015) Evaluation of liquefaction potential of soil based on shear wave velocity using multi-gene genetic programming. Handbook of Genetic Programming Applications Part II:309–343
13. Samui P, Sitharam TG (2011) Machine learning modelling for predicting soil liquefaction susceptibility. Natural Hazards and Earth Sciences 11:1–9
14. Samui P (2007) Seismic liquefaction potential assessment by using Relevance Vector Machine. Earthq Eng Eng Vib 6(4):331–336
15. Sun Y, Kamel MS, Wang Y (2006) Boosting for learning multiple classes with imbalanced class distribution. In: Proceedings data mining ICDM'06. Sixth international conference on data mining, Hong Kong, pp 592–602
16. Weiss GM, Provost F (2003) Learning when training data are costly: The effect of class distribution on tree induction. Journal of Artificial Intelligence Research 19:315–354
17. Yuan B, Liu W (2012) A measure oriented training scheme for imbalanced classification problems. In: Proceedings 15th international conference on new frontiers in applied data mining, PAKDD'11, vol 7104. Springer, Berlin, Heidelberg, pp 293–303

18. Zitzler, E., Laumanns, M., Thiele, L.: Spea2: Improving the strength Pareto evolutionary algorithm, Technical Report 103, Computer Engineering and Networks Laboratory (TIK), Swiss Federal Institute of Technology (ETH) Zurich, (2001).
19. Zong W, Huang GB, Chen Y (2013) Weighted extreme learning machine for imbalance learning. Neurocomputing 101:229–242

Seismic Pullout Capacity of Strip Plate Anchors Embedded in Sandy Slopes

S. Khuntia and J. P. Sahoo

Abstract For strengthening the foundations subjected to uplift loads mostly in the case of transmission towers, bridges and buried pipelines, etc., plate anchors are often used and their design must ensure considerable safely against the resistance of supporting soil mass to uplift failure. In the present study, the pullout capacity of horizontal strip anchors embedded in sandy soil with sloping surface has been obtained under seismic conditions. The computations have been carried out by using lower bound limit analysis with finite elements in conjunction with linear programming. The uplift capacity is expressed in terms of a dimensionless seismic uplift factor due to the contribution of soil self-weight with the inclusion of pseudo-static seismic body forces. The analysis has been performed for different values of (i) slope angle varying from 15° to 35°, (ii) embedment ratio of the anchor plate varying from 1 to 7, (iii) angle of internal friction of soil ranging from 20° to 40°, and (iv) horizontal seismic acceleration coefficient (K_h) ranging from 0 to 0.3. With pullout load normal to the anchor plate, the solutions have been obtained for three different orientations of the anchor plate, i.e., horizontal, parallel to the sloping face and vertical. It is observed that for the given magnitudes of embedment ratio and angle of internal friction of soil mass, the pullout capacity reduces with an increase in the angle of slope, inclination of anchor plate and seismic acceleration coefficient. With the increase in the roughness of anchor plate, the pullout capacity is found to be increasing significantly.

Keywords Anchor · Finite elements · Seismic · Slope · Uplift capacity

1 Introduction

Anchors are often provided to resist the uplift force on different structures such as transmission towers, dry docks and pipelines under water, etc., subjected to

S. Khuntia
National Institute of Technology, Rourkela, Odisha 769008, India

J. P. Sahoo (✉)
Indian Institute of Technology, Uttar Pradesh, Kanpur 208016, India
e-mail: jpsahoo@iitk.ac.in

T. G. Sitharam et al. (eds.), *Earthquake Geotechnics*, Lecture Notes in Civil Engineering 187, https://doi.org/10.1007/978-981-16-5669-9_4

uplift forces. Depending upon the location of structures, anchors are embedded also embedded in the sloping ground, where the behavior of anchor against uplift forces is different from that buried in soil with horizontal ground surface. Numerous investigations have been performed for finding the uplift resistance of plate anchors embedded in horizontal ground surface with or without seismic loading by different researchers [1–11]. But the influence of slope on the uplift capacity has been studied by few researchers [12–17]. Based on the upper bound limit analysis with rigid block failure mechanism, Kumar [5] has developed closed form solutions for seismic uplift capacity of horizontal plate anchors buried in sand. Choudhury and Subba Rao [13] obtained the seismic uplift capacity factors of horizontal strip anchors embedded in an inclined ground surface using limit equilibrium method with pseudo-static body force. Yu et al. [17] provided solutions for pullout capacity for inclined strip plate anchors in sandy slopes using upper bound approach. However, limited studies seem to be reported to predict the uplift capacity of anchors buried in a frictional soil with sloping ground surface in the presence of seismic body forces. In the present analysis, it is aimed to obtain solutions for finding the uplift capacity of strip plate anchors for three anchor orientations (i.e., horizontal, parallel to sloping face and vertical) embedded in frictional soil with sloping ground surface in the presence of seismic loading. The solutions were obtained on the basis of lower bound limit analysis in combination with finite elements and linear optimization techniques.

2 Statement of the Problem

A rigid strip plate anchor of width B and infinitesimal thickness is embedded at a depth H in a sloping ground of frictional soil with slope angle β as illustrated in Fig. 1a. The anchor plate is embedded in the sloping ground at different orientations with an angle of inclination of α with the vertical. The soil medium is assumed to be perfectly plastic, and it obeys the Mohr–Coulomb failure criterion with an associated flow rule. The unit weight and angle of internal friction of soil are denoted as γ and ϕ, respectively. In general, the magnitude of the uplift resistance is not affected by the roughness of the anchor plate for horizontal anchor with horizontal ground surface [2]. However, the uplift resistance of horizontal anchor plate embedded in sloping ground has been reported to be lower for smooth anchor plate as compared to that of rough interface condition [17]. The analysis was performed in the present analysis for smooth anchor-soil interface condition in order to obtain the results on the safer side.

It is intended to determine the seismic uplift resistance P_{us} per unit length of anchor plate at the ultimate shear failure of soil mass, which is expressed in terms of a dimensionless uplift factor as follows

$$P_{us} = \gamma \, B \, H \, F_{\gamma s} \tag{1}$$

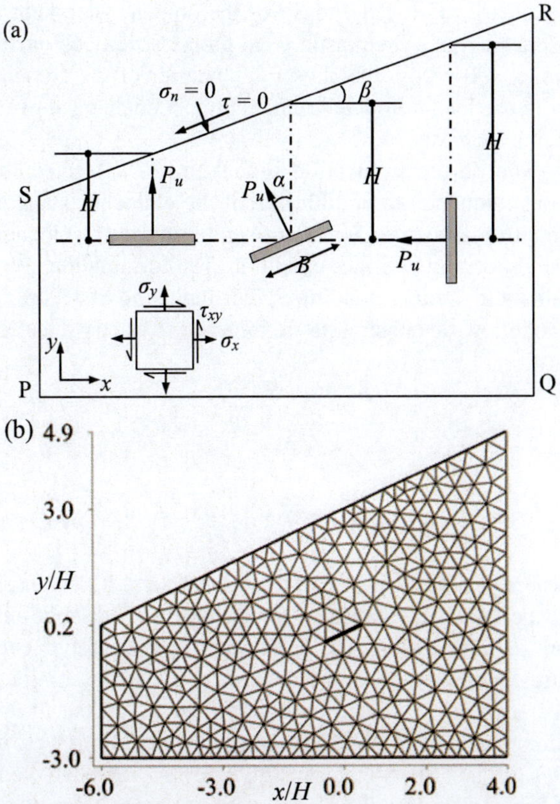

Fig. 1 **a** Problem definition; **b** a typical finite element mesh along with stress boundary conditions

Here, $F_{\gamma s}$ is the seismic uplift factor due to self-weight of soil in the presence of seismic loading.

3 Finite Element Lower Bound Limit Analysis

The computations have been performed using finite element approach based on the lower bound theorem of limit analysis proposed by Sloan [18]. The problem domain PQRS as shown in Fig. 1a was chosen for performing the analysis and was discretized in to three noded triangular elements by writing codes in MATLAB [19]. A typical finite element mesh for the problem considered is shown in Fig. 1b. Each node is involved with three basic unknowns, namely, horizontal normal stress (σ_x), vertical normal stress (σ_y) and the shear stress (τ_{xy}) on the horizontal/vertical plane; the positive directions of σ_x, σ_y and τ_{xy} are indicated in Fig. 1a. By integrating the normal stresses along the interfaces of anchor and surrounding soil, the magnitude of the seismic uplift collapse load (P_{us}) for the anchor plate is determined. The boundaries QR, RS and SP of the domain were kept at far away from the anchor plate such that the plastic yield points developed during the analysis are contained quite well within the chosen boundaries of the domain, and the magnitudes of the collapse load computed remain almost unchanged even if the size of the domain is being increased.

The objective function is maximized subjected to satisfaction of (i) equality constraints due to equilibrium of the elements, statically admissible stress discontinuities, and stress boundary conditions and (ii) inequality constraints due to yield condition and interface condition. The formulation given by Sloan [18] is applicable for static loading conditions. For inclusion of seismic body forces in the analysis, the following constraints are imposed to satisfy the element equilibrium conditions.

$$\frac{\partial \sigma_x}{\partial x} + \frac{\partial \tau_{xy}}{\partial y} = -\gamma K_h \tag{2a}$$

$$\frac{\partial \tau_{xy}}{\partial x} + \frac{\partial \sigma_y}{\partial y} = \gamma \tag{2b}$$

where γ is the unit weight of soil mass and K_h is horizontal seismic coefficient.

For attaining statically admissible stress discontinuities, lines of discontinuities are permitted along the interfaces of adjacent elements, and the shear and normal stresses along these interfaces are imposed to be continuous. The ground surface (RS) is free from stresses (i.e., normal stress (σ_n) and shear stress (τ) were specified as zero along this surface). No stress boundary conditions were imposed along the boundaries PQ, QR and SP. All the constraints are expressed in the form of matrices and vectors. After constituting the global matrices and vectors, the magnitude of ultimate uplift load is maximized subject to the set of all the constraints. A necessary pre-processing computer code was generated in MATLAB [19] for performing the analysis, and optimized solutions were obtained with the help of linprog an optimization toolbox available in MATLAB [19].

4 Results and Discussions

The seismic uplift factors are presented as a function of normalized embedment depth ($H/B = 1$ to 7), angle of slope ($\beta = 15°, 25°, 35°$), internal friction angle of soil ($\phi = 20°, 30°, 40°$) and horizontal seismic coefficient ($K_h = 0, 0.1, 0.2, 0.3$). The following observations have been made for horizontal anchor ($\alpha = 0°$), anchor parallel to sloping face ($\alpha = \beta$) and vertical anchor ($\alpha = 90°$) which are presented in Figs. 2, 3 and 4, respectively. The variation of seismic uplift factor with H/B for different combinations of β, ϕ and K_h is presented in Figs. 2, 3 and 4. From Figure, it is clearly seen that for higher embedment ratios (H/B), uplift capacity factors are higher as expected. For example, referring to Figs. 2e, 3e and 4e, for the case of $\phi =$

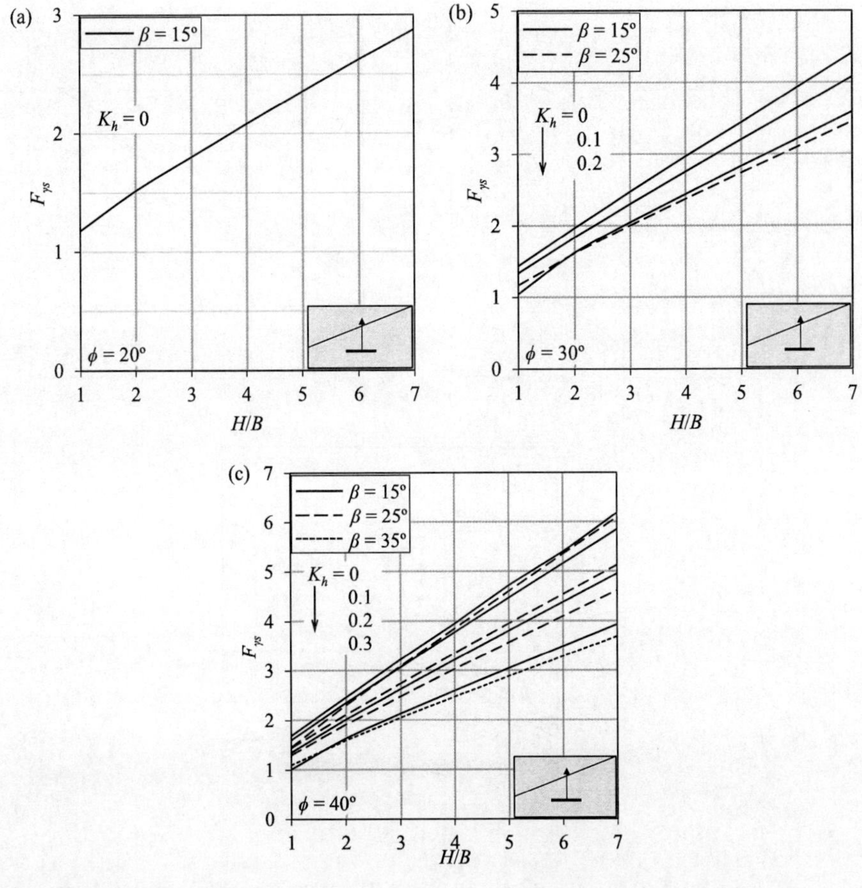

Fig. 2 For horizontal anchor ($\alpha = 0°$), variation of $F_{\gamma s}$ with H/B for different values of β, K_h **a** $\phi = 20°$, **b** $\phi = 30°$, **c** $\phi = 40°$

40°, $\beta = 25°$, $K_h = 0.2$, as H/B changes from 2 to 7, the increase in $F_{\gamma s}$ values are 2.65, 2.45, 2.6 times, respectively, for $\alpha = 0°$, $\alpha = \beta$ and $\alpha = 90°$, respectively. The rate of reduction in $F_{\gamma s}$, due to horizontal seismic acceleration coefficients is higher for lower values of H/B. For example, from Fig. 2e, for the case of $\phi = 40°$, $\beta = 25°$ as K_h changes from 0.0 to 0.2, the reductions in $F_{\gamma s}$ values for $H/B = 1$ and 7 are found to be 40 and 36%, respectively. Similar trends can be observed for anchor parallel to sloping ground and vertical anchor. From all figures, it is clearly seen that with the increase of slope angle (β) and horizontal seismic acceleration coefficient (K_h), there is a significant reduction in the magnitude of $F_{\gamma s}$. For higher values of K_h, the slope becomes stable at high magnitude of angle of internal friction (ϕ). The magnitude of $F_{\gamma s}$ is found to be maximum for $\alpha = 90°$, that is, when the plate becomes vertical. It can be observed from Fig. 5, the magnitude of $F_{\gamma s}$ is higher for rough anchor when compared with smooth anchor plate for $\alpha = 0°$ and 90°,

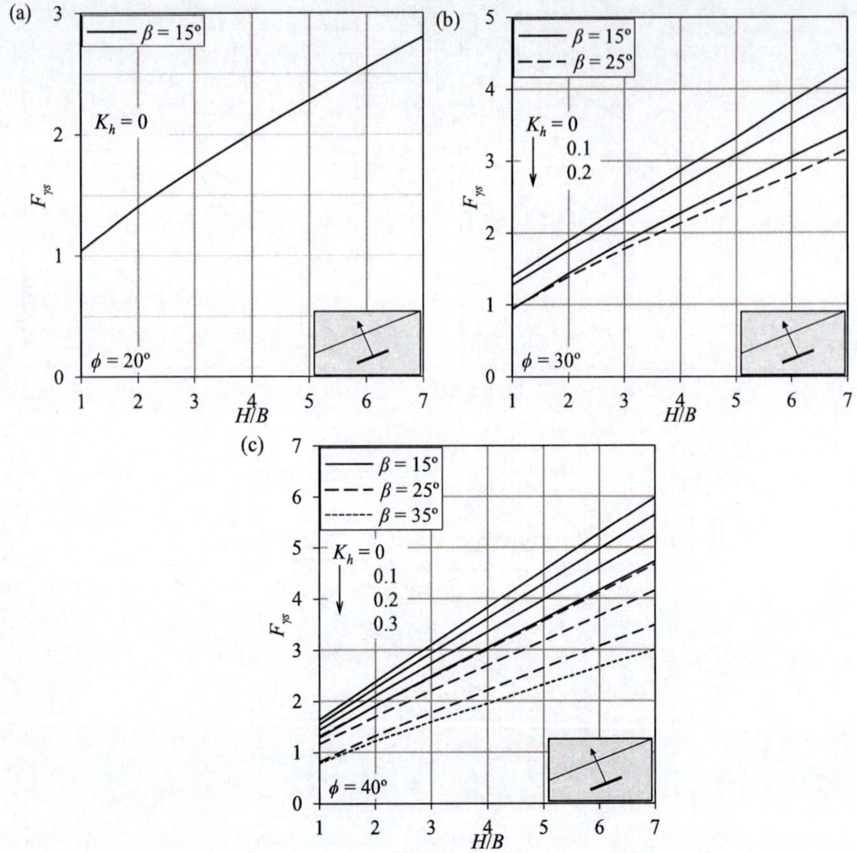

Fig. 3 For anchor placed parallel to sloping face ($\beta = \alpha$), variation of $F_{\gamma s}$ with H/B for different values of β and K_h **a** $\phi = 20°$, **b** $\phi = 30°$, **c** $\phi = 40°$

whereas there is no significant effect of anchor roughness observed when anchor placed parallel to the sloping surface.

5 Comparisons

In Fig. 6a, for horizontal anchor plates, the values of $F_{\gamma s}$ with changes in horizontal seismic coefficient (K_h) for $\phi = 30°$ and 40° and $H/B = 4$ and 7 were compared with the numerical results of Kumar [5] using the upper bound theorem of limit analysis and Bhattacharya and Kumar [20] by using lower bound finite element limit analysis. As expected, the present values of $F_{\gamma s}$ are found to be a little lower than the upper bound solutions provided by Kumar [5] and closely matching with the lower bound

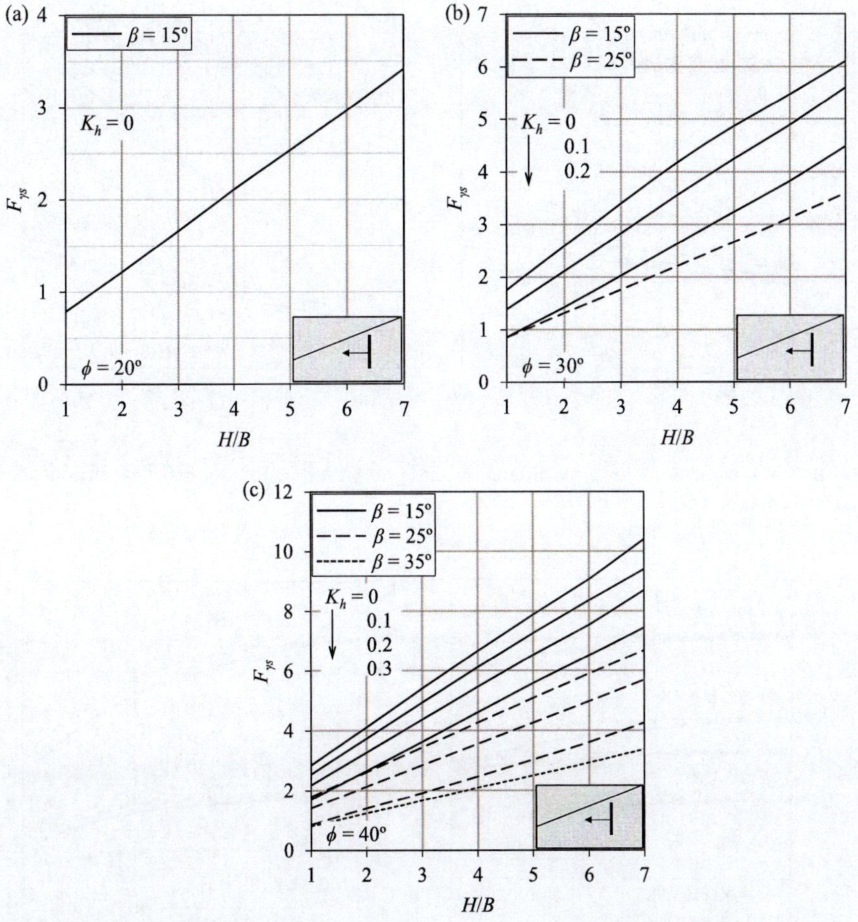

Fig. 4 For vertical anchor ($\alpha = 90°$), variation of $F_{\gamma s}$ with H/B for different values of β and K_h **a** $\phi = 20°$, **b** $\phi = 30°$, **c** $\phi = 40°$

solutions of Bhattacharya and Kumar [20]. The variation of $F_{\gamma s}$ with slope angle (β) is compared with the rigid block upper bound solutions of Yu et al. [17] as presented in Fig. 6b. It needs to be mentioned that the solutions of Yu et al. [17] are available only for static condition. The results of Yu et al. [17] are slightly higher than the present solutions.

6 Conclusions

By using the lower bound theorem of the limit analysis in combination with finite elements and second-order cone programming, the seismic pullout capacity of strip

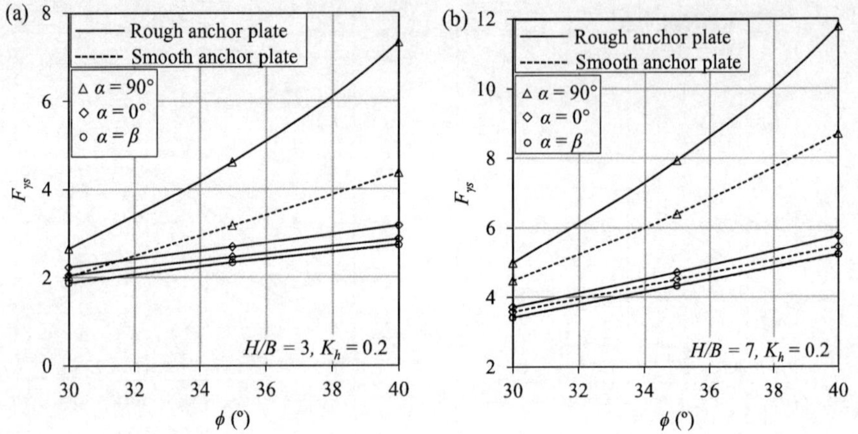

Fig. 5 Variation of $F_{\gamma s}$ with ϕ for different values of α with $\beta = 15°$, $K_h = 0.2$ for smooth and rough anchor for **a** $H/B = 3$, **b** $H/B = 7$

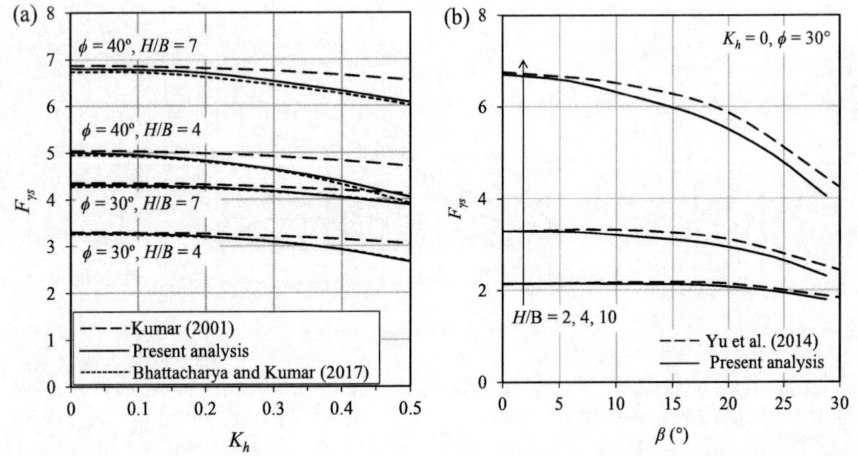

Fig. 6 Comparison of pullout capacity of horizontal anchor embedded in **a** horizontal ground with different combination of K_h, H/B and ϕ; **b** sloping ground for different values of β and H/B

plate anchor embedded in sand under the influence of seismic loading has been determined for three different orientations (i.e., horizontal, parallel to slope and vertical) of the plate with different combinations of H/B, β, ϕ and K_h. The pullout resistance has been found to be maximum when the anchor plate is vertically oriented. On the other hand, for horizontal and vertical orientation of the plate, the pullout resistance becomes minimum for smooth anchor plate and negligible effect of anchor roughness observed for the case of anchor parallel to the sloping surface ($\beta = \alpha$).

The pullout resistance increases quite significantly with an increase in the values of both H/B and ϕ.

References

1. Meyerhof GG, Adams JI (1968) The ultimate uplift capacity of foundations. Can Geotech J 5(4):225–244
2. Rowe RK, Davis EH (1982) The behaviour of anchor plates in sand. Geotechnique 32(1):25–41
3. Murray EJ, Geddes JD (1987) Uplift of anchor plates in sand. J Geotech Eng 113(3):202–215
4. Rao KSS, Kumar J (1994) Vertical uplift capacity of horizontal anchors. J Geotech Eng 120(7):1134–1147
5. Kumar J (2001) Seismic vertical uplift capacity of strip anchors. Geotechnique 51(3):275–279
6. Merifield RS, Lyamin AV, Sloan SW (2005) Stability of inclined strip anchors in purely cohesive soil. J Geotech Geoenviron Eng 131(6):792–799
7. Merifield RS, Sloan SW (2006) The ultimate pullout capacity of anchors in frictional soils. Can Geotech J 43(8):852–868
8. Kumar J, Kouzer KM (2008) Vertical uplift capacity of horizontal anchors using upper bound limit analysis and finite elements. Can Geotech J 45(5):698–704
9. Ghosh P (2009) Seismic vertical uplift capacity of horizontal strip anchors using pseudo-dynamic approach. Comput Geotech 36(1–2):342–351
10. Pain A, Choudhury D, Bhattacharyya SK (2016) Seismic uplift capacity of horizontal strip anchors using a modified pseudodynamic approach. Int J Geomech 16(1):04015025
11. Ganesh R, Khuntia S, Sahoo JP (2018) Seismic uplift capacity of shallow strip anchors: A new pseudo-dynamic upper bound limit analysis. Soil Dyn Earthq Eng 109:69–75
12. Kumar J (1997) Upper bound solution for pullout capacity of anchors on sandy slopes. Int J Numer Anal Meth Geomech 21(7):477–484
13. Choudhury D, Rao KSS (2004) Seismic uplift capacity of strip anchors in soil. Geotech Geol Eng 22(1):59
14. El Sawwaf MA (2007) Uplift behavior of horizontal anchor plates buried in geosynthetic reinforced slopes. Geotech Test J 30(5):418–426
15. Ganesh R, Sahoo JP (2016) Uplift capacity of horizontal strip plate anchors adjacent to slopes considering seismic loadings. Soils Found 56(6):998–1007
16. Khuntia S, Sahoo JP (2018) Vertical uplift capacity of strip plate anchors embedded horizontally adjacent to cohesive-frictional soil slopes. Int J Geomech 18(3):06018001
17. Yu SB, Merifield RS, Lyamin AV, Fu XD, Fu D (2014) Kinematic limit analysis of pullout capacity for plate anchors in sandy slopes. Struct Eng Mech 51(4):565–579
18. Sloan SW (1988) Lower bound limit analysis using finite elements and linear programming. Int J Numer Anal Meth Geomech 12(1):61–77
19. MATLAB R2015b [Computer software]. MathWorks, Natick, MA
20. Bhattacharya P, Kumar J (2017) Seismic pullout capacity of inclined anchor plates in sand. Geotech Geol Eng 35(2):679–692

Role of Borehole Information in Minimizing Uncertainties in Surface Wave Testing

Aniket Desai, Alla Kranthikumar, and Ravi S. Jakka

Abstract Seismic surface wave methods have become widely adopted methods to acquire the shear wave velocity (V_s) profile of soil over the last two decades. However, the results of these methods are highly dependent on the interpretation and judgment of the person. When the dispersion curve obtained by an MASW test is inverted, it provides numerous V_s profiles equivalent to the dispersion curve. This generates indecisiveness about the actual V_s profile of the soil. Hence, a need has arisen to develop some ways to reduce the uncertainties in the MASW test. In this study, numerical simulation of the MASW test was carried out, and its results were processed with and without borehole data to study the reduction in uncertainties due to the borehole data. The results demonstrate that the availability of borehole information drastically reduces the uncertainties in MASW testing, thereby increasing the accuracy and reliability of the surface wave methods.

Keywords Seismic site characterization · MASW method · Rayleigh wave propagation · A-priori information

1 Introduction

Seismic site characterization is considered a task of prime importance in geotechnical earthquake engineering. It is required for many crucial operations such as seismic site response analysis, site classification, seismic microzonation. To carry out site characterization, the most popular methods are the surface wave methods currently. Among them, the spectral analysis of surface waves (SASW) method was popular earlier [1]. Currently, the majorly adopted method is the multichannel analysis of surface waves (MASW) [2, 3]. These two are non-invasive methods that provide the shear wave velocity (V_s) profile of the soil. They are based on the dispersion phenomenon of Rayleigh waves. An elaborate description of the surface wave analysis methods is available in [4].

A. Desai · A. Kranthikumar · R. S. Jakka (✉)
IIT Roorkee, Roorkee, Uttarakhand 247667, India
e-mail: ravi.jakka@eq.iitr.ac.in

T. G. Sitharam et al. (eds.), *Earthquake Geotechnics*, Lecture Notes in Civil Engineering 187, https://doi.org/10.1007/978-981-16-5669-9_5

Although the surface wave methods have become so popular, there are some limitations present in the results of these tests. These tests provide the dispersion curve (plot between Rayleigh wave phase velocity and frequency) of the soil. This curve is then inverted to retrieve the V_s profile of the soil. However, the inversion process is non-unique; i.e., several profiles are generated from a single dispersion curve. Hence, the actual V_s profile of soil can't be ascertained with full confidence. It affects further analyses such as site response analysis [5, 6]. Overall, these methods highly rely on the initial definition of V_s and layer thickness by the user. Due to these factors, there is a dire need to have some ways to reduce these uncertainties or account for them in further analyses.

To reduce these uncertainties, one way can be the use of borehole information along with the MASW test. If the information about the number of soil layers and layer thicknesses becomes available along with the MASW test, it may help in generating better MASW results. However, it is an extremely important task to determine how much the borehole information can become useful to reduce the uncertainties in the MASW test. In this study, numerical simulations are used to investigate the extent up to which borehole information can help to minimize the uncertainties in the MASW test. This information can become critical for people working on MASW or any further analyses.

There are some field tests available to obtain the borehole information about the soil. The tests such as standard penetration test (SPT), cone penetration test (CPT) can provide data about the number of soil layers and the layer thicknesses. The effect of a-priori information on the inversion of surface waves has been investigated earlier [7, 8]. It was found that subsurface investigations can help in constraining the surface wave inversion in a better way. In a research work that provides the guidelines on surface wave analysis, it is mentioned that any a-priori information available is highly important for surface wave analysis [9]. However, there is no particular study that examines how much the borehole information can help in reducing the uncertainties in an MASW test. In the current study, such an investigation has been carried out. This study focuses on quantifying the reduction of uncertainties after the use of borehole information along with the MASW test.

2 Model Description

ABAQUS [10] software, which is based on the finite element method, was used for simulating the MASW test. The soil model was taken as an axisymmetric model having a size of 400^*400 m. The dimensions were kept higher to avoid wave reflections and refractions. At the left and right hand boundary, horizontal displacement was restricted and at the bottom, horizontal and vertical displacement was restricted. The finite element mesh of the model is shown in Fig. 1. The criteria given by Kuhlemeyer and Lysmer [11] were used to select the mesh size. For the soil, a synthetic V_s profile was considered as shown in Table 1.

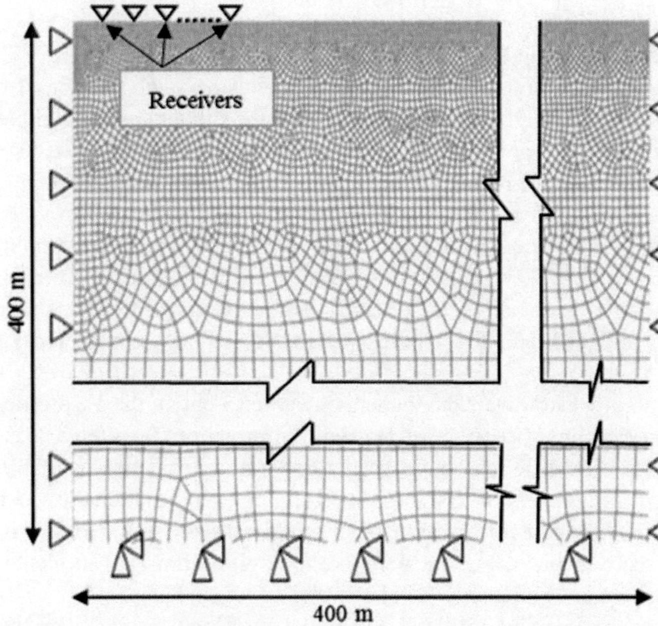

Fig. 1 Finite element mesh of the model

Table 1 Considered shear wave velocity profile in the model

Layer	Thickness (m)	V_s (m/s)	Density (kg/m^3)	ν (Poisson's ratio)
1	5	180	1700	0.3
2	7	240	1750	0.3
3	12	300	1800	0.3
Half-space	–	760	2200	0.24

3 Methodology and Results

The dropping of the hammer in the MASW test was simulated using a modified Gabor wavelet based on Colombero et al. [12]. At the location of loading, a velocity time history was applied as a point load. The equation of velocity time history is given below.

$$v(t) = C_b * \beta * t^{\gamma *} \exp\left[-\left(\frac{2^*\pi^*t}{T_s^*\alpha}\right)^2\right] * \cos(\frac{2*\pi^*t}{T_s*\alpha}),\ 0 \le t \le 1.2T_s \qquad \text{(1)}$$
$$= 0, \qquad \text{otherwise}$$

In the above equation, C_b is the momentum of the load, T_s is the time period of the load, t is the time, and α, β, and γ are constants. The parameters in the equation were assumed as per Colombero et al. [12]. The hammer weight and height of fall were considered 40 kg and 2 m, respectively.

Figure 2 shows the applied input velocity time history. The sampling frequency was kept at 500 Hz. After the load was applied and the simulation was completed, the velocity time histories were collected at 72 locations which represent the receivers in an MASW test. The inter-receiver spacing was kept at 1 m, and the source-receiver distance was kept at 2.5 m. These output time histories were used to generate the dispersion curve for the considered soil profile. Geopsy [13] software, which utilizes the f-k method [14, 15], was used to obtain the dispersion curve from the available velocity time histories. Figures 3 and 4 show the generated dispersion image using Geopsy and the finally extracted dispersion curve for the considered condition, respectively.

The next step was to invert the dispersion curve to obtain the V_s profile. In this procedure, to examine the effect of borehole information, 3 different cases were considered. The first case was such that no borehole information available, and it was assumed during inversion that there are 3 soil layers present. In the second case, again there is no borehole information, but 5 soil layers were assumed. The inversion process was carried out using the neighborhood algorithm [16] implemented by Dinver framework in the Geopsy [13] software.

For the parameterization of the ground model, the V_s values in the top two layers were allowed to vary from 50 to 500 m/s. In the 3rd and 4th layers (for case 2), this range was 200 to 600 m/s. In the bottom-most layer, it was 200 to 1100 m/s for case 1 and 300 to 1100 m/s for case 2. The ranges of the bottom depth of all the layers were selected based on the minimum and maximum wavelengths (λ_{min} and λ_{max}) values and the recommendations by Foti et al. [9]. The ranges of V_p and Poisson's ratio were kept 200–5000 m/s and 0.2–0.49, respectively, for all the layers in both cases. The

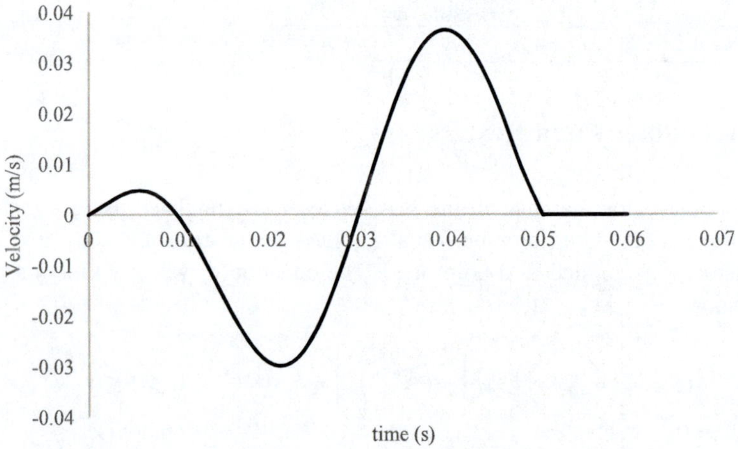

Fig. 2 Applied velocity time history: modified gabor wavelet

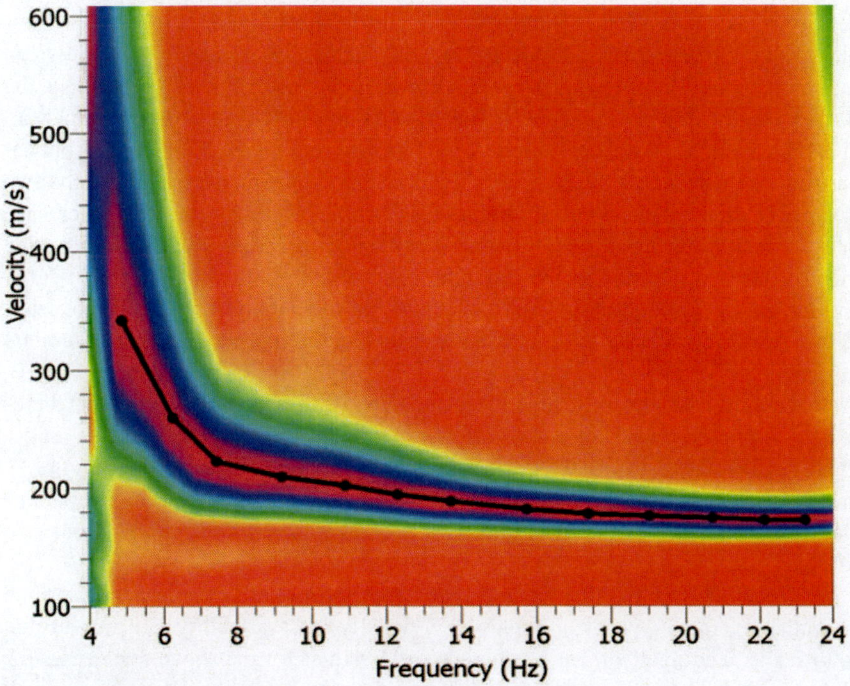

Fig. 3 Dispersion image generated from the numerical simulation of MASW

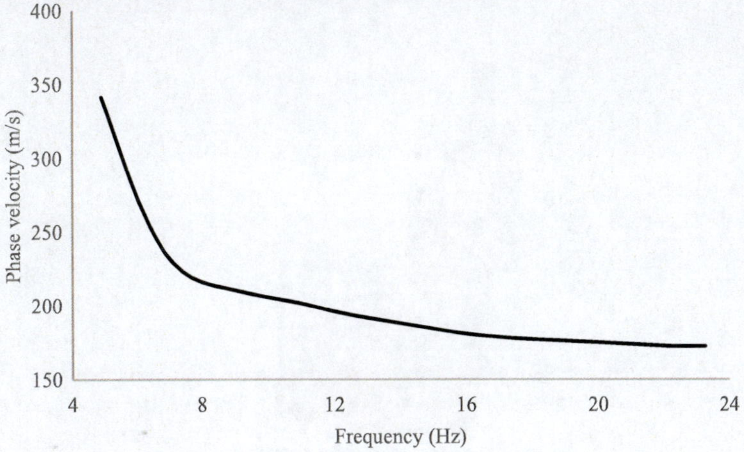

Fig. 4 Extracted dispersion curve of the considered V_s profile

density varied from 1700 to 2000 kg/m^3 in all the upper layers; and in the bottom-most layer, its range was 1800–2000 kg/m^3. The choice of density and Poisson's ratio ranges were loosely based on the recommendations of Foti et al. [9]. Also, the density values were chosen to be increasing with depth as suggested by Ivanov et al. [17]. For a single parameterization, 5 runs (5 different seeds) were carried out for ensuring the robustness of the results. Figures 5 and 6 show the results for the two cases considered above. A total of 10,000 V_s profiles were generated for each case out of which those having misfit value less than 0.1 are shown in the figures. The misfit value is a measure of the difference between the observed dispersion curve and the theoretical dispersion curve of the V_s profiles obtained from the inversion [13]. A lower misfit value indicates a better match of the experimental and theoretical dispersion curves.

In the third case, it was assumed that the borehole information is available; therefore, it is known that there are 4 soil layers, and the layer thicknesses are as per Table 2. This data was utilized during the inversion process. Figure 7 shows the V_s profiles obtained after inversion in the presence of borehole information. The parameterization selected for the inversion, in this case, was similar to the previous two cases.

In the above figures, it is visible that for the third case (where borehole data is available), the V_s profiles display significantly less variability. To present this statistically, a comparison has been made of V_s values for the above three cases for all the layers. They are presented using box plots in Figs. 8, 9, 10 and 11 which show the distribution of V_s values. In the plots, the middle horizontal line inside the colored rectangles indicates the median value. The bottom and top-end line of

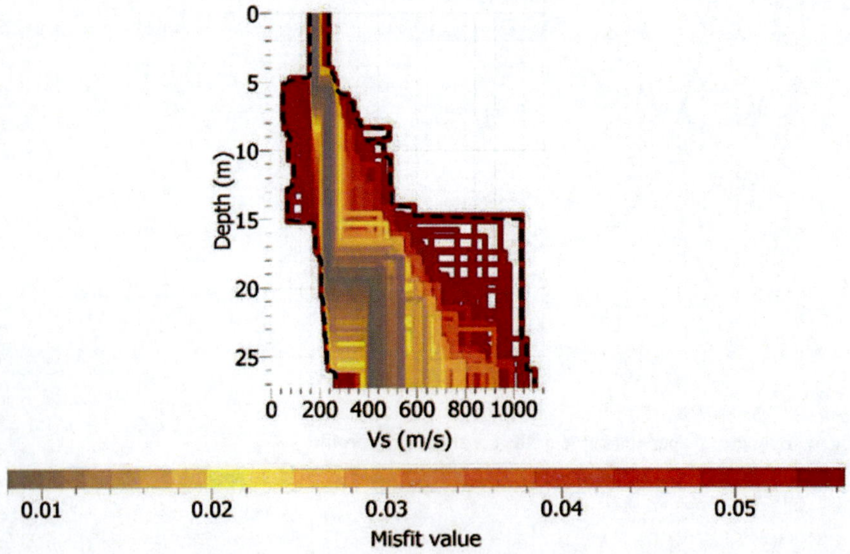

Fig. 5 V_s profiles obtained, while assuming 3 soil layers (without borehole data)

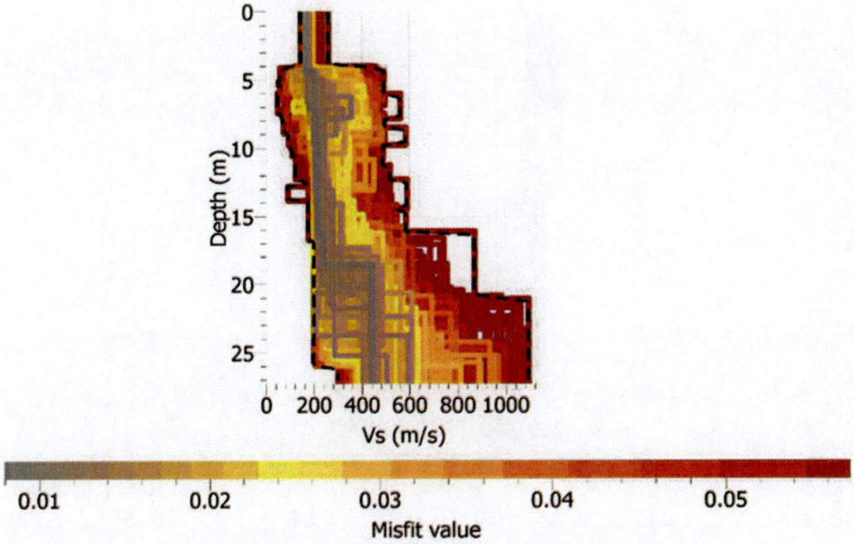

Fig. 6 V_s profiles obtained, while assuming 5 soil layers (without borehole data)

Table 2 Borehole information considered

Soil layer	Thickness (m)
Soft clay	5
Loose sand	7
Stiff clay	12
Rock	-

the colored rectangles signify the lower and upper quartile values. The start (at the lower end) and the end (at the upper end) of the whisker display the minimum and the maximum values of the distribution. The outlier values are neglected here. The median, minimum and maximum values of V_s for all the layers have been compared for the three cases in Table 3.

In Figs. 8, 9, 10 and 11, it is easily visible that in case 3 when borehole information is used, the variability of V_s values for all the layers is very less compared to the previous two cases. Also, Table 3 provides the range of V_s values from inversion, giving a firm indication that the range of V_s values is quite narrow in case 3 compared to the other two cases. For example, in the layer 1, the minimum and maximum values of Vs for the 3 cases are 170 and 197; 159 and 199; and 173 and 179 m/s, respectively. Similar phenomenon is happening for each of the layers present. Also, with no borehole data, the layer thicknesses are assumed as a range of values, which generated many values of thicknesses, causing significant variability in the inverted profiles. On the other hand, with borehole data (in case 3), the layer thicknesses and number of layers would be already known. Because of this, all the inverted profiles

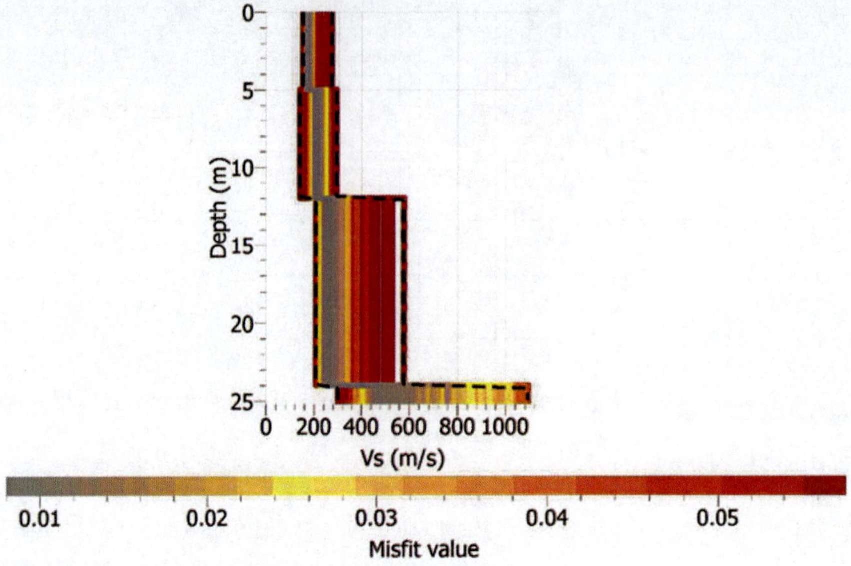

Fig. 7 V_s profiles obtained, while assuming 4 soil layers (Using the borehole data)

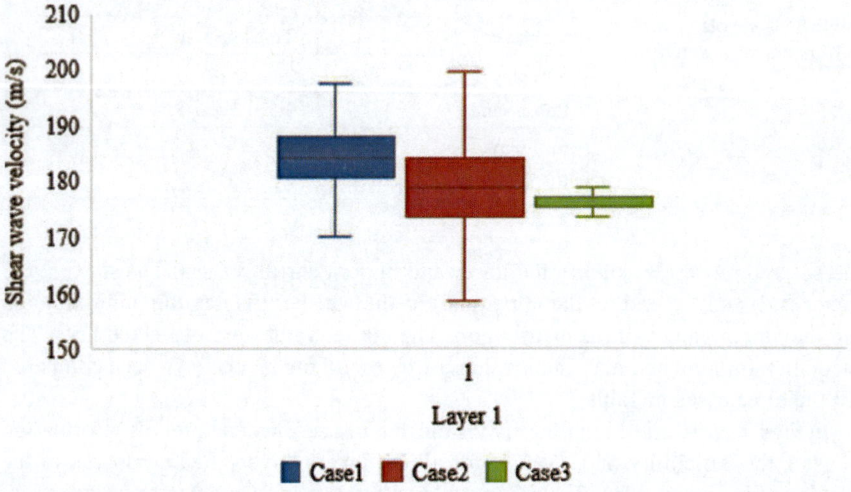

Fig. 8 Box plots showing a comparison of variability in V_s values in layer 1 for the three cases. Case 1: Assumption of 3 layered V_s profile, no borehole data; Case 2: Assumption of 5 layered V_s profile, no borehole data and Case 3: Using borehole data, 4 layered V_s profile considered

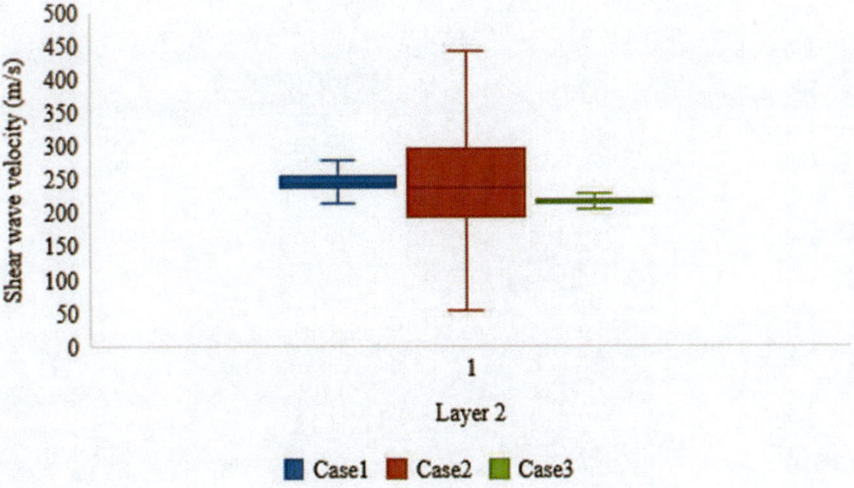

Fig. 9 Box plots showing a comparison of variability in V_s values in layer 2 for the three cases. Case 1: Assumption of 3 layered V_s profile, no borehole data; Case 2: Assumption of 5 layered V_s profile, no borehole data and Case 3: Using borehole data, 4 layered V_s profile considered

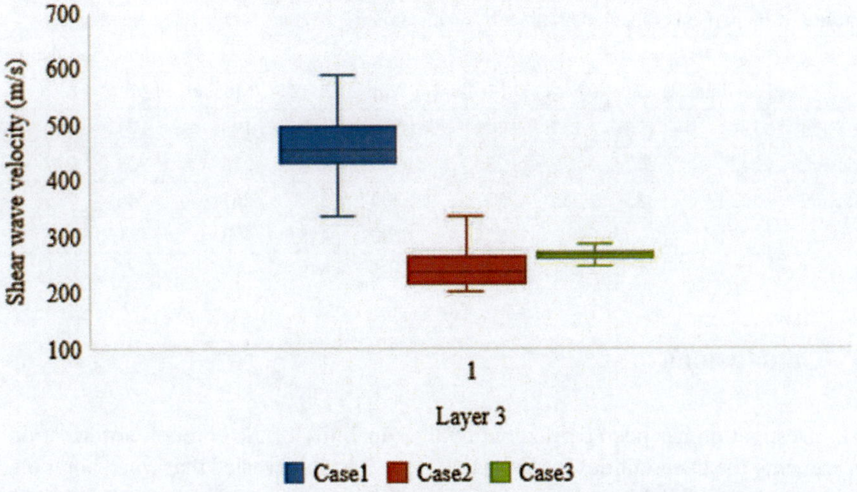

Fig. 10 Box plots showing a comparison of variability in V_s values in layer 3 for the three cases. Case 1: Assumption of 3 layered V_s profile, no borehole data; Case 2: Assumption of 5 layered V_s profile, no borehole data and Case 3: Using borehole data, 4 layered V_s profile considered

have the same values of the layer thicknesses in all inverted models of case 3. These all inferences strongly confirm that the availability of borehole information aids in drastically improving the confidence in the obtained V_s profiles from the MASW test.

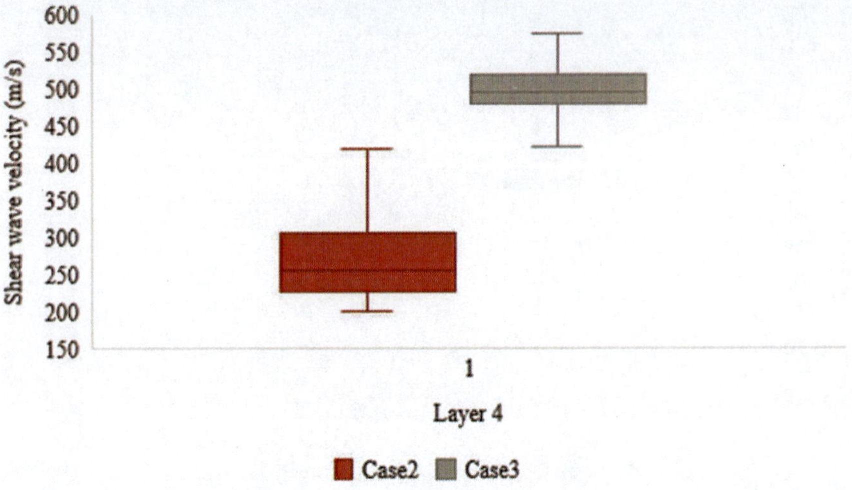

Fig. 11 Box plots showing a comparison of variability in V_s values in layer 4 for the two cases having more than 3 layers. Case 2: Assumption of 5 layered V_s profile, no borehole data and Case 3: Using borehole data, 4 layered V_s profile considered

Table 3 Comparison of median, minimum and maximum V_s values for the three cases

	Case 1			Case 2			Case 3 (With a-priori data)		
	Median	Min	Max	Median	Min	Max	Median	Min	Max
Layer1	184	170	197	179	159	199	175	173	179
Layer2	243	212	277	236	52	442	216	203	227
Layer3	452	335	585	237	200	336	267	246	286
Layer4	–	–	–	254	200	418	493	421	573
Layer5	–	–	–	460	300	678	–	–	–

4 Conclusions

An investigation has been carried out on the capability of the borehole information in reducing the uncertainties in the MASW test. Results revealed that when borehole information is available along with the MASW test, there is a huge reduction in the variability of the V_s profiles representing the site. This indicates that the actual V_s profile can be better represented if the borehole data is also available during surface wave testing. Hence, it can be stated that borehole information can prove as a huge asset in constraining the V_s profiles from the MASW test. In this way, the uncertainties in surface wave testing are significantly reduced when the borehole information is used for V_s profile generation from the surface wave tests.

Acknowledgements The authors wish to thank Science and Engineering Research Board (SERB), Department of Science and Technology (DST), Government of India, for providing financial support for carrying out the current research (Project Grant Code No. SB/FTP/ETA-164-2014).

References

1. Nazarian S, Stokoe KH (1986) Use of surface waves in pavement evaluation. Transportation Research Record 1070, TRB, National Research Council. pp 132–144
2. Park CB, Miller RD, Xia J (1999) Multi-channel analysis of surface waves. Geophysics 64(3):800–808
3. Xia J, Miller RD, Park CB (1999) Estimation of near-surface shear-wave velocity by inversion of Rayleigh waves. Geophysics 64(3):691–700
4. Socco LV, Foti S, Boiero D (2010) Surface-wave analysis for building near-surface velocity models—established approaches and new perspectives. Geophysics 75(5):75A83–75A102
5. Boaga J, Vignoli G, Cassiani G (2011) Shear wave profiles from surface wave inversion: the impact of uncertainty on seismic site response analysis. J Geophys Eng 8:162–174
6. Roy N, Jakka RS, Wason HR (2013) Effect of surface wave inversion non-uniqueness on 1-D seismic ground response analysis. Nat Hazards 68(2):1141–1153
7. Foti S, Sambuelli L, Socco VL, Strobbia C (2003) Experiments of joint acquisition of seismic refraction and surface wave data. Near surface geophysics 1(3):119–129
8. Wood CM, Wotherspoon LM, Cox BR (2015) Influence of a-priori subsurface layering data on the development of realistic shear wave velocity profiles from surface wave inversion. In: Proceedings 6th international conference on earthquake geotechnical engineering, Christchurch, pp 1–9
9. Foti S, Hollender F, Garofalo F, Albarello D, Asten M, Bard P-Y, Comina C, Cornou C, Cox B, Di Giulio G, Forbriger T, Hayashi K, Lunedei E, Martin A, Mercerat D, Ohrnberger M, Poggi V, Renalier F, Sicilia D, Socco V (2017) Guidelines for the good practice of surface wave analysis: a product of the InterPACIFIC project. Bull Earthq Eng 16(6):2367–2420
10. Dassault Systemes (2015) Abaqus. Retrieved from http://www.3ds.com/products-services/simulia/products/abaqus/
11. Kuhlemeyer RL, Lysmer J (1973) Finite element method accuracy for wave propagation problems. J Soil Mech Found Div 99(SM5):421–427
12. Colombero R, Kontoe S, Foti S, Potts DM (2015) Numerical modeling of drop load tests. Soil Dyn Earthq Eng 77:279–289
13. Wathelet M (2008) An improved neighborhood algorithm: parameter conditions and dynamic scaling. Geophys Res Lett 35:L09301. https://doi.org/10.1029/2008GL033256
14. Capon J (1969) High resolution frequency—wavenumber analysis. Proc IEEE 57:1408–1418
15. Yilmaz O (1987) Seismic data processing: society of exploration geophysicists, Tulsa, Oklahama, USA
16. Sambridge M (1999) Geophysical inversion with a neighbourhood algorithm: I searching a parameter space. Geophys J Int 138:479–494. https://doi.org/10.1046/j.1365-246X.1999.00876.x
17. Ivanov J, Tsoflias G, Miller R, Xia J (2009) Practical aspects of MASW inversion using varying density. In: 22nd EEGS symposium on the application of geophysics to engineering and environmental problems, european association of geoscientists and engineers, pp 171–177

Engineering Geological Mapping for Seismic Hazard Assessment of the Gandhinagar City, Western India

Vasu Pancholi, V. K. Dwivedi, B. Sairam, Pallabee Choudhury, Sumer Chopra, and N. Y. Bhatt

Abstract Gandhinagar is the capital of Gujarat state in western India and a centre of intense economic and urban development activity, which requires a constant need for development and expansion. The city, having a total area of 57 km^2, is situated on the bank of the river Sabarmati. Geologically, the city is covered by alluvium deposits, and the top layer of 80–100 m is composed of loose sandy and soft clayey soils. Most of the urban activities in the city are located on these subsoil strata. A geotechnical database is prepared at various locations of the city along with the geological and geomorphological details. In the framework of this study, 14 boreholes have been drilled, and detailed mapping of the city is done according to its lithology, soil properties, standard penetration test (N value) and shear wave velocity (Vs). With the use of the average shear wave velocity, the amplification factor and peak ground acceleration (PGA) at surface are estimated and further used for the assessment of seismic hazard. The seismic hazard analysis is done by using published ground motion prediction equations (GMPEs) developed for the Gujarat region. Four published GMPEs are used in a logic tree approach by taking magnitude 6.0 as the scenario earthquake for near-field case. The distance of the source from each borehole is calculated and used in GMPEs. Two scenario earthquakes representing Mw7.7, 2001 Bhuj earthquake and M6.5 from Narmada-Son region, are also considered as far-field earthquakes. From the detailed analysis, soil maps, SPT N value maps, Vs map and PGA map are prepared and placed on Geographic Information system (GIS) platform to interpret the complex data for reliable interpolation. The study shows that the area in the central and south-western parts of the city are more prone to high ground shaking and related damage during a major earthquake in Gujarat. The results can be considered to guide further urban development and possible infrastructural development of the city.

V. Pancholi (✉) · V. K. Dwivedi · B. Sairam · P. Choudhury · S. Chopra
Institute of Seismological Research, Gandhinagar 382009, India
e-mail: vasupancholi@isr.res.in

N. Y. Bhatt
M. G. Science Institute, Gujarat, Ahmedabad 380015, India

T. G. Sitharam et al. (eds.), *Earthquake Geotechnics*, Lecture Notes in Civil Engineering 187, https://doi.org/10.1007/978-981-16-5669-9_6

Keywords Seismic hazard · Engineering geology · Standard penetration test · Shear wave velocity

1 Introduction

Seismic hazard always attracts much attention in the field of the earth science and any other geologic studies. Gujarat, western India, has recently experienced a major earthquake in 2001. The Bhuj earthquake (2001) drew the attention of society towards the safeguard and different safety measures to be taken into account at the time such major earthquake to minimise the hazard caused by it. As Gandhinagar is the capital city, the economy and other social prospects of the state depend on its advancement and safety. In order to carry out seismic hazard analysis and its mitigation, geotechnical studies play vital role all around. The Gandhinagar City is very important part of the Gujarat State as it is the capital of the state and growing rapidly with modern infrastructures. The city area is about 57 km^2 (10×4.5 km), lying around 23.2°N 72.6°E (or 23°11'–23°15' and 72°37'–72°40') (Fig. 1). Most of the area falls in Survey of India Topographical Sheet 46 A/12. Elevation is 80–100 m, with higher elevation in the NE and general slope towards west and south. Population is about 2.67 lakhs [2]. Annual rainfall is 70 cm and groundwater potential yield are 10–25 L/sec with water table at 15–20 m depth. The area has small oil fields of Vavol and

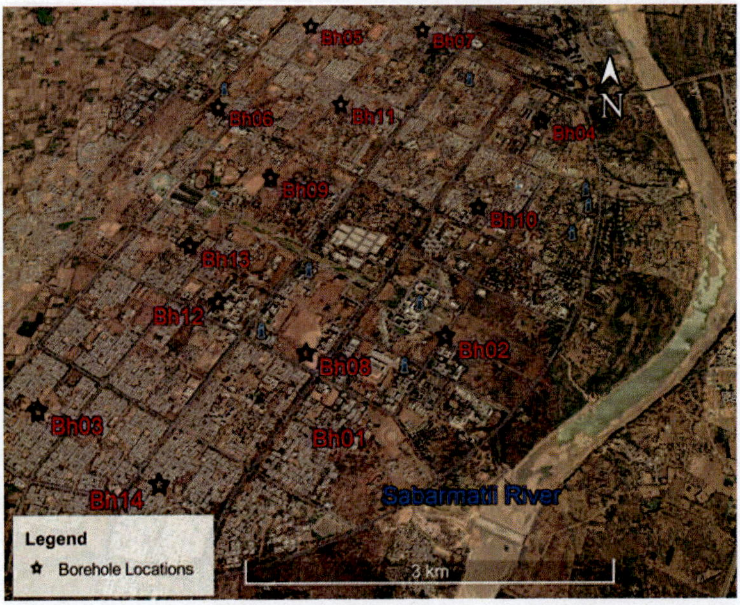

Fig. 1 Borehole location map

Table 1 Stratigraphy of gandhinagar area

Period	Epoch	Lithology	Thickness
Quaternary	Holocene	Sand, silt, clays and gravels	80 – 100 m ~ 80 m near Gandhinagar
	Pleistocene	Yellow and grey clays, coarse sands, gravel and kankar	300 m
Tertiary	Pliocene	Claystone, sandstone and conglomerate	200 m
	Miocene	Ferruginous sandstone, conglomerate and grey clay	400 m
	Oligocene	Grey shale, sandy shale and argillaceous sandstone	~ 600 m
	Eocene	Black shale, carbonaceous shale	430 m
	Paleocene	Volcanic conglomerates	+1000 m
Mesozoic	Cretaceous	Deccan trap	+4000 m

Indroda. The region has experienced several earthquakes of magnitude ≤ 6 in the past.

The city of Gandhinagar is part of north Gujarat alluvial tract and situated in tectonically controlled, about 50 km wide, Cambay Basin, which is characterized by two northwest-trending boundary faults named as East Cambay Fault (lying close to Gandhinagar) and West Cambay Fault. The basin is filled up with 300–400 m thick Quaternary sediments and about 3 km or more Tertiary sediments constituting alluvial plains of Mainland Gujarat. The Quaternary sediments are of fluvial and Aeolian origin. The detail stratigraphy [11] is given in Table 1.

There is not much seismicity seen around Gandhinagar City. Eight shocks of magnitude 2.7–5 have been registered within 80 km radius of the city. In the historical past, Ahmedabad located 15 km south-west of Gandhinagar experiences 3 earthquakes in 1840, 1843 and 1864 of magnitude 4.6, 3.7 and 5.0, respectively.

The state capital Gandhinagar located on the right (west) bank of Sabarmati River falls in zone III where intensity VIII from a local earthquake (of magnitude 6) or intensity VII from a regional earthquake can be expected that can damage single to multistory buildings. Thus, there is a need for engineering geological mapping of the area. The investigations include geological, geotechnical, and geophysical measurements, which yielded SPT N value, shear wave velocity, amplification maps as well as PGA using different GMPEs for near and far-field earthquake source.

The seismic hazard of Gujarat region has been addressed by number of workers. Some seismic hazard studies have been done in Gujarat with the deterministic seismic hazard assessment (DSHA) and probabilistic seismic hazard assessment (PSHA) by researchers. For Gujarat, DSHA had been done by Parvez et al. [15], Chopra et al. [5], Kolathayar et al. [8], Mohan [13], Choudhry et al. [6], Magrin et al. [9] and Parvez et al. [16], while PSHA of Gujarat had been done by Mandal et al. [10], National Disaster Management Authority [14], Chopra and Choudhury [4], Joshi et al. [7],

Rout et al. [18]. In this study, the city of Gandhinagar is selected for study which is almost equidistant from the epicentre of Bhuj earthquake (2001) and lies consecutive north to Ahmedabad City where much destruction was recorded that time. In this paper, we estimate the seismic hazard possibilities with the help of the geotechnical investigations.

2 Methodology

Geotechnical studies were carried out in detail through drilling of 14 boreholes. A total of 8 boreholes of 50 m depth and 6 boreholes of 35 m depth were drilled (Fig. 1). The SPT tests were carried out in all these boreholes for N-value at every 3 m interval. Disturbed/undisturbed soil samples were collected alternately at every 1.5 m interval and their soil properties were measured. Further, the N value was corrected by the correction factors like (a) Overburden Pressure (CN), (b) Hammer energy (CE), (c) Borehole diameter (CB), (d) the presence or absence of liner (CS), (e) Rod length (CR) and (f) Fine content Seed et al. [20, 21]; Youd et al. [23]; Cetin et al. [3], Skempton [22] and Pearce and Baldwin [17]. The shear wave velocity with the help of multi-channel analysis of surface waves (MASW) was performed at the different locations nearer to the borehole locations Sairam et al. [19]. After performing the test for the index properties of the soil, it is found that the area is mostly covered by the sandy soil and at some places, the area is having the silty sand type of soil. Based on that, the empirical correlation is carried out with the help of 33 pairs of the SPT-Vs at different locations in two categories, i.e. (1) for sandy soil and (2) for all soil. Further, the seismic hazard analysis is done both for near and far-field scenarios, by using published ground motion attenuation equations (GMPE) developed for Gujarat region. For near-field, a scenario of M6.0 earthquake was considered and east marginal fault as capable fault. Two scenario earthquakes equivalent of Mw7.7, 2001 Bhuj earthquake and M6.5 Narmada-Son region are also considered as far-field. PGAs are calculated at 14 boreholes at base rock for all the scenarios. The Vs30 values are used for the calculation of the amplification factor (AF) to estimate the peak ground accelerations (PGA) at surface. The GMPEs of Mandal et al. [10], NDMA [14], Chopra and Choudhury [4] and Joshi et al. [7] are in the form of

$$\ln(\mathrm{PGA}) = -7.9527 + 1.4043M - \ln(rjb2 + 19.832)1/2 - 0.0682S \quad (1)$$

$$\begin{aligned}\ln(\mathrm{PGA}) = {} & -4.7653 + 1.3935M - 0.0162M2 - 0.002R \\ & - 1.4183\ln(R + 0.0187e0.9899M) + 0.1243\log(R)f0 + \ln(0.3596)\end{aligned} \quad (2)$$

$$\log(\mathrm{PGA}) = -3.8017 + 0.5568M - 0.00929R - 0.1125S \quad (3)$$

$$\ln(\text{PGA}) = -2.56 + 1.17M - 0.015R - 0.001 * \ln(E + 15) \quad (4)$$

where $f0 = \max[\ln(r/100),0]$, M is the magnitude, *rjb* is the closest distance to the surface projection of the fault rupture in km, R is the hypocentral distance in km, E is the epicentral distance in km. *S* is considered according to the local geology and taken 0 for rock and 1 for soil sites. Equations 1 and 2 provide PGA at Vs = 760 m/s and $Vs \geq 1500$ m/s, respectively. Equation (5) gives the ground motion at soil or rock.

Evaluation of the site effect is very important in the prediction of strong motion based on GMPE. One of the methods usually used in site effect evaluation is based on the correlation between the site effects and Vs30 Borcherdt et al. [1]; Midorikawaet al. [12]. Midorikawa et al. [12] provided the following formulation for determining the acceleration amplification factor

$$\log \text{ARA} = 1.35 - 0.47 \log(\text{AVs30}) \quad (5)$$

where ARA is the amplification for the bedrock of $Vs = 745$ m/s and AVs30 is the average shear wave velocity to a depth of 30 m. Borcherdt and Fumal [1] have provided acceleration-dependent amplification factors (AF) estimated through geotechnical and geophysical investigations at a number of strong-motion sites in the USA. He provided amplification factors for short and mid-period ranges for different shear wave velocities with respect to base rock of different shear wave velocities. The short period (0.3 s) AF with respect to base rock of shear wave velocity 1378 m/s is given by

$$AF = (1378/\text{Vs30})0.326 \quad (6)$$

where Vs30 is the average shear wave velocity to a depth of 30 m.

3 Result and Discussion

The Gandhinagar City falls in zone III of the seismic zoning map of India (BIS) where the maximum expected magnitude of an earthquake is 6.0. Gandhinagar City is the capital city of Gujarat, and it is having special importance due to the rapid development and infrastructural growth. We carried out all the field and laboratory analyses very precisely and prepared maps on the GIS platform so they can be helpful for further infrastructural development. In the study area, arenaceous and argillaceous soils are alternatively deposited with varying thickness. Layers of clayey soil are thin and found at a shallower depth at most of the places. The northern and central part of the city comprises a mixture of silt and sand with a thin layer of clay. In the eastern city, fine-grained soil is found in the upper layer. At some places of southern part of the city, another layer of sandy soil is present up to borehole termination depth. The sub-surface litho-assemblage of Gandhinagar City is shown in Fig. 2. The standard penetration test was performed at an interval of 3 m till the termination depth. The $(N1)_{60}$ maps were prepared for 3, 6, 9 and 12 m depth. At the depth of 3 m, the values are ranging up to a maximum range of 50. Values from 30–40 cover a major part of the city which includes central and some northern part of the study area. Values from 15 to 30 are seen majorly in the eastern (Sector-16, 17, 19, 22, 23) and south-western part (Infocity) of the study area. Values greater than 45 are seen as some patchy outcrops northern and western part of the study area. The area around Sector-22 is having $(N1)_{60}$ less than 15 shows the highest possibility of the hazard in the study area as it indicated very poor compaction of the soil (Fig. 3a). At the depth of 6 m, the values are ranging up to a maximum range of 100 (Refusal). This depth

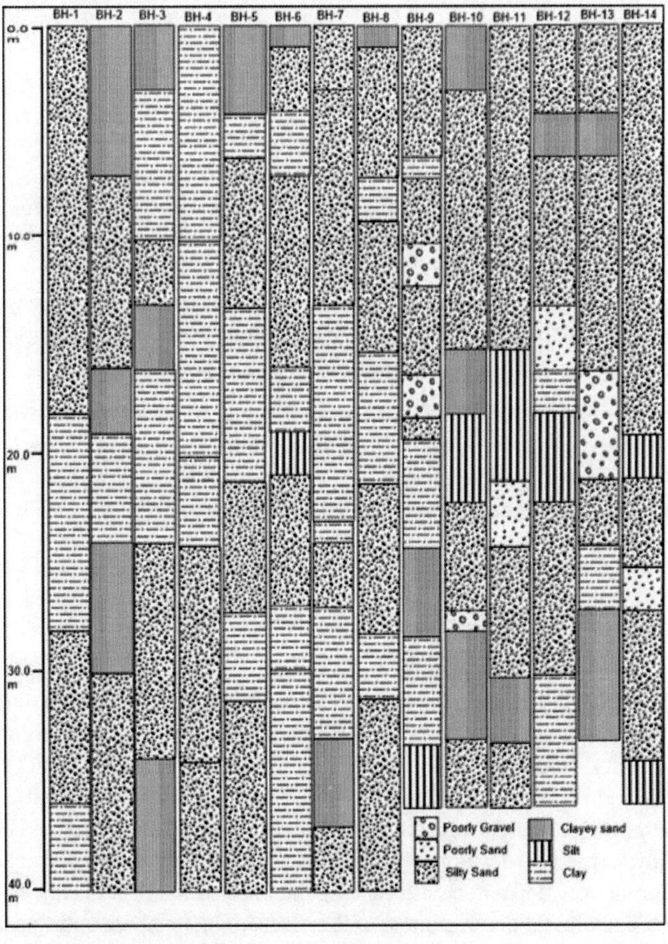

Fig. 2 Litho-assemblage of study area

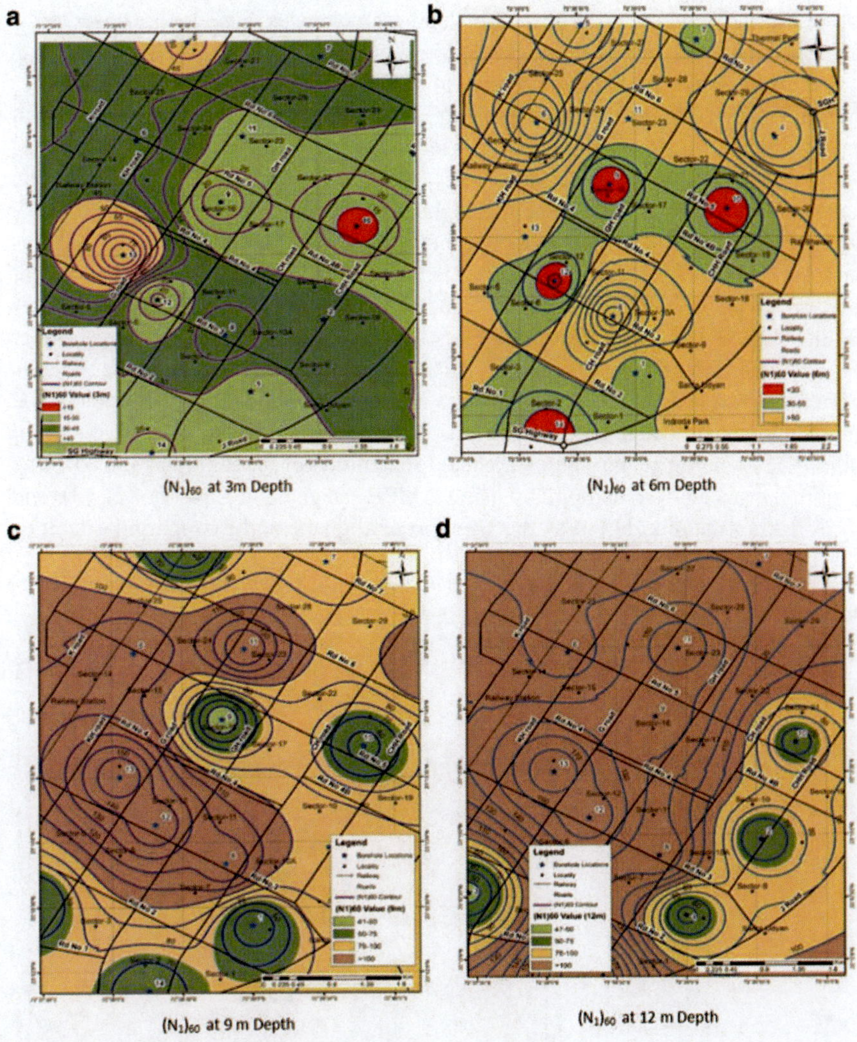

Fig. 3 **a** $N_1(60)$ at 3 m depth, **b** $N_1(60)$ at 6 m depth, **c** $N_1(60)$ at 9 m depth, **d** $N_1(60)$ at 12 m depth

is very important when the plan to build more than ten-storey building. Values from 0 to 30 are very less and are seen in the scattered form at the location of Sector-16, 22 and the southern part of the sector 2 and sector 21. A major part of the city is covered by value ranging from 50 to 100 $(N1)_{60}$. Here, at 6 m depth the central and south-western part of the study area is having $(N1)_{60}$ in between 30 and 50 (Fig. 3b). At the depth of 9 m, the values are ranging up to a maximum range of 100 (Refusal). The lowest value (40–50) is seen in very less area around sector-16. Other than that, most of the area is having $(N1)_{60} > 50$. Highest values are observed in the western part of the study area (Fig. 3c). At 12 m depth, same ranges of values are observed as in the previous category. In this category, most of the area is having $(N1)_{60}$ more than 50 which clearly indicated the presence of stiff soil at the depth of 12 m (Fig. 3d).

The Vs_{30} in the study area ranges from 330 to 405. In addition, the average shear wave velocity in the top 30 m (Vs_{30}) is used in the NEHRP Provisions and the 1997 Uniform Building Code to separate sites into classes for earthquake engineering design. Within the study area, the green area, covered by boreholes 1, 9, 12 and 14, shows the lowest shear wave velocity and comes under site class D (stiff soil) as per the NEHRP classification. The area is having low value due to the presence of soft soil layers (having low SPT- N value) at shallow depth. The other boreholes having $Vs_{30} \geq 360$ m/s come under class C (very dense soil and soft rock) as per NEHRP classification Fig. 4. The amplification soil rating is given based on the Vs_{30} value as shown in Table 2. The study area comes under the site class B and class C soil category. The Gandhinagar City falls in zone III of the seismic zoning map of India (BIS) where the maximum expected magnitude of an earthquake is 6.0. The seismic hazard analysis is done by using GMPEs as given in Eqs. (1), (2), (3), and (4) by taking magnitude 6.0 as the scenario earthquake and east marginal fault as

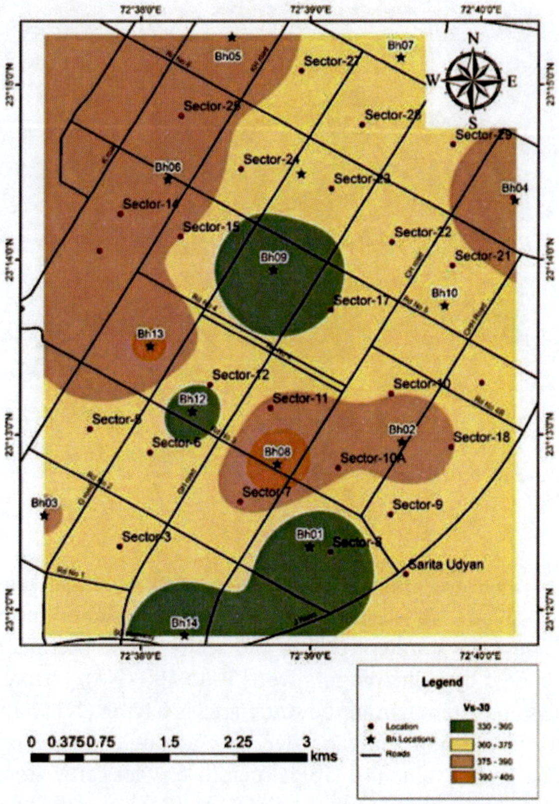

Fig. 4 Shear wave velocity (Vs30) Map

Table 2 Amplification rating and corresponding soil category label

Borehole No	V_S30	Amplification rating (Finn 1991)	Soil category label
01	334	Moderate	C
02	384	Low	B
03	341	Moderate	C
04	384	Low	B
05	359	Moderate	C
06	427	Low	B
07	342	Moderate	C
08	374	Low	B
09	269	Moderate	C
10	316	Moderate	C
11	325	Moderate	C
12	307	Moderate	C
13	308	Moderate	C
14	309	Moderate	C

capable fault for generating near-field earthquake [5]. The focal depth is considered as 10 km. The distances of the sources from every borehole are calculated and used in GMPEs. Two scenario earthquakes equivalent of Mw7.7, 2001 Bhuj earthquake and M6.5 Narmada-Son region are also considered as far-field [5] and included in the hazard analysis. In the first place, PGAs are estimated using Eqs. (1) to (4) for a scenario earthquake of M6.0, accounted for the near-field earthquake. The distances from boreholes (BH1 to BH14) are calculated, and PGAs are estimated accordingly. It may be noted that Chopra and Choudhury [4] provided the ground motion for the rock and soil sites as they used only recorded data. There is no mention about the local geology in the study done by Joshi et al. [7]. Mandal et al. [10] and NDMA [14] provided ground motions at $Vs = 760$ and $Vs \geq 1500$ m/s, respectively. Thus, a direct comparison between the PGAs estimated with the four GMPEs is not much realistic. Hence, an amplification factor needs to be multiplied with the acceleration values given by Mandal et al. [10] and NDMA [14] to arrive at ground motion at the surface. There is no such data for the Gujarat region. So, following Midorikawaet al. [12] (Eq. 5) and Borcherdt [1] (Eq. 6), the amplification factors are calculated with respect to the base rock of Vs 745 m/s and 1378 m/s, respectively, using the shear wave velocities obtained at the 14 boreholes. The shear wave velocities at all the 14 boreholes are estimated till the depth of 30 m using SPT N value. The Vs30 values are used in the GMPEs of Mandal et al. [10] and NDMA [14] to calculate the PGAs at the surface. The GMPE of Chopra and Choudhury [4] was considered for soil sites and GMPE of Joshi et al. [7] considered as it is. The variation of PGAs estimated using four GPMEs with distance is shown in Fig. 5. The PGAs for a near-field earthquake vary from 50 to 80 cm/s^2 as estimated by Mandal et al. [10] and Joshi et al. [7] and from 150 to 250 cm/s2 as estimated Chopra and Choudhury [4]

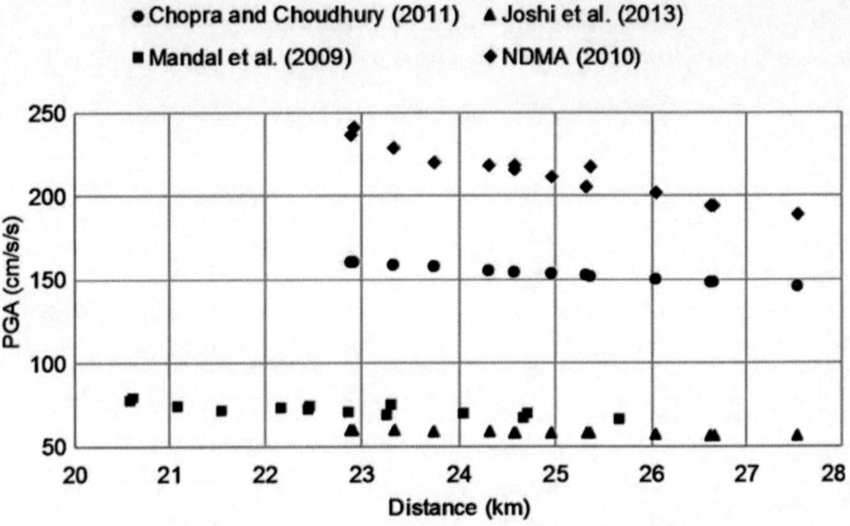

Fig. 5 Surface PGAs for a scenario earthquake of magnitude 6.0 estimated by four GMPEs at different boreholes. The distance represents the hypocentral distance except for Mandal et al. [10] where it represents the fault distance

and NDMA [14], at a distance range 20–28 km. It is evident from Fig. 5 that Mandal et al. [10] and Joshi et al. [7] provide low ground motion values while NDMA [14] provides high values. Chopra and Choudhury [4] provide PGA values consistent with the BIS recommended value for zone III. To account for the far-field earthquake, two sources were considered (i) a scenario earthquake of M6.5 from Narmada-Son region (distance ~ 180 km) and (ii) a scenario earthquake of M7.7 from the Kachchh region (distance ~ 240 km). Chopra and Choudhury [4] relation is valid for magnitude up to 6.0, and Joshi et al. (2013) relation is valid up to a distance 120 km. In view of this, for both the far-field scenarios GMPEs of NDMA [14] and Mandal et al. [10] were considered to calculate the PGAs. Here also, using the amplification factors obtained using Eqs. 5 and 6, the PGAs at 14 boreholes are estimated from GMPEs of Mandal et al. [10] and NDMA [14] for scenario earthquakes of M6.5 and M7.7. Figure 6a and b shows the PGAs at 14 boreholes estimated using both the GMPEs. It is observed in Fig. 6b that the PGA values at Narmada region do not differ much between the two GMPEs. The PGAs are in the range 22–28 cm/s^2. On the contrary, for the Kachchh region, the two GMPEs give many different values of PGA. The GMPE of Mandal et al. [10] gives higher values of PGA for Kachchh region, being in the range 95–105 cm/s^2, than the GMPE of NDMA [14], the value is in the range 60–65 cm/s^2. Next, to follow a unified approach for all the three scenarios, GMPEs of Mandal et al. [10] and NDMA [14] are used and a logic tree was constructed for the GMPEs. To construct the GMPEs, Mandal et al. [10] used only recorded data and NDMA [14] used simulated data, generated using region-specific parameters. In view of this, weights of 0.6 and 0.4 were assigned to relationships of Mandal et al.

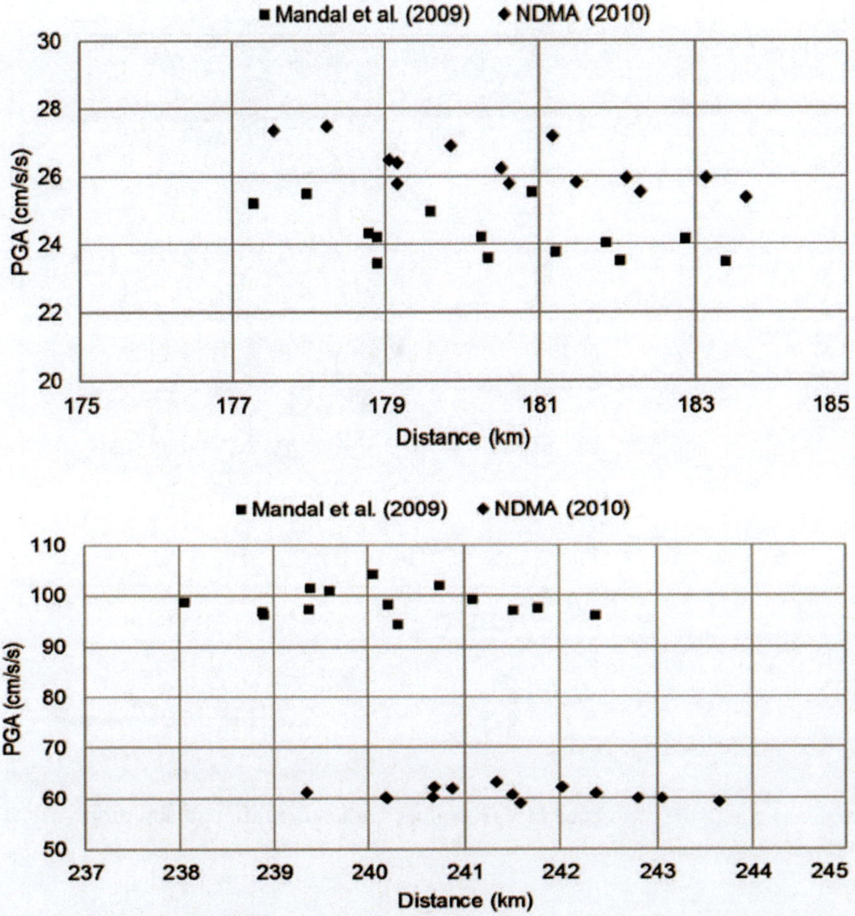

Fig. 6 PGA (in cm/s2) at the surface for scenario earthquakes of **a** magnitude 6.5 **b** magnitude 7.7 estimated by two GMPEs. The distance represents fault distance and epicentral distance for Mandal et al. [10] and NDMA [14], respectively

[10] and NDMA [14], respectively. Three PGA maps were thus prepared for these three scenarios of M6.0, M6.5, M7.7 as shown in Figs. 7, 8 a and b, respectively. It is observed from Fig. 7 that PGAs for a near-field earthquake in the study area vary between 115 and 145 cm/s^2. Within the study area, the PGA values at boreholes 5,6, 7, 11 and 13 are lowest, being in the range 115–125 cm/s2 and highest PGA values are observed at BH 1, 2 and 14, the value ranging from 136 to 145 cm/s^2. Figure 8 a and b reveals that within the study area, BH 1, 9, 12, 14 show higher PGA values for both the scenarios, M6.5 and M7.7. The values being in the range 26–30 and 86–90 cm/s^2, respectively, for M.6.5 and M7.7. Rest of the boreholes show PGA values of 20–25 and 80–85 cm/s^2, respectively, for M.6.5 and M7.7. The difference in PGA

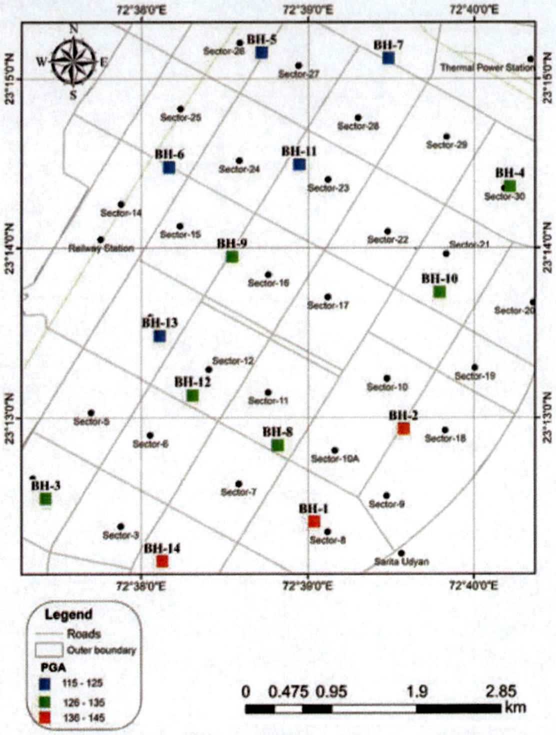

Fig. 7 Weighted PGAs (in cm/s^2) estimated using GMPEs of Mandal et al. [10] and NDMA [14] at 14 boreholes for a scenario earthquake of M6.0

values is due to the difference in Vs30 values, indicating different soil properties at different boreholes.

4 Conclusions

Gandhinagar City is consisting silty sand at most of the part. The soil comes under stiff category from 12 m depth at almost all area. According to NEHRP soil classification based on Vs30, the study area falls under site class C (very dense soil and soft rock) and site class D (stiff soil). The relatively threatening area of Gandhinagar City is Sectors 2, 12 and 16 in terms of Seismic hazard. The study may helpful for further urban infrastructural development of the city.

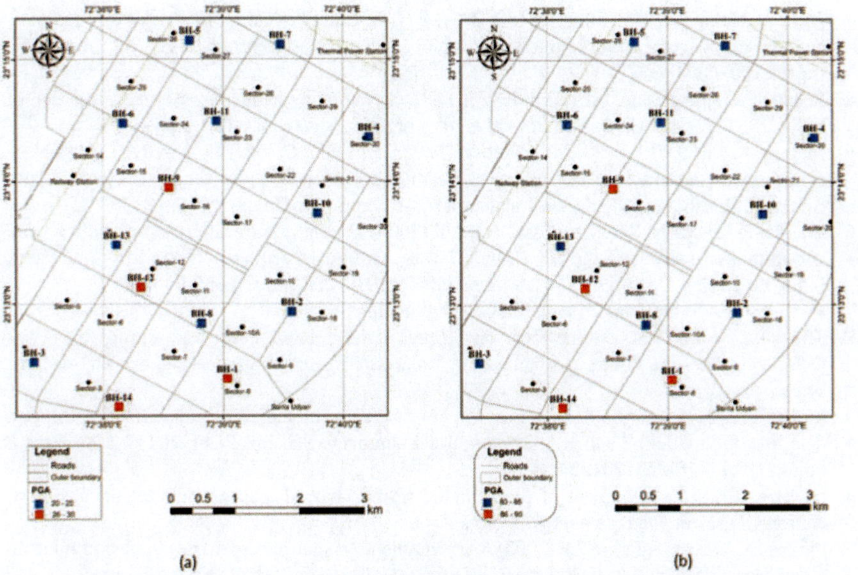

Fig. 8 Weighted PGAs (in cm/s^2) at all 14 boreholes estimated using GMPE of Mandal et al. [10] and NDMA [14] for a scenario earthquake of **a** M6.5 from Narmada region and **b** M7.7 from Kachchh region

Acknowledgements We are thankful to Director General, Institute of Seismological Research, Gandhinagar, for his support and encouragements. We acknowledge the department of science and technology government of Gujarat for financial support.

References

1. Borcherdt RD, Fumal TE (2002) Shear-wave velocity compilation for Northridge strong-motion recording sites. vol 2. US Department of the Interior, US Geological Survey.
2. Census of India (2011) Provisional population totals. Registrar General and Census Commissioner of India, Ministry of Home Affairs, New Delhi, India
3. Cetin KO, Seed RB, Der Kiureghian A, Tokimatsu K, Harder LF, Kayen RE, Moss, RES (2004) Standard penetration test-based probabilistic and deterministic assessment of seismic soil liquefaction potential. J Geotech Geoenviron Eng. https://doi.org/10.1061/(ASCE)1090-0241(2004)130:12(1314)
4. Chopra S, Choudhury P (2011) A study of response spectra for different geological conditions in Gujarat, India. Soil Dyn Earthquake Eng 1551–1564. https://doi.org/10.1016/j.soildyn.2011.06.007
5. Chopra S, Kumar D, Rastogi BK, Choudhury P, Yadav RBS (2012). Deterministic seismic scenario for Gujarat region, India. Natural Hazards 517–540. https://doi.org/10.1007/s11069-011-0027-y
6. Choudhury P, Chopra S, Roy KS, Sharma J (2016) Ground motion modelling in the Gujarat region of Western India using empirical Green's function approach. Tectonophysics. 675:7–22. https://doi.org/10.1016/j.tecto.2016.02.027

7. Joshi A, Kumar A, Mohan K, Rastogi BK (2013) Hybrid attenuation model for estimation of peak ground accelerations in the Kutch region, India. Natural Hazards 249–269. https://doi.org/10.1007/s11069-012-0524-7
8. Kolathayar S, Sitharam TG, Vipin KS (2012) Deterministic seismic hazard macrozonation of India. 1351–1364. J Earth Syst Sci. https://doi.org/10.1007/s12040-012-0227-1
9. Magrin A, Parvez IA, Vaccari F, Peresan A, Rastogi BK, Cozzini S, Mir RR (2016) Neo-deterministic definition of seismic and tsunami hazard scenarios for the territory of Gujarat (India). In Earthquakes and their impact on society Springer, Cham, pp 193–212
10. Mandal P, Kumar N, Satyamurthy C, Raju IP (2009) Ground-motion attenuation relation from strong-motion records of the 2001 Mw 7.7 Bhuj earthquake sequence (2001–2006), Gujarat, India. Pure Appl Geophys 451–469. https://doi.org/10.1007/s00024-009-0444-y
11. Merh SS (1995) Geology of Gujarat. GSI Publications
12. Midorikawa S (1994) Site effects on strong-motion records observed during the 1987 Chiba-ken-toho-oki, Japan earthquake. In Proceedings ninth japan earthquake engineering symposium. pp 85–90
13. Mohan K (2014) Seismic-hazard assessment in the kachchh region of gujarat (India) through deterministic modeling using a semi-empirical approach. Seismol Res Lett 117–125. https://doi.org/10.1785/0220120123
14. NDMA (2011) Development of probabilistic seismic hazard map of india technical report. National Disaster Management Authority
15. Parvez IA, Vaccari F, Panza GF (2003) A deterministic seismic hazard map of India and adjacent areas. Geophys J Int 489–508. https://doi.org/10.1046/j.1365-246X.2003.02052.x
16. Parvez IA, Magrin A, Vaccari F, Ashish MirRR, Peresan A, Panza GF (2017) Neo-deterministic seismic hazard scenarios for India—a preventive tool for disaster mitigation. J Seismol 1559–1575. https://doi.org/10.1007/s10950-017-9682-0
17. Pearce JT, Baldwin JN, Vaughn JD (2005) In: Liquefaction susceptibility mapping. St. Louis, Missouri and Illinois . William Lettis & Associates, pp 42
18. Rout MM, Das J (2018) Probabilistic seismic hazard for the Himalayan region using kernel estimation method (zone-free method). Natural Hazards. 967–985. https://doi.org/10.1007/s11069-018-3336-6
19. Sairam B, Rastogi BK, Agrawal S, Chauhan M, Bhonde U (2011)Seismic site characterization using Vs30 and site amplification in Gandhinagar region, Gujarat, India. Current Sci 100(5):754–761
20. Seed HB, Idriss IM, Arango I (1983) Evaluation of liquefaction potential using field performance data. J Geotech Eng 458–482. https://doi.org/10.1061/(ASCE)0733-9410(1983)109:3(458)
21. Seed HB, Tokimatsu K, Harder LF, Chung RM (1985) Influence of SPT procedures in soil liquefaction resistance evaluations. J Geotech Eng 1425–1445. https://doi.org/10.1061/(ASCE)0733-9410(1985)111:12(1425)
22. Skempton AW (1986) Standard penetration test procedures and the effects in sands of overburden pressure, relative density, particle size, aging and over consolidation. Geotechnique 425–447. https://doi.org/10.1680/geot.1986.36.3.425
23. Youd TL, Idriss IM, Andrus RD, Arango I, Castro G, Christian JT, Stokoe KH (2001) Liquefaction resistance of soils: a Summary report from the 1996 NCEER and 1998 NCEER/NSF workshops on the evaluation of liquefaction resistance of soils. J Geotech Geoenviron Eng 297–313. https://doi.org/10.1061/(ASCE)1090-0241(2001)127:10(817)

Linear and Nonlinear Site Response Study of Soil-tire Crumb Mixture

P. S. Naresh Dixit and Mohammed Zubair

Abstract Unrecycled tire waste is an enormous global problem because of its non-biodegradability, flammability, and chemical composition. The production and disposal of tires have been on the rise ever since. Due to a shortage of natural resources and increase in waste disposal cost, waste tires in the form of crumbs are widely used as a lightweight material for the backfill in embankment construction. Due to the inherent energy absorption and damping behavior of rubber, research on sand-tire crumb mixtures has shown that there is increase in shear strength and energy absorption capacity. But, very few tests have been performed on soil-tire crumb mixtures. The primary objective of our study involves covering this gap by determining the shear strength and energy absorption characteristics of soil-tire crumb mixture (STCM) by performing direct shear and unconsolidated undrained (UU) triaxial test on disturbed soil samples collected from various site locations, by varying percentage of tire crumbs. Site response studies are done using DEEPSOIL software using both linear and nonlinear methods. Based on the result obtained, it is observed that there is an increase in shear strength in the range of 20–30%, increase in energy absorption 15–20%, and reduction in PGA by 10–15% for the optimum percentage of tire crumb which is site-specific.

Keywords Tire crumbs · Soil · PGA · Shear strength · Energy absorption

1 Introduction

Over the years, earthquakes have caused a tremendous amount of damage to life and property. Though earthquakes do not cause the damage directly, the destruction caused to man-made infrastructure and edifices eventually causes the death and loss of property. While the direct costs of it itself amount to a large sum every year, earthquake's indirect economic damage caused for the further years is also detrimental. As per The Oklahoma Economist [1], due to the 48 earthquakes of magnitude 5.5 or more

P. S. Naresh Dixit (✉) · M. Zubair
Department of Civil Engineering, CMR Institute of Technology, Bangalore 560037, India
e-mail: naresh.d@cmrit.ac.in

T. G. Sitharam et al. (eds.), *Earthquake Geotechnics*, Lecture Notes in Civil Engineering 187, https://doi.org/10.1007/978-981-16-5669-9_7

which have occurred in developed countries since 1985, the estimate of economic damage amounts to about more than $232 billion (in 2015 dollars). In addition, about 800 thousand people have died due to earthquakes from 2000 to 2015, as per Statista Research Department [2]. The current seismic resistance methods are expensive and are not affordable by all. They currently amount to about 15–20% of the total project cost. Contemporaneously, the disposal of used tires has been a growing environmental issue. Tires take up a large volume of an already limited landfill space. They also trap and on incineration produce obnoxious gases like methane, which could potentially be flammable [3]. Tires once used up by a consumer cannot be re-used. They are, thus, discarded, and as their volume is quite high, they consume a lot of space. In 2008, around one billion tires were being produced globally each year [4], with an estimated further four billion already in landfills [5]. To tackle this issue of disposal, with a sustainable future in mind, research is being done to utilize the waste tires. Due to the inherent energy absorption and damping behavior of rubber, research has shown that waste tire in the form of crumbs can act as quite an effective base isolator [6]. Waste tire crumbs in mixture with soil have also been used as a lightweight backfill material and in retaining structures [7, 8].

Dynamic studies have also been carried out on soils reinforced with tire crumbs. When a retaining wall which was subjected to sinusoidal dynamic excitations, it was observed that the horizontal displacement and incremental lateral earth pressures were significantly reduced when the retaining wall was included with tire chips as a cushion in between the wall and the fill [9]. With the help of a finite element program, another study found that the structural responses in terms of acceleration and inter-story drift of a low to medium rise building could be reduced by 40–60% [6]. Moreover, when sand-tire mixtures were subjected to undrained monotonic and cyclic triaxial tests, it was found that the undrained shear strength of the specimens tested increased with the inclusion of oven-dried, large-sized tire chips along with an improvement in liquefaction resistance [10].

Furthermore, site characterization and site response studies have been carried out [11]. Seismic hazard evaluation studies have also been carried out. It was observed that due to the conditions of the soil, large modification of wave amplitudes is observed resulting in higher peak ground acceleration when compared to rock-level acceleration [12]. Combining these aspects of using tire crumbs in the form of mixture with soil along with the application of seismic response studies present, certain tests were, firstly, performed on soil-tire crumb mixtures to estimate its shear strength and energy absorption capacity. In addition, linear and nonlinear site response study and analysis for each of the considered sites and its corresponding optimum soil-tire crumb mixtures were carried out, simulating ground conditions. This analysis was done with the help of a software, DEEPSOIL [13]. DEEPSOIL is a 1D equivalent linear and nonlinear site response analysis platform.

2 Methodology

Site selection was based on the borehole data which were collected from various government and other private agencies. Out of the available data, sites with normal standard penetration test (NSPT) value greater than 35 along with sand percentage up to 55% to a depth of 3 m were selected. Five sites were shortlisted—AECS layout A-block (Site 1), Kadugodi (Site 2), AECS layout D-block (Site 3), AECS layout (Site 4), and AECS layout (Site 5). Disturbed samples were collected from all the sites in re-sealable polythene bags to maintain its in situ moisture content. Tire crumbs were obtained from a local industry. The size of the tire crumbs was less than 4.75 mm. To confirm it, the tire crumbs were also sieved along with soil through a 4.75 mm sieve. Mixed-size (tire buff) tire crumbs were selected for further testing as uniformly sized tire crumbs showed little to no improvement in soil strength in the initial, preliminary tests.

Test samples were then prepared. The in situ density of all the soil samples was determined by core cutter method [14]. Tire crumbs were mixed in increments of 2.5% up to 10%. This was done in accordance with previous studies on sand, soil, and tire crumb mixtures. Furthermore, sieve analysis was carried out to obtain the grain-size distribution. D50 value was determined from the grain size distribution curve. The soils were then graded. Standard proctor compaction test was performed on the soil samples to obtain its optimum moisture content (OMC) [15]. On its determination, test samples for unconsolidated undrained triaxial test were prepared. This was done by hand mixing of required tire crumbs and soil, along with the addition of water equal to the OMC of soil. Then, direct shear and unconsolidated undrained triaxial tests were performed on STCM for the prepared samples. The dimension of the shear box was 60 mm × 60 mm × 30 mm. Testing was done under normal stresses of 50,100, and 200 kPa as per IS:2720 (Part 13)—1986 [16]. Triaxial test was performed to obtain the stress versus strain curves. These curves were further used to find the total energy absorption capacity of the STCM samples. The energy absorption capacity was determined by finding the area under the stress versus strain curves. The area under the stress–strain curve up to a given value of strain is the total mechanical energy per unit volume, which the material consumed while straining to that value [17], and is given by:

$$\mathrm{EA} = \int_0^{\epsilon} \sigma(\epsilon)\mathrm{d}\epsilon \tag{1}$$

where $\sigma(\epsilon)$ is stress as a function of strain and EA is the energy absorption capacity.

2.1 DEEPSOIL

After the required tests were performed and the corresponding data was determined, its analysis was done using DEEPSOIL software. DEEPSOIL is capable of performing linear and equivalent linear and linear and nonlinear analysis in frequency and time domains, respectively. Post-analysis, the output data can be exported conveniently into a Microsoft Excel file for our further reference. For our study, linear and nonlinear time domain analyses were performed. This can be opted in the first screen in DEEPSOIL and is shown in Fig. 1. With the help of the borehole data previously collected, soil profiles were made. Soil profiles of different layers with their respective properties can be made in the next screen under analysis tab. The depth of each soil layer, its unit weight, shear modulus, and damping ratio were entered for each layer considered. This is shown in Fig. 2. Condition of no ground water table was selected as the ground water table was assumed to be below the soil profile considered. The unit weight for each layer was obtained from the borehole data collected. The shear modulus for each layer was determined using the relation given by Anbazhagan and Sitharam [19], given by:

$$G_{\max} = 24.28\, N^{0.55} \tag{2}$$

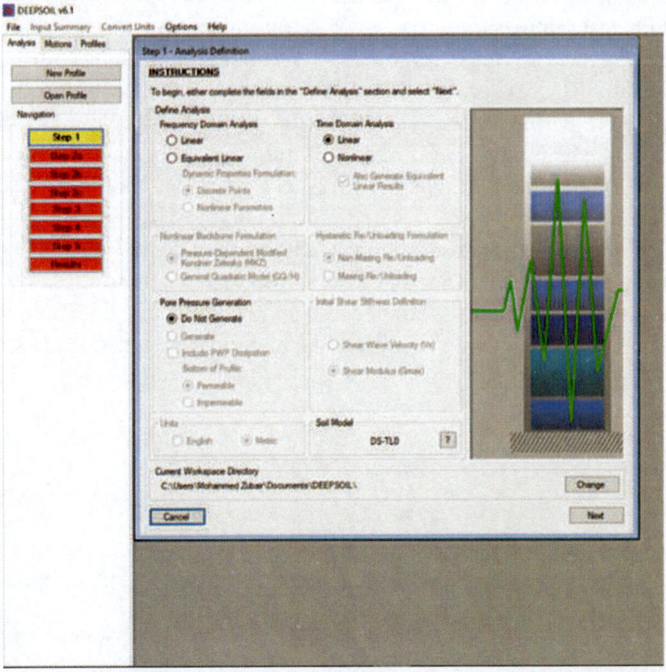

Fig. 1 DEEPSOIL interface

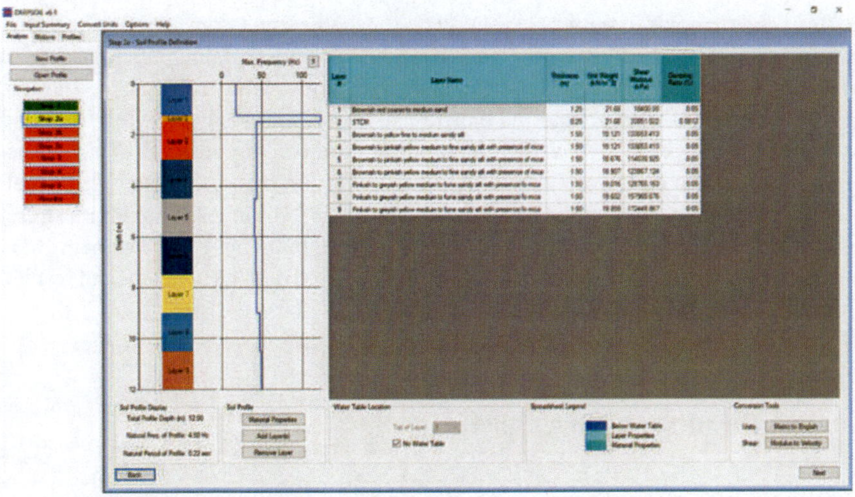

Fig. 2 DEEPSOIL interface showing elastic half-space

where G_{max} is the maximum shear modulus and N is the SPT "N" value.

The shear modulus of the STCM layer was determined using the direct shear curve, where the slope of the straight portion of the stress versus strain curve gives the shear modulus. This was done to the curve of the optimum percentage of tire crumbs soil. Optimum percentage of tire crumbs was selected by considering all the considered parameters, including shear strength and energy absorption capacity, of the improved STCM mixture. The damping ratio for each layer was set to 0.05% except the STCM layer considered. The damping ratio for the STCM layer was obtained as per the increase in energy absorption capacity of soil void of tire crumbs and that of soil with the optimum percentage of tire crumbs. The unit weight of STCM was kept equal to the pure soil as the analysis was done by weight batching.

Under the motions tab, one can select the type of input motion to provide for further analysis. The input motions provided to the elastic half-space were Koyna earthquake which had occurred at Koynanagar, Maharashtra, on December 11, 1967 [18]. The magnitude 6.6 shock hits with a maximum Mercalli intensity of eight, amounting to a death count of about 180 people. This type of earthquake in the Deccan Plateau possessed an intensity which has an occurrence period of 100 years. Hence, two versions of this earthquake were used for analysis—once which was not scaled down and the other which was scaled down to an intensity equivalent to an occurrence period of 30 years. This scaled down earthquake would probably occur in the lifespan of a building. The peak ground acceleration (PGA) of the input motions used was 0.36 and 0.1 g for the non-scaled down and the scaled down input motion, respectively. This made sure that both a higher intensity and a lower intensity strong ground motions were considered for analysis.

In regard to elastic half-space properties, Bangalore bedrock properties were considered. The shear modulus was kept to 5000 kPa and the unit weight to 160

kN/m^3. The damping ratio was kept to 20% for the bedrock considered. The input motion was applied to all the soil layers. Response spectra damping ratio of 5% was provided, followed by a frequency-independent damping. Time history interpolation method was done linearly, and the elastic half-space was analyzed. In the output page, one can check the time history plots, stress–strain plots, and spectral plots. The output was then exported to Excel for further comparison and analysis. The Excel output file provides peak ground acceleration (PGA, g), strain (%), maximum stress ratio (ratio of shear stress to effective vertical stress), maximum displacement (M), maximum strain, and other data for each layer. Required data were taken, and results were concluded.

3 Test Results and Discussion

Various experimental tests were performed on the soil samples. These experiments ranged right from sieve analysis up to unconsolidated undrained (UU) triaxial test. Furthermore, software analysis was performed on the soil profiles with the help of DEEPSOIL.

3.1 Sieve Analysis

Sieve analysis was performed on all the soil samples which were obtained from the shortlisted sites. Grain size distribution curves were plotted, and their respective D50 values were determined. In addition, the soils were classified. The grain size distribution curves are shown in Fig. 3.

The D50 values, along with uniformity coefficient (Cu), coefficient of curvature (Cc), and the percentage of coarse-grained particles of soils from the sites under consideration are as shown in Table 1.

The D10, D30, D60, and D50 values for soil samples of the considered sites were determined from the gradation curves. Using these values, the uniformity coefficient (Cu), coefficient of curvature (Cc), and the coarse-grained content (%) were determined. Based on the values obtained, the soils were classified as shown in Table 1.

3.2 Direct Shear Test

Direct shear test was performed on all the soil samples with varying percentages of tire crumbs. Figure 4 shows a plot of variation of stress versus normal pressures for different percentages of tire crumbs for Site 1 soil sample. From the figure, the variation of maximum stress values at a particular pressure can be observed. It can be

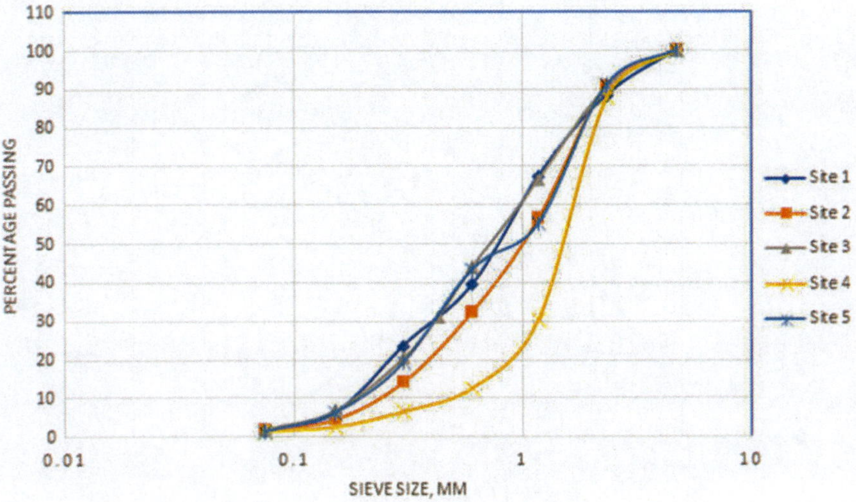

Fig. 3 Sieve analysis results

Table 1 D50, Cu, and Cc values of soils

Site	D50	Cu	Cc	Coarse content (%)	Soil classification as per unified soil classification system
Site 1	0.79	5.59	1	51.26	SP—poorly graded sand
Site 2	1	7.12	0.66	51.28	SW—well-graded sand
Site 3	0.71	5.39	0.96	51.82	SP—poorly graded sand
Site 4	1.4	3.2	1.51	69.53	SP—poorly graded sand
Site 5	0.95	6.5	0.62	44.81	SP—poorly graded sand

inferred that when the percentage of tire crumbs was 5%, the values of stress increased at all pressures. This increase was seen to reach its peak by 27.13% at 100 kPa normal pressure. Hence, 5% was selected as the optimum percentage of tire crumbs for Site 1. In a similar way, the variation of stress versus normal pressures was determined for all the five sites under consideration. While at few percentages, the stress value was seen to decrease, and an appreciable increase in stress was also observed at certain percentages. This increase was observed consistently for all the values of normal pressures. The optimum percentage of tire crumbs was, thus, determined to be 5%, 7.5%, 10%, 2.5%, and 2.5% for Sites 1, 2, 3, 4, and 5, respectively. The maximum increase in stress value was observed to be 27.13%, 22.81%, 11.52%, 23.68%, and 7.37% for Sites 1, 2, 3, 4, and 5, respectively.

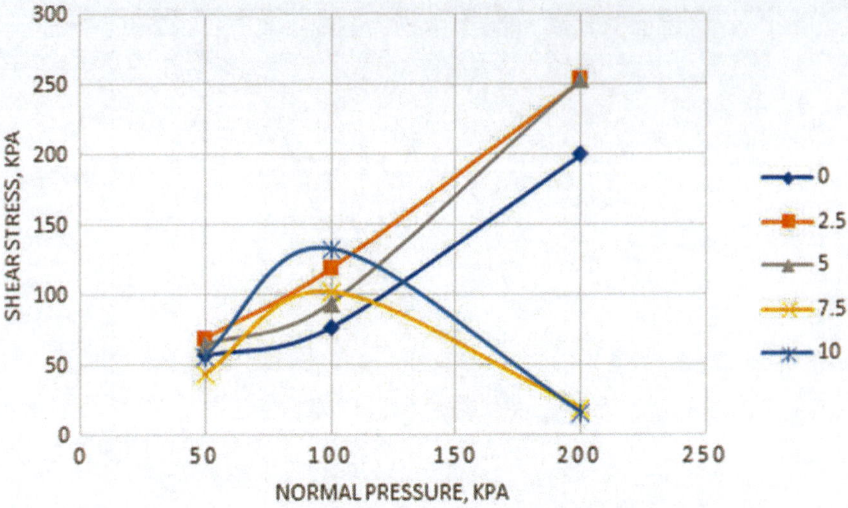

Fig. 4 Variation of stress versus normal pressures for different percentages of tire crumbs for Site 1 soil sample

3.3 *Unconsolidated Undrained Triaxial Test*

Unconsolidated undrained triaxial test was performed on all samples of soil-tire crumb mixtures for varying percentages of tire crumbs.

Figure 5 shows variation of deviatoric stress versus pressures for varying percentages of tire crumbs for Soil 1 soil sample. The influence of tire crumbs can be noticed by the significant increase in deviatoric stress value at 10% tire crumbs at 100 kPa confining pressure. But, there is an equally significant drop when the deviatoric stress reached its zenith at the same percentage of tire crumbs at 200 kPa confining pressure, thereby rejecting 10% as being the optimum. Considering overall improvement in strength, 5% tire crumbs was selected as the optimum as the increase was consistent at all confining pressures, with the maximum being 43.47% at 200 kPa confining pressure. This increase was also observed at 5% tire crumbs in direct shear test for Soil 1.

On a similar note, for Sites 2, 3, 4, and 5, the maximum increase in deviatoric stress value was observed at optimum percentage of tire crumbs determined in direct shear test. The maximum increase in deviatoric stress value for Sites 2, 3, 4, and 5 was observed to be 52.65%, 86.36%, 55.62%, and 25.19%, respectively.

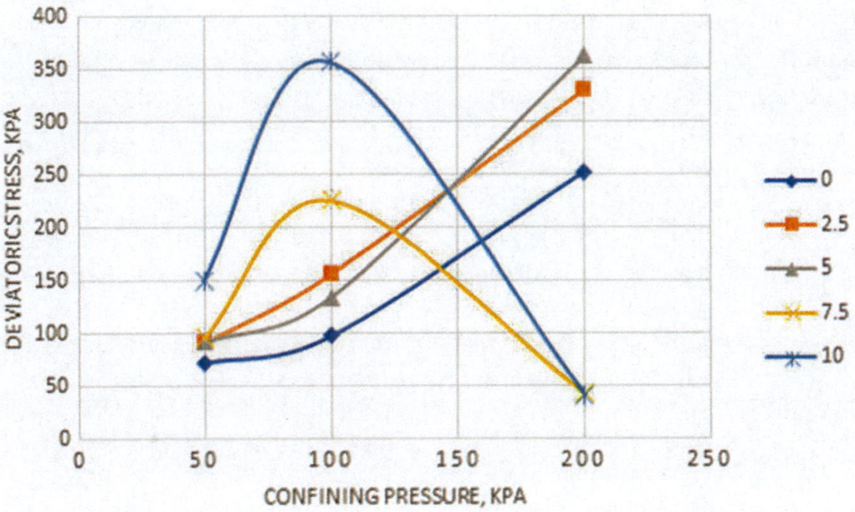

Fig. 5 Variation of deviatoric stress versus confining pressures for varying percentages of tire crumbs for Soil 1 soil sample

3.4 Energy Absorption Capacity

The energy absorption capacity was determined by finding the area under the stress versus strain curves obtained from UU triaxial tests. The area under the stress–strain curve up to a given value of strain is the total mechanical energy per unit volume which the material consumed while straining to that value. Table 2 shows the energy absorption values for all the sites considered for the study.

It can be observed that the addition of tire crumbs did cause a significant increase in the energy absorption capacity of tire crumbs. This increase was observed to be in the optimum percentage of tire crumbs which was determined with the help of direct shear and UU triaxial tests. For Site 1, there is an overall maximum increase in energy absorption capacity at all confining pressures observed at its optimum percentage of 5% tire crumbs. A maximum increase of 62.38% is observed at a confining pressure of 200 kPa. Similarly, the maximum increase for Sites 2, 3, 4, and 5 was observed to be 23.96%, 81.96%, 331.39%, and 54.94%, respectively.

3.5 DEEPSOIL Analysis

Soil profiles were generated with the help of the borehole data previously collected. Different properties for the elastic half-spaces were assigned as explained under Sect. 2.1. For easier understanding and analyzes, the different layers of soil were numbered and named as Layer 1, Layer 2, and so on. Layer 1 consisted of the top

Table 2 Energy absorption (kJ/m^3) values for different sites for varying percentages of tire crumbs

Material	Pressure (kPa)	Site 1	Site 2	Site 3	Site 4	Site 5
Soil	50	11.0915	1.67777	11.3185	0.92842	2.17
	100	9.11306	2.07577	22.9851	0.82245	1.803
	200	17.8712	5.14877	27.7431	0.08248	0.3637
Soil + 2.5%	50	8.3136	10.3273	14.681	0.59642	2.0459
	100	14.2994	6.8074	14.6586	0.32565	2.1769
	200	20.4996	5.53761	15.9281	0.3559	0.5635
Soil + 5.0%	50	17.3144	9.73345	4.19716	0.08157	1.3821
	100	10.7521	7.31931	6.30232	0.18977	1.2557
	200	29.0199	9.6219	12.5823	0.26144	1.1777
Soil + 7.5%	50	13.6032	10.312	21.5314	0.25605	2.0436
	100	31.9539	8.35287	10.3378	0.68375	0.6272
	200	6.03928	6.38228	14.1482	0.38438	1.3886
Soil + 10%	50	15.8558	6.17253	15.4262	0.7979	1.5601
	100	37.499	1.79361	22.2137	0.35563	1.1445
	200	4.40239	3.13397	50.4806	0.28345	1.3204

layer of the soil. Layer 2 constituted the STCM layer. This made sure that the STCM layer under a depth in accordance with a foundation of a small to medium rise building. Further layers constituted the lower layers of soil. Linear and nonlinear analysis was carried out on the soil profiles.

3.5.1 Linear Analysis

Site 1 analyses results are tabulated and shown in Table 3. The effective stress (kPa), PGA (g), maximum displacement (m), and maximum stress ratio for Site 1 for two different input motions are shown below. It is observed that the effective vertical stress at Layer 1 reduced significantly on the use of 0.25 m of STCM layer for both the input motions. With the STCM layer of thickness 1.25 m, the decrease in effective stresses in improved soil was found to be 88.37%, with the decrease in PGA and displacement being 5.03% and 6.2%, respectively. At 1.25 m thickness, there was a decrease in stress ratio by 50.8%. Considering the overall improvement compared to other depths of STCM used, 0.25 m was the optimum.

The effective stress (kPa), PGA (g), maximum displacement (m), and maximum stress ratio for Sites 2, 3, 4, and 5 are shown in Table 4. Since there is little to no change observed in the soil layers present below the STCM layer, only the top two layers have been considered for analysis and shown in the same table. At 1.25 m thickness of STCM in Site 2 analysis, there was a reduction in PGA and maximum displacement by 4.59% and 5.66%, respectively, observed. A decrease in effective stresses in improved soil by 83.37%, along with a 21.9% increase in stress ratio, was

Table 3 Site 1 DEEPSOIL linear analysis output

Site-specific ground motion—0.36 g						Site-specific ground motion—0.1 g					
Depth of STCM layer	Layer	Effective stress (kPa)	PGA (g)	Maximum displacement (m)	Maximum stress ratio	Depth of STCM layer	Layer	Effective stress (kPa)	PGA (g)	Maximum displacement	Maximum stress ratio
0 m	Layer 1	16.25904	0.64197	0.0050387	0.618227	0 m	Layer 1	16.259	0.17965	0.00141	0.173001
	Layer 2						Layer 2				
	Layer 3	46.85795	0.62818	0.0048804	0.609331		Layer 3	46.858	0.17579	0.001366	0.170511
	Layer 4	104.2204	0.60058	0.0045583	0.597109		Layer 4	104.22	0.16806	0.001276	0.167091
	Layer 5	191.9265	0.46178	0.0028136	0.522913		Layer 5	191.927	0.12922	0.000787	0.146329
0.25 m	Layer 1	2.709285	0.63903	0.0050615	1.22578	0.25 m	Layer 1	2.70929	0.17882	0.001416	0.343015
	Layer 2	18.96828	0.63906	0.0050596	0.611762		Layer 2	18.9683	0.17883	0.001416	0.171191
	Layer 3	46.85795	0.60095	0.0046457	0.584411		Layer 3	46.858	0.16817	0.0013	0.163538
	Layer 4	104.2204	0.57551	0.0043443	0.569974		Layer 4	104.22	0.16105	0.001216	0.159498
	Layer 5	191.9265	0.45075	0.0026789	0.497877		Layer 5	191.927	0.12613	0.00075	0.139323

Table 4 Sites 2, 3, 4, and 5 DEEPSOIL linear analysis output

	Site-specific ground motion—0.36 g						Site-specific ground motion—0.1 g					
	Depth of STCM layer	Layer	Effective stress (kPa)	PGA (g)	Maximum displacement	Maximum stress ratio	Depth of STCM layer	Layer	Effective stress (kPa)	PGA (g)	Maximum displacement	Maximum stress ratio
Site 2	0 m	Layer 1	13.97919	0.60942	0.0027589	0.591886	0 m	Layer 1	13.9792	0.17054	0.000772	0.165629
		Layer 2						Layer 2				
	1.25 m	Layer 1	2.329391	0.5814	0.0026027	0.721624	1.25 m	Layer 1	2.32939	0.1627	0.000728	0.201935
		Layer 2	16.3086	0.58108	0.0025995	0.557303		Layer 2	16.3086	0.16261	0.000727	0.155952
Site 3	0 m	Layer 1	15.81654	0.64088	0.0049977	0.616949	0 m	Layer 1	15.8165	0.17934	0.001399	0.172643
		Layer 2						Layer 2				
	1 0.25 m	Layer 1	2.635555	0.59869	0.0046059	1.16418	1.25 m	Layer 1	2.63555	0.16753	0.001289	0.325777
		Layer 2	18.45209	0.59876	0.0046045	0.593644		Layer 2	18.4521	0.16755	0.001288	0.166121
Site 4	0 m	Layer 1	11.84178	0.62654	0.0046027	0.599729	0 m	Layer 1	11.8418	0.17533	0.001288	0.167824
		Layer 2						Layer 2				
	0.25 m	Layer 1	9.868547	0.58939	0.0042962	0.556545	0.25 m	Layer 1	9.86855	0.16493	0.001202	0.15574
		Layer 2	21.71032	0.58319	0.0042241	0.574823		Layer 2	21.7103	0.1632	0.001182	0.160855
Site 5	0 m	Layer 1	12.9367	0.6313	0.004716	0.605578	0 m	Layer 1	12.9367	0.17666	0.00132	0.169461
		Layer 2						Layer 2				
	1.25 m	Layer 1	2.155677	0.64978	0.0051382	1.50638	1.25 m	Layer 1	2.15568	0.18183	0.001438	0.421535
		Layer 2	15.09235	0.64996	0.0051373	0.608359		Layer 2	15.0924	0.18188	0.001438	0.170239

also observed. And, 1.25 m was, therefore, selected as the optimum depth of STCM for Site 2. The reduction in PGA value for Site 3 on the usage of 1.25 m thick layer of STCM was 6.7%. The decrease in effective stresses in the improved soil was found to be 83.37%, with the decrease in PGA and displacement being 6.58% and 7.84%, respectively. In addition, there was also an 88.7% increase in stress ratio. Hence, 1.25 m was considered optimum.

The decrease in effective stresses in the STCM soil was found to be 83.37%, with the decrease in PGA and displacement being 2.49% and 2.89%, respectively, for Site 4, when the thickness of STCM was 0.25 m. There was also a 107.07% increase in stress ratio. Hence, the optimum thickness of STCM layer was selected to be 0.25 m In Site 5 soil analysis, on the usage of 1.25 m thick STCM, there was little to no difference observed in the PGA and maximum displacement values. However, there was a significant reduction in the effective stress values. The optimum depth of STCM for Site 5 was, hence, concluded to be 1.25 m. With the STCM layer of thickness 1.25 m, the decrease in effective stresses in improved soil was found to be 83.3367%, with the decrease in PGA and displacement being 6.21% and 8.76%, respectively. In addition, an increase of stress ratio by 127.2% was observed. Hence, 1.25 m was selected as the optimum.

3.5.2 Nonlinear Analysis

Nonlinear time domain analysis was carried out on the soil profiles made on DEEPSOIL. Table 5 shows the DEEPSOIL analysis output for Site 1 soil sample. The effective stress (kPa), PGA (g), maximum displacement (m), and maximum stress ratio for Site 1 for two different input motions are shown.

It is observed that similar to linear analysis, there is a significant reduction in effective stress for both the input motions when the STCM layer used is 1.25 m thick. With the STCM layer of thickness 1.25 m, the decrease in effective stresses in the STCM soil was found to be 59.6%, with the decrease in PGA and displacement being 4.3% and 5.66%, respectively. In addition, a decrease of stress ratio by 4.56% was observed. Considering the overall improvement in performance of soil to the input motions, 1.25 m was selected as the optimum depth of STCM. The effective stress (kPa), PGA (g), maximum displacement (m), and maximum stress ratio for Sites 2, 3, 4, and 5 are shown in Table 6. Since there is negligible change observed in the soil layers present below the STCM layer, only the top two layers have been considered for analysis and shown in the same table.

The effective stress (kPa), PGA (g), maximum displacement (m), and maximum stress ratio for Site 2 for two different input motions are shown. A reduction in PGA and displacement by 5.66% and 8.90%, respectively, was observed. Furthermore, a 10.80% decrease in soil strain was also observed. Considering the improvement in performance of soil to the input motions, 1.25 m was selected as the optimum depth of STCM. Site 3 soil sample analysis results are tabulated and shown. A decrease in PGA and displacement by 7.1% and 9.76%, respectively, were observed when the

Table 5 Site 1 DEEPSOIL nonlinear analysis output

Site-specific ground motion—0.36 g						Site-specific ground motion—0.1 g					
Depth of STCM layer	Layer	Effective stress (kPa)	PGA (g)	Maximum displacement (m)	Maximum stress ratio	Depth of STCM layer	Layer	Effective stress (kPa)	PGA (g)	Maximum displacement	Maximum stress ratio
0 m	Layer 1	16.26	0.163	0.0018	0.003	0 m	Layer 1	16.259	0.5	0.009	0.011
	Layer 2						Layer 2				
	Layer 3	46.86	0.16	0.00176	0.0067		Layer 3	46.858	0.5	0.009	0.026
	Layer 4	104.2	0.155	0.00166	0.0131		Layer 4	104.22	0.49	0.008	0.058
	Layer 5	191.9	0.122	0.00107	0.0239		Layer 5	191.93	0.4	0.006	0.124
1.25 m	Layer 1	2.709	0.161	0.00179	0.0011	1.25 m	Layer 1	2.7093	0.49	0.008	0.004
	Layer 2	18.97	0.161	0.0017	0.0113		Layer 2	18.968	0.49	0.008	0.049
	Layer 3	46.86	0.154	0.00166	0.0063		Layer 3	46.858	0.49	0.008	0.025
	Layer 4	104.2	0.149	0.00157	0.0124		Layer 4	104.22	0.48	0.008	0.055
	Layer 5	191.9	0.12	0.00101	0.0224		Layer 5	191.93	0.4	0.005	0.114

Table 6 Sites 2, 3, 4, and 5 DEEPSOIL nonlinear analysis output

	Site-specific ground motion—0.36 g						Site-specific ground motion—0.1 g					
	Depth of STCM layer	Layer	Effective stress (kPa)	PGA (g)	Maximum displacement	Maximum stress ratio	Depth of STCM layer	Layer	Effective stress (kPa)	PGA (g)	Maximum displacement	Maximum stress ratio
Site 2	0 m	Layer 1	13.9792	0.1614	0.00107	0.1597	0 m	Layer1	13.9792	0.5301	0.00578	0.5250
		Layer 2						Layer 2				
	1.25 m	Layer 1	2.3294	0.1551	0.00100	0.2317	1.25 m	Layer 1	2.3294	0.5022	0.00529	0.9224
		Layer 2	16.3086	0.1551	0.00101	0.1534		Layer 2	16.3086	0.5027	0.00529	0.5237
Site 3	0 m	Layer 1	15.8165	0.1632	0.00180	0.159	0 m	Layer 1	15.8165	0.5045	0.00883	0.4992
		Layer 2						Layer 2				
	1.25 m	Layer 1	2.6356	0.1524	0.00164	0.3425	1.25 m	Layer 1	2.6355	0.4700	0.007862	0.4411
		Layer 2	18.4521	0.1525	0.00164	0.1653		Layer 2	18.4521	0.4709	0.00786	0.5335
Site 4	0 m	Layer 1	11.8412	0.1539	0.00179	0.1491	0 m	Layer 1	11.8418	0.4490	0.00957	0.4393
		Layer 2						Layer 2				
	0.25 m	Layer 1	9.8685	0.1446	0.00164	0.1357	0.25 m	Layer 1	9.8685	0.4268	0.00844	0.3839
		Layer 2	21.7103	0.1443	0.00162	0.1570		Layer 2	21.7103	0.4302	0.008378	0.4805
Site 5	0 m	Layer 1	12.9367	0.1611	0.00168	0.1571	0 m	Layer 1	12.9367	0.5049	0.00816	0.4967
		Layer 2						Layer 2				
	1.25 m	Layer 1	2.1556	0.1598	0.00180	0.4553	1.25 m	Layer 1	2.1557	0.4719	0.00836	1.5841
		Layer 2	15.0924	0.1597	0.00180	0.1593		Layer 2	15.0923	0.4724	0.00836	0.4872

STCM was 1.25 m thick. A 7.78% decrease in stress ratio was also observed. Hence, 1.25 m was selected as the optimum.

In Site 4, at 0.25 m thick STCM, the decrease in effective stresses in the STCM induced soil was found to be 42.8%, with the decrease in PGA and displacement being 5.8% and 12.63%, respectively. A 5.4% decrease in stress ratio was also observed. Considering the improvement in performance of soil to the input motions, 0.25 m was selected as the optimum depth of STCM.

The decrease in effective stresses in the STCM reinforced soil in Site 5 sample was found to be 62.4%, with the decrease in PGA and displacement being 3.1% and 12.5%, respectively. There was also a 5.1% decrease in stress ratio. Hence, 1.25 m was selected as the optimum depth of STCM.

4 Conclusion

Based on all the experiments performed, analysis conducted, and the results obtained, the following may be concluded:

1. There was a definite improvement in soil properties like shear strength, energy absorption, and ductility of the soil on the addition of tire crumbs to the soil.
2. The improvement was site-specific—different tire crumbs percentages improved different soils
3. There was a definite decrease in effective stress (kPa), PGA (g), and maximum displacement (m), which are highly desirable, on the usage of STCM at foundation depth.
4. The optimum percentage of tire crumbs for Site 1 was found to be 5%, and on performing analysis on DEEPSOIL, the optimum thickness of the STCM layer was found to be 0.25 m. Hence, the best improvement of the soil foundation would be to provide 0.25 m thick of STCM layer which has 5% tire crumbs.
5. Similarly, the optimum thickness of STCM layer and percentage of tire crumbs which would most improve the soil were found to be 1.25 m with 7.5% tire crumbs for Site 2, 1.25 m with 10% tire crumbs for Site 3, 0.25 m with 2.5% tire crumbs for Site 4, and 1.25 m with 2.5% tire crumbs for Site 5, respectively.
6. The improvement of soil in all aspects does display the importance of this study, by providing an affordable and economic energy absorption and base isolation technique for an earthquake resistant structure.

4.1 Future Scope

This study deals with improving shear strength and energy absorption characteristics of a specific site and its application in vibration isolation for shallow foundations. It can further be done for deeper foundations. Analytical study can be extended further to create simulation models for better results. Variation can also be done in the percentage of tire crumbs to a finer extent to obtain more accurate variations.

References

1. https://www.kansascityfed.org/publications/research/oke/articles/2016/economic-damage-large-earthquakes
2. https://www.statista.com/statistics/263108/global-death-toll-due-to-earthquakes-since-2000/
3. Price W, Smith ED (2006) Waste tire recycling: environmental benefits and commercial challenges. Int J Environ Technol Manag 6(3–4):362–374
4. Martínez, JD, Neus P, Ramón M, Tomás G, Victoria NM, Maria MA (2013) Waste tyre pyrolysis—a review. Renew Sustain Energ Rev 23:179–213
5. World Business Council for Sustainable Development, Managing End-of Life Tires. Available at https://www.wbcsd.org/contentwbc/download/1105/14236
6. Tsang HH, Lo SH, Xu X, Neaz Sheikh M (2012) Seismic isolation for low-to-medium-rise buildings using granulated rubber–soil mixtures: numerical study. Earthquake Eng Struct Dyn 41(14):2009–2024
7. Lee JH, Salgado R, Bernal A, Lovell CW (1999) Shredded tires and rubber-sand as lightweight backfill. J Geotech Geoenviron Eng 125(2):132–141
8. Cecich V, Gonzales L, Hoisaeter A, Williams J, Reddy K (1996) Use of shredded tires as lightweight backfill material for retaining structures. Waste Manag Res 14(5):433–451
9. Reddy SB, Krishna AM (2019) Mitigation of dynamic loading effects on retaining walls using recycled tire chips. In: Soil dynamics and earthquake geotechnical engineering. Springer, Singapore, pp 141–148
10. Hong Y, Yang Z, Orense RP, Lu Y (2015, November) Investigation of sand-tire mixtures as liquefaction remedial measure. In: Proceedings of the 10th Pacific conference on earthquake engineering, November, Sydney, Australia
11. Anbazhagan P, Sitharam TG (2008) Site characterization and site response studies using shear wave velocity. J Seismol Earthquake Eng 10(2):53–67
12. Anbazhagan P, Thingbaijam KKS, Nath SK, Kumar JN, Sitharam TG (2010) Multi-criteria seismic hazard evaluation for Bangalore city, India. J Asian Earth Sci 38(5):186–198
13. http://deepsoil.cee.illinois.edu/
14. IS: 2720-29 (1975) Methods of test for soil-part XXIX: Determination of dry density of soils in-place
15. Code, I. S. IS 2720 (Part 7)–1980. Determination of water content, dry density relation using light compaction (second revision)
16. IS: 2720, Part–13. (1986). Methods of tests for soils: direct shear test
17. Roylance D (2001) Stress-strain curves. Massachusetts Institute of Technology study, Cambridge
18. Gupta H, Narain H, Rastogi BK, Mohan I (1969) A study of the Koyna earthquake of December 10, 1967. Bull Seismol Soc Am 59(3):1149–1162
19. Anbazhagan P, Parihar A, Rashmi HN (2012) Review of correlations between SPT N and shear modulus: a new correlation applicable to any region. Soil Dyn Earthq Eng 36:52–69

Effects of Air Intrusion Technique on Permeability of Liquefiable Soil

Punit Bhanwar, Shivam Thakur, and Trudeep N. Dave

Abstract Earthquakes are one of the most devastating natural disasters, which may lead to infrastructural damages like slope failure, ground subsidence, and liquefaction of soils. As soil liquefaction causes soil to lose its ability to support structures rested upon it, several ground improvement techniques were devised in the past to minimize its repercussions. The present study highlights a novel technique of intruding air into a saturated liquefiable soil to change its permeability. This air intrusion is achieved by mixing oxygen-generating eco-friendly chemical, i.e., sodium percarbonate in definite proportion by weight to alter the degree of saturation thereby reducing permeability. This experimental study using conventional constant head permeability test setup accentuated that mixing of 1% of sodium percarbonate can reduce soil permeability by about 90% and consequently improve liquefaction resistance. The test findings also substantiated that oxygen bubbles so formed in admixed samples were stable under a given head for a considerable time.

Keywords Liquefaction mitigation · Air intrusion technique · Permeability test · Optimum sodium percarbonate

1 Introduction

In the event of an earthquake causing significant shaking motion and displacement of the ground, various infrastructures fail to sustain such excitation. This is majorly attributed to a loss in strength of supporting soil or ground beneath these structures. Soil liquefaction is one of the reasons for such failure. In this phenomenon, loose and saturated soils due to a sudden change in soil's stress condition and pore-pressure development start to behave like a liquid. This effectively causes a loss in bearing capacity and stiffness of the soil. Generally, soils having a particle size in the range of 0.002–2.36 mm (silt and sands) are mainly susceptible to this hazard [30]. The devastating effects of soil liquefaction on infrastructure are quite apparent from

P. Bhanwar · S. Thakur · T. N. Dave (✉)
Institute of Infrastructure Technology, Research and Management, Ahmedabad, India
e-mail: trudeepdave@iitram.ac.in

T. G. Sitharam et al. (eds.), *Earthquake Geotechnics*, Lecture Notes in Civil Engineering 187, https://doi.org/10.1007/978-981-16-5669-9_8

Table 1 Classification of soil liquefaction mitigation techniques

Techniques	Mechanism of improvement		
	Dissipation of pore pressure	Densification of soil structure	Reinforcement of soil structure
	Induced partial desaturation [34]	Vibro-compaction [21]	Permeation grouting [8]
	Earthquake PVDs [24]	Vibro-replacement [25]	Deep soil mixing [27]
	Electro-osmotic consolidation [9]	Dynamic compaction [20]	Jet grouting [19]
	Microbial-induced desaturation [23]	Compaction grouting [2]	Passive site remediation [7]
		Blasting compaction [22]	Microbial-induced calcite precipitation [31]

the long-prevailing history, i.e., Alaska and Nigata earthquake (1964), El Sentro earthquake (1979), Kobe (1995), to recent ones in Kocaeli (1999), and Bhuj (2001). Settlement or tilting of the building, rupture, and cracking at joints of buried utilities, lateral spreading of slopes are repercussions of construction in such soils.

It is always preferred to avoid such calamitous losses of public infrastructure and human lives associated by ground improvement of liquefiable soils. Geotechnical professionals generally employ methods that substantially improve critical soil aspects like pore-pressure dissipation, state of packing, and reinforcement compatibility. As these elements directly affect soil performance in the event of liquefaction [10, 29], several soil liquefaction mitigation techniques based on improving the same have been developed. Some of them are listed below (see Table 1).

Among these techniques, induced partial desaturation due to its adaptability in variable site geometric conditions and cost-effectiveness promises to be a novel sustainable solution for liquefaction mitigation. It relies on the principle of dissipating excess pore water pressure of fully saturated liquefiable soil by desaturating it via air or gas intrusion. Initial studies on improving the strength of liquefiable soils by foresaid mechanism started back in 1978 [18], where it was found that a reduction in the degree of saturation from 100 to 97.8% can lead to more than 30% increase in liquefaction strength. Similar kinds of studies [3, 34] also confirmed that the resistance to liquefaction increases approximately two times as much as that of fully saturated samples when the degree of saturation drops to 90%. It was the pioneering work of [32], which boosted much clarity on inducing partial saturation in the sands without disturbing the in situ stress conditions and density of the sand skeleton. Studies were then strived to investigate the behavior of partially saturated sands in cyclic loading conditions [6].

1.1 Air Intrusion by Sodium Percarbonate

In the past, researchers have employed electrolysis and drainage-recharge mechanism to partially desaturate fully saturated liquefiable sands [32]. It was found that these mechanisms were not effective in reducing a substantial amount of saturation and were slow in action. In another study, a chemical sodium perborate was utilized [6] to achieve partial desaturation, which was later declared "carcinogenic, mutagenic, or toxic for reproduction" as per [4] in the year 2009. Efforts are also endured to achieve desaturation via microbial activity, but then processes so involved in it require a specific skill set and sophisticated technologies [23, 29].

A chemical performing similar actions of producing oxygen gas bubble upon reaction with water is found to be sodium percarbonate ($Na_2CO_3.1.5H_2O_2$). It is white solid in appearance while chemically crystalline, hygroscopic, and water-soluble in nature [5]. Sodium percarbonate is industrially produced by crystallizing sodium carbonate with hydrogen peroxide maintaining proper control of the pH and concentrations. It is also an environment-friendly, non-toxic, non-flammable chemical that readily decomposes into oxygen, water, and natural soda ash when dissolved in water [11].

$$(Na_2CO_3.1.5H_2O_2) \rightarrow 1.5H_2O_2 + CO_3^{-2} + 2Na^{+1}$$
$$1.5H_2O_2 \rightarrow 1.5H_2O + 0.75O_2$$

As it can be seen from the above reactions, the intermediate product of the reaction is hydrogen peroxide (H_2O_2), which is the actual source of oxygen bubbles. Variation in permeability during liquefaction is directly attributed to excess pore-pressure ratio sustained [26]. Substantial increase in soil permeability and associated large settlement during liquefaction may be controlled by inducing partial saturation. Thus, studying changes occurring in permeability of a saturated liquefiable sand can be an indirect approach to determine improvement in soil's liquefaction resistance. The present study aims to assess sodium percarbonate for its effectiveness on air intrusion capabilities in the fully saturated sand. The said effectiveness is comprehended on the basis of changes occurring in the permeability of treated liquefiable sands. Further, the studies were endeavored to find out the optimum sodium percarbonate content to achieve desired partial saturation.

2 Material Overview

The sand used in the current study was procured from Pinal Corporation, Ahmedabad. According to the manufacturer, it was designated as Grade-II [12] sand having a particle size ranging between 1.0 and 0.5 mm. However, for detailed insight on soil, more geotechnical tests like particle size distribution, specific gravity, relative density, and permeability were also carried out. As per USCS classification, the

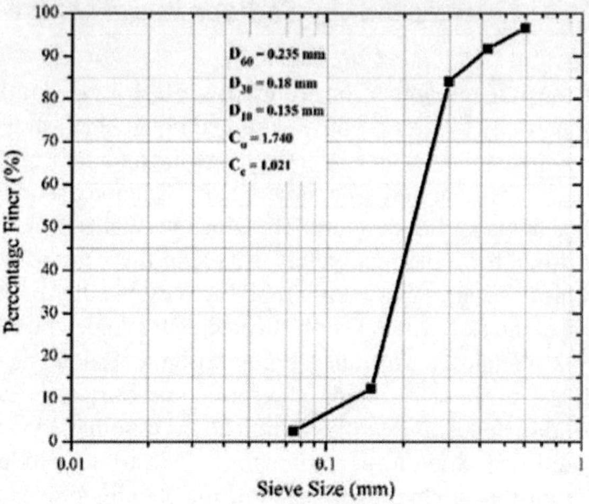

Fig. 1 Particle size distribution curve of Grade-II sand

Table 2 Geotechnical properties of Grade-II sand

Properties[1]	Grade-II sand
Specific gravity (G)	2.62
Effective diameter (D_{10})	0.135 mm
D_{60}	0.235 mm
D_{30}	0.18 mm
Coefficient of uniformity (C_u)	1.740
Coefficient of curvature (C_c)	1.021
Maximum void ratio (e_{max})	0.93
Minimum void ratio (e_{min})	0.63
Constant head permeability[2] (k)	4×10^{-4} cm/s

[1]Geotechnical properties conforming to IS 2720 (Part-3, Part-4, Part-14, Part-17) [13–15]
[2]Based on constant head permeability test [16]

sand was characterized as poorly graded sand, i.e., SP (as shown in Fig. 1). Sodium percarbonate was procured from Ravi Marketing, an Ahmedabad-based chemical supplier. A brief summary of all geotechnical properties found in the laboratory is presented in Table 2.

3 Experimental Program

In order to perceive the ability of sodium percarbonate to reduce the substantial amount saturation percentage, tests for measuring changes in the permeability of

Fig. 2 Schematic for constant head permeability test

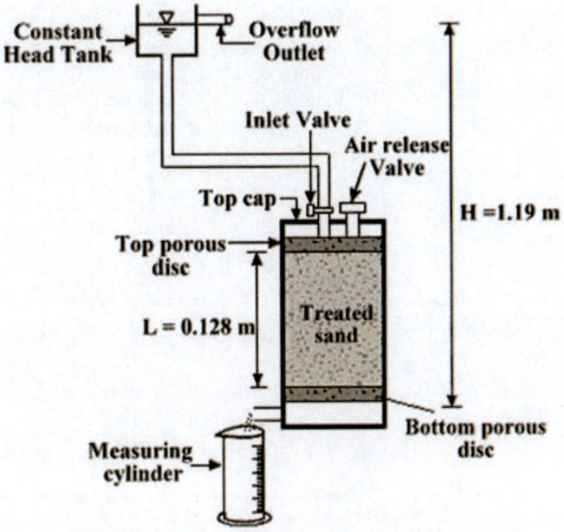

given sand at various chemical contents were performed. Specimens for permeability tests were prepared by adding sodium percarbonate in percentage by weight of dry virgin sand. These percentages were varied in order of 0.5%, 1%, and 2%, respectively. After which, it was ensured that proper mixing of sodium percarbonate and dry sand had been achieved.

The treated sand was then deposited in permeameter using dry pluviation method [1, 17]. Finally, treated sand sample was subjected to constant head permeability test, as indicated in Fig. 2. It must be noted that for the entire duration of the test, the outlet valve was kept open. This was done in order to determine sustainability of oxygen bubbles under given head.

4 Results and Discussion

In order to evaluate the efficacy of sodium percarbonate as an air intrusion admixture, a series of constant head tests were performed, and results of these tests are presented in Table 3 (Fig. 3).

The coefficient of permeability of virgin liquefiable sand was found to be 4.4 $\times$ 10^{-3} cm/s. This permeability was considered to be as baseline permeability for further comparative analysis. From Table 3, it is evident that with the addition of 0.5% sodium percarbonate, the permeability is reduced by three-fourth of the permeability of virgin soil and found to be 1.1 $\times$ 10^{-3} cm/s. The observed reduction in permeability was achieved within 40 min from start of test. This may be due to limited amount of oxygen bubbles formed owing to the lesser amount of sodium percarbonate and, hence, a lesser amount of void spaces occupied by oxygen bubbles. However, addition

Table 3 Permeability of sand at different sodium percarbonate concentrations

Head (m)	Time (min)	Permeability of virgin sand k_0 (cm/s)	Permeability of sand with 0.5% sodium percarbonate $k_{0.5}$ (cm/s)	Permeability of sand with 1% sodium percarbonate k_1 (cm/s)	Permeability of sand with 2% sodium percarbonate k_2 (cm/s)
1.19	5	4.5×10^{-4}	1.8×10^{-3}	1.4×10^{-3}	4×10^{-4}
	10	4.4×10^{-3}	1.3×10^{-3}	6×10^{-4}	3×10^{-4}
	20	4.4×10^{-3}	1.2×10^{-3}	4×10^{-4}	3×10^{-4}
	40	4.4×10^{-3}	1.1×10^{-3}	4×10^{-4}	2×10^{-4}
	80	4.4×10^{-3}	1.1×10^{-3}	4×10^{-4}	2×10^{-4}

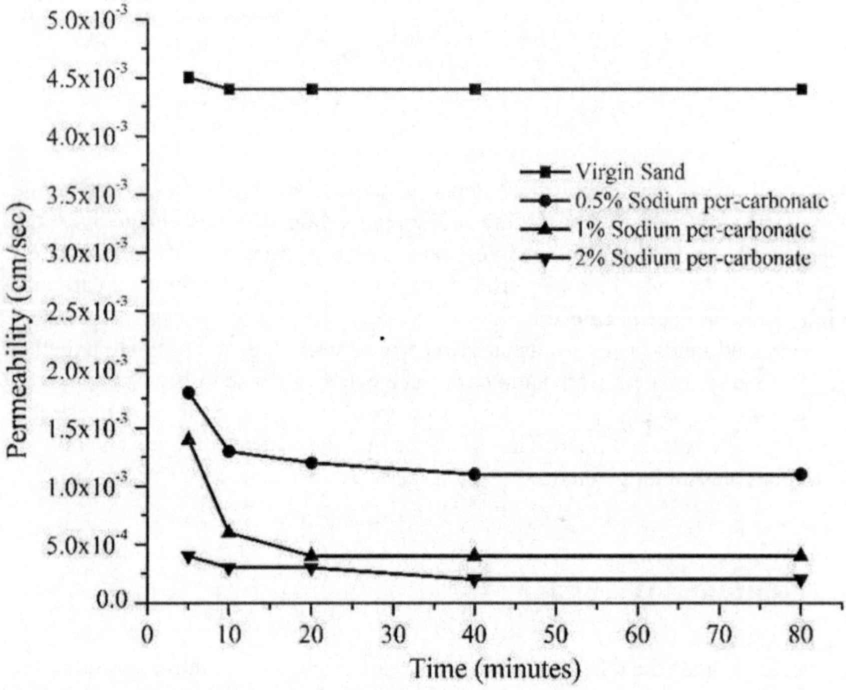

Fig. 3 Comparative illustration of permeability change at different chemical concentration

of 1% sodium percarbonate reduced permeability to 4×10^{-4} cm/s which was about 90% less as compared to virgin sand. Further, to achieve a constant flow, it required 50% less time as compared to that with 0.5% admixture. It is anticipated that relatively larger volume of voids in this case were predominantly occupied by oxygen bubbles generated during chemical reaction of sodium percarbonate with water. The trend of further reduction in permeability was also witnessed at a 2% concentration level of sodium percarbonate. In this case, the permeability was about 5% of that of the virgin soil. This may be attributed to large void volume occupied by air bubbles generated during chemical reactions and, hence, longer travel paths to be followed by unreacted water to drain out. Mainly, decrease in permeability of virgin sand due to air intrusion by sodium percarbonate would make the soil partially saturated and, hence, increase the cyclic resistance ratio. This is validated with respect to findings of [28, 33], which reports an increase in cyclic resistance of partially saturated sands occurring upon a decrease in permeability.

5 Summary and Conclusions

In the present study, an attempt is made to ascertain the effectiveness of an eco-friendly chemical, i.e., sodium percarbonate for air intrusion capability in the potentially liquefiable sand. The said effectiveness was measured in terms of changes occurring in permeability upon addition of 0.5, 1, and 2% by weight of a dry liquefiable Grade–II sand. A constant head permeability test deployed to see such permeability changes on these treated sands fetched following conclusions:

- Reduction in permeability for all admixed sands was significant, i.e., 75%, 90%, and 95%, respectively, for 0.5%, 1%, and 2% addition when compared to the permeability of virgin sand. The time required to achieve such reductions were 40, 20, and 40 min, respectively.
- As for the entire duration of the test, the outlet valve was kept open, and the permeability value of admixed samples remained unchanged. This confirms that bubbles generated upon the reaction of sodium per carbonate with pore water were stable under a head of 1.19 m.
- All of the above-indicated results substantiate that sodium percarbonate is a potent alternative for air intrusion in liquefiable sands. Also, an optimal concentration of 1% by weight of dry sand can be regarded as a desirable amount for air intrusion from efficient functionality and economy point of view.
- Air intrusion by sodium percarbonate resulted in substantial reduction in soil permeability causing soil partially saturated and can increase the liquefaction resistance of the sand. The liquefaction resistance of sand taken in the study increased as it was evident with observed reductions in permeability due to air intrusion by addition of sodium percarbonate.

Although a simplified version of air intrusion is considered in this study, research should be ventured to reinforce the field-scale application of said mechanism.

References

1. Amini F, Qi GZ (2000) Liquefaction testing of stratified Silty sands. J Geotech Geo-environ Eng ASCE 126(3):208–217

2. BDJ Team 2: Chemical used in beauty salon teeth whitening banned by EU.15075. https://doi.org/10.1038/bdjteam.2015.75 (2015)
3. Boulanger RW, Hayden RF (1995) Aspects of compact grouting of liquefiable soils. Geotech Eng ASCE 121(12):844–855
4. Chaney R (1978) Saturation effects on the cyclic strength of sands. Earthquake Eng Soil Dyn 1
5. Craig WJ (1999) Applications of hydrogen peroxide and its derivatives. Royal Soc Chem. ISBN 0-85404-536-8
6. Eseller B, Yegian MK, Alshawabkeh A, Gokyer S (2013) Liquefaction response of partially saturated sands: experimental results. ASCE J Geotech Geo-environ Eng 139(6):863–871
7. Gallagher PM (2000) Passive site remediation for mitigation of liquefaction risk. Dissertation, Virginia Polytechnic Institute and State University
8. Hayashi K, Zen K, Yamazaki A (2000) A field test on a new chemical grouting method to improve the liquefaction resistance of sandy layers beneath the existing structure. In: Proceedings of international symposium on coastal geotechnical engineering in practice, vol 1, A.A. Balkema, Rotterdam, the Netherlands, pp 291–297
9. Hocking G, Geoffrey HC (2012) Soil liquefaction prevention by electro-osmosis and an in situ method to quantify a soil's tendency to liquefy. Ground Modif Seismic Mitig 337–342. https://doi.org/10.1061/40864(196)45
10. Huang Y, Yu M (2013) Review of soil liquefaction characteristics during major earthquakes of the twenty first century. Nat Hazards 65(3):2375–2384
11. Human and environmental risk assessment on ingredients of European household cleaning products Sodium percarbonate (CAS No. 15630-89-4) August, 2002 Available on website http://www.heraproject.com
12. Indian Standard 650-1991 (Re-affirmed 2008): Specification for Standard Sand for Testing of Cement, Bureau of Indian standards, New Delhi
13. Indian Standard 2720-1990 Part 3—Section 1 (Re-affirmed 2002): Method of test for soils, Determination of specific gravity, Bureau of Indian standards, New Delhi
14. Indian Standard 2720-1985 Part 4 (Re-affirmed 2006): Method of test for soils, Grain size analysis, Bureau of Indian standards, New Delhi
15. Indian Standard 2720-1983 Part 14 (Re-affirmed 2006): Method of test for soils, Determination of density index of cohesionless soils, Bureau of Indian standards, New Delhi
16. Indian Standard 2720-1986 Part 17 (Re-affirmed 2002): Method of test for soils, Laboratory determination of permeability, Bureau of Indian standards, New Delhi
17. Ladd RS (1974) Specimen preparation and liquefaction of sands. J Geotech Geo-environ Eng ASCE 100(10):1180–1184
18. Martin GR, Liam Finn WO, Seed HB (1978) Effects of system compliance on liquefaction tests. J Geo-tech Eng Div ASCE 104(GT4):463–479. Proc. Paper 13667
19. Martin JR II, Olgun CG (2006) Liquefaction mitigation using jet-grout columns-1999 Kocaeli earthquake case history. Ground Modif Seismic Mitig. https://doi.org/10.1061/40864(196)47
20. Meyer ME (2001) Liquefaction mitigation at JFK airport using dynamic compaction. In: Proceedings for 2001: a geo-odyssey (Geotechnical Special Publication No. 113). ASCE, Virginia Tech, Blacksburg, VA, 9–13 June 2001
21. Mitchell JK (2008) Mitigation of liquefaction potential of Silty sands. From research to practice in geotechnical engineering. ASCE, pp 433–451
22. Narin WA, Mitchell JK (1995) New insights into explosive compaction of loose, saturated, Cohesionless soils. In: Conference soil improvement for earthquake hazard mitigation, ASCE, pp. 51–65. Geotechnical Special Publication no. 49
23. O'Donnell ST, Bruce ER, Kavazanjian EJ (2017) MIDP: liquefaction mitigation via microbial denitrification as a two-stage process I. J Geotech Geo-environ Eng 143(12):04017094
24. Rollins KM, Goughnour RR, Anderson JKS, Wade SF (2004) Liquefaction hazard mitigation by prefabricated vertical drains. In: International conference on case histories in geotechnical engineering, vol 4

25. Seed HB, Booker JR (1977) Stabilization of potentially liquefiable sand deposits using gravel drains. ASCE J Geotech Eng Div 103(7):757–768
26. Shahir H, Pak A, Taiebat M, Jeremić B (2012) Evaluation of variation of permeability in liquefiable soil under earthquake loading. Comput Geotech 40:74–88
27. Tanaka T, Yasuda S, Murasawa Y, Konishi T, Uchiyama J (2003) Mitigation of subsoil liquefaction by columnar deep mixing. JSCE J Earthquake Eng 27. Paper No. 210 (in Japanese)
28. Thevanayagam S, Ecemis N (2008) Effects of permeability on liquefaction resistance and cone resistance. Geotech Earthquake Eng Soil Dyn IV:1–11
29. Towhata I (2008) Mitigation of liquefaction-induced damage. In: Geotechnical earthquake engineering. Springer, Berlin, pp 588–642
30. Tsuchida H (1970) Prediction and countermeasure against liquefaction in sand deposits. In: Abstract of the Seminar of the Port and Harbour Research Institute, vol 3.1–3.33. Japanese Ministry of Transport, Yokosuka, Japan
31. Whiffin VS, Paassen LA, Harkes MP (2007) Microbial carbonate precipitation as a soil improvement technique. Geomicrobiology J 24(5):417–423
32. Yegian MK, Eseller-Bayat EA, Alshawabkeh S (2007) Induced-partial saturation for liquefaction mitigation: experimental investigation. J Geotech Geo-environ Eng 133(4):372–380
33. Yegian M, Alshawabkeh A (2013) Gas delivery system to provide induced partial saturation through solute transport and reactivity for liquefaction mitigation. U.S. Patent Application 13/969, 885, filed December 19
34. Yoshimi Y, Yanaka K, Tokimatsu K (1989) Liquefaction resistance of partially saturated sand. J Soils Found 29

A Study on Effect of 12 mm Polypropylene Fiber on Shear Strength Characteristics of Fine Sand

D. Neeraj Varma and T. Vamsi Nagaraju

Abstract Fiber is a non-definite failure plane reinforcing material used for enhancing the engineering properties of various building materials. This paper presents the effect of 12 mm polypropylene fiber on shear strength characteristics of fine sand. Direct shear tests are conducted on fine sand reinforced with varying percentages of randomly distributed fiber, from 0 to 1.50% at an increment of 0.25% (by dry weight of soil). With increase in fiber content, angle of shearing resistance increased. Development of apparent cohesion is witnessed from the test data due to the formation of thin films of fiber. Maximum shear strength of fine sand observed at 1.0% of fiber content. In addition, the fiber effect on permeability characteristics is also ascertained through constant head permeability test, and reduction in coefficient of permeability is observed with fiber inclusion. Thus, increases the seepage resistance of the fine sand.

Keywords Fine sand · Fiber · Shear stress · Angle of shearing resistance · Apparent cohesion · Fiber film

1 Introduction

Infrastructure is a major assert for any country to compete with other challenging countries. The progress of infrastructure mainly depends on the available resources and economy. For a fast developing country like India, it is necessary to utilize all the available resources economically. With a rapid growth in construction industry, the availability of material with good engineering properties is being declined. As a result, the cost of materials increasing drastically. From the geotechnical engineering perspective, foundation materials such as coarse sand and moorum come under the same category due to their superior engineering properties. In the current scenario,

D. Neeraj Varma (✉)
PhD Scholar, Department of Civil Engineering, NIT-Rorkela, Rourkela, Odisha, India

Present Address:
T. Vamsi Nagaraju
M-Tech, Department of Civil Engineering, NIT-Rourkela, Rourkela, Odisha, India

T. G. Sitharam et al. (eds.), *Earthquake Geotechnics*, Lecture Notes in Civil Engineering 187, https://doi.org/10.1007/978-981-16-5669-9_9

for effective utilization of low-cost material, it is necessary to use existing materials with less engineering properties through proper stabilization techniques.

Fine sand is an abundant material, available in various places around the globe. Due to the poor gradation and inadequate engineering properties, fine sand is not utilized as foundation material [1, 2]. Conventional stabilization techniques using cementitious products are effectively used for improving the engineering properties of soil [3–5]. However, the usage of cementitious products, commercially available cement, emits greenhouse gases, which is an environmental concern [6]. Hence, researchers are working on new stabilization techniques by considering both environment and economy aspects.

Reinforcement with fiber is one of the new stabilization techniques adopted for strengthening materials. Researchers are working with variety of synthetic and natural fibers for stabilizing various types of soil [7–12]. Fiber reinforcement imparts stability to soil through the development of tensile forces [7]. As granular soils are weak in tension, the engineering properties can be enhanced by reinforcing with randomly distributed fiber. One should not contemplate randomly distributed fiber reinforcement as reinforced earth in which reinforcement in the form of sheets, strips, grids, etc., placed in predetermined positions. Unlike reinforced earth, randomly distributed fiber will not show any definite failure plane [13, 14] and offer strength isotropy [15]. Hence, in the present study, the shear strength characteristics of fine sand was improved by reinforcing with randomly distributed 12 mm polypropylene fiber.

2 Materials and Methods

For the present study, soil collected from Suryalanka, Bapatla, Andhra Pradesh. About 99% of soil passed through 425 micron IS sieve and retained on 75 micron IS sieve. Most of the soil particles were fine in nature and not suitable for foundation material. Therefore, the collected soil was best suited for fiber reinforcement. As per IS Soil Classification System [16], the collected soil sample was classified as poorly graded sand (SP) with a specific gravity of 2.62. Various laboratory tests were conducted to determine properties of fine sand, and the same were presented in Table 1.

Commercially available 12 mm Recron 3S polypropylene fiber was used for reinforcing the fine sand. The properties of fiber were presented in Table 2. Fine sand was mixed with varying percentages of fiber (0.25, 0.50, 0.75, 1.00, 1.25, and 1.50%) by dry weight of soil. Initially, optimum moisture content (OMC) and maximum dry density (MDD) of fiber reinforced fine sand reinforced was determined by performing IS light compaction test as per IS 2720 part 7, 1980 [17]. The compaction test results were presented in Table 3. With increase in fiber content, the maximum dry density decreased and optimum moisture content increased.

Fiber-reinforced soil specimens are compacted at respective OMC and MDD in shear box of 60 mm × 60 mm × 60 mm size, and direct shear test was conducted

Table 1 Engineering properties of collected soil sample

S. No.	Engineering properties	Values
1	Compaction characteristics	
	(a) Optimum moisture content	11.2%
	(b) Maximum dry density	15.6 kN/m^3
2	Shear strength parameters	
	(a) Cohesion	0
	(b) Angle of internal friction	23°
3	Coefficient of permeability	2.67×10^{-3} cm/s

Table 2 Properties of Recron 3S polypropylene fiber

S. No.	Property	Value	Units
1	Diameter	0.04	mm
2	Cut length	12	mm
3	Aspect ratio	0.0033	–
4	Specific gravity	0.91	–
5	Tensile strength	4000–6000	kN/m^2
6	Melting point	>250°	C
7	Elongation	60–90	%
8	Shape of fiber	Triangle shape	–

Table 3 Variation of compaction characteristics with varying fiber content

S. No.	*F*(%)	OMC (%)	MDD (kN/m^3)
1	0	11.2	15.6
2	0.25	12.4	15.52
3	0.50	13.4	15.50
4	0.75	14.8	15.46
5	1.00	16	15.40
6	1.25	17	15.30
7	1.50	18.2	15.24

Note F(%)—Percentage of fiber by dry weight of soil

on soil specimens as per IS 2720 part 13, 1986 [18]. The shear load was applied at a rate of 1.25 mm/min until the specimen failed or 20% of strain was reached, whichever was earlier. Direct shear tests carried out for different fiber contents (% *F* = 0.25, 0.50, 0.75, 1.00, 1.25, and 1.50%) under different normal stresses (0.5, 1.0, and 2.0 kg/cm^2).

For accessing the permeability characteristics of fiber-reinforced soil, constant head permeability test was conducted on soil specimens prepared at OMC and MDD as per IS 2720 part 17, 1986 [19]. For achieving uniform saturation, water was allowed to flow through soil specimen for two hours prior to the test.

3 Results and Discussions

3.1 Shear Strength Parameters

Typical results of direct shear test conducted on fiber-reinforced fine sand were given in Table 4. The angle of shearing resistance (ϕ) of fine sand increased with increase in fiber content. Many researchers also reported a similar trend in enhancement of shear strength parameters [9, 11, 12, 20]. Marandi et al. [10] reported that the shear zone and surface failure direction of soil was effected by the tensile properties of fiber. With an increase in fiber content, the interaction of fiber with soil grains increased through surface friction. Therefore, under the combined action of surface friction and tensile resistance of the fiber, the angle of shearing resistance of fine sand increased. The tensile force developed in the fiber under the action of normal stress was represented in Fig. 1b. The maximum shearing resistance of fiber-reinforced fine sand was encountered at 1.0% fiber content. Thereafter, with increase in fiber content, the portion of soil grains were replaced by the fiber. The behavior of reinforced soil is categorically governed by the fiber, resulting in the reduction of angle of shearing resistance. Further, in the presence of moisture content, thin films of fiber

Table 4 Variation of shear parameters with varying fiber content

S. No.	$F(\%)$	Angle of shearing resistance (ϕ) (°)	Apparent cohesion (kN/m^2)
1	0	23	0
2	0.25	37	0
3	0.50	40	9
4	0.75	40	29
5	1.00	45	32
6	1.25	42	28.5
7	1.50	41	35.5

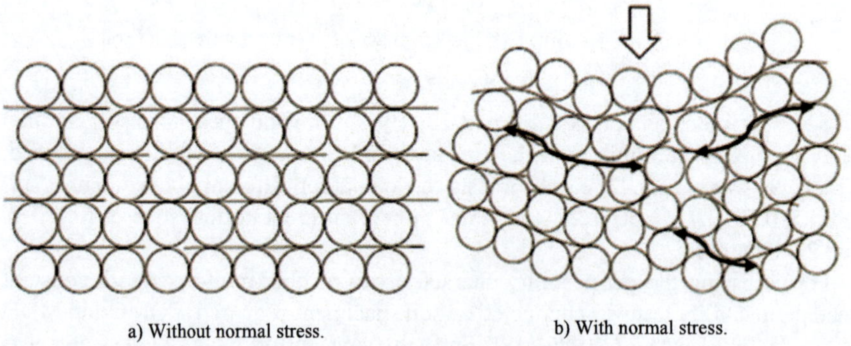

Fig. 1 Effect of normal stress on tensile resistance of fiber

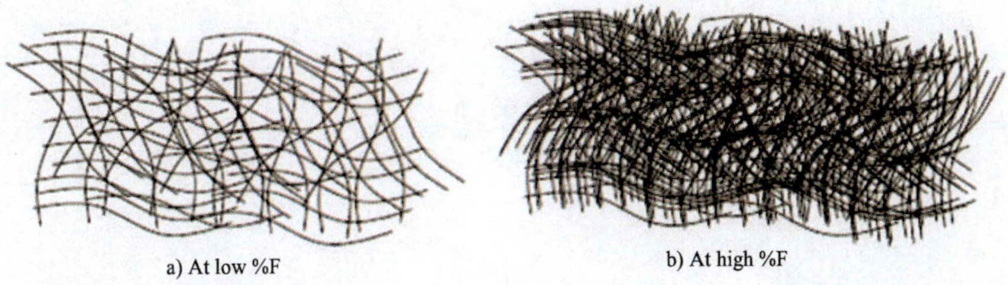

Fig. 2 Thin film of fiber

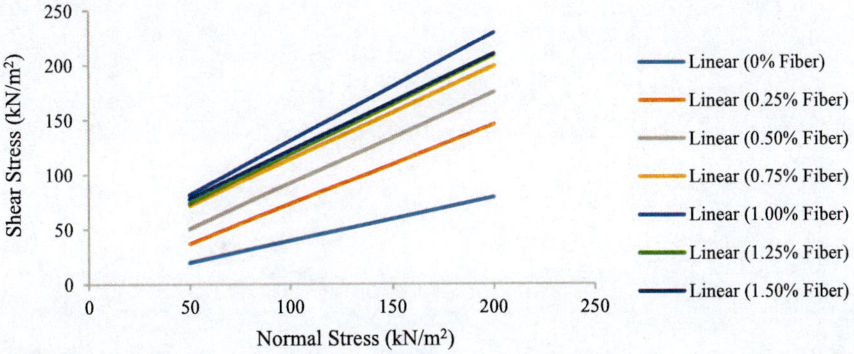

Fig. 3 Failure envelope of fiber-reinforced fine sand

(shown in Fig. 2) were developed between soil grains causing superficial force of attraction between soil grains. The resultant force of attraction was represented as apparent cohesion. At low fiber content (i.e., 0.25 and 0.5%), apparent cohesion is not predominant due to the insufficiency in fiber content for producing superficial force (Fig. 2a). With increase in fiber content, apparent cohesion increased with increase in fiber content and maximum of 35.5 kN/m^2 observed at a fiber content of 1.5%. In contrast, Darvishi and Erken [12] reported a decrease in the apparent cohesion beyond 1.0% of fiber addition. As fiber and fine sand are inert in nature, the development of apparent cohesion is constrained to certain limit only. For most of the fiber contents, apparent cohesion was in the range of 28–35.5 kN/m^2. The typical failure envelopes for varying percentages of fiber were presented in Fig. 3.

3.2 *Shear Stress and Ductile Behavior of Fine Sand*

The variation in shear stress for various proportions of fiber at each normal stresses (i.e., 50, 100, and 200 kN/m^2) presented in Figs. 4, 5, and 6. With increase in fiber content, the peak stress of fine sand increased under the action of tensile forces in

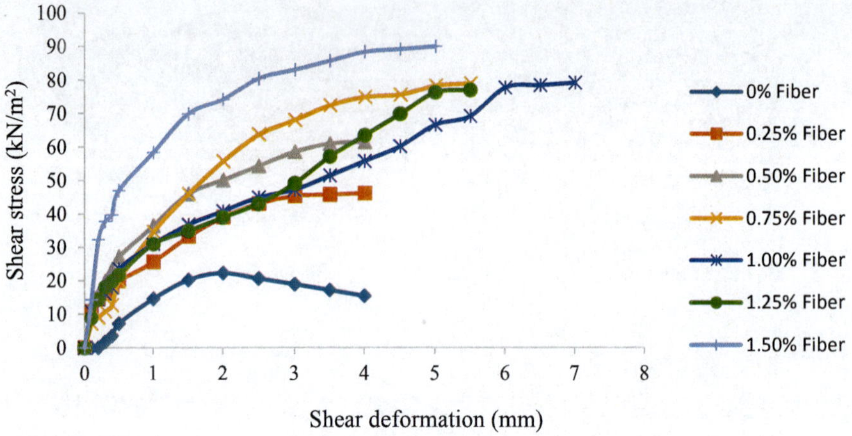

Fig. 4 Shear deformation vs. shear stress plot at 50 kN/m^2 normal stress

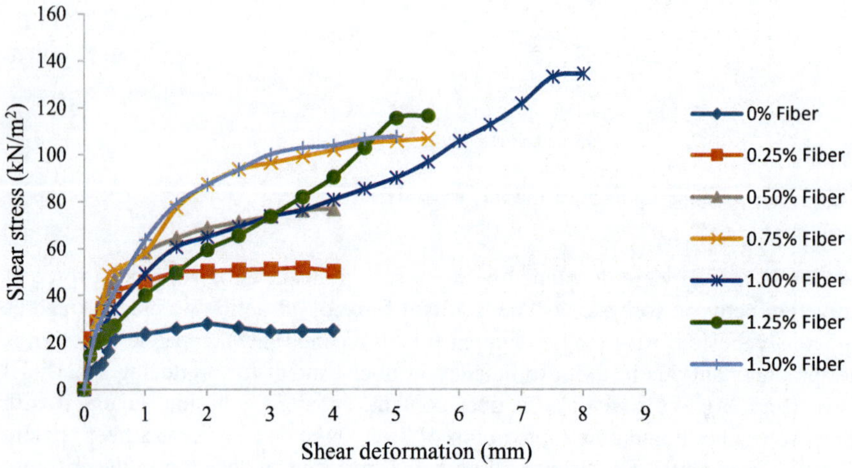

Fig. 5 Shear deformation vs. shear stress plot at 100 kN/m^2 normal stress

the soil–fiber matrix (as shown in Fig. 1b) and surface friction. For most of the fiber-reinforced specimens subjected to varying normal loads, strain-hardening behavior was noticed. Consoli et al. [21] reported that the longer fiber exhibits strain-hardening behavior and shorter fiber exhibits strain-softening behavior. Heineck et al. [22] reported that the strain-hardening behavior was not only controlled by fiber length but also with aspect ratio. However, the variation in aspect ratio is not in the scope of this study. Further, the addition of fiber increased the failure strain in fine sand, resulting in induction of ductile behavior in fine sand [10].

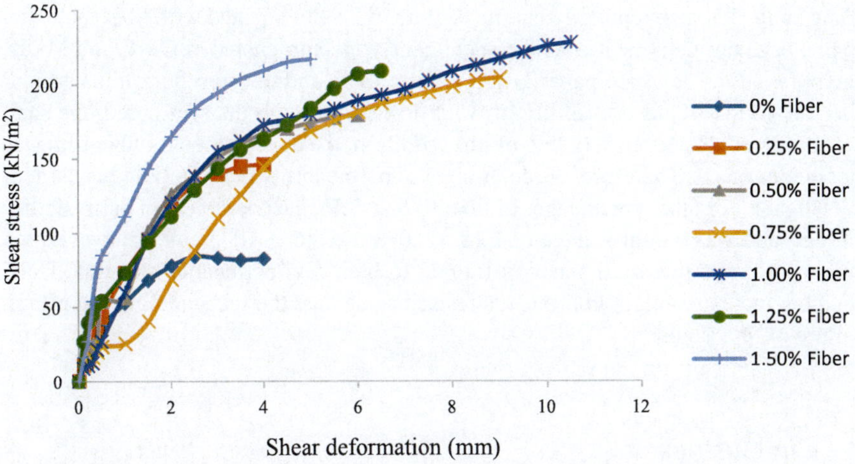

Fig. 6 Shear deformation vs. shear stress plot at 200 kN/m^2 normal stress

3.3 Permeability Characteristics

The coefficient of permeability (k) of fine sand reinforced with varying percentages of fiber is shown in Fig. 7. A significant decrease in k value was observed up to 0.5% of fiber content. With increase in fiber content from 0.5 to 0.75%, the decrease became insignificant. With further increase in fiber content form 0.75 to 1.5%, a slight increase in permeability was noticed. Irrespective of fiber percentage, the coefficient of permeability of fiber-reinforced fine sand was less than that of unreinforced soil.

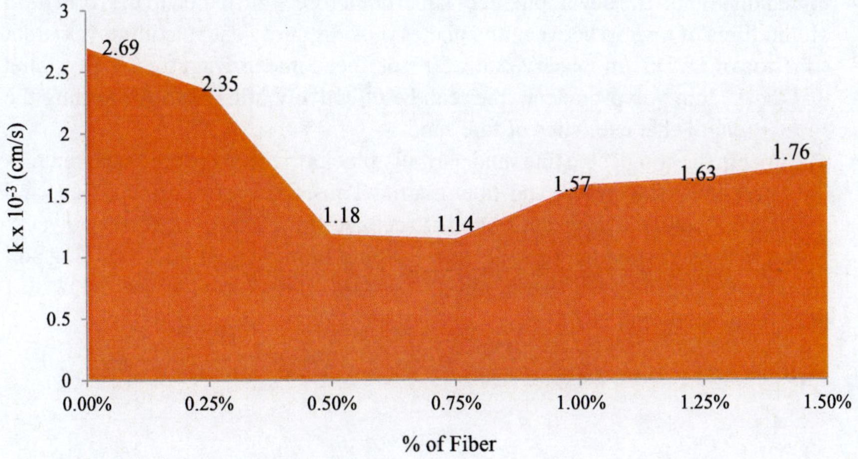

Fig. 7 Variation in k (cm/s) with fiber content

Yang et al. [23] also reported that the decline in k value of sand with fiber inclusion is higher under dense state. As the specimens were compacted at OMC and MDD, the pores in the sand are partially filled with fibers and develop fiber films around the soil grains. Thus, the ability of water to flow through the reinforced fine sand was hindered. However, at 0.25% of fiber addition, the development of fiber film was not predominant. Therefore, the coefficient of permeability (k) at initial fiber content was higher. For other percentages of fiber (0.5–1.5%), the coefficient of permeability of fine sand was within a range of 1.18 × 10–3 to 1.76 × 10–3 cm/s. However, the fiber-reinforced fine sand was constrained to be a low permeable material. Thus, the fiber inclusion might improve the resistance against the seepage flow and piping failure of fine sand.

4 Conclusions

From the results and discussion cited above, the following conclusions were adopted:

- With increase in 12 mm polypropylene fiber content, the maximum dry density decreased and optimum moisture content increased. These results are likely attributed to the resistance offered by the fiber under compaction energy which prevents denser packing and increases the void ratio in the sand-fiber mix.
- The addition of high tensile strength fiber in fine sand increased the interaction between grains through friction. Under the mutual action of tensile force and friction, the angle of shearing resistance (ϕ) of soil increased. The peak value of ϕ is observed at 1.0% of fiber. In the similar manner, the improvement in shear stress and ductile behavior of fine sand are observed with fiber inclusion.
- Under the action of moisture content, apparent cohesion was developed in reinforced fine sand. The development of superficial forces attributed to the formation of thin films of fiber in between the planes of soil grains. The maximum apparent cohesion of 35.5 kN/m^2 observed at 1.50% of fiber content. Therefore, it was stated that the 12 mm polypropylene fiber can be effectively utilized for increasing the shear strength characteristics of fine sand.
- The fiber in the compacted fine sand partially blocks the seepage pores and reduces the water flow through the sand-fiber matrix. Thus, the fiber-reinforced fine sand might have potential resistance against seepage flow. However, further investigation is to be carried out to analyze the piping failure effect under varying soil characterizes such as void ratio, initial moisture content, etc., to use the best of fiber-reinforced sand.

References

1. Mola-Abasi H, Shooshpasha I (2016) Influence of zeolite and cement additions on mechanical behaviour of sandy soil. J Rock Mech Geotech Eng 8:746e752
2. Nader A, Masoud M (2018) Improvement of geotechnical properties of silty sand soils using natural pozzolan and lime. I J Geo-Eng 9:4
3. Huang JT, Airey DW (1998) Properties of an artificially cemented carbonate sand. J Geotech Geoenviron Eng 24(6):492–499
4. Consoli NC, Fonseca AV, Cruz RC, Heineck KS (2009) Fundamental parameters for the stiffness and strength control of artificially cemented sand. J Geotech Geoenviron Eng 135(9):1347–1353
5. Consoli NC, Festugato L, da Rocha CG, Cruz RC (2013) Key parameters for strength control of rammed sand–cement mixtures: influence of types of Portland cement. Constr Build Mater 49:591–597
6. Andrew RM (2018) Global CO_2 emissions from cement production. Earth Syst Sci Data 10:195–217
7. Ranjan G, Vasan RM, Charan HD (1996) Probabilistic analysis of randomly distributed fiber-reinforced soil. J Geotech Eng 122:419–426
8. Santoni ZL, Tingle JS, Webster SL (2001) Engineering properties of sand-fiber mixtures for road construction. J Geotech Geoenviron Eng 127(3):258–268
9. Ibraim E, Fourmont S (2006) Behaviour of sand reinforced with fibres. In: Geotechnical symposium in Roma, 16 and 17 March 2006
10. Marandi SM, Bagheripour MH, Rahgozar R, Zare H (2008) Strength and ductility of randomly distributed palm fibers reinforced silty-sand soils. Am J Appl Sci 5(3):209–220. ISSN 1546-9239
11. Eldesouky HM, Morsy MM, Mansour MF (2015) Fiber-reinforced sand strength and dilation characteristics. Ain Shams Eng J 2090–4479
12. Darvishi A, Erken A (2018) Effect of polypropylene fiber on shear strength parameters of sand. In: Proceedings of the third world congress on civil, structural, and environmental engineering (CSEE'18), Hungar
13. Carmine GG (2016) Failure mechanisms of geosynthetic-reinforced wall with modular blocks: from centrifuge test to numerical analysis. In: Sixth European geosynthetics congress, At Ljubljana, Slovenia
14. Carlos DM, Pinho-Lopes M, Lopes ML (2016) Effect of geosynthetic reinforcement inclusion on the strength parameters and bearing ratio of a fine soil. Procedia Eng 143:34–41
15. Babu GLS, Chouksey SK (2010) Model for analysis of fiber-reinforced clayey soil. Geomech Geoeng Int J
16. IS 1498-1970 Classification and identification of soils for general engineering purposes [CED 43: Soil and Foundation Engineering]. BIS, New Delhi
17. IS 2720 (Part 7): 1991, Method of testing for soil, Part 7: Determination of water content-dry density relation using light compaction [CED 43: Soil and Foundation Engineering]. Bureau of Indian Standards, New Delhi
18. IS 2720-13 (1986): Methods of test for soils, Part 13: Direct shear test [CED 43: Soil and Foundation Engineering]. BIS, New Delhi
19. IS 2720-17 (1986): Methods of test for soils, Part 17: Laboratory determination of permeability [CED 43: Soil and Foundation Engineering]. BIS, New Delhi
20. Consoli NC, Consoli BS, Festugato L (2013) A practical methodology for the determination of failure envelopes of fiber-reinforced cemented sands. Geotext Geomembr 41:50–54
21. Consoli NC, Festugato L, Heineck KS (2009) Strain-hardening behaviour of fibre-reinforced sand in view of filament geometry. Geosynth Int 16(2):109–115
22. Heineck KS, Coop MR, Consoli NC (2005) Effect of microreinforcement of soils from very small to large shear strains. J Geotech Geoenviron Eng 131(8):1024–1033
23. Yang KH, Adilehou WM, Jian ST, Wei SB (2017) Hydraulic response of fibre-reinforced sand subject to seepage. Geosyn Int. 1072-6349 © 2017 Thomas Telford Ltd

Local Site Effects Influence on Earthquake Early-Warning Parameters

A. Mugesh, Aniket Desai, Ravi S. Jakka, and Kamal

Abstract Earthquake early-warning (EEW) system is a valuable device to detect an impending earthquake in real time and helps in the mitigation of seismic risk. The technology of EEW is used for calculating the size of an earthquake based on the first three seconds of recorded P-wave ground motion at the given site. The issues of precision are still challenging in this system. In this paper, an endeavor is made to show the influence of local site effects on earthquake early-warning parameters such as average ground motion period (τ_c) and the peak displacement amplitude (P_d) in the initial three seconds of the motion records. The analysis is carried out using the ground motion records of the Japanese dataset. The vertical component of the acceleration waveforms is used to determine the earthquake early-warning parameters. The comparison is made between the variations of τ_c and P_d parameters with the magnitude τ_c, epicentral distance, and average shear wave velocity for the surface motions. The results indicated that the local site effects significantly influence the earthquake early-warning parameters.

Keywords Earthquake early-warning parameters · Magnitude · Average shear wave velocity · Borehole motion · And surface motion

1 Introduction

In the current scenario, the earthquake early-warning system plays a vital role in the prediction of the most damaging earthquake, but still, it is in developing stage. There were many countries like Japan, Taiwan, and Southern California that provide an early warning during an earthquake was in good agreement with an source parameters of

A. Mugesh (✉) · R. S. Jakka · Kamal
Centre of Excellence in Disaster Mitigation and Management, IIT Roorkee, Roorkee 247667, India

Kamal
e-mail: kamalfes@iitr.ac.in

A. Desai
Department of Earthquake Engineering, IIT Roorkee, Roorkee-247667, India

T. G. Sitharam et al. (eds.), *Earthquake Geotechnics*, Lecture Notes in Civil Engineering 187, https://doi.org/10.1007/978-981-16-5669-9_10

the earthquake. But for providing the early warning, there were many parameters used like predominant period, cumulative average velocity, average ground motion period, peak displacement amplitudes, etc. Researchers in this field are still developing new applications for improving the accuracy of the early-warning period in the earthquake early-warning system. In the past and present, there were many early-warning signal parameters are found they are ground motion average period τ_c and peak displacement amplitude P_d and it shows a good improvement for measurement of the earthquake size. During the origin of the earthquake, the early-warning waveform of three seconds time window was taken into account to determine the early-warning parameters like τ_c and P_d. In Taiwan, the experiment conducted for the onsite early-warning approach suggested that both τ_c and the peak displacement amplitude were possible to give a reliable threshold warning during the earthquake and stated that those given parameters had the complement usage for the regional approach in Taiwan [1]. Wu and Zhao [2] clearly explained the earthquake early-warning practice, and peak displacement amplitude provides a more advanced warning period from the three seconds of the early-warning signal. They have shown the magnitude determination from the P_d parameter with lesser uncertainty, and this parameter is used to reduce the false alarms caused by the small and moderate earthquake [3]. For many practical early warning purposes, the considered magnitude was greater than 6 is resulted in a betterment way during the warning period compared with lower size of the magnitude and it suggested that the future work like uncertainties findings in the estimation of magnitude and intensity from the initial P-wave motion. Authors [4, 5] suggested that the relationship between τ_c and M shows a better fit in the regression analysis, and they were used the high-pass filter with a cutoff frequency of 0.075 Hz for removing the noise data in the collected waveform. Wu et al. [6] shows the linear relationship of the τ_c and P_d parameters with the function of the magnitude in the range of 4–8 by using the regression analysis with a better fit of lesser standard deviation. Bhardwaj [7] concluded that the magnitude obtained from the τ_c of the initial P-wave provided a satisfactory solution. The comparison was made between the τ_c and P_d with the function of the magnitude by using linear regression analysis with the considered Indian dataset [8]. The logarithmic τ_c with magnitude shows a better correlation with a positive slope in regression analysis [9]. This study deals with, by comparing the early warning parameters with epicentral distance and the average shear wave velocity were shown the influence of local site effects in the initial P-wave of three seconds time window.

2 Data Analysis

This study considered the bedrock and surface vertical component waveform of the seven individual station records with the single earthquake event. The data were collected in a Japanese strong motion database in KIK-net sensors installed in both borehole and the surface at the sites with a magnitude of 6.7 within the covered epicentral distance of 18 to 66 km. In the considered seven-station records, each

Table 1 List of stations and earthquakes used in this study

Eventname	Station	Latitude	Longitude	Magnitude	Epicentral distance (km)
Eastern an central Iburi	HDKH04	42.5126	142.0381	6.7	18
	IBUH02	42.8714	142.1285		20
	HDKH03	42.5934	142.3521		22
	HDKH01	42.7031	142.2296		26
	IRKH02	43.2204	141.6523		30
	IBUH01	42.8739	141.8191		44
	SRCH09	43.0587	141.8063		66

station has two sets of motion that is bedrock motion and the surface motion, so it covered 14 vertical component of the P-wave motion. The site characterization details were also collected from the KIK-net Web site, it includes the soil profile data for each station, and these collected soil profile data are used to classify the sites based on average shear wave velocity for each site. The earthquake event details are given in Table 1.

3 Methodology

In this study, the waveform data of the vertical component of bedrock and surface acceleration recorded by the KIK-net seismic networks of the National Research Institute for Earth Science and Disaster Prevention (NIED-Japan) are used. The selected stations are near to the epicenter, and the collected data for each station show an uncorrected value of the acceleration records, so here linear baseline correction is carried out by using SeismoSignal software for the betterment of the results. The value of τ_c and P_d has been determined at the level of bedrock and the surface motion for each station record. Then, the site classification is made for the selected sites based on average shear wave velocity followed by NEHRP [10].

4 τ_{c} and P_{d} Method

Toward finding the magnitude of the earthquake, it is essential to describe whether the occurrence of slip motion is stationary or it is still developing, which is usually revealed in the period of the primary motion. The minor events produce a short period, and the extended events yield a longer period of the initial motions, respectively. Normally, the slip motion is a difficult measurement, and also, a large event often begins with a short period, followed by a long period motion. Accordingly, it is significant to define the average period during the first motion. τ_c is a measurement of the average period of the ground motion with a three-second time window of P waves developed [1]. τ_c is determined as

$$\tau_c = \frac{2\pi}{\sqrt{r}}$$

where $r = \frac{\int_0^{\tau_0} \dot{u}^2(t)\mathrm{d}t}{\int_0^{\tau_0} u^2(t)\mathrm{d}t}$, τ_0 = duration of the time window (generally 3 s),

$\dot{u}(t)$ = velocity and $u(t)$ = vertical displacement obtained from ground motion record on double integration.

P_d is the peak displacement amplitude in the first three seconds after the arrival of P waves, it is the robust measurement for determining the magnitudes of earthquakes, and it has practical application in the earthquake early-warning systems [2].

5 Results and Discussions

5.1 *Variation of τ_c in bedrock and Surface motion*

The motion at the base of the soil deposit (also the top of the rock) is known as bedrock motion, and the free surface in the top of the soil deposit is known as surface motion. The acceleration data were collected from the Kiban Kyoshin network (KIK-net) strong motion seismograph network, which consists of pairs of seismographs installed in a borehole and surface at a different site. In the borehole and the surface, the recorded motion was considered as a bedrock and surface motion for the analysis. The collected acceleration data are not about the zero axis line, so here the baseline correction analyses were carried out in the SeismoSpect software v18, and these records were integrated to obtain velocity and displacement. A bandpass second-order Butterworth filter with cutoff frequency 0.07–30 Hz was applied on the record to remove the low-frequency drift in the velocity and displacement signal. The average ground motion period (τ_c) and the peak displacement amplitude (P_d) for the bedrock and the surface motion estimated with the selected time window of the 3 s.

In Fig. 1, the comparison is made for logarithmic τ_c of bedrock and surface motion with different sites for the magnitude of 6.7. The surface motion early-warning parameter shows higher than the bedrock motion parameter for all sites. It represents that the surface amplification is higher than the bedrock amplification due to softer soil conditions in all sites.

The logarithmic parameter τ_c increased with increasing epicentral distance. Figure 2 represents that the surface motion parameter is high compared to the bedrock motion parameter, and it reveals that with the increasing epicentral distance, the surface motion parameter is increasing. And also, it shows that the surface motion has a higher difference with a distance greater than 35 km due to site effects.

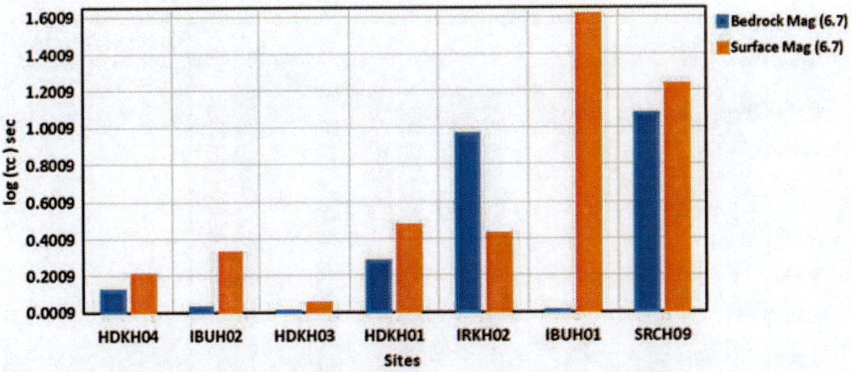

Fig. 1 τ_c recorded at different sites for the considered earthquake of magnitude 6.7

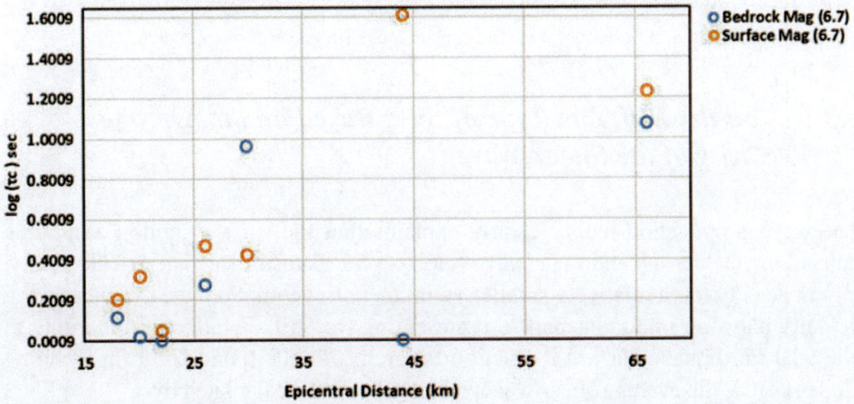

Fig. 2 Variation of $\log(\tau_c)$ with epicentral distance

5.2 *Variation of P_d in Bedrock and Surface Motion*

The logarithmic peak displacement amplitude of the early-warning parameter P_d is compared with the different sites with a magnitude of 6.7 in Fig. 3. It shows that most of the site surface motion parameter is lower than the bedrock motion parameter, the observation made in P_d has very low-level amplitudes that were represented in negative log parameter, and the positive P_d parameter reflected as a higher amplitude due to the considerable soil amplification. The site IBUH01 has a higher surface motion parameter than the bedrock motion parameter, and it may due to high-level amplification of the motion.

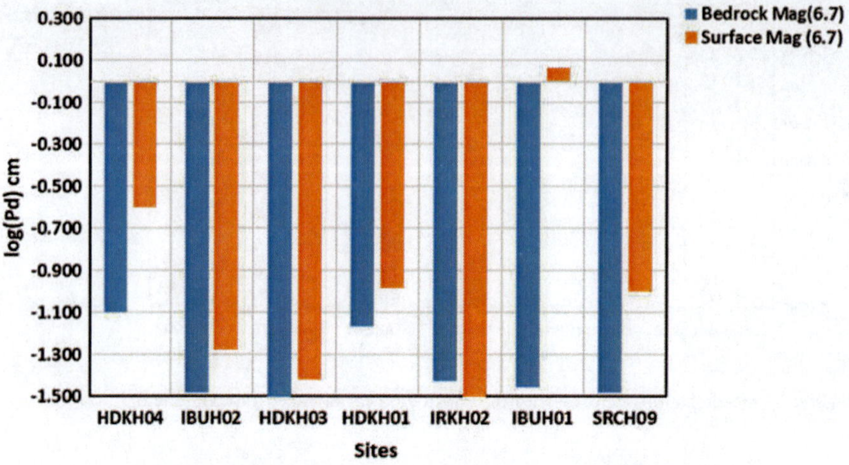

Fig. 3 P_d recorded at different sites for the considered earthquake of magnitude 6.7

5.3 τ_c on the Different Type of Sites Based on the Average Velocity of the Shear Wave

The wave propagation theory clearly explains that the surface motion amplitude depends on the density and shear wave velocity of the subsurface material. Generally, in-situ density has a relatively smaller variation with depth. So, here the shear wave velocity plays an important part in representing the soil site effects. According to National Earthquake Hazard Reduction Program (NEHRP) and Uniform Building Code (UBC), the average shear wave velocity has been calculated by

$$\text{Average Shear wave velocity}(V_s) = \frac{\sum \text{di}}{\sum \frac{di}{Vi}}$$

di—thickness of soil layers (m)

Vi—shear wave velocity of the soil layers (m/s) (Table 2).

5.3.1 Influence of Average Shear Wave Velocity on τ_c and P_d for Surface Motion

The average shear velocity is one of the most vital properties in the site effect characterization. In Fig. 4, the logarithmic parameter of τ_c is compared with the function of average shear wave velocity and it states that log τ_c is decreased with increasing average shear wave velocity, and also it shows the site which has the lowest shear wave velocity, there the τ_c parameter shows the highest value in the surface motion.

Table 2 Site Classification based on Average Shear-wave Velocity

Site	Average Shear Wave Velocity (m/s)	Site Class	Site Type
HDKH01	426	C	Very dense soil or soft rock
HDKH03	732	C	Very dense soil or soft rock
HDKH04	509	C	Very dense soil or soft rock
IBUH01	408	C	Very dense soil or soft rock
IBUH02	733	C	Very dense soil or soft rock
IRKH02	588	C	Very dense soil or soft rock
SRCH09	221	D	Stiff soil

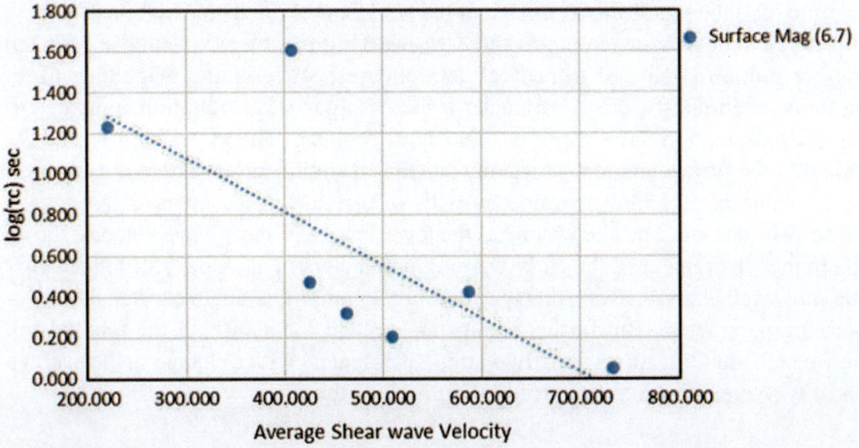

Fig. 4 Variation of τ_c with average shear wave velocity of the site

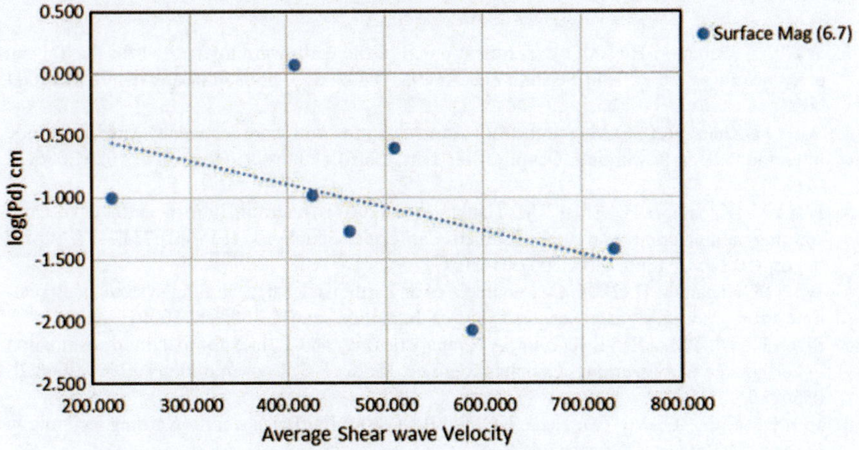

Fig. 5 Variation of P_d with average shear wave velocity of the site

The logarithmic parameter of P_d is compared with the function of average shear wave velocity (Fig. 5). The above mentioned graphical representation explains that the log P_d is decreased with increasing average shear wave velocity, and also it shows the site which has the lowest shear wave velocity, there the P_d parameter shows the highest value in the surface motion.

6 Conclusions

This work endeavored to study the effects of local site conditions on earthquake early-warning parameters of the bedrock and the surface motion. The sites considered in this study show the increment in the early-warning parameters τ_c and P_d for the surface motion compared to bedrock motion irrespective of the magnitude of an earthquake. Similarly, the τ_c parameter is increasing in surface motion compared to the bedrock motion with respective epicentral distance. The increase in τ_c and P_d values in the surface motion is certainly due to soil amplification. There is a decrease in the value of τ_c and P_d parameters with an increase in the average shear wave velocity of the site. The site which has the lowest average shear wave velocity shows the highest increment of the early-warning parameters of τ_c and P_d. It reflects that the sites which have the soft soil type show higher amplification in earthquake early-warning parameters. But further studies are needed for justifying the relationship between the modifications of early-warning parameters with a change in the average shear wave velocity and the epicentral distance of the site.

References

1. Wu YM, Kanamori H (2005) Experiment on an onsite early warning method for the Taiwan early warning system. Bull Seismol Soc Am 95(1):347–353. https://doi.org/10.1785/0120040097
2. Wu YM, Zhao L (2006) Magnitude estimation using the first three seconds P-wave amplitude in earthquake early warning. Geophys Res Lett 33(16):4–7. https://doi.org/10.1029/2006GL026871
3. Wu YM, Kanamori H, Allen RM, Hauksson E (2007) Determination of earthquake early warning parameters, τ_c and P_d, for southern California. Geophys J Int 170(2):711–717. https://doi.org/10.1111/j.1365-246X.2007.03430.x
4. Wu YM, Kanamori H (2008) Development of an earthquake early warning system using real-time strong motion signals. Sensors 8(1):1–9. https://doi.org/10.3390/s8010001
5. Shieh JT, Wu YM, Allen RM (2008) A comparison of τ_c and T_p max for magnitude estimation in earthquake early warning. Geophys Res Lett 35(20):1–5. https://doi.org/10.1029/2008GL035611
6. Wen KL, Shin TC, Wu YM, Hsiao NC, Wu BR (2009) Earthquake early warning technology progress in Taiwan. J Disaster Res 4
7. Bhardwaj R (2012) Analysis of Tauc (t_c) and P_d attributes for earthquake early warning in India. In: 15th world conference on earthquake engineering (15WCEE)

8. Shanyou L, Jindong S (2008) A new magnitude estimation method based on predominant period and peak amplitude. 14 WCEE. ftp://128.46.154.21/spujol/Andres/files/S05-03-012.PDF
9. Zhang J, Li H (2018) Testing and comparison of three frequency-based magnitude estimating parameters for earthquake early warning based events in the Yunnan region, China in 2014. J Geophys Eng 15(3):971–979. https://doi.org/10.1088/1742-2140/aaa85d
10. NEHRP Recommended seismic provisions for new buildings and other structures FEMA P-750/2009 Edition

Application of Scrap Tyre as Backfill in Retaining Wall: A Review

Sadia Kantroo and Furkhanda Khalid Khan

Abstract Every year, waste materials are expanding and leading to myriad environmental issues. Recycling of these materials is the safest way of their disposal. Scrap tyre is one of those wastes, and its disposal is a major trouble we have been facing. A new geoenvironmental approach suggested the use of waste tyre in several civil engineering applications. As tyre aggregates possess less unit weight and are flexible in nature, thus, it tends to reduce the dynamic earth pressure and lateral displacement of retaining wall effectively. This makes them a suitable backfill material. This paper aims at reviewing the enhancement in the properties of different types of soil by the inclusion of waste tyre rubber and confirmation of its suitability as a retaining wall backfill. One more property that is mandatory in the backfill material of retaining wall is permeability, and due to this reason, sand has always been a preferential material. Research has been carried out to replace sand by other materials that are cheap and easily available. This study also provides insight into the micro- and macro-level studies of effect on the properties of sand by the inclusion of rubber. It can be summarized that the addition of waste tyre leads to improvement in the properties of soil, contributes to the waste management burden, and also gives a cost-effective solution for a backfill.

Keywords Scrap tyre · Sand · Retaining wall

1 Introduction

1.1 Background

At present, the disposal of waste is a growing problem worldwide. Scrap tyre is one of those waste materials whose disposal is a great challenge in the present scenario. The amount of waste tyre is increasing at a rapid rate because of the increasing number of vehicles. In 2017, more than 2.9 billion of yearly waste tyre generation was recorded [1]. Tyre waste is non-biodegradable, highly durable, and consumes a

S. Kantroo (✉) · F. K. Khan
Lovely Professional University, Jalandhar, Punjab 144411, India

T. G. Sitharam et al. (eds.), *Earthquake Geotechnics*, Lecture Notes in Civil Engineering 187, https://doi.org/10.1007/978-981-16-5669-9_11

valued space in the environment because of its low compressibility [2]. In order to avoid these complications, recycling is the promising approach. In 2016, global tyre recycling value was recorded as USD 0.95 billion and is expected to increase by 2.1% in upcoming period [3]. Waste tyre has several beneficial applications in engineering projects [4] such as in road construction, ground erosion control, lightweight backfill in retaining wall, thermal insulation, drainage material in landfills, in bridge abutments, and light-rail construction, as a reinforcement to improve bearing capacity in asphalt mixture and cement concrete [5, 6]. The retaining structures associated with civil engineering are frequently influenced by static and dynamic loading. With growing seismic activities, it is becoming necessary to find earthquake mitigating techniques and building of earthquake-resistant structures. The working of retaining wall in the presence of seismic and static load relies on the soil and its ability to perform as a backfill, and more often, granular soil is chosen as a backfill material. Studies have been conducted where sand and rubber are mixed and used as backfill material [4, 7–9]. Scrap rubber is light in weight and compressible in nature [10–12] which makes it suitable for backfill material. Various researchers conducted triaxial tests [4, 7, 8] and the direct shear test [9] to understand the stress–strain behaviour of waste tyre with sand, and results revealed the improvement in shear strength and angle of shearing resistance in sand by the addition of rubber tyre [6, 13–17]. Performance of waste tyre varies with its size and shape also. If tyre waste is in crumb form, it tends to reduce shear strength of soil [18, 19] although improves ductility [19]. Most of the engineers found maximum shear strength at 35% rubber content [16, 20, 21] when shredded tyre is used as backfill material in retaining wall; it is responsible for the reduction of stress on the wall and reduces dimensions of retaining wall [22, 23].

In both dynamic and static condition, retaining wall backfill with tyre aggregates is economical compared to the traditional granular backfill of soil due to minimum requirement of concrete in retaining wall construction, dimensions, amount of backfill material, and volume of excavation. Numerical simulations also reported similar conclusion as reported above [12, 24, 25].

2 Effect on the Properties of Soil by Mixing Waste Tyre

In the last two decades, the use of scrap tyre as an admixture is accepted widely and found useful in numerous applications. Scrap tyre shows a significant impact on the geomehanical properties of sand. Researchers have observed that by mixing rubber tyre with sand, internal friction angle and shear strength show significant increase Figures 1 and 2 illustrates that at 40% of shredded tyre shear strength and friction angle are maximum [13]. Kheira et al. reported increase in shear strength at 30% rubber percentage in sand and suggested 20% rubber as optimal for improving friction angle when examined for chlefs river sand [17]. Other properties that are modified by admixing rubber are dilatancy, effects of confinement, relative density, and initial tangent modulus [6]. Mashiri et al. observed three behavioural zones of

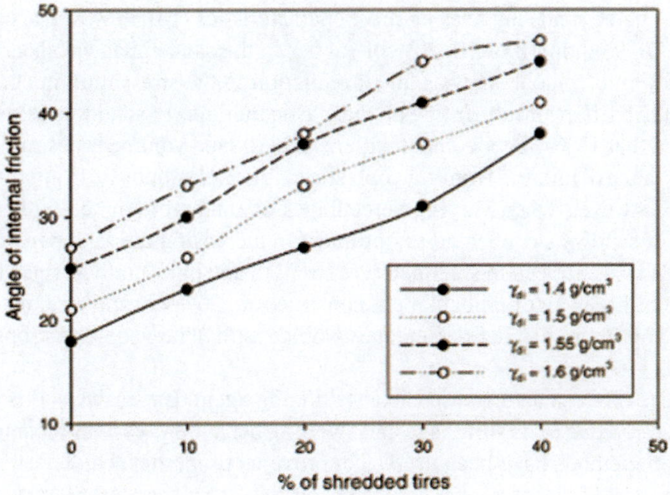

Fig. 1 Effect of percentage of shredded tyre on friction angle

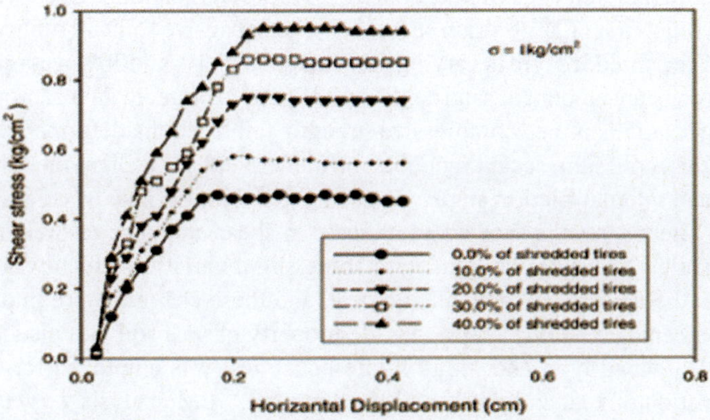

Fig. 2 Effect of percentage of shredded tyre on shear strength

sand and tyre mixture. In the first zone, sand acted as a matrix material, in the second zone, sand and tyre chips both form the matrix and show substantial advancement in shear strength and decrease in dilatancy, and in third zone, rubber alone forms the skeleton and showed similar behaviour with that of sand with a huge decrease in the dilatancy and initial tangent modulus. Segregation of material was also noticed in this zone due to empty voids in case of a rubber matrix. An optimum percentage of the waste tyre was found to be 35% for the improvement of sand properties. At this percentage, dilatency reduced and shear strength improved. The factors that influence shear strength of sand–rubber mixture (SRM) include content of rubber, normal stress, and mixture unit weight [6]. Mohamad et al. proposed that the utilization of

10% rubber in the sand enhances its shear characteristics [14]; however, according to Al-Rkaby 2019, addition of 10–20% of rubber in the sand enhanced shear strength and 30–50% tyre content shows a quick reduction in the stress ratio in comparison with sand [15]. Effect of rubber tyre on the elastic modulus of sand was assessed, and it was noted that 15% rubber content with respect to sand volume is desirable to alter the performance of mixture from elastoplastic to linear elastoplastic to approximately linear [8]. However, when varying percentages of sand were mixed with shredded tyre, 30% of sand was confirmed as optimal. On increasing the sand percentage, an upward trend was noticed in shredded tyre strength and in its unit weight. Moreover, in the case of highway or embankment construction, 20% of sand tyre mixture was found to be optimum. Higher percentages would result in large deformations that are not suitable [16].

The waste tyre can also be used as a stabilizing agent. Expansive soil is the most problematic soil due to its shrinkage and swelling behaviour. Various techniques and stabilization methods have been used to improve its properties. Dunham-Friel et al. investigated the effect of rubber inclusion on the expansive soil at varying mean stresses, i.e., 50,100, and 200 kPa. The results of the research indicated that rubber inclusion reduced the possibility of phase transformation and eradicates undrained stability. Further increment in rubber content for the aforementioned effective stresses resulted in reduction of large strain stiffness in the expansive soil and rubber mixture [26]. However, shredded tyre in varying percentages 0, 5, 10, and 20% was mixed with peat soil at constant amount of sand content 400 kg/m^3. The geotechnical properties of peat soil such as unconfined compressive strength, failure strain, deformability index, bulk modulus, cohesion, secant modulus, brittleness index, resilient modulus, shear modulus, and internal friction angle were estimated with the help of UCS and direct shear test. The test results showed an increase in shear modulus, resilient modulus, elastic modulus, and bulk modulus with some slight variation which was due to a decrease in the uniformity of stabilized peat, and these changes were improved by growing percentages of tyre chips. Ductile property of peat soil was also improved by adding shredded tyre, and X-ray fluorescence test was employed for chemical characterization of peat and sandy soil; furthermore, SEM analysis was conducted. The analysis characterizes untreated peat as loose bound fibres with organic content and as compact matrix with less voids [27]. Scrap tyre rubber improves the strength and dynamic behaviour of clayey soil and can be used as a reinforcing element [28]. The effect of tyre chip inclusion was also seen on cohesive clay soil by performing grain size analysis, Atterberg limit, direct shear test, permeability, and compaction tests. All the outcomes indicated the suitability of tyre chips mixed with clayey soil as fill material in highway embankments, bridge abutments, and backfills in retaining wall. It is suggested to use 20% of tyre chips if coarser and 30% if finer, above the ground water table. There use has been restricted in areas where drainage is required and sand or gravel is recommended in place of clay [29].

Scrap tyre can be reused in different forms, shapes, and sizes. Tyre crumb is one of the forms of machine granulated tyre. For any type of soil, 20% inclusion of tyre crumb gives maximum friction angle and decreases dry unit weight, whereas 10% inclusion and 30% inclusion decrease friction angle except in fine soil. In case of

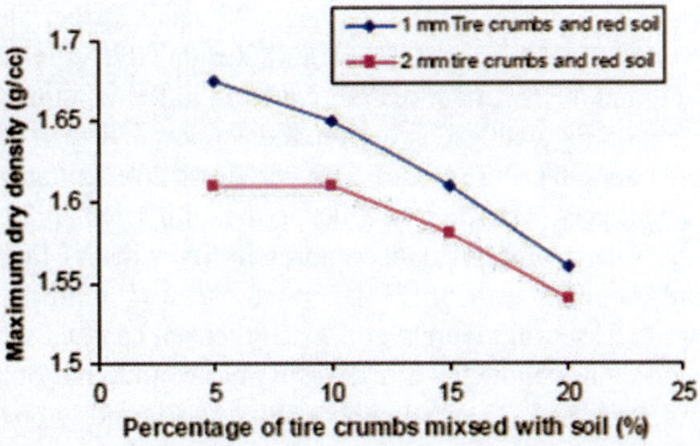

Fig. 3 Variation of max dry density with the percentage of tyre crumbs from compaction test

sandy soil, dilation is observed to be low [2]; unlike other forms of waste tyre, crumb reduces maximum and minimum density, angle of shearing resistance, and max dry density of sand. Tyre crumbs do not effect OMC. However, maximum dry density decreases with an increase in tyre crumb as depicted in Fig. 3 with the help of a graph plotted between maximum dry density g/cc and percentage of tyre crumbs mixed with soil (%).

When it comes to acting as a base, seismic isolation material tyre crumb proves to be supportive. Its application can be seen by filling the trench below the foundation by tyre crumbs–soil mixture [18]. Furthermore, tyre crumb is also witnessed to be responsible for the decrease of shear strength which is opposite to the behaviour shown by tyre shred and tyre chip. Addition of tyre crumb increases axial strain with respect to peak deviatoric stress. This property of tyre crumb is applicable where ductility in the structure is required and ductility is related to the axial strain [19]. Tyre crumb application can also be seen as an infill material in geocells where it gives higher pressure resistance as compared to that of sand but shows slightly higher horizontal displacement in comparison with the sand infilled retaining wall [30].

3 Use of Scrap Tyre in Retaining Wall

Retaining wall is earth retaining structure that plays a vital role in public life These structures are subjected to varied loading condition, and their working under seismic and static loading is based on the type of backfill soil, and mostly, clean granular cohesionless materials are preferred as backfill; however, alternate lightweight backfill materials that are used mostly include geofoam, geogrid, polystyrene foam, and shredded tyre chips [10, 31–33]. Lot of studies have been conducted to observe behaviour of retaining wall when using shredded tyre as a backfill [23, 34, 35]. Cecich et al. carried out research on uniformly graded shredded tyre with unit weight ranging within 5.51–5.86 kN/m^3 which is a smaller value when comparing to the unit weight of sand. Hydraulic conductivity was found to be 0.03 cm s^{-1}, cohesion 147 lbs ft-2, and internal friction angle 27°, and further, these properties were used for the designing purposes of retaining wall in which the backfill comprised of shredded tyre, and the results revealed reduction in the volume of backfill. These dimensions of retaining wall lead to a safer and economic design curtailing overturning, sliding, and strength failure risks as compared to the retaining wall backfilled with sand [23]. The properties of scrap tyre like permeability, density, shear strength, specific gravity, lateral earth pressure, and compressibility make it a favourable material for subgrade and retaining wall fill [34]. Tyre waste with sand is suitable as thermal backfills due to lesser unit weight and lesser thermal conductivity as compared to other soils. Xiao et al. conducted a test of thermal conductivity on the mix of sand–tyre shred and pure sand. Thermal conductivity was found to be minimum at 40% volumetric mixing ratio than that of sand. Larger sized tyre shreds resulted in higher thermal conductivity [36]. Recently, a new type of reinforcement element prepared with the help of scrap tyre called as geoscrap tyre was used in a mechanically stabilized earth walls as a transverse member for enhancement of pullout resistance. The pullout test results inferred the better performance of geo scarp tyre as compared to geosynthetics [37].

Numerical studies were carried out by many researchers on retaining wall with rubber tyre as backfill. Numerical study of cantilever retaining wall backfilled with varying STC mixtures was carried out by Abdullah et al. The conclusion illustrated the reduction in total displacement, vertical displacement, lateral displacement, max bending moments, and max shear force. Optimum tyre chips and sand ratio were found to be 66.54/33.46 by weight or 81/19 by volume [38]. Shrestha et al. [25] employed Plaxis software to compare the performance of tyre aggregate backfill and granular soil backfill under static and dynamic loading conditions. It was observed that tyre aggregate induces low S.F and B.M throughout the stem of retaining wall when compared with granular soil backfill as illustrated in Fig. 4, whereas at the footing backfill of sandy soil shows less shear force developed by the granular soil backfill. B.M was found to be more near the toe and less near heel in the tyre aggregate backfill when compared with granular soil. Singh and Chauhan employed OPTUM G2 (finite element-based numerical tool) to study the effect of waste tyre when using as a backfill material waste tyre used as backfill 8 m high wall on total lateral earth pressure From the results of this, it has been inferred that inclusion of waste tyre as a backfill material significantly reduces total lateral earth pressure on the wall within the range of 60–64% in comparison with the wall backfilled with the soil [12].

Gravel can also be used as backfill material in the gabion-type retaining wall when combined with a waste tyre. This was proved when an artificial scenario under artificial rainfall of about 0.8 mm/h was set up. It was observed that the slope was stable and horizontal movements negligible [38]. Industrial wastes like quarry waste is also used as backfill material in retaining wall with varying percentages of a waste tyre in which an addition of 30% granulated rubber decreased the unit weight of quarry waste from 21.8% to 14.1 KN/m^3 and the angle of shearing resistance improved by 14° [39].

The use of tyre rubber also attracts the attention of researchers throughout the world due to its ductile performance under dynamic loads. A model of 600 mm high

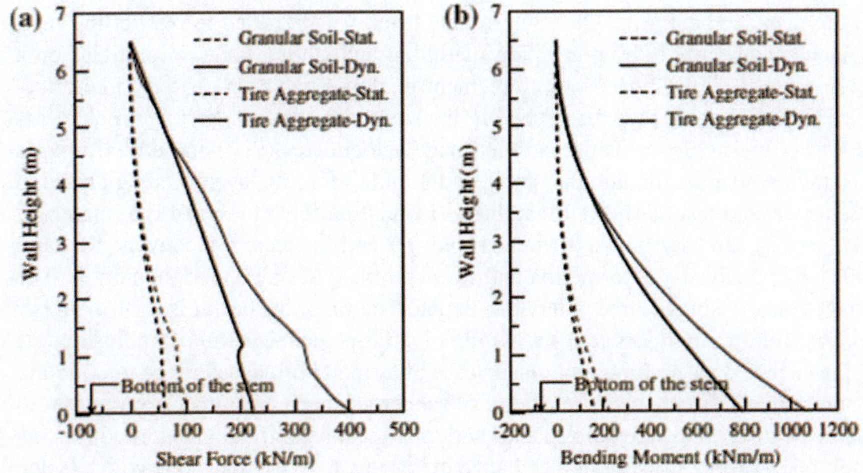

Fig. 4 **a** Maximum shear force envelop along the stem and **b** maximum bending moment along the stem

retaining wall was set up in the laboratory by Reddy and Krishna with the inclusion of recycled tyre in the backfill and analysed under dynamic loading generated by the shaking table test. The dynamic responses were measured by pressure sensors, displacement transducers, and accelerometers. It is noted that the involvement of tyre chips showed decrease of 60–75% in horizontal displacement and that of 68–83% in incremental earth pressure compared with the control test wall [11]. Djadouni et al. employed cyclic triaxial test under controlled strain condition to calculate shear moduli and also damping ratios of rubber and sand particles in which the materials were taken of the same size. Varied percentages of rubber were taken into account 10, 30, and 50% by weight. It was noted that 10% rubber in the sand–rubber mix (SRM) gives desirable static and dynamic properties static and can be beneficially used in low-rise buildings for seismic base isolation. The test mentioned earlier was conducted under saturated conditions and with 2 mm size sand and rubber particles. The reason for using 2 mm size sand and rubber particles was a requirement of uniform and homogenous mixture to cut out segregation, damping and to curtail problems due to the size of particles in the triaxial test [40].

Very few studies have been conducted until now to understand the particle-level mechanical properties of sand rubber mixtures. Recently, Gong et al. [41] employed a 3D discrete element method to compute contact forces by simulation of triaxial test on large particle comprising sand–rubber mixtures (SRMs). The numerical test results indicated that the mixing of rubber particles declined to contact normal anisotropy towards the loading direction. The contact between rubber and rubber forms weak force networks; however, they contribute in strong force chains when rubber content is increased with this results in the formation of strong force chains that develops uniform forces in the sand–rubber mixture. Rubber changes the strain localization in the displacement area. On a microscale level, it is believed to restrict the rearrangement of particle by prevention of rolling and sliding between particles on a microscale. Using rubber as reinforcement in slopes has been proved to be effective. Hajiazizi et al. [42] researched in his study the effect of horizontal elements of waste tyre in slopes using particle image velocimetry (PIV) and digital images The factors like length, number, and location of reinforcing layers were analysed in this paper, and results showed that the horizontal reinforced layer of tyre increased the bearing capacity and settlement about 3.5 and 3 times, respectively Khatami [43] while analysing deformation and stress profiles uses a trapdoor apparatus. The images clearly shows three reductions in deformation of the backfills at 30% rubber and also minimum shear strain was observed at this percentage. Rubber inclusion was believed to reduce arching modulus in case of a small normalised range of structure displacements. Mechanical behaviour of rubber has been examined from micro- to macro-scale with different host sands and found that peak friction angle reduces the addition of rubber because of the change in inter-particle friction angle which is due to the relocation of particles mostly observed in dilative sands. When sand is smooth and unbreakable, angle of friction increases due to increase in friction at the surface of sand and rubber particles; however, in case sand is rough and breakable, adding rubber results in reduction of angle of friction because of the weak frictional forces between sand and rubber particles then pure sand interface [44].

4 Conclusion

This paper was aimed at providing an overview of the use of waste tyre in modifying the properties of clayey soil, sandy soil, peat, and expansive soil. These modifications can prevail their use in vast areas by improving their quality despite of their inherent limitations that hinders their use. Moreover, it also turns industrial waste into a useful material that can be utilized in constructing civil structures. The paper focuses on the utilization of waste tyre as a backfill in retaining wall. As waste tyre is responsible for increasing shear strength and permeability, thus, it makes it an appropriate backfill material. Its flexible behaviour prevents the dynamic loads to cause damage to the structure. The literature review revealed that waste tyre successfully reduces lateral thrust, lateral displacement of retaining wall, and also reduces the requirement of complex designing and dimensions, thus making structure economic. Recently, researchers are conducting micro- to macro-level study of backfill material, and much more needs to be done in the future. Also, more types of waste materials need to be investigated and examined in order to analyse their applicability as a backfill material.

References

1. Raffoul S, Garcia R, Escolano-Margarit D, Guadagnini M, Hajirasouliha I, Pilakoutas K (2017) Behaviour of unconfined and FRP-confined rubberised concrete in axial compression. Constr Build Mater 30(147):388–397
2. Bosscher PJ, Edil TB, Kuraoka S (1997) Design of highway embankments using tire chips. J Geotech Geoenviron Eng 123(4):295–304
3. Global Tire Recycling Market Share, Size, Trends Analysis | Forecast 2024, (n.d.). https://www.goldsteinresearch.com/report/global-tire-recycling-industrymarket-trends-analysis. Accessed April 8, 2019
4. Lee JH, Salgado R, Bernal A, Lovell CW (1999) Shredded tires and rubber-sand as lightweight backfill. J Geotech Geoenviron Eng 125(2):132–141
5. Lepcha KH, Agnihotri AK, Priyadarshee A, Yadav M (2014) Application of tire chips in reinforcement of soil? A review. J Civil Eng Environ Technol 1(5):51–53
6. Mashiri MS, Vinod JS, Sheikh MN, Tsang HH (2015) Shear strength and dilatancy behaviour of sand–tyre chip mixtures. Soils Found 55(3):517–528
7. Masad E, Taha R, Ho C, Papagiannakis T (1996) Engineering properties of tire/soil mixtures as a lightweight fill material. Geotech Test J 19(3):297–304
8. Ehsani M, Shariatmadari N, Mirhosseini SM (2017) Experimental study on behavior of soil-waste tire mixtures. Scientia Iranica 24(1):65–71
9. Yang S, Lohnes RA, Kjartanson BH (2002) Mechanical properties of shredded tires. Geotech Test J 25(1):44–52
10. Reddy SB, Krishna AM (2015) Recycled tire chips mixed with sand as lightweight backfill material in retaining wall applications: an experimental investigation. Int J Geosyn Ground Eng 1(4):31
11. Reddy SB, Krishna AM (2017) Tyre chips as compressible inclusions in earth-retaining walls. Proc Inst Civil Eng Ground Improv 170(3):137–148
12. Singh V, Chauhan VB (2019) Reduction of dynamic lateral earth pressure on retaining wall using waste tyre
13. Attom MF (2006) The use of shredded waste tires to improve the geotechnical engineering properties of sands. Environ Geol 49(4):497–503
14. Mohamad ET, Latifi N, Marto A, Moradi R, Abad SV (2013) Effects of relative density on shear strength characteristics of sand-tire chips mixture. Electron J Geotech Eng 18:623–632
15. Al-Rkaby AH (2019) Strength and deformation of sand-tire rubber mixtures (STRM): an experimental study. Studia Geotechnica et Mechanica 10:1 (ahead-of-print)
16. Youwai S, Bergado DT (2003) Strength and deformation characteristics of shredded rubber tire sand mixtures. Can Geotech J 40(2):254–264
17. Kheira N, Ahmed A (2019) A direct shear investigation on the determination of the shearing resistance of reinforced soil with waste rubber. In: Recent advances in geo-environmental engineering, geomechanics and geotechnics, and geohazards. Springer, Cham, pp 295–299
18. Anbazhagan P, Mamatha M, Soumyashree P, Sushyam N, Bharatha TP, Vivekan RW (2011) Laboratory characterization of tyre crumbs soil mixture for developing low cost damping materials. Int J Earth Sci Eng 4(6):63–66
19. Neaz Sheikh M, Mashiri MS, Vinod JS, Tsang HH (2012) Shear and compressibility behavior of sand–tire crumb mixtures. J Mater Civ Eng 25(10):1366–1374
20. Bergado DT, Youwai S, Rittirong A (2005) Strength and deformation characteristics of flat and cubical rubber tyre chip–sand mixtures. Geotechnique 55(8):603–606
21. Zornberg JG, Cabral AR, Viratjandr C (2004) Behaviour of tire shred sand mixtures. Can Geotech J 41(2):227–241
22. Tweedie JJ, Humphrey DN, Sandford TC (1998) New England transportation consortium. In: Tire chips as lightweight backfill for retaining walls—phase II. New England Transportation Consortium
23. Cecich V, Gonzales L, Hoisaeter A, Williams J, Reddy K (1996) Use of shredded tires as lightweight backfill material for retaining structures. Waste Manag Res 14(5):433–451

24. Ahmed I (1993) Laboratory study on properties of rubber-soils
25. Shrestha S, Ravichandran N (2018) Performance of retaining wall backfilled with tire aggregate under static and dynamic loading conditions: conventional designs and finite element simulations. Int J Geotech Eng 17:1–3
26. Dunham-Friel J, Carraro JA (2011) Shear strength and stiffness of expansive soil and rubber (ESR) mixtures in undrained axisymmetric compression. In: Geo-frontiers 2011: advances in geotechnical engineering 2011, pp 1111–1120
27. Saberian M, Mehrinejad Khotbehsara M, Jahandari S, Vali R, Li J (2018) Experimental and phenomenological study of the effects of adding shredded tire chips on geotechnical properties of peat. Int J Geotech Eng 12(4):347–356
28. Akbulut S, Arasan S, Kalkan E (2007) Modification of clayey soils using scrap tire rubber and synthetic fibers. Appl Clay Sci 38(1–2):23–32
29. Cetin H, Fener M, Gunaydin O (2006) Geotechnical properties of tire-cohesive clayey soil mixtures as a fill material. Eng Geol 88(1–2):110–120
30. Kolathayar S, Kumar CR (2019) Potential of tire waste as infill material in geocells for soil retention systems. In: Geo-congress 2019: geoenvironmental engineering and sustainability, 2019 Mar 21. American Society of Civil Engineers, Reston, VA, pp 193–201
31. Reddy SB, Krishna AM (2013) Numerical simulations of earth-retaining structures using EPS geofoam inclusions. In: Indian geotechnical conference 2013
32. Voottipruex P, Bergado DT, Tanchaisawat T (2010) Lightweight recycled geomaterials reinforced with geogrids. Proc Inst Civil Eng Ground Improv 163(1):13–21
33. Inglis D (1996) Basement wall with seismic earth pressures and novel expanded polystyrene foam buffer layer. In: Proceedings of 10th annual symposium on earth retention system, Canadian Geotechnical Society, Vancouver, Canada, pp 1–10
34. Humphrey DN, Sandford TC (1993) Tire chips as lightweight subgrade fill and retaining wall backfill. In: Proceedings of the symposium on recovery and effective reuse of discarded materials and by-products for construction of highway facilities, 1993 Oct 19. US Department of Transportation, Federal Highway Administration, pp 5–87
35. Kaneda K, Hazarika H, Yamazaki H (2007) The numerical simulation of earth pressure reduction using tire chips in backfill. In: Proceedings of the international workshop on scrap tire derived geomaterials—opportunities and challenges, Yokosuka, Japan, 2007 Nov 8, pp 245–251
36. Xiao Y, Nan B, McCartney JS (2019) Thermal conductivity of sand-tire shred mixtures. J Geotech Geoenviron Eng 145(11):06019012
37. Tajabadipour M, Dehghani M, Kalantari B, Lajevardi SH (2019) Laboratory pullout investigation for evaluate feasibility use of scrap tire as reinforcement element in mechanically stabilized earth walls. J Clean Prod 237:117726
38. Abdullah A (2019) Potential of tyre derived geomaterial as alternative material in gabion type retaining wall. Int J Integr Eng 11(6):274–284
39. Ambarakonda P, Mohanty S, Shaik R (2019) Utilization of quarry waste and granulated rubber mix as lightweight backfill material. J Hazardous Toxic Radioactive Waste 23(4):06019001
40. Djadouni H, Trouzine H, Gomes Correia A, Miranda TF (2019) 2D numerical analysis of a cantilever retaining wall backfilled with sand–tire chips mixtures. Eur J Environ Civ Eng 27:1–7
41. Gong L, Nie L, Xu Y, Wang H, Zhang T, Du C, Wang Y (2019) Discrete element modelling of the mechanical behaviour of a sand-rubber mixture containing large rubber particles. Constr Build Mater 30(205):574–585
42. Hajiazizi M, Mirnaghizadeh M, Nasiri M (2019) Experimental study of waste tire-reinforced sand slope. Int J Mining Geo-Eng 53(2):183–191
43. Khatami H, Deng A, Jaksa MB (2019) Passive arching in rubberised sand backfills. Canadian Geotech J
44. Li W, Kwok CY, Sandeep CS, Senetakis K (2019) Sand type effect on the behaviour of sand-granulated rubber mixtures: integrated study from micro-to macro-scales. Powder Technol 15(342):907–916

Filter Paper Calibration Using Osmotic Coefficients to Measure Total Soil Suction

P. B. Pande, S. R. Khandeshwar, and S. P. Bajad

Abstract Suction is the pivotal parameter in arena of unsaturated soil. Filter paper method is the easy and cost-effective method for quantifying suction in which the calibration of filter paper is indispensable. The calibration curve with respect to total suction is impractical for measuring suctions a smaller than 1000 kPa. Thus, the distinct calibration curves are needed for measuring matric suction. Filter paper procedure is suitable up to the development of calibration curve in concern with total suction but difficult pertaining to development of matric suction calibration curve. Normally the pressure plate and pressure membrane were preferred for developing calibration curves, nevertheless these methods are having some limitations such as time consuming, skill and unavailability of equipment. The objective of this paper is to sidestep those limitations by adopting the prescribed procedure of establishing both curves required for measuring total as well as matric suction from filter paper technique. In this paper, the calibration equation for quantifying total suction was recognized from the wetting calibration curve achieved for Whatman No. 42 filter papers and matched with calibration equations developed by several other researchers. The total suction calibration equation was so selected that percentage error between obtained and selected must be less than 5%. Then, the drying calibration curve developed by the same researcher for the same filter paper was used for matric suction assessment of clayey soil used in this study. The procedure is recommended to reduce the efforts and additional time required for developing drying calibration curves. This technique can be appropriate in case of unavailability of equipment required for constructing drying calibration curve.

Keywords Filter paper calibration · Osmotic coefficients · Suction

---

P. B. Pande (✉) · S. R. Khandeshwar
Yeshwantrao Chavan College of Engineering, Nagpur, India

S. P. Bajad
Regional Board of Technical Education, Nagpur, India

T. G. Sitharam et al. (eds.), *Earthquake Geotechnics*, Lecture Notes in Civil Engineering 187, https://doi.org/10.1007/978-981-16-5669-9_12

List of Notation

ψ_t is the total suction (kPa)
w_f is the gravimetric filter paper water content (%)
w_m is the gravimetric water content of soil (%)
γ_d is the dry unit weight of soil (KN/m^3)

1 Introduction

Suction is the important parameter in geotechnical engineering intended for shear strength, bearing capacity and stability. Measuring suction is not possible without essential equipment. In unsaturated soil mechanics, the filter paper method is well known for its simplicity and cost-effectiveness [1–3]. ASTM D 5298-94 recommended the standard procedure for assessing soil suction by means of filter paper method [4]. Filter paper calibration is the important part of filter paper technique which governs soil suction.

Generally the filter paper is calibrated using the relative humidity values readily available in the literature. Relative humidity values are also suggested in ASTM 5298-10 for calibrating filter paper. In this paper, the osmotic coefficient values are used for calibrating filter paper. The easiness is compared with the relative humidity values for calibrating filter paper for measuring total suction. This paper also attempted to extend the filter paper calibration procedure suggested in ASTM D5298-10. The osmotic coefficient suggested in various literatures is included for calibrating the Whatman 42 filter paper and analyzed the logarithmic as well as exponential calibration equations. The validation of calibration equations was confirmed by measuring the soil suction developed in this study. The total suction of soil is measured to demonstrate the practical applicability of osmotic coefficients used. The precision in total suction estimated is perceived with the method adopted in this study.

There are still controversies identified associated with the calibration methods [16]. Developing wetting calibration curve using salt solution is easy because salt required for wetting process is easily available. Suction plate and pressure plate are equipment for developing drying calibration curve which are not easily available [7]. Due to non-availability and lack of knowledge about this equipment, it is difficult to perform drying process for calibrating filter paper. In this research, an attempt has made for measuring matric from readily available drying calibration curves. These curves are obtained for the same filter paper, i.e., Whatman 42 by developing the wetting calibration curves with the help of same salt solution [5].

Fig. 1 Suction measurement using filter paper method [7]

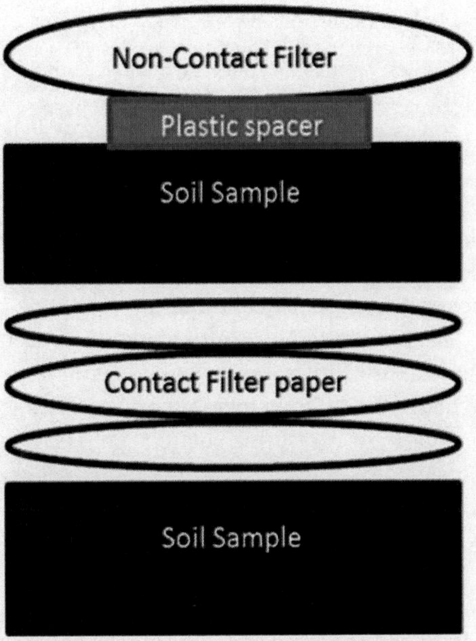

2 Literature Survey

2.1 Filter Paper Method

Filter paper method is one of the prominent and result oriented techniques out of other indirect suction measuring technique. Since it is simple and cost-effective, its reapplication is possible and thereby accurate result can be attained. This technique can be utilized to assess both total as well as matric suction. The filter paper is in direct contact with soil in case of total suction assessment and hanged over the soil in case of matric suction. Total and matric suction are measured in terms of vapor transfer and liquid transfer, respectively, (Fredlund and Rahardjo 1993). The filter paper technique suggested in ASTM D5298 and Bulut (1996) is followed in this research work (Fig. 1).

2.2 Calibration of Filter Paper

The filter paper process is a simple means of quantifying suction only the calibration part is time consuming and somewhat tiresome, nonetheless it is the essential part of this method. The equipment like pressure plate and pressure membrane was generally used for constructing the drying filter paper calibration curves. In pressure plate type devices the filter papers holds large extent of water at low suctions during the

Table 1 Equipment required for calibration of filter paper with respect to suction range [19]

Equipment	Suction (kpa)
Suction plate	10, 20 and 30
Pressure plate	70,100,200,300 and 400
Salt solution	500, 1000, 2000, 3000, 4000 and 5000

drying or wetting process. At lower suction range, pressure plate type devices should be investigated consciously. Table 1 represents the suction range with respect to equipment used for calibration.

In this study, Nacl salt solution has been used for constructing wetting calibration curve of filter paper which is commonly applicable for assessing total suction. Whatman 42 filter paper is the most commonly used filter paper which is calibrate and used for measuring suction. The thermodynamic relationship and osmotic coefficients are used to establish wetting calibration curve which helps to quantity the total suction, ψ_t (or osmotic suction). Combination of procedures suggested in [2, 6] was followed for calibrating the filter paper.

Vapor transfer mechanism is preferred to calibrate the filter paper used to assess the total suction. The water transfers through the vapor absorption by putting the filter paper above salt solution. Cleaned tweezers and hand gloves are required for performing the calibration procedure. The filter papers are hanged to the inverted plastic cones, glass crocks, and containers should not be handled without gloves, otherwise filter papers may absorb more water due to oiliness.

Further the procedure followed in the presented research work discussed. Initially the representative filter papers are dried by keeping it in to oven at 105 °C. The salt solution is made by adding required quantity of NaCl into water, and the concentration of solution is ranging from 0.003 to 2.7 molality [13]. The solution was prepared in 50 ml distilled water and dispensed into small the glass (nonreactive container). The precaution has been taken during placing two papers over the salt solution with the help of cone. The jar containing complete arrangement was wrapped and kept in constant temperature container for seven days period of equilibration. The tolerance for temperature during equilibration period was ± 0.10 C which may affect the suction result [1]. The moisture content of filter papers was obtained after seven days by removing it immediately rapid to sidestep the evaporation laboratory air. The oven drying method was used for assessing water content in which the weighing balance with nearest precision of 0.0001 g. Then, the papers were placed in aluminum cans and kept for oven drying for overnight. After oven drying the filter paper reweighed by placing cans over aluminum block for 30 s to assess the filter paper water content.

2.3 Filter Paper Calibration

The thermodynamic relationship and relative humidity is used for assessing the total suction (ψ_t, kPa) which is represented by Kelvin's equation (Eq. 1).

$$\psi_t = \frac{RT}{V} \ln\left(\frac{P}{P_0}\right) \tag{1}$$

Lang (1967) suggested the formula for the Osmotic coefficient [9] represented as (Eq. 2).

$$\emptyset = \frac{\rho_w}{vmw} \ln\left(\frac{P}{P_0}\right) \tag{2}$$

In this study, the amalgamation of Eqs. (1) and (2) is utilized for assessing the osmotic suctions (Eq. 3) with respect to different salt solutions as presented in Table 2.

$$\psi_\pi = -vRTm\emptyset \tag{3}$$

In the various literatures the Osmotic coefficients are already portrayed [12] for different salt solutions. In this study, the osmotic coefficients for only NaCl salt solutions are depicted and represented in Table 2.

The calibration curve are mostly established from the osmotic coefficient of sodium chloride (NaCl). The osmotic suction is governed by its osmotic coefficient, and therefore, the osmotic suction values are determine from the typically used osmotic coefficient. Table 2 represents the osmotic suction value for sodium chloride salt solution concentrations ranging from 0.05 to 2.0.

Table 2 Sodium chloride concentration, osmotic coefficients for sodium chloride salt solutions [12] and osmotic suction (ψ_π) for 25 °C [21]

Molality	Sodium chloride in (gms) in 1000 ml of water	Sodium chloride in (gms) in 150 ml of water	Osmotic coefficients at 25 °C for NaCl (Hamer 1972)	Total suction (ψ_π)	
				Hamer (1972)	Lang (1967)
0.05	2.9221	0.4383	0.94357	211	234
0.1	5.8443	0.8766	0.93250	418	462
0.2	11.6886	1.7533	0.92387	827	915
0.3	17.5328	2.6299	0.92123	1237	1368
0.4	23.3771	3.5066	0.92000	1648	1823
0.5	29.2214	4.3832	0.92224	2062	2281
0.6	35.0657	5.2599	0.92300	2480	2744
0.8	46.7542	7.0131	0.92900	3328	3682
0.9	52.5985	7.8898	0.93350	3762	4158
1	58.4428	8.7664	0.93600	4191	4640
1.4	81.8199	12.2730	0.95300	5974	6623
1.6	93.5085	14.0263	0.96487	6892	7652
2	116.8856	17.5328	0.98400	8812	9780

2.4 Filter Paper Calibration Curve

Calibration curve with respect to filter paper governs the precision of soil suction. It is the curve obtained between osmotic suction and filter paper moisture content. Each set of filter paper required to calibrate separately during particular experimental work. The osmatic suction (ψ_t) and filter paper water content (w_f) relationship give the calibration curve. The calibration curve of filter paper is generally divided in two sections. Wetting calibration curves are one which is used for assessing the total suction. The filter papers are normally suspended over the salt solution. The drying calibration curves used for measuring matric suction and generally use the pressure plate or membrane apparatus. The thermodynamic relationship of relative humidity and total suction is used to develop the calibration curve for this salt solution with different concentration.

Initially, the filter papers were kept in oven at 40 °C which may be nearly equal to room temperature. To maintain cleanness, filter papers were handled with uncontaminated tweezers and by wearing gloves. Usually, Whatman 42 and Schleicher and Schuell No. 589 filter papers are used for performing the filter paper method which are most commonly used filter papers. Each filter paper is having single calibration curve. It may be either drying or wetting [16, 17].

The filter paper technique is accomplished in corporation with procedures suggested in ASTM D 5298 and Bulut [11]. The 55 mm diameter Whatman No. 42 filter paper weighed on around 0.2 g. is used according to the ASTM D 529. The osmotic suction of sodium chloride solution was used having concentrations from 0.05 to 2.0. Initially, dried filter paper was put over inverted cone inserted in salt solutions which results wetting calibration curve. The jars were sealed properly to make it airtight. Two week period was considered as period for equilibration for wetting curve. The sealed containers containing filter papers and different salt solution were kept in a constant temperature container for the period of two weeks [1, 5, 14]. Fluctuation in temperature of constant temperature container may give false results [14, 15]. Hence, the constant temperature of 25 °C was considered during the equilibration period [4]. The condensations and evaporations were sidestepped by minimizing the variation which should not less than ± 0.10 °C [1].

The procedure discoursed above is for calibrating the curve which outlays the equation for measuring total suction. The filter paper moisture content and total suction values with respect to salt solution were used to obtain the calibration curve was developed as shown in Fig. 2. The calibration curve is represented by a straight line and expressed by Eq. 4.

$$\begin{aligned} &\text{For } w_f <= 36\% \\ &\psi = 5.4816 - 0.0816 w_f, \quad R^2 = 0.9501 \end{aligned} \tag{4}$$

The equations concerning calibration curves for Whatman 42 filter papers established by various researchers are presented in Table 3 [12]. The calibration equation developed in this research work is compared with the equations established by the

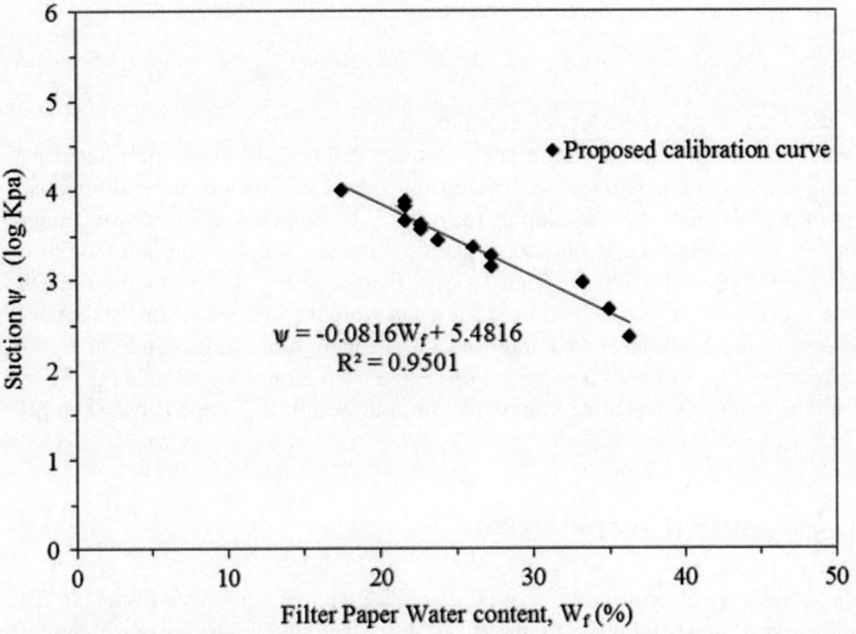

Fig. 2 Calibration curve for Whatman 42 filter paper

Table 3 Calibration equations for Whatman 42 filter paper [5, 20]

References	Water content (%)	Calibration equations
[9]	$W_f < 45.3$	$\psi = 5.777 - 0.06w_f$
[2, 11]	$W_f < 45.3$	$\psi = 5.327 - 0.0779w_f$
[17]	$W_f < 47$	$\psi = 4.945 - 0.0673w_f$
[4]	$W_f < = 47$	$\psi = 5.201 - 0.062w_f$
Author (2018)	$W_f < 36$	$\psi = 5.4816 - 0.0813w_f$

various other researchers summarized Table 3. The closely matched equation with respect to *a* and *b* parameters was comprised for comparison. The equations involved are the calibration equations for measuring total suction and those are developed from the wetting calibration method for the filter paper moisture content less than 47%.

ASTM D 5298 [2] adopted the equations proposed in [11] which was developed using the data originally presented in [9].

For $w <= 47\%$

$$\psi = 5.201 - 0.062w_f, \quad R^2 = 0.976 \tag{5}$$

For $w > 47\%$

$$\text{For } w > 47\%$$
$$\psi = 2.909 - 0.021w_f \tag{6}$$

where, R^2 = Coefficient of correlation.

In current practices, for evaluating matric suction by use of same filter paper, filter paper would be required to develop new matric calibration curve (drying calibration curve). Instead of developing the new calibration curve for measuring matric suction, the similar total calibration equation is chosen which is similar to the equation developed in this study. Bichalo's [4] equation (Eq. 5) is matched with the developed equation (Eq. 4) and used for calculating the matric suction. Filter paper water contents were measured and used for calculation of total and matric suction by arranging the soil and filter papers inside the glass jar as illustrated in Fig. 1. The detail procedure is explained in next section followed from [7] and is based on [2].

3 Soil Suction Measurement

The soil test is performed on the soil specimens identically prepared from clay. The similar procedure is followed for the all specimen testing. The sieving of soil through 2 mm sieve is systematically mixed with water the essential to maintain unit weight of soil sample. The soil–water mixture was then set aside the tank for 24 h exclusively made for compacting the soil sample. After achieving the hydraulic equilibrium, the soil water mixture was compacted by use of mechanized compaction equipment developed during this study [18]. The water content averagely maintained during compaction is 36%. The consistency of the soil in terms of liquid limit and plastic limit is observed as 68.5% and 29.1%, respectively, and the specific gravity is 2.61. The optimum moisture content of soil exhibits is 25% and the maximum dry unit weight ($\Upsilon_{d\max}$) 14.4 kN/m^3; also the compaction curve is illustrated in Fig. 3.

Previously, some researchers had measured the suction of highly plastic clay suction using Whatman 42 filter paper [8, 20]. The soil samples are extracted of size 55 mm diameters and height 150 mm, and the tests were performed for determining total and matric suction. The dried 55 mm diameter filter paper is sandwiched between the soil specimen compacted at known initial moisture content. The filter paper is removed from the soil sample, and moisture content of filter paper was quickly determined with precaution. The total and matric suction was then measured by putting the water content of filter paper in the respective calibration equations. The total suction was measured from @@(4) obtained from the calibrati on curve developed in this study, and matric suction was measured from the (6) taken from [4] whose total suction curves were very similar.

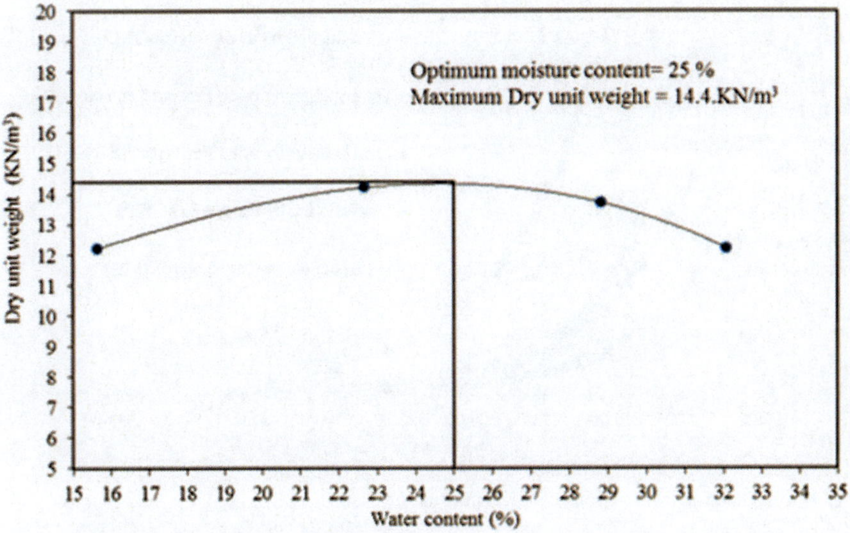

Fig. 3 Compaction curve obtained for soil

4 Results and Discussion

The equation developed by Bicalho et al. [4] was found to be much closed to the proposed calibration equation (Table 3).

The measured total suction values using Eq. (4) obtained from the curve shown in Fig. 2 was compared with the total suction measured from the readily available equations in literature summarized in Table 3. The calibration curve developed with the help of filter paper water content (w_f) and total suction (kPa) is presented in Fig. 4. This curve the equations which helps in measuring the total suction. The same equation was compared with the equations enlisted in Table 3. The coefficient of determination (R^2) values for the curves obtained from the proposed and other included equations was averagely appeared to be 0.99. Out of all the curves, the total suction curve obtained from proposed calibration equation curve represented by solid filled dots found to be much closer to [4] and ASTMD 5298 [2]. Authors have compared four total calibration curves proposed in various literatures [4, 9, 17]. The method adopted only covers of the total suction measurement by use of gravimetric water contents of filter paper ($16\% < w_f < 43\%$). To interpret the measured total suction error has been calculated and explored the well agreed calibration curve with the proposed one. Total suction estimations from the calibration equation of ASTMD 5298 [2] and the proposed calibration curve are differed by 24%. The errors observed from Fawcett et al. [9] and Leong et al. [17] were found to be −77% and 104%, respectively, which was larger as compared to [4] and ASTMD 5298 [2]. The difference is −6% when compared with [4] and found near to the values measured by proposed one.

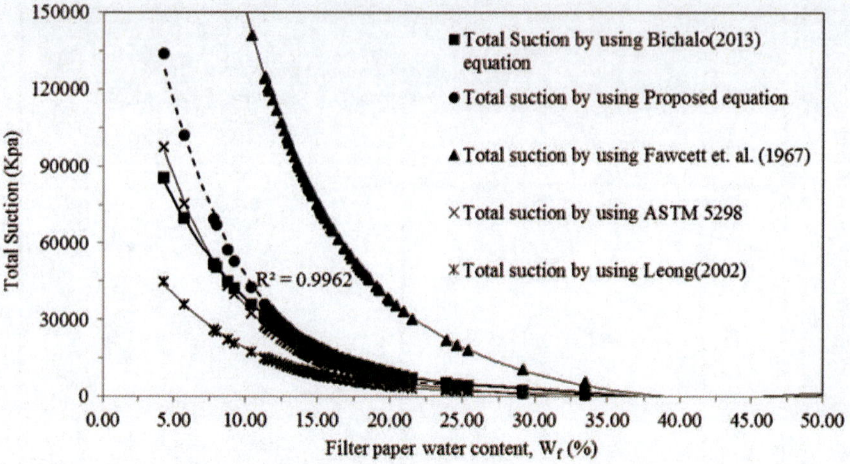

Fig. 4 Total suction from various calibration equations

The results of total suction from various equations at various gravematric water content (%) for the soil are plotted and presented in Figs. 5, 6, 7 and 8. Total suction was plotted against the gravimetric water content to compare the data obtained from the test performed. The proposed equation and four wetting calibration equations are compared and studied from the plotted curves. Soil was tested at average dry density of 13.09 KN/m^3 and 36% moisture content. Each figure shows data for filter paper method performed and developed water retention curve. The SWCC can be well-defined by total suction and gravimetric water content. Figure 7 shows the trend line obtained from Fawcett (1967) which is observed somewhat above the trend line observed in proposed Fig. 5, wherein the trend line obtained from ASTM D 5398 [2] (Fig. 8) is observed. Figures 5 and 6 show trend lines that are at same level which are obtained from proposed. From the above results and discussion, it has been observed

Fig. 5 Total suction values with gravematric moisture content (%) from proposed equation

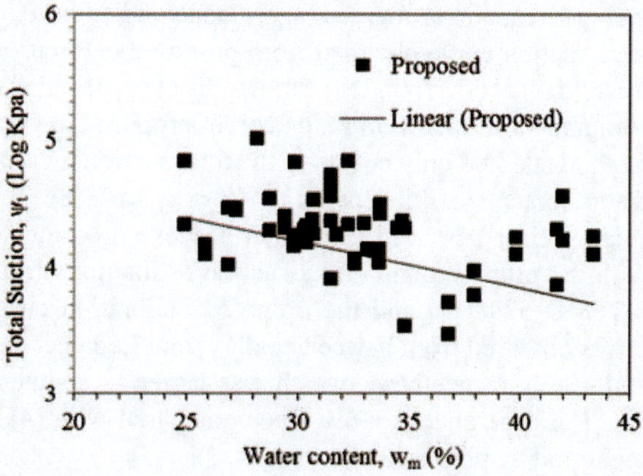

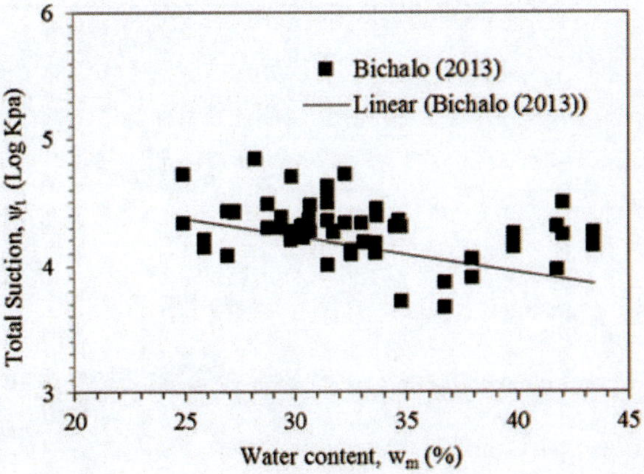

Fig. 6 Total suction values with gravematric moisture content (%) from [4] equation

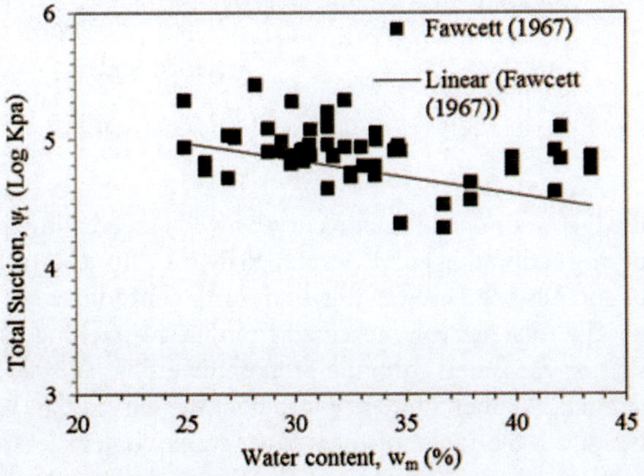

Fig. 7 Total suction values with gravematric moisture content (%) from Fawcett [9]

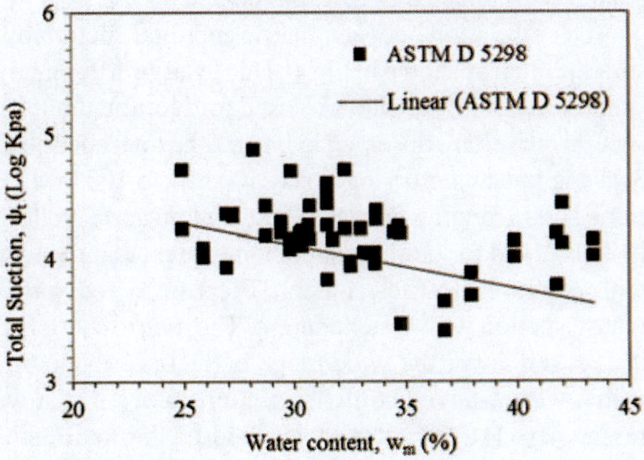

Fig. 8 Total suction values with gravematric moisture content (%) from [2]

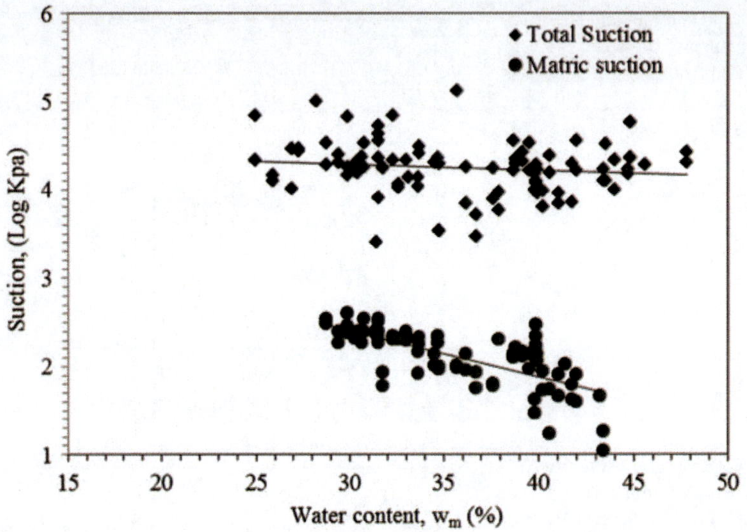

Fig. 9 Relation between suction (total suction and matric suction) versus gravimetric water content

that [4] wetting calibration curve is well agreed with the proposed curve. Therefore, drying calibration curve developed by Bicalho et al. [4] for measuring matric suction of soil has been used in this study with conformity.

The total suction measured from the [4] wetting calibration curve and the total suction measured from the proposed wetting calibration curve is found to be very similar. Another similarity was both the curves that were developed for Whatman No. 42 in the range of filter paper water content less than 47%. Therefore, matric suction values of compacted clayey soil can be evaluated by the drying calibration equation established by Bicalho et al. [4].

The correlation between matric and total suction by considering water content of soil is plotted as shown in Fig. 9. Due to large difference in the range of total and matric suction logarithmic scale was used for plotting the relationship. The total and matric suction curves are observed in linear. The total suction assessment is ranging between 3.41 log kpa and 5.01 log kpa (2500 kPa to 100,000 kPa) and the matric suction for same soil is ranging between 2.00 log kpa and 2.60 log kpa (10–500 kPa). The trend line observed for both the suction indicated that the total suction decreases gradually with increase in water content. The compacted soil is showing linear decrease in matric suction with water content. The matric suction significantly decreases though it does not cover the wide range of suction. A mixture of matrix and total suction which was observed at higher suction values which was also observed by the other researchers [10]. Scattering in trend of the total and matric suction comprehends nonlinear relationship amongst filter paper gravimetric moisture content and suction (Log).

5 Conclusions

Calibration curve can be developed by use of osmotic coefficient of salt solutions. Combination of procedure suggested in [2] and Bulut [6] was successfully implemented to develop the calibration curve. Calibration equation of total suction can be utilized to utilize the matric suction calibration equation by checking the similarity with already available total suction equations. The percentage error should be within 5–10%. This study reduces the efforts and additional time required for developing drying calibration curves. The satisfactory results of matric suction are more reliable and can be utilized for establishing the soil water retention curves that were obtained. This paper proposed methodology to expedite the filter paper technique which depends on trial and error.

References

1. Al-Khafaf S, Hanks R (1974) Evaluation of the filter paper method for estimating soil water potential. Soil Sci 117(4):194–199
2. ASTM Standard D5298–94 (1997) Standard test method for the measurement of soil potential (suction) using filter paper annual book of ASTM standards, ASTM International, West Conshohocken. 04(09):157–162
3. ASTM Standard D 2216–98 (1998) Test method for laboratory determination of water (moisture) content of soil and rock. Annual book of ASTM standards, ASTM International, West Conshohocken. 04(08):188–191
4. Bicalho K, Cupertino K, Bertolde A (2013) Evaluation of the suction calibration curves for Whatman 42 filter paper. Advances in Unsaturated Soils, Taylor & Francis Group, London, pp 225–230. ISBN 978-0-415-62095-6
5. Bicalho K (2007) Filter paper method of soil suction measurement. In: Proceedings of 13th Panamerican conference on soil mechanics and geotechnical engg, Isla Margarita, Venezuela, pp 215–219
6. Bulut R (1996) A re-evaluation of the filter paper method of measuring soil suction. Civil Engineering Master of Science, Thesis, Civil Engineering Graduate Faculty of Texas Tech University
7. Bulut R, Lytton R, Wray W (2001) Suction measurements by filter paper method. Am Soc Civil Eng Geotech 115:243–261
8. Chandler R, Gutierrez C (1986) The filter paper method of suction measurements. Geotechnique 36:265–268
9. Fawcett R, Collis-George N (1967) A filter-paper method for determining the moisture characteristics of soil. Aust J Exp Agric Anim Husb 7:162–167
10. Fredlund D, Rahardjo H, Fredlund M (2012) Unsaturated soil mechanics in engineering practice. John-Wiley and Sons, New York
11. Greacen E, Walker G, Cook P (1987) Evaluation of the filter paper method for measuring soil water suction. In: International conference on measurement of soil and plant water status, pp 137–143
12. Hamblin A (1981) Filter-paper method for routine measurement of field water potential. J Hydrol 53:355–360
13. Hamer W, Wu Y (1972) Osmotic coefficients mean activity coefficients of uni-univalent electrolytes in water at 25 °C. J Phys Chem Ref Data 1(4):1047–1099
14. Houston S, Houston W, Wagner A (1994) Laboratory filter paper suction measurements. Geotech Test J 17(2):185–194

15. Krahan J, Fredlund D (1971) On total, matric and osmotic suction. Soil Sci 114(5)
16. Lang A (1967) Osmotic coefficients and water potentials of sodium chloride solutions from 0 to 40 °C. Aust J Chem 20:2017–2023
17. Leong E, He L, Rahardjo H (2002) Factors affecting the filter paper method for total and matric suction measurements. Geotech Test J 25(3):322–333
18. Nam S, Gutierrez M, Diplas P, Petrie J (2009) Comparison of testing techniques and models for establishing the SWCC of riverbank soils. Eng Geol 110(1–2):1–10
19. Oliveira O, Marinho F (2006) Evaluation of filter paper method. Unsaturated soils ASCE, pp 1845–1851
20. Oloo S, Fredlund D (1995) Matric suction measuring in an expansive soil subgrade in Kenya. In: Proceedings of 1st international conference on unsaturated soils, Paris
21. Pande P, Khandeshwar S, Bajad S, Raut S (2019) Calibration of filter paper for measuring total soil suction. SN Appl Sci 1:803

Performance of Pavement Sections Constructed Using RAP with Geocell as Reinforcement

Mohan Badiger, Rahul Raju Bhatakande, K. H. Mamatha, and S. V. Dinesh

Abstract Nowadays, non-availability and/or lack of superior quality materials is posing a threat for the construction of structures such as bridges, roadways, buildings, dams and so on. A well-planned road network will enable accessibility to remote areas and establish connectivity from one place to the other. Pavement as a structure consists of multi-layers constructed using soil, aggregates, asphalt/cement concrete, and the requirement of these materials is huge as pavements extend over thousands of kilometres. In the present era, the engineers are forced to think of the sustainable materials without compromising on the quality, and in this connection, several industrial by-products such as fly ash, rice husk ash and steel slag and waste materials such as vehicular tyres, C&D waste and over burnt bricks are being utilized for construction. But, in India, the rate of use of these sustainable materials in construction is not popular owing to the fact that the utilization of these materials demands highly skilled manpower, advanced equipment, etc. Thus, there is a need to explore the performance of such wastes in pavement sections. In this study, model pavement sections were constructed with clayey subgrade which is generally not suitable for construction and among the wastes available; RAP was considered. The model pavement section comprised of subgrade, base and unpaved surfacing simulating unpaved low volume roads. To enable the construction of pavement section with black cotton soil and RAP, geocell was considered as reinforcement. The model pavement sections were tested under repetitive loading. During testing, the pressure at subgrade level and deformation at the surface were monitored and recorded. The test results indicated that the geocell reinforced pavements with RAP in the unbound granular layer perform similar to that of the conventional pavement section. The study shows the possible way of constructing sustainable pavements with RAP and geocell as reinforcement. The study also showed that the pavement thickness could be reduced. This will lead to reduction in construction cost.

Keywords Weak subgrade · RAP · Geocell · Pavement performance

---

M. Badiger · R. R. Bhatakande · K. H. Mamatha (✉) · S. V. Dinesh
Department of Civil Engineering, Siddaganga Institute of Technology, Tumkur, Karnataka 572103, India
e-mail: mamathakh@sit.ac.in

T. G. Sitharam et al. (eds.), *Earthquake Geotechnics*, Lecture Notes in Civil Engineering 187, https://doi.org/10.1007/978-981-16-5669-9_13

1 Introduction

In the recent years, the increased infrastructural development in the construction sector leads to the scarcity of virgin aggregates. The natural resources are on the verge of depletion and creating an alarming situation for the engineers to seek for an alternative sustainable solution. On the other hand, large amount of reclaimed asphalt pavement (RAP) is produced during reconstruction and rehabilitation of roads. The need to find appropriate disposal locations is a major concern particularly in urban and semi-urban areas. RAP contains well-graded aggregates along with aged bitumen. In this connection, recycling program offers a viable solution by partially supplementing the present material requirement. Utilization of recycled materials for construction offers considerable savings in construction cost and time. In addition, subgrades on black cotton soils are considered as problematic for road construction due to their high swelling and shrinkage characteristics. Whether it is a temporary access road or a permanent road built over a weak subgrade, a large deformation of the subgrade can lead to deterioration of the paved or unpaved surface. This will lead to increased maintenance cost of the roads. To minimize these problems, thickness of the road section needs to be increased, good quality construction materials need to be used or subgrade needs to be improved by any suitable technique. Nowadays, geosynthetics is being effectively used in various construction practices. Geosynthetics placed either on top of the subgrade or within the base course layer works with the soil and granular material to create a reinforced section through separation, confinement and/or reinforcement functions. Several studies in the past have shown that geosynthetics can extend the service life of pavements [1–12], reduce base course thickness for a given service life and traffic [1, 2, 5, 12–14] and delay rutting development [3, 12, 15]. Among various forms of geosynthetics, geocells are being widely used owing to its three-dimensional structure. The term geocell refers to a polymeric, honeycomb-like cellular material. The structure of these cells interconnected by joints to form a cellular network could be used for the confinement of soil. Geocells completely encase the infill material and provide all-round confinement, thus preventing the lateral spreading of the infill material. Because of this, the geocell reinforced layer acts as a stiff mat, distributing the load over a much larger area of the subgrade soil [15]. This helps in reducing vertical and lateral deformations of the foundation soil to a large extent besides increasing the overall bearing capacity of the foundation soil [12, 15, 16]. The ultimate bearing capacity of the clay bed reinforced with the combination of bamboo cell and bamboo grid was about 1.3 times higher than the geocell and geogrid reinforced clay beds [17]. The bearing pressure of the jute geocell reinforced soil was about 3.5 times higher than that of the unreinforced soil [18]. In this study, an attempt is made to evaluate the effectiveness of utilization of RAP as filler material inside the geocells in the model pavement sections built in the laboratory. The geocell was placed in the granular base layer. The model pavement section consists of subgrade and granular layer and was subjected to repeated load tests.

2 Materials

In this study, three types of soils were used for the construction of model pavement sections. For the construction of subgrade, black cotton soil was used and was collected from Hiriyur, Chitradurga District, Karnataka. For the construction of improved subgrade (in case of conventional model pavement section) and sacrificial layer, red soil and gravel was collected locally. For the construction of conventional granular base layer, aggregates of different sizes (i.e. 40 mm down, 12.5 mm down and 4.75 mm down) were used. The aggregates were collected from a nearby quarry. RAP used for the construction of granular base layer was collected from a road site located in the rural area near Tumkur where the stretch of road was strengthened by removing the asphalt layer. To reinforce this granular base layer constructed with RAP, geocell was used as reinforcement and was procured from Strata Geosystems Private. Ltd., New Delhi.

The engineering properties of soils and aggregates were tested as per the relevant IS guidelines. The properties of soils are presented in Table 1. Figure 1 shows the grain size distribution of all the three soils. It was found that the black cotton soil fails to meet the requirements of subgrade in terms of minimum maximum dry unit weight (i.e. 16.5 kN/m^3) and soaked CBR (i.e. 2%) of subgrade material according

Table 1 Engineering properties of soils

Tests	Subgrade	Improved subgrade	Sacrificial layer
Specific gravity	2.71	2.65	2.65
Soil classification			
IS soil classification	CH	SM	SM
HRB classification	A-7-5	A-2-6	A-3
Atterberg's limits			
Liquid limit (%)	67	40	15
Plastic limit (%)	33	29	NP
Plasticity index (%)	34	11	–
Standard proctor test			
Max. dry unit weight (kN/m^3)	13.3	19.62	20.4
OMC (%)	31	13.2	11.2
California bearing ratio (%)			
Unsoaked	6	23	61
Soaked	<2	14	32
Free swell index (%)	40	–	–

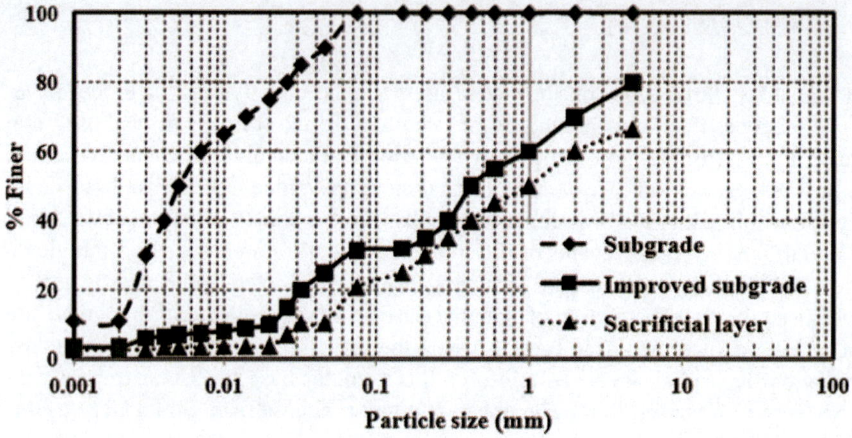

Fig. 1 Grain size distribution curves of soils

to Ministry of Road Transport and Highways (MoRT&H) [19] and Ministry of Rural Development (MORD) [20] guidelines for low volume roads application. Thus, the soil forms a weak subgrade. The red soil used for improved subgrade possesses a soaked CBR of 14% and satisfies the minimum CBR requirement of 10% for improved subgrade application as per IRC:SP:72 [21] guidelines. Figure 2 shows the gradation requirements of sacrificial layer as per MORD [20], and it was found that the gradation of sacrificial layer confirms to the requirement specified by MORD [20].

The physical and mechanical properties of aggregates were tested as per relevant IS codes, and the test results are presented in Table 2. The aggregates were found

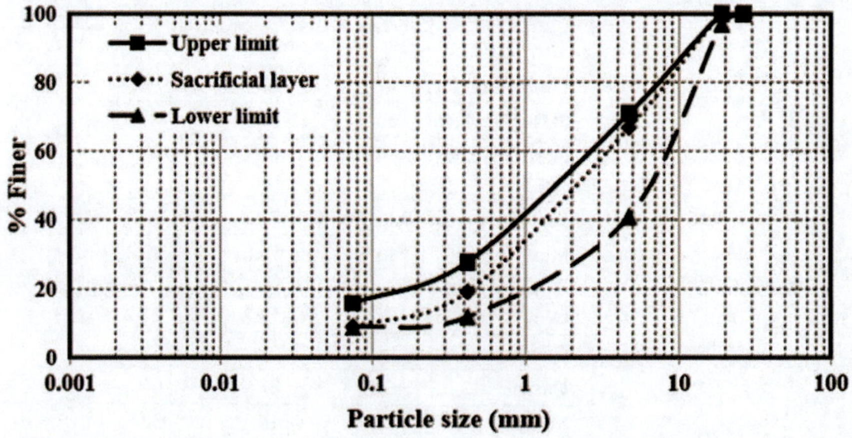

Fig. 2 Gradation curve of soil used for sacrificial layer confirming to MORD (2014)

Table 2 Physical and mechanical properties of aggregates

S. No	Test	12.5 mm down	40 mm down	MoRT&H (2013) specifications	Remarks
1	Specific gravity	2.62	2.63	2.5–3.2	Acceptable
2	Aggregate crushing value (%)	26	26	Max. 30%	
3	Aggregate impact value (%)	27	27	Max. 30%	
4	Los Angeles abrasion (%)	28	28	Max. 40%	
5	Combined indices (%)	32	32	Max. 35%	
6	Water absorption (%)	0.17	0.21	Max. 2%	

to satisfy the requirements specified by MoRT&H [19]. Individual gradation was carried out for each of the fractions by considering the 5000 gms, and then combined gradation was carried out to design a mix satisfying the gradation requirements of granular base of grading A of MORD [20] guidelines. Figure 3 shows the gradation curve of the designed conventional granular base mix along with the upper and lower limit values suggested by MORD [20]. The designed conventional granular base mix was found to possess a maximum dry unit weight of 22.13 kN/m^3 and an optimum moisture content of 6% under modified Proctor condition. The mix showed a soaked CBR of 45% under soaked test condition corresponding modified proctor condition.

The collected recycled asphalt pavement (RAP) material was first broken into smaller fractions manually using rammer and then processed through jaw crusher

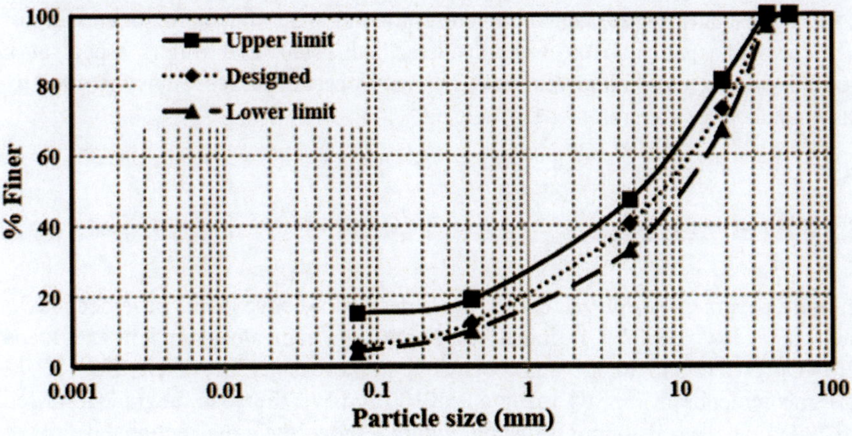

Fig. 3 Gradation curve of designed conventional base mix confirming to Grading A of MORD (2014) guidelines

Fig. 4 RAP **a** in its original form and **b** after processing through jaw crusher

Table 3 Properties of geocell

Property	Value
Cell size (mm)	259 × 224
Nominal sheet thickness (mm)	Min. 1.52
Weld spacing (mm)	356
Cell depth (mm)	150
Strip thickness (mm)	1.53
Cell seam strength (N)	2310
Density (g/cm^3)	0.935–0.965

for further breaking down. Fractional distillation was carried out on the RAP to extract the binder and the average binder content was found to be 4%. The RAP was found to possess a maximum dry unit weight of 19.81 kN/m^3 and an optimum moisture content of 5.62% under modified proctor condition. It exhibited a soaked CBR of 30% under soaked condition corresponding to modified proctor condition.

Geocell of aspect ratio 0.67 was considered in this study as the higher aspect ratio; geocell yields greater improvement [12]. The properties of geocell as provided by the manufacturer are presented in Table 3.

3 Pavement Design

Initially, pavement design was carried out to design the conventional pavement thickness as per IRC: SP:72 [21] guidelines. For design, equivalent standard axel loads (ESAL) of 10,000 to 30,000 was considered. For a soaked CBR of 2%, IRC: SP:72 [21] recommends an improved subgrade of 100 mm over the weak subgrade followed by 200 mm base and 50 mm surfacing. Figure 5 shows the cross-sectional details of the model pavement sections considered.

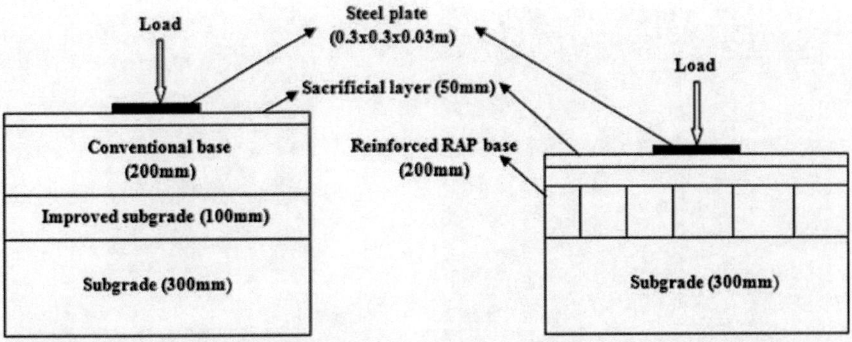

Fig. 5 Cross section of unreinforced and geocell reinforced model pavement sections

Table 4 Experimental program

Section No	Description	Designation
1	Weak subgrade (300 mm) + Improved subgrade (100 mm) + Base (200 mm) + Surfacing (50 mm)	UR
2	Weak subgrade (300 mm) + Geocell reinforced base with RAP (200 mm) + Surfacing (50 mm)	RE (RAP)

4 Experimental Program

Two model pavement sections were considered in this study. Among the two, one is conventional pavement section as detailed in Sect. 3, and the other corresponds to reinforced pavement section. In the reinforced section, the improved subgrade was eliminated and RAP was used as base material. Table 4 shows the experimental program and associated designations considered.

5 Construction of Model Pavement Sections, Experimental Set up and Testing Procedure

Figure 6 shows the schematic representation of stage construction adopted during the construction of model pavement sections and experimental set up.

In this study, a steel tank of size 2 m × 2 m × 2 m was used to construct the test sections. A steel plate of size 0.3 m × 0.3 m × 0.03 m was used to transfer the load to the test section. Stage construction was adopted in the model pavement construction and detailed in the following paragraphs.

Black cotton soil subgrade of thickness 300 mm was constructed and it was compacted manually using a rammer. A relative compaction of 95% was maintained throughout for the black cotton soil subgrade. Over the compacted subgrade, 100 mm

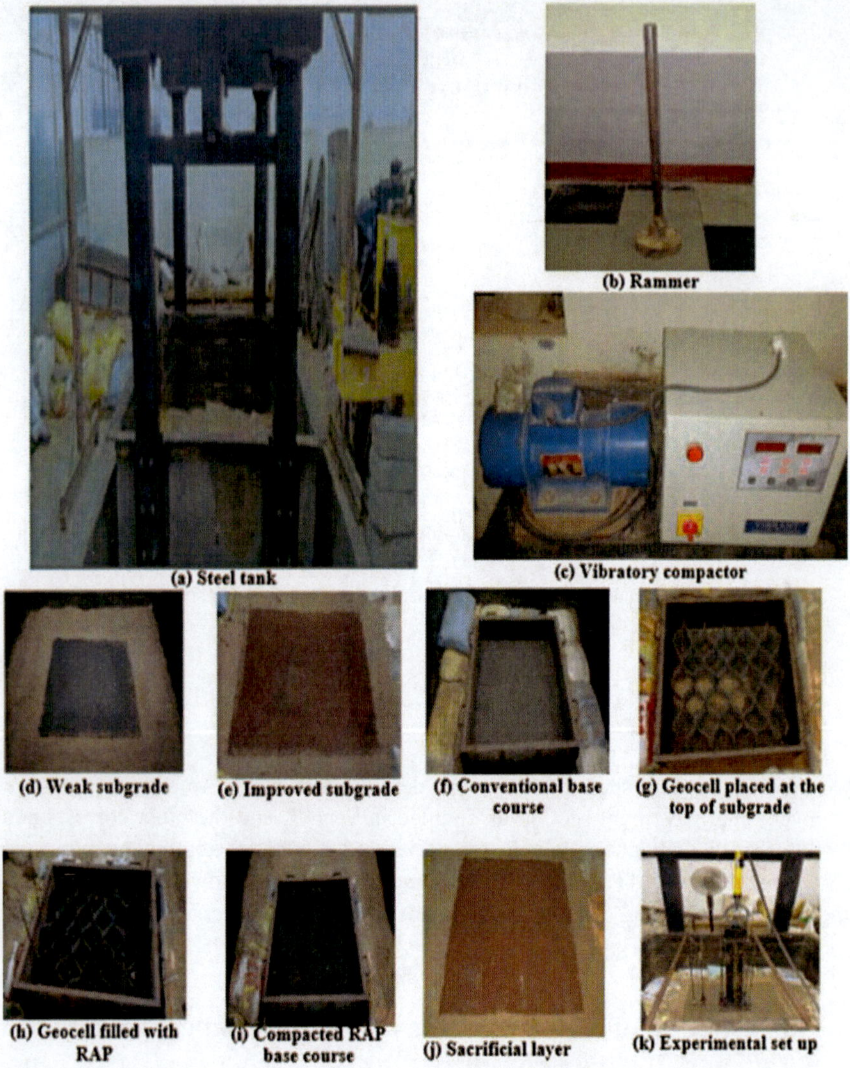

Fig. 6 Schematic representation of stage construction adopted during the construction of model pavement sections and experimental set up.

thick improved subgrade was prepared using vibratory plate compactor. A 200 mm thick conventional base layer was constructed over the prepared improved subgrade. On the prepared base layer, a sacrificial layer of 50 mm thick was prepared. In case of geocell reinforced RAP base section, subgrade was laid and compacted similar to unreinforced section. Over the prepared subgrade, geocell was laid and filled using RAP followed by the preparation of sacrificial layer. Improved subgrade, conventional base, reinforced RAP base and sacrificial layer were compacted using vibratory plate compactor which operates at 20, 40, 60 and 80 Hz frequency. A frequency of 80 Hz was considered for the compaction of all these layers. A relative compaction of greater than 98% was maintained throughout for all these layers.

Dial gauges were used to record the deformation of the test plate and the surface profile. Four dial gauges were positioned on the test plate to record its deformation during testing. Average of these four dial gauges was considered as deformation of the test plate. Two dial gauges were positioned at a distance of 30 and 40 cm on one side from the centre of the test section to record the surface profile of the test section.

The setup used for the repeated load test was assembled as per IS: 5249 [22] guidelines. The steel plate was placed at the centre of the prepared surface layer in the model tank. Initially, the dial gauges were set to zero. A seating load of 7kN was applied to create a pressure of 7 kg/cm^2 with respect to the area of the plate (i.e. 30 cm × 30 cm) and was maintained constant throughout the test. When there is no further settlement, the dial gauges were set to zero again. The test sections were loaded manually using a hydraulic jack of 100 kN capacity with a least count of 1 kN. To simulate vehicular loading, haversine loading was considered. For testing, a maximum pressure of 560 kPa was adopted. The tests were terminated at 500 load cycles. At the end of each load cycle, the dial gauge readings after loading, unloading and rest period were recorded. These readings were used in the determination of total, plastic and elastic deformations of the test section.

6 Results and Discussions

Figures 7, 8 and 9 show the comparison of total, plastic and elastic deformation of conventional and geocell reinforced RAP pavement sections, respectively. The plastic settlement forms more than 90% of the total settlement. The pavement sections under consideration are unpaved roads and such roadways fails predominantly by

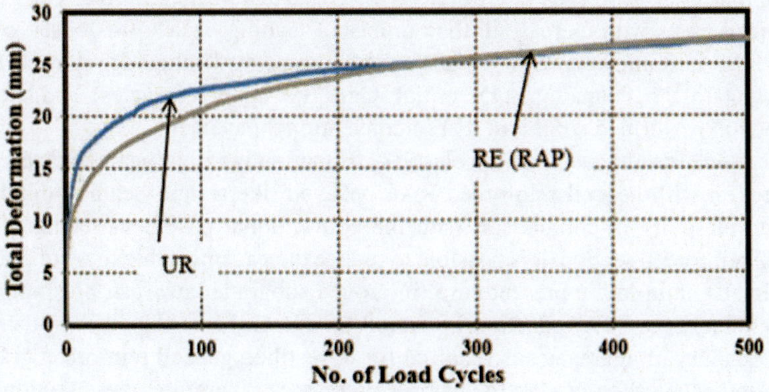

Fig. 7 Comparison of total deformation of unreinforced and geocell reinforced test sections

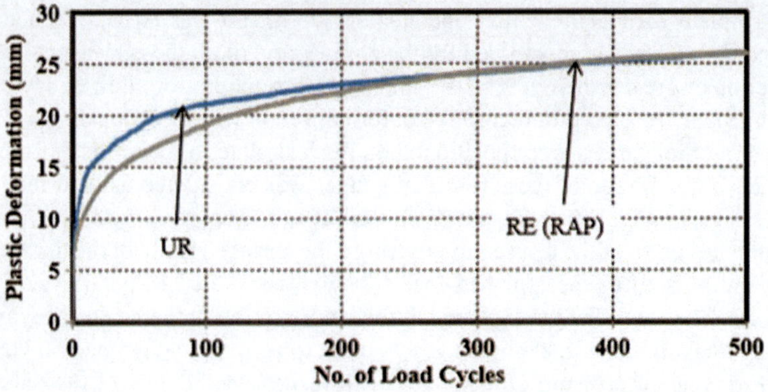

Fig. 8 Comparison of plastic deformation of unreinforced and geocell reinforced test sections

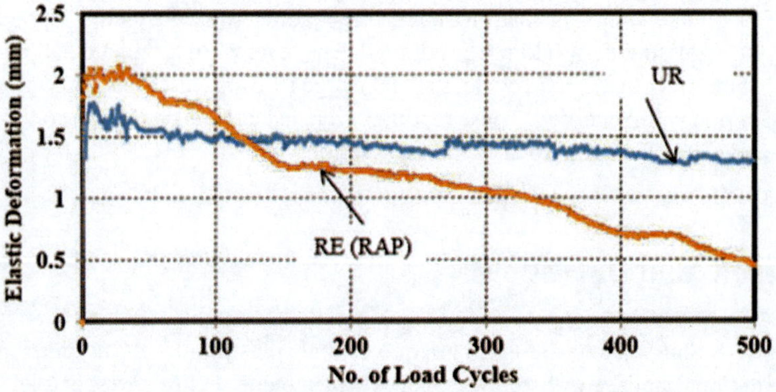

Fig. 9 Comparison of elastic deformation of unreinforced and geocell reinforced test sections

rutting. The total and plastic deformations increased in both unreinforced and geocell reinforced cases with increase in the number of loading cycles. It is observed that when RAP is reinforced with geocell, with no improved subgrade, showed lesser plastic and total deformation at the initial stages (i.e. up to 200 cycles), and later, the test section performed similar to that of conventional pavement.

The influence of geocell on the elastic deformation is insignificant. However, the test section with geocell reinforced RAP behaved like a rigid beam with slightly reduced elasticity in comparison with the conventional pavement section. In the geocell reinforced section, it is possible to reduce the pavement thickness of granular layer by 100 mm in the present case (improved subgrade layer can be eliminated) which amounts to 20% reduction in overall thickness. From this, it is evident that lesser quantity of aggregates is required for RAP-filled geocell reinforced granular layers and enables to preserve the natural aggregates. Therefore, these sections will be economical for the same performance and sustainable because of the use of RAP.

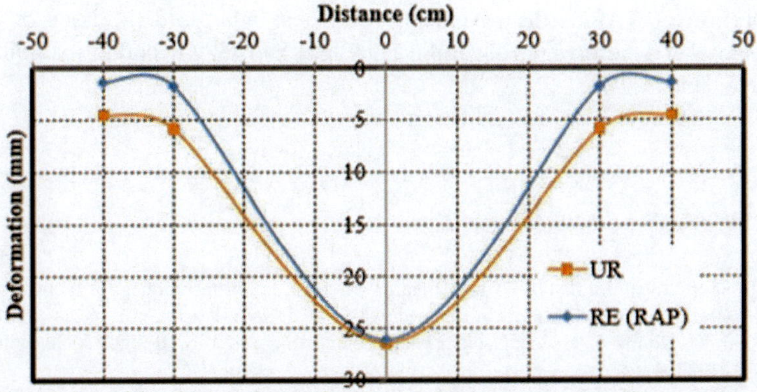

Fig. 10 Surface profiles at the top of the pavement section for conventional and geocell reinforced test sections at the end of 500 load cycles

Figure 10 shows the surface profiles at the top of pavement section for conventional and geocell reinforced RAP base test sections at the end of 500 load cycles. It is observed that when RAP is reinforced with geocell with no improved subgrade showed similar deformation behaviour compared to the conventional test section at the centre of the test section (point of loading) and lesser deformation was observed at a distance of 30 and 45 cm because of the rigidity of the geocell reinforced layer. Geocell reduces lateral spreading and improves the stiffness by providing lateral support. Geocell will reduce the pressure on the subgrade by distributing the load to a larger area.

7 Conclusions

Based on the repeated load test conducted under laboratory condition, the following conclusions are drawn.

1. The use of RAP as filler in the geocell reinforced layer in the granular base layer shows the same performance in terms of plastic and total deformation as that of a conventional pavement layer with provision of improved subgrade on top of weak black cotton soil subgrade.
2. The provision of geocell reinforced pavement layer acts as a rigid beam at the top of weak subgrade and reduced lateral flow of granular layer; as result, the surface heaving at locations away from loading is less compared to conventional pavement layer. This will reduce the overall surface rutting and enhance the riding quality of pavements.
3. The study has shown the possibility of reduction in pavement thickness to the extent of 20% in case of RAP-filled geocell reinforced layers in granular base layer compared to conventional pavement sections on weak subgrade.

4. The use of RAP in the pavement section has addressed the issue of RAP disposal and preservation of natural aggregates and helps in building sustainable pavements.

References

1. Webster SL, Watkins JE (1977) Investigation of construction techniques for tactical bridge approach roads across soft ground. Report no. S-77–1, USAE Waterways Experiment Station
2. Giroud JP, Noiray L (1981) Geotextile reinforced unpaved road design. J Geotech Eng 107(9):1233–1254
3. Potter JF, Currer EWH (1981) The effect of a fabric membrane on the structural behaviour of a granular road pavement. RRRL Report, Report no. LR996, Crowthorne, UK
4. Haliburton TA, Barron JV (1983) Optimum method for design of fabric reinforced unsurfaced roads. Transportation Research Record, 916, Transportation Research Board, National Research Council, Washington D C 26–32
5. Love (1984) Model testing of geogrids in unpaved roads. Thesis submitted to University of Oxford, England
6. Lawson CR (1992) Some examples of separation geotextiles under road pavements. Proc Inst Civ Eng Transp 95(3):197–200
7. Austin DN, Coleman DM (1993) A field evaluation of geosynthetic reinforced haul roads over soft foundation soils. Proceedings of Geosynthetics, Industrial Fabrics Association International, St. Paul, Minn, pp 65–80
8. Al-Qadi IL, Brandon TL, Valentine RJ, Smith TE (1994) Laboratory evaluation of geosynthetic reinforced pavement sections. Transportation Research Record, 1439, Transportation Research Board, National Research Council, Washington, D C, pp 25–31
9. Fannin RJ, Sigurdsson O (1996) Field observations on stabilization of unpaved roads with geosynthetics. J Geotech Eng 122(7):544–553
10. Hufenus R, Rueegger R, Banjac R, Mayor P, Springman SM, Bronnimann R (2006) Full scale field tests on geosynthetic reinforced unpaved roads on soft subgrade. Geotext Geomembr 24(1):21–37
11. Madhavi Latha G, Nair AM, Hemalatha MS (2010) Performance of geosynthetics in unpaved roads. Int J Geotech Eng 4:151–164
12. Mamatha KH, Dinesh SV (2017) Performance evaluation of geocell reinforced pavements. Int J Geotech Eng 12(6):546–555
13. Bush DI, Jenner CG, Bassett RH (1990) The design and construction of geocell foundation mattresses supporting embankments over soft ground. Geotext Geomembr 9(1):83–98
14. Babu S, Pawan Kumar GL (2012) An approach for evaluation of use of geocells in flexible pavements. In: Proceedings of Indian geotechnical conference, Delhi, December 13th–15th, Paper No. E502, (2012)
15. Mamatha KH, Dinesh SV (2017b) Evaluation of strain modulus and deformation characteristics of geosynthetic reinforced soil—aggregate system under repetitive loading. Inter J Geotech Eng 1–10
16. Hedge A, Sitharam TG (2016) Behaviour of geocell reinforced soft clay bed subjected to incremental cyclic loading. Geomech Eng 10(4):405–422
17. Hedge A, Sitharam TG Experimental and analytical studies on soft clay beds reinforced with bamboo cells and geocells. Int J Geosynthet Ground Eng 13(1)
18. Surya M, Anisha S, Kailash S, Navaneeth M, Vignesh R (2019) Performance assessment of square footing on jute geocell-reinforced sand. Int J Geosynthet Ground Eng 5(25):1–10
19. MoRT&H: Ministry of road transport and highways: Specifications for roads and bridges. In: 5th Edition, Indian Roads Congress, New Delhi (2013)

20. MORD: Ministry of rural development: Specifications for rural roads. The Indian Roads Congress, New Delhi (2014)
21. IRC: SP: 72: Guidelines for the design of flexible pavements for low volume rural roads. The Indian Roads Congress, New Delhi (2015)
22. IS 5249: Determination of dynamic properties of soil-method of test. 2nd Edition, Bureau of Indian Standards, New Delhi (1992)

Dynamic Properties of Soils—A Stress Path Approach

Mutyala Gowtham Kumar and Gali Madhavi Latha

Abstract The present study focuses on variation of dynamic properties of soils with change in stress path. Cyclic triaxial test, which is the most common laboratory test used to evaluate the dynamic properties of soils, suffers several limitations. Two major limitations are its inability to introduce continuous rotation of principal stress axes and the difference in simulated stress path when compared to vertically propagating shear waves. Present study focuses on quantifying one of these limitations, effect of change in stress path on the dynamic properties of soils, including shear modulus and damping ratio. Also, cyclic triaxial tests are created and performed in a way to simulate more realistic stress paths through slow dynamic loading. To keep the stress path vertical, minor principal stress is decreased with the increase in major principal stress by reducing the confining pressure. These tests showed that hysteresis loops became considerably broader with the stress path getting vertical, shear modulus decreased by about 25–45% and damping ratio increased by about 40–70% compared to conventional cyclic triaxial test. These results proved that the shear modulus is overestimated, and damping ratio is underestimated in a conventional triaxial test with inclined stress path, which does not actually represent a field scenario. The new tests with vertical stress path provide more realistic estimates of dynamic soil properties.

Keywords Stress path · Dynamic properties · Cyclic triaxial test · Shear modulus · Damping ratio

1 Introduction

Response of a soil due to any vibration is mainly influenced by the dynamic properties of soil. Among these properties, shear modulus and damping ratio of soils are the two most important parameters that are required for ground response analysis or design of foundations subjected to cyclic loads. Further, these properties are the

M. G. Kumar (✉) · G. M. Latha
Indian Institute of Science, Bangalore, India
e-mail: gowthamm@IISc.ac.in

T. G. Sitharam et al. (eds.), *Earthquake Geotechnics*, Lecture Notes in Civil Engineering 187, https://doi.org/10.1007/978-981-16-5669-9_14

basic input parameters for any numerical model that simulates dynamic response of soil. Accuracy of these numerical models mainly depends on the accuracy with which these parameters are determined from field or laboratory tests. Among the laboratory tests used to estimate the dynamic properties of soils, cyclic triaxial test is the widely used test because of its relative simplicity in applying cyclic loading conditions on soil and ease of parameter control in tests. In laboratory tests, dynamic properties of soils are influenced by many parameters such as the strain rate, number of loading cycles, confining pressure, relative density, over consolidation ratio, frequency of loading and type of soil. Many earlier researchers have studied the variation of damping ratio and shear modulus with different parameters. Ishihara [1], Hardin and Drnevich [2], Kramer [3] and Seed and Idriss [4] have shown that dynamic properties of soil depend on various parameters like type of soil, relative density, plasticity index, effective confining pressure, over consolidation ratio, frequency of loading and number of cycles.

In a cyclic triaxial test, shear modulus is represented by the inclination of hysteretic cyclic loading loop, and damping ratio is controlled by the breadth of the hysteresis loop. Most of the earlier researchers assumed a symmetric hysteresis loop while computing the dynamic soil properties from a triaxial test. Hardin and Drnevich [2] have shown experimentally that shear modulus decreases and damping ratio increases with increase in strain amplitude, considering symmetrical hysteresis loop. However, the hysteresis loop is seldom symmetric in triaxial tests because of inherent nonlinearity, non-homogeneity and anisotropy of soils. Hence, considering the realistic asymmetric hysteresis loop for the estimation of dynamic soil properties is very important. Kumar et al. [5] showed that by considering asymmetric hysteresis loop, damping ratio has reached a peak value at a specific strain and then decreased.

Shear strength parameters and dynamic parameters estimated from a triaxial test are greatly influenced by the stress path. The stress path in a triaxial test does not represent the true field conditions as described by many researchers. Kokusho [6] concluded that care must be taken in triaxial testing since it imposes conditions which are quite different from vertically propagating shear waves. Limited research has been done on the effect of stress paths on dynamic properties. Sayao and Vaid [4] showed that a part of strain induced by vertically propagating shear waves is due to principal stress axis rotation that cannot be simulated in a triaxial test, which is one of the limitations of triaxial testing.

The concepts of stress paths related to field and cyclic triaxial conditions are discussed here in detail. Figure 1a shows the condition of a dry soil element in field when a single cycle of vertically propagating shear wave (Fig. 1b) passes through it. Here, σ_v and σ_h represent vertical and horizontal stresses on the soil element, respectively, and τ represents the shear stress amplitude. In these figures, four different stages (1, 2, 3 and 4) of the soil element when a vertically propagating shear wave is passing through it are shown. Stages 1 and 3 represent 'at rest' earth condition while stages 2 and 4, respectively, represent the stages when positive and negative peaks of a shear wave reach the element. Mohr circles drawn for these four stages are shown in Fig. 1c. It can be observed that the Mohr circle expands and contracts about the center O, representing different stages. Figure 1d shows the stress paths in a t–s plot,

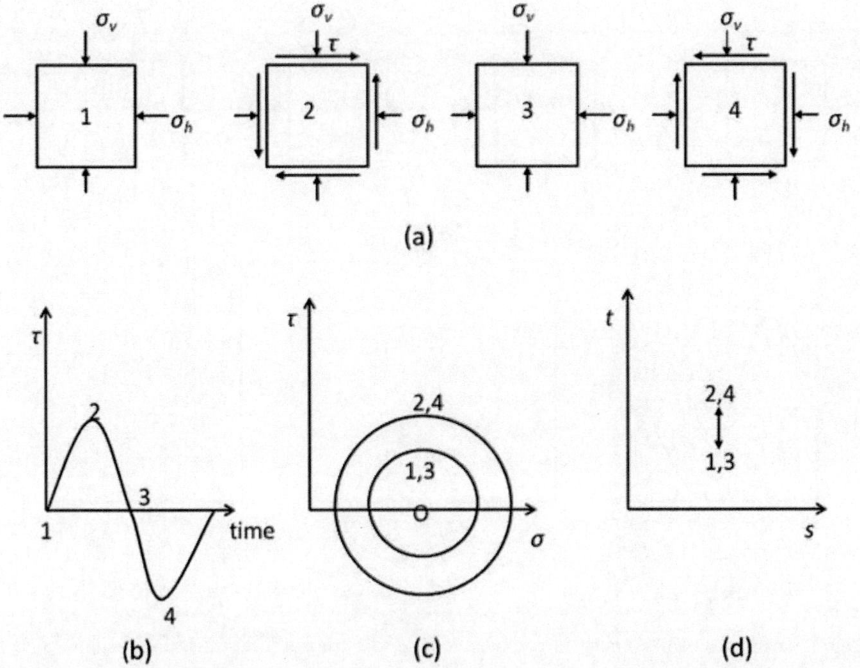

Fig. 1 Representation of shear loading on a soil element in field. **a** Shearing stages, **b** shear wave, **c** Mohr circles, **d** stress paths

where t is the deviatoric stress given by $(\sigma_1 - \sigma_3)/2$, s is the mean principal stress given by $(\sigma_1 + \sigma_3)/2$, σ_1 and σ_3 representing major and minor principal stresses, respectively.

Though the cyclic triaxial test should ideally simulate the above field conditions, the Mohr circles and stress paths in a cyclic triaxial test are quite different as explained in Fig. 2. The four different stages of shear loading on a soil element in a cyclic triaxial test are shown in Fig. 2a. Here, σ_1^1, σ_1^2, σ_1^3 and σ_1^4 represent axial stresses at stages 1, 2, 3 and 4 of an element, respectively, under normal triaxial loading. The cyclic variation of normal stress is shown in Fig. 2b, where axial stress (σ_1) is sinusoidal and cell pressure (σ_3) is constant. Stage 1 represents the initial condition of the soil element after the application on all-round confining pressure σ_3. Stage 2 represents the stage when the axial stress reaches a positive peak σ_1^2, stage 3 represents the condition when the axial stress σ_1^3 drops to the confining pressure. Stage 4 represents the condition where the axial stress reaches the negative peak σ_1^4. Note that the cell pressure (σ_3) is constant in all stages and unlike actual field condition shown in Fig. 2; here, Mohr circle expands and contracts about different centers $O1$ and $O2$ (Fig. 2c), resulting in inclined stress paths as shown in Fig. 2d.

Above explanations establish the clear difference in the cyclic loading pattern of a soil element in a triaxial test and in field. The rotation of principal stresses as happens

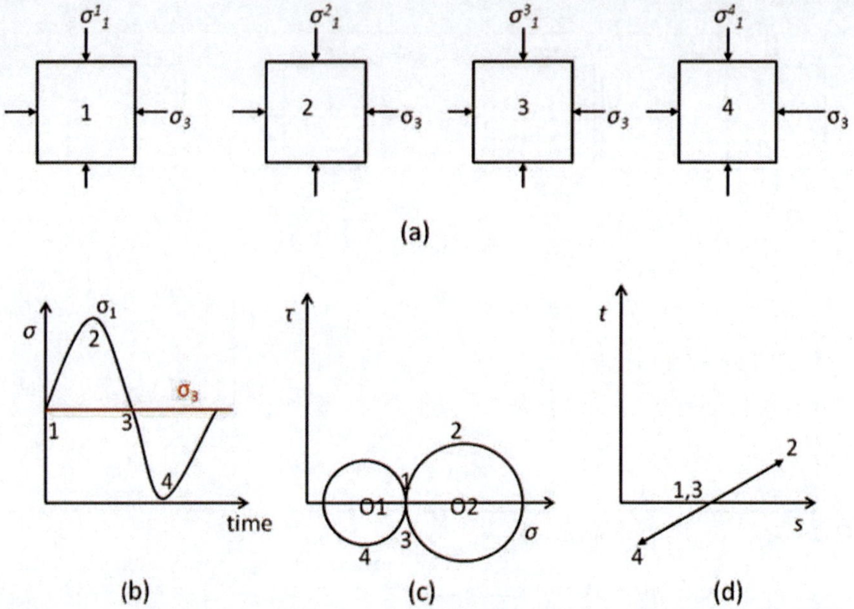

Fig. 2 Loading on soil element in a conventional cyclic triaxial test. **a** Triaxial loading stages, **b** loading cycle, **c** Mohr circles, **d** stress paths

in the field cannot be simulated in cyclic triaxial tests, but stress paths as in field can be simulated. In order to simulate the stress paths created by vertically propagating shear waves in a conventional cyclic triaxial test, cell pressure should be controlled to match the change in axial stress. For example, when the axial stress increases, the cell pressure should be correspondingly decreased and vice versa. To authors' knowledge, no such experiments were reported in literature. The present study tries to simulate the field conditions closely in a cyclic triaxial test by controlling the cell pressure. These tests are referred to as advanced cyclic triaxial tests in this paper.

Figure 3 explains the condition of a soil element in an advanced cyclic triaxial test. The four stages of advanced cyclic triaxial loading are shown in Fig. 3a. The loading cycle is shown in Fig. 3b, where the axial stress is sinusoidal and cell pressure is also sinusoidal. Here, σ_1^1, σ_1^2, σ_1^3 and σ_1^4 represent axial stresses at stages 1, 2, 3 and 4 of an element, respectively, and σ_3^1, σ_3^2, σ_3^3 and σ_3^4 represent confining pressures at stages 1, 2, 3 and 4 of an element, respectively, under advanced triaxial loading. Stage 1 represents the initial condition of the soil element after the application on all-round confining pressure σ_3^1. Stage 2 represents the stage when the axial stress reaches a positive peak σ_1^2 and the cell pressure reaches a negative peak σ_3^2, stage 3 represents the condition when the axial stress drops to σ_1^3 and confining pressure increases to σ_3^3, both matching at this stage, as shown in Fig. 3c. Stage 4 represents the condition where the axial stress reaches the negative peak σ_1^4 and cell pressure reaches the positive peak σ_3^4. Unlike the conventional cyclic triaxial test, cell pressure

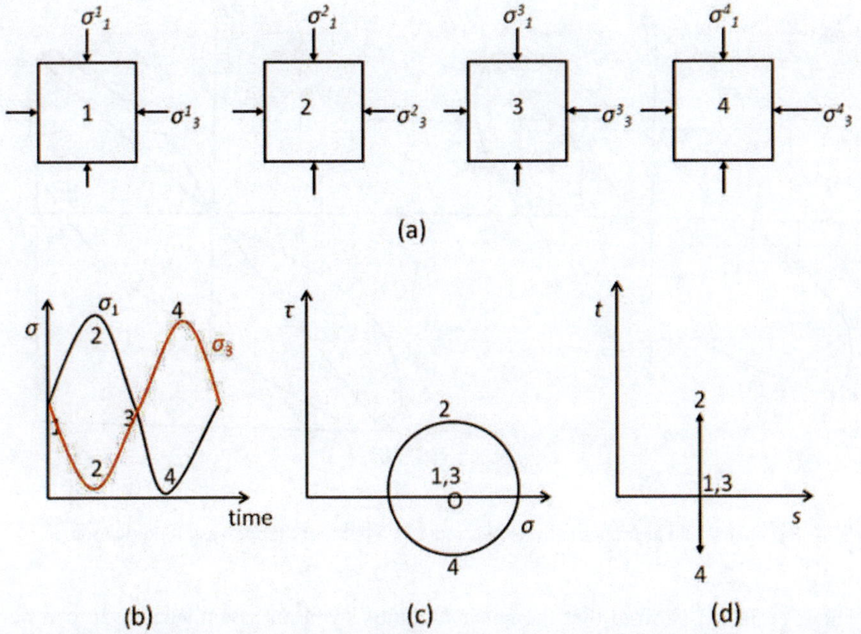

Fig. 3 Loading on soil element in an advanced cyclic triaxial test. **a** Triaxial loading stages, **b** loading cycle, **c** Mohr circles, **d** stress paths

also varies sinusoidally in this test. Though the condition is not exactly same as the condition created by a vertically propagating shear wave, the Mohr circle expands and contracts about a constant point *O*, the resultant stress paths closely simulating the field conditions as shown in Fig. 3d.

Present study evaluates the dynamic soil properties through conventional and advanced cyclic triaxial tests to understand the effect of variation of stress paths on shear modulus and damping ratio of the soils.

2 Shear Modulus and Damping Ratio in Triaxial Tests

Cyclic loading on soils results in hysteresis loop. For a symmetric hysteresis loop shown in Fig. 4a, estimation of shear modulus and damping ratio is well described in literature [1, 2, 4, 6–8]. For a symmetric loop, elastic modulus is computed as the slope of the line *oa* in Fig. 4a, and shear modulus is then computed from this value using Poisson's ratio of the soil. Damping ratio is determined as ($\Delta A/4\pi A_r$), where ΔA is the total area of the loop *abcdef*, which represents the energy dissipated in a cycle and A_r is the area of the shaded triangle *oag* in Fig. 4a.

Hysteresis loops are seldom symmetric, and hence, the shear modulus and damping ratio cannot be determined using the method specified for symmetric loops.

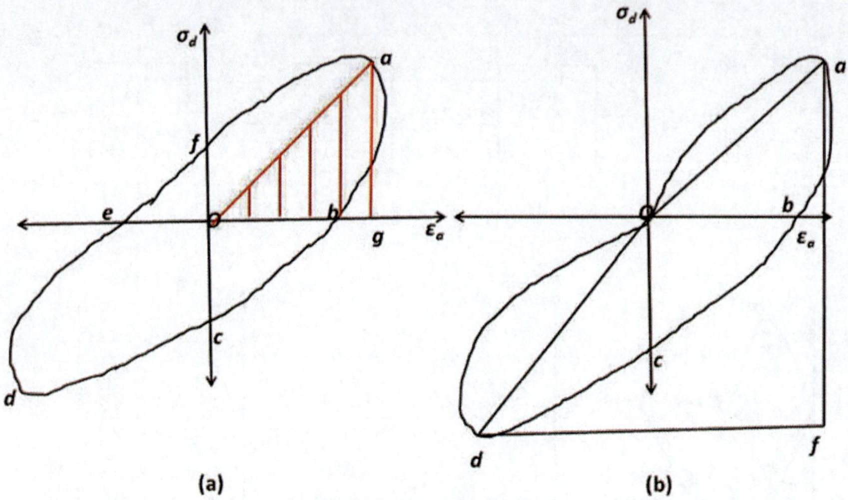

Fig. 4 Symmetric and asymmetric hysteresis loops. **a** Symmetric loop, **b** asymmetric loop

Kumar et al. [5] showed that the shear modulus from an asymmetric loop can be computed using the average elastic modulus (E) and Poisson's ratio (μ), using the expression, $G = E/2(1 + \mu)$. Average elastic modulus is the average of the slope of the lines *oa* and *od* representing compression and tension of the sample. A modified method given by Takeji Kokusho [6] is considered for computing damping ratio for asymmetric loops. Damping ratio is determined as $\Delta A/\pi A$, where ΔA is the total area of the loop *oabcdo*, which represents the energy dissipated in a cycle and A is the area of the trapezium *oafd* in Fig. 4b, which represents the maximum elastic energy stored.

3 Soil Used in Test

Sand used in this study is angular, with sieve analysis showing D10 of 0.08 mm, D30 of 0.16 mm and D60 of 0.3 mm. The coefficients of uniformity and curvature are computed as 3.6 and 1.1, respectively, classifying the soil as poorly graded sand with letter symbols SP as per the unified soil classification system. The maximum and minimum dry unit weights of the sand are 20.23 kN/m^3 and 14.93 kN/m^3, respectively.

4 Cyclic Triaxial Tests

The fully automated cyclic triaxial test setup (GDS make) used in this study can test specimens with diameter range of 38–300 mm and maximum cell pressure of 1 MPa. A 10 kN load frame fitted with a cyclic actuator which can apply a cyclic loading of frequency 0–10 Hz is used. Cyclic triaxial tests were done on remolded dry specimens of 70 mm diameter and 140 mm height. Dry tamping method was used for preparing the sample in five layers. All specimens were prepared to a relative density of 70%, which corresponds to the unit weight of 18.3 kN/m^3.

Both conventional and advanced cyclic triaxial tests were performed on sand specimens. In conventional cyclic triaxial tests, cell pressure was kept constant and axial stress was varied sinusoidally. In advanced cyclic triaxial tests, both axial stress and cell pressure were varied sinusoidally such that a vertical stress path was created as explained earlier. Stress paths in a conventional triaxial test are inclined on *t*–*s* plot, whereas they are vertical lines parallel to *t* axis with a constant value on *s* axis in case of an advanced triaxial test. So, to make stress paths vertical in a cyclic triaxial test, the *s* coordinate (mean principal stress) is kept constant throughout the test. When the axial stress is increased in a test, cell pressure is decreased by the same amount so that mean stress is constant. Similarly, when the axial stress is decreased, cell pressure is increased by the same amount.

5 Results and Discussions

Figure 5 shows the variation of axial stress and cell pressure with time in conventional and advanced cyclic triaxial tests. In the conventional cyclic triaxial test shown in Fig. 5a, cell pressure (σ_3) is constant at 70 kPa and axial stress (σ_1) varied sinusoidally from 70 kPa with an amplitude of 30 kPa. This is the regular stress controlled cyclic triaxial test (case 1), where the cell pressure is kept constant, and deviator stress is varied cyclically. In the advanced cyclic triaxial test (case 2) shown in Fig. 5b, axial stress and cell pressure are varied simultaneously, their values varying between 70 and 100 kPa, and the mean stress remained constant at 65 kPa. Since time period of both sine waves representing axial and confining stresses are the same and are 180° out of phase, the mean stress remained constant in this triaxial test. Figure 6 shows the variation of mean stress for conventional and advanced cyclic triaxial tests. Hence, the objective of keeping mean stress constant throughout the test is achieved. The resulting stress paths from conventional and advanced tests can be seen in Fig. 7. Stress path for conventional tests is inclined and in case of advanced tests, the stress path is similar to the stress path caused by vertically propagating shear wave. Variation of hysteresis loop for the two cases can be seen in Fig. 8.

Given the fact that the stress path in case of an advanced cyclic triaxial test is similar to the stress path under a vertically propagating shear wave in field, the shear modulus and damping ratio estimated from advanced cyclic triaxial test are more

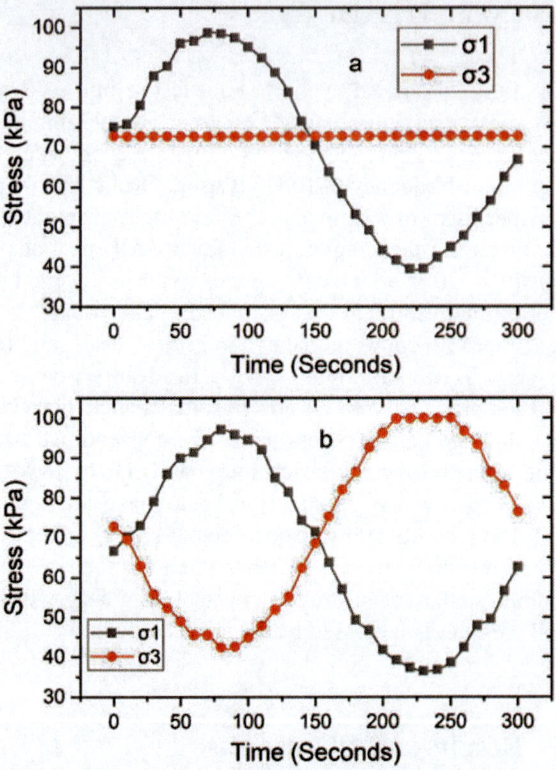

Fig. 5 Comparison of axial stress and confining pressure in conventional and advanced cyclic triaxial tests, **a** conventional, **b** advanced

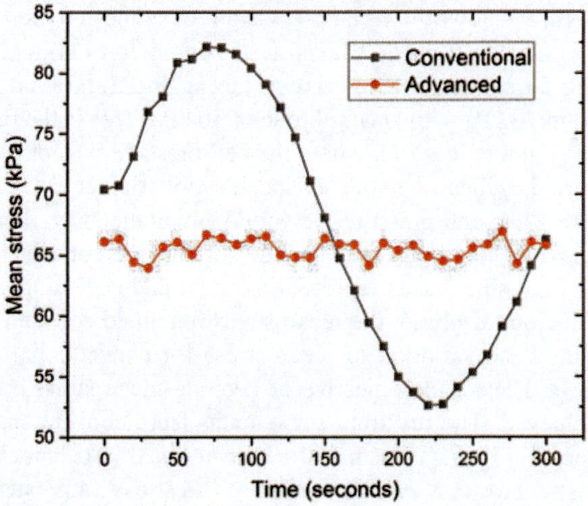

Fig. 6 Variation of mean principal stress in conventional and advanced cyclic triaxial tests

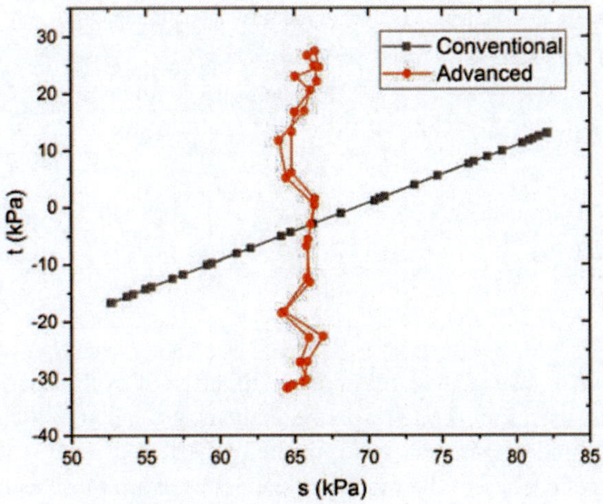

Fig. 7 Stress paths in conventional and advanced cyclic triaxial tests

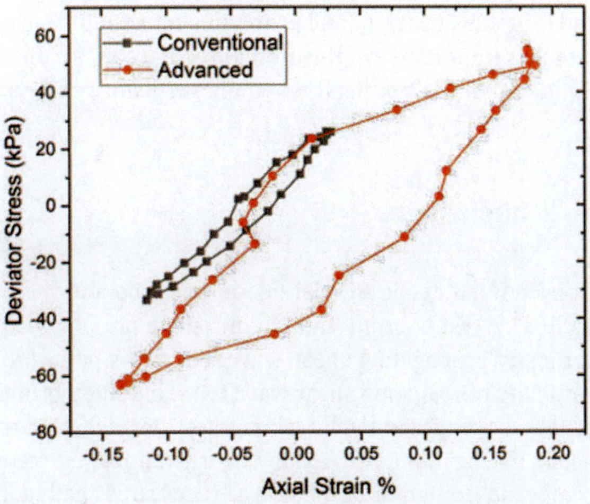

Fig. 8 Hysteresis loops for conventional and advanced cyclic triaxial tests

realistic. To establish the applicability of the results for tests with multiple loading cycles, tests were conducted for four cycles with a period of 5 min each, and the third cycle is considered for the evaluation of dynamic properties.

Calculation of shear modulus (G) and damping ratio (ε) for conventional and advanced cyclic triaxial tests for the hysteresis loops from single cycle and multiple cycle triaxial tests are shown in Table 1. A Poisson's ratio of 0.25 is assumed for the sand.

Results from both the tests established that sand behaved differently with different stress paths. Table 1 shows that in conventional triaxial test, G is seriously overestimated and damping ratio is underestimated. Use of dynamic properties obtained from conventional cyclic triaxial tests for the design of foundations subjected to

Table 1 Calculation of shear modulus and damping ratio

Parameter	Single cycle		Multiple cycles	
	Conventional	Advanced	Conventional	Advanced
ΔA	1.12	10.46	1.09	7.604
A	3.03	20.64	2.92	12.20
Elastic modulus (kPa)	68,875	38,440	75,125	56,845
G (kPa)	27,550	15,376	30,050	22,738
ε (%)	11.7	16.13	11.8	19.8

cyclic loads can result in overestimation of cyclic bearing capacity. If the properties are used for ground response analysis, ground amplifications and displacements will be underestimated. Also, in the numerical programs to simulate soils subjected to cyclic loads, if the input parameters used are those estimated from the conventional cyclic triaxial tests, serious mismatch with the actual behavior can be expected. In the design codes, such limitations are not clearly laid out and most of the designs and programs use dynamic properties estimated from cyclic triaxial tests. This study provides important insights into the actual stress paths in a cyclic triaxial test and the limitation of the test itself when the verticality of stress paths is not followed.

6 Conclusions

Conventional cyclic triaxial test imposes conditions which are quite different from field. This test not only fails to introduce principal stress axis rotation as in case of vertically propagating shear wave, but stress paths are quite different from those in vertically propagating shear wave too. This study brings out this important limitation of the conventional cyclic triaxial test through plots of hysteresis loops and stress paths. Further, an advanced cyclic triaxial testing method is presented, which overcomes the limitation of the non-vertical stress path and rotation of principal axes in a triaxial test and succeeds in creating a condition close to the realistic stress path as in case of a vertically propagating shear wave in field. Results showed that the conventional cyclic triaxial test overestimates the shear modulus shear modulus by about 25–45% and underestimates the damping ratio by about 40–70% compared to the advanced cyclic triaxial test.

References

1. Ishihara K (1996) Soil behaviour in earthquake geotechnics. Oxford Science Publications
2. Hardin BO, Drnevich VP (1972) Shear modulus and damping in soils: measurement and parameter effects. J Soil Mech Found Div ASCE 6:603–624
3. Kramer SL (1996) Geotechnical earthquake engineering. Prentice Hall, New Jersey

4. Sayao ASF, Vaid YP (1989) Deformations due to principal stress rotation. In:12th International proceedings on conference on soil mechanics and foundation engineering, vol 1. Rio de Janeiro, Brazil, pp 107–110
5. Kumar SS, Krishna AM, Dey A (2017) Evaluation of dynamic properties of sandy soil at high cyclic strains. Soil Dyn Earthq Eng 99:157–167
6. Kokusho T (1980) Cyclic triaxial test of dynamic soil properties for wide strain range. Soils Found 20(4):1–18
7. Seed HB, Idriss IM (1970) Soil moduli and damping factors for dynamic response analyses. Report No. EERC 70–10. Berkeley, Earthquake Engineering Research Centre, University of California, California
8. Saglam S, Bakir BS (2014) Cyclic response of saturated silts. Soil Dyn Earthq Eng 61:164–175

Bottom Ash Stabilized Subgrade Soil Admixed with Sugarcane Bagasse Ash

Prathik Anand Krishnan, V. Pradeep Gokul, B. Adithya, and Anil Kumar Sharma

Abstract Expansive soils are popularly known as black cotton soils in India. It shows very high swelling and shrinking characteristics. Hence, it becomes necessary to stabilize these soils for infrastructural development such as for the construction of roads. For the purpose of full or partial replacement of traditional stabilizers, recent research in the field of geotechnical engineering focuses more on the use of cheap and locally available industrial wastes as stabilizing agents. In this study, it is aimed to evaluate the possible utilization of industrial wastes like bottom ash (BA) which is a by-product obtained from thermal power plant, effectively in, along with sugarcane bagasse ash (SCBA) as stabilizing agent for a locally available expansive soil. Various engineering property tests like proctor compaction, California Bearing Ratio (CBR), unconfined compressive strength (UCS) and split tensile strength were conducted for soil–SCBA–BA mixes in different proportions. BA was varied from 0 to 40% and SCBA was varied 0–9%. A small amount of cement (3%) was also added to promote pozzolanic reaction. It was observed that the engineering properties of the mixes depend mainly on the amount of BA and SCBA as well as curing period. Microstructural and mineralogical studies were also conducted to check the effectiveness of the proposed stabilizing agent at nano-scale. This experimental investigation suggests the possibility of using BA and SCBA in combined form to get maximum benefits in soil stabilization. By promoting these materials, the number of costly additives like cement and lime can be reduced significantly which will provide an economical and environmental friendly solution.

Keywords Soil stabilization · Sugarcane bagasse ash (SCBA) · Pozzolans · Bottom ash (BA)

P. A. Krishnan
Atkins India Pvt. Ltd., Bangalore, India

V. Pradeep Gokul · B. Adithya
Department of Civil Engineering, Amrita School of Engineering, Amrita Vishwa Vidyapeetham, Amrita University, Coimbatore, India

A. K. Sharma (✉)
Department of Civil Engineering, National Institute of Technology Patna, Ashok Rajpath, Patna, Bihar 800005, India
e-mail: aks.ce@nitp.ac.in

T. G. Sitharam et al. (eds.), *Earthquake Geotechnics*, Lecture Notes in Civil Engineering 187, https://doi.org/10.1007/978-981-16-5669-9_15

1 Introduction

Environmental pollution arising due to disposal of wastes from industrial activities is among the major problems faced by developing countries like India. There is a great concern about the effect of waste generated from various industries on health and environment. At present, there is a great need to focus on the reuse and recycling of waste for civil engineering applications [1]. The conventional approaches for utilizing agricultural waste, industrial waste and municipal waste are not sufficient to solve the disposal problems. One of the ways to establish sustainable development is by approaching 100% full utilization of all types of industrial and agricultural wastes. Industrial waste utilization in civil engineering applications have been found to be effective in accelerating the attempts towards sustainable goal [2]. Industrial wastes like fly ash, bottom ash, blast furnace slag, rice husk ash, silica fume etc., have been successfully implemented in geotechnical engineering applications such as fill material, binder, subgrade stabilizer and embankment material [3–5]. Recent trend of research work in the field of geotechnical engineering focused more on the study of cheap and locally available industrial and agricultural wastes as stabilizing agents for the purpose of full or partially replacement of traditional stabilizers. These wastes are increasingly becoming a focus for researchers because of the enhanced pozzolanic capabilities. They investigated on the use of fly ash for the stabilization of typical soils [6]. They observed that engineering properties of the stabilized soil and were capable of taking higher load coming from structures. They concluded that using fly ash is an effective way for modifying the soil properties.

Researchers also tried to improve shear strength of soft clay soil stabilized with silica fume in the presence of lime. The experiments showed that silica fume and lime mix increased the shear strength and angle of internal friction of the soil [7].

There are several other industrial wastes products used as stabilizers for improving the bearing capacity and strength properties of the soil [8–10]. It has been found that mixing different wastes as binder/stabilizer is more effective than individual materials [11, 12].

This study explores the use of sugarcane bagasse ash (SBA) and bottom ash (BA) to stabilize and improve the properties of subgrade soil. Sugarcane straw is a waste product from sugar industry and is produced in very large quantities in India. India is the second-largest sugarcane producing country in the world. Attempts have been made for utilisation of sugarcane straw ash which possess high pozzolanic activity [13]. BA is a by-product obtained from thermal power plant and can be effectively used in geotechnical application.

Laboratory tests were conducted in the geotechnical laboratory with the collected soil sample to classify the soil to evaluate its physical and engineering properties and to study the compaction characteristics. This elaborates the various physical and engineering properties of sub-grade soil namely natural moisture content, specific gravity, liquid limit, plastic limit, shrinkage limit, grain size distribution, optimum moisture content, maximum dry density, unconfined compressive strength and CBR strength.

2 Materials and Methodology

2.1 Materials

Black cotton soil was collected from a construction site in Periyanayakampalaym, Coimbatore in the Tamil Nadu state of India. Bottom ash and sugarcane bagasse ash were procured from Neyveli Lignite Corporation Limited (NLC), Tamil Nadu. Sugarcane bagasse ash (SBA) was collected from Amrita Recycle Centre (ARC) in the Amrita University campus, Coimbatore, where it is used as fuel. These raw materials were sieved through 4.75 mm sieve and the particle size distribution of soil and SBA was determined (see Figs. 1 and 2). All the physical properties of the soil were determined conforming to IS: 2720 in Table 1. The specific gravity of BA and

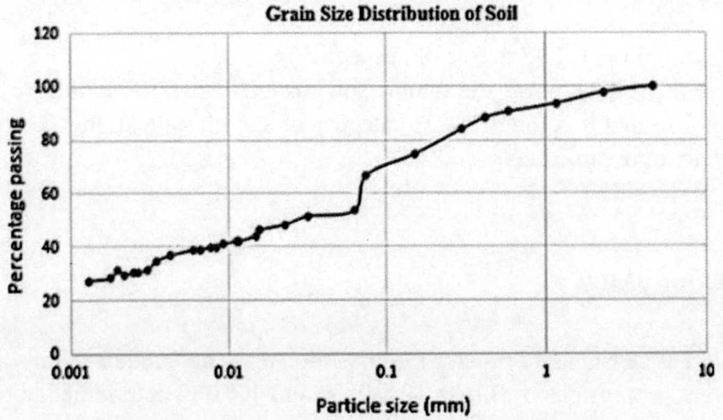

Fig. 1 Particle size distribution curve of soil

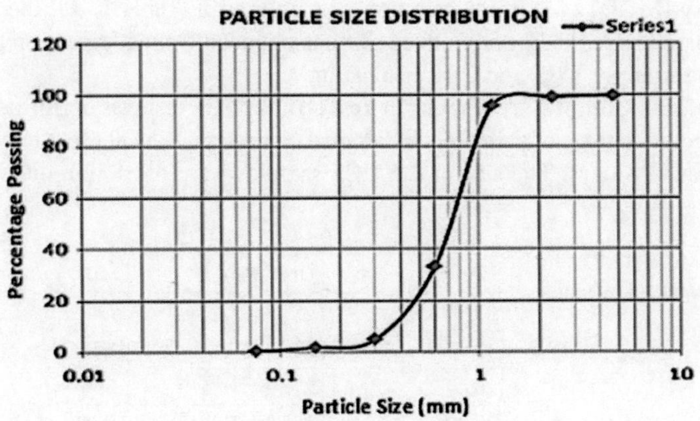

Fig. 2 Particle size distribution curve of bottom ash

Table 1 Physical properties of soil sample

Liquid limit (%)	63.37
Plastic limit (%)	31.72
Plasticity index (%)	31.65
Optimum moisture content (%)	28.73
Maximum dry density (g/cc)	1.42
Free swell index (%)	77.78
Specific gravity	2.33
C_u	2.00
C_c	1.13
UCS, kPa	163.39
CBR (unsoaked) @ 2.5 mm	2.78%
CBR (unsoaked) @ 5 mm	%

SCBA was found to be 2.43 and 1.88, respectively.

Based on the A-line graph, the sample soil has LL = 63.37% and PL = 31.72%. We can confirm that it belongs to CH category of soil classification. CH stands for high plasticity, inorganic clay.

2.2 *Methodology*

Compaction Test: Standard proctor's compaction tests are conducted on soil mixed with different proportions of BA (0, 10, 20, 30 and 40%) to determine the optimum moisture content and maximum dry density of soil sample. Mini compaction test apparatus developed by Sridharan and Sivapullaiah [14] was used to study the compaction characteristics. The optimum moisture content (OMC) and maximum dry density (MDD) of soil–BA mixtures are reported in Table 2. All the samples were prepared as per their optimum conditions obtained from compaction tests and were then tested for UCS and CBR behaviour.

Unconfined Compressive Strength Test (UCS): UCS test was performed on the different combinations of soil–BA–SCBA mixtures after curing at room temperature for 7, 14 and 28 days. The size of the UCS samples were of 38 mm diameter and

Table 2 OMC and MDD for varying percentage of bottom ash

BA in %	OMC (%)	MDD (g/cc)
0.00	28.73	1.42
10.00	23.73	1.53
20.00	21.12	1.56
30.00	18.82	1.64
40.00	23.32	1.54

height of 76 mm and were prepared by compacting to their respective MDD and OMC obtained from mini compaction test. The test was then conducted according to the IS: 2720 (Part 10):1991.

Split Tensile Test: Split tensile test is a simple and reliable method of measuring the tensile strength of cohesive soils compacted at dry of optimum. In this, test samples of various mix were prepared and split tensile test was carried out according to ASTM D3967-16. The dimensions of this test are similar to the UCS test samples.

California Bearing Ratio (CBR) Test: CBR is the prime factor which determines the thickness of each pavement layer in the design of pavement. California Bearing Ratio test was performed on the soil sample mixed with BA and SCBA. The soaked CBR value is determined by subjecting the specimen in the mould for four days soaking; then, it is cured in moist condition for 3 days, and the water is drained 15 min prior to the testing of the specimen. Unsoaked CBR value is determined by subjecting the specimen in the mould for seven-day air curing. The CBR test was conducted as per IS 2720 (Part 16)-1987.

XRD and SEM Analysis: X-ray diffraction analysis was carried out to learn about the mineralogical changes occurring due to pozzolanic activity in the stabilized soil. XRD diffraction patterns were analysed using software X'pert Highscore. The scanning electron microscope (SEM) is a nano-scale level investigation to capture the microstructural changes during the stabilization of the soil. SEM analysis was carried out on the stabilized soil samples at different curing periods.

3 Results and Discussions

3.1 Unconfined Compressive Strength Test (UCS)

Unconfined compressive strength (UCS) tests were performed on the soil sample stabilized with varying percentages of BA such as 0, 10, 20, 30 and 40% combinations. Table 3 shows the UCS results of soil + BA combinations. We can see from UCS of the soil gradually increases from 0 to 30% and then it decreases. This may be due to loss of cohesive property of soils with higher amount of BA. Hence, the optimum value of BA is to be considered as 30% for further studies. Thereafter, different percentages of SCBA, i.e. 5, 7 and 9%, were added to the soil–BA mixture

Table 3 UCS of soil/bottom ash mixtures

Bottom ash (BA) (%)	UCS (kPa)
0	163
10	170
20	172
30	175
40	107

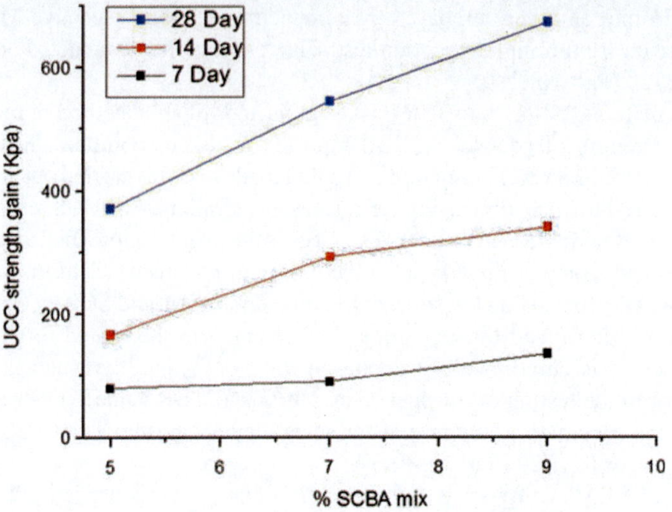

Fig. 3 Variation of UCS at different percentage of SCBA at different curing periods

to check the UCS behaviour. A small amount of cement, i.e. 3%, was also added to enhance the pozzolanic characteristics.

The variation of UCS for the soil mixed 30% BA with the different percentages of SCBA (see Fig. 3). We can observe the increase in strength with respect to increase in curing period as well as with the increase in the SCBA proportion.

SCBA is very fine particle and possesses pozzolanic characteristics which helps in filling the voids and improving the strength characteristics. Moreover, the strength increases as a result of the pozzolanic reaction and the formation of cementitious compounds, such as CSH. Since the soil properties are getting altered and its gaining strength with increase in curing period, soil is getting more stabilized and can be used for various construction purposes. The mechanism of strength gain be more clearly understood with the help of XRD and SEM analysis.

3.2 CBR Test

The CBR variation at both soaked and unsoaked conditions were determined (see Fig. 4). From this Figure, it was observed that the CBR increases with the increase in SCBA content. At the SCBA content of 9%, it shows the highest CBR value for both unsoaked and soaked conditions. The CBR under soaked condition shows lesser values as compared to unsoaked condition sample. Since the strength is reduced upon saturation, soaked CBR test is conducted to check the performance of the mix in the worst likely condition, that is, when the soil is in complete saturated condition. We can observe from the graphs that 9% SCBA mix yields maximum CBR value as

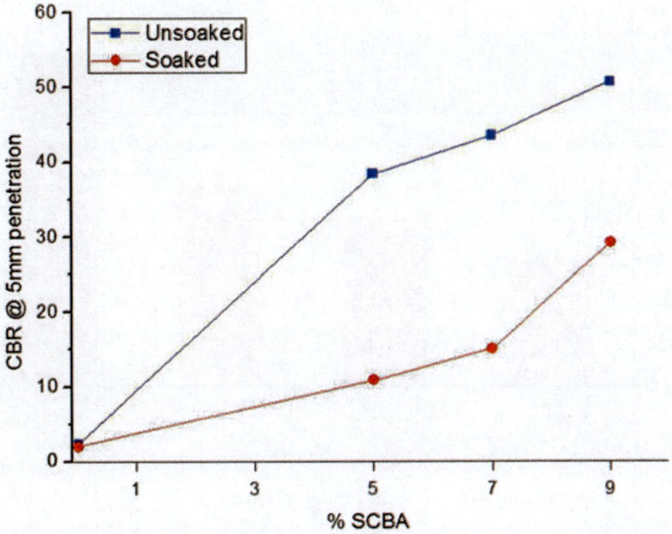

Fig. 4 Variation of CBR at different percentage of SCBA

compared to other mixes. The mixture containing soil with 30% BA, 9% SCBA and 3% cement gave the best result from the CBR studies. It can be inferred that the soil load carrying capacity has significantly improved.

3.3 Split Tensile Test

The behaviour of tensile strength of the stabilized soil with 30% BA and different percentages of SCBA (see Fig. 5). The trend is similar to the previous results increases with increase in SCBA content. It has been found with the addition of 9% SCBA, the tensile strength has increased by 2.5 times the tensile strength of pure soil sample. This indicates that there is a significant improvement in the tensile strength of the soil. The increase in tensile strength shows that the stabilized subgrade soil would be able to take up the tensile stresses coming on them.

3.4 XRD and SEM Analysis

XRD diffraction pattern of 7 day and 28 day cured soil sample treated with 30% BA, 9% SCBA and 3% cement was determined (see Fig. 6). From these XRD analysis, we can observe that formation of cementitious compounds like CSH and CASH can easily be detected from the diffraction pattern. This indicates the fact that with

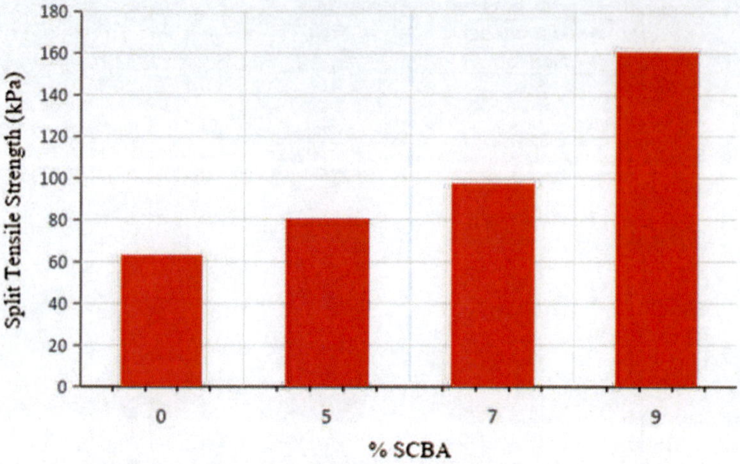

Fig. 5 Variation of split tensile strength at different percentage of SCBA

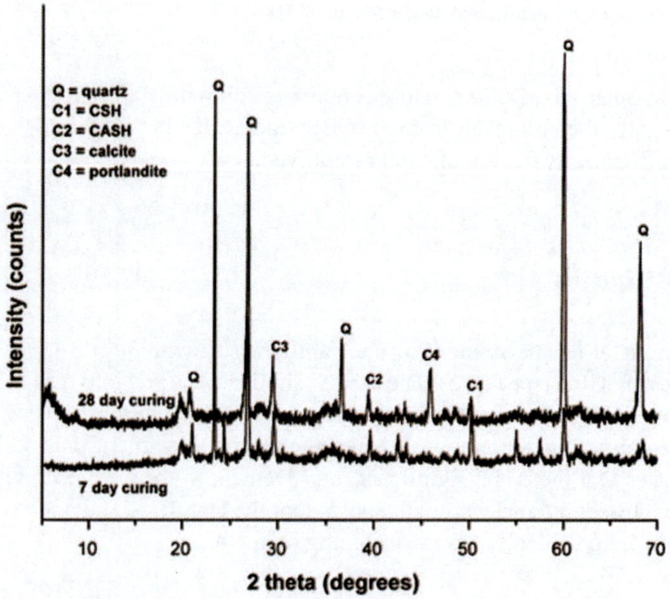

Fig. 6 XRD analysis of soil containing 30% BA, 9% SCBA and 3% cement

increase of curing period, the sample undergoes pozzolanic reaction which leads to the formation of cementitious compounds.

Microstructural study was carried out to check the changes occurring in the microstructure of the stabilized samples after curing period. The stabilized soil sample with 30% BA and 9% SCBA and 3% cement cured for 7 days and 28 days was

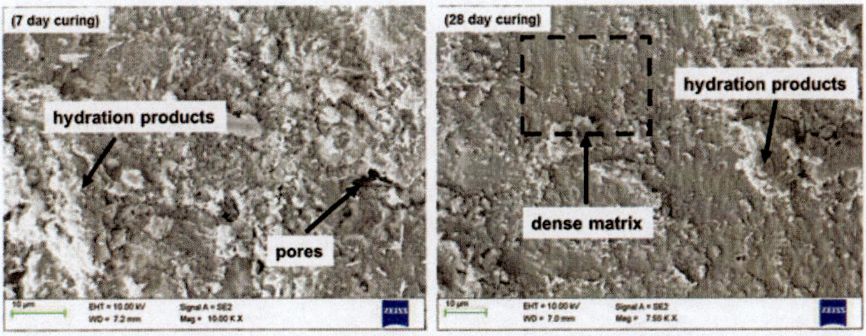

Fig. 7 SEM microstructure of soil containing 30% BA, 9% SCBA and 3% cement

determined (see Fig. 7). The images show a very dense like microstructure, usually observed in stabilized soil samples. We can clearly see that all those pores have been completely filled, and a dense structure is obtained. This also provides indications about the existence of pozzolanic activity happening between the stabilizer and the soil. The pozzolanic reaction results in the formation of cementitious compounds thereby increasing the strength and improving the denseness of microstructure of the stabilized soil.

4 Conclusion

The main aim of this research was to check the feasibility of using combination of BA and SCBA as stabilizer for expansive soil. From the test results, the following conclusions can be drawn:

- The maximum dry density increased with increase in BA content up to 30%, while it decreased with further addition.
- The UCS of the soil with the addition of up to 30% BA content showed slight improvement, while it showed reduction in strength with the addition of 40% of BA.
- Addition of SCBA had a great impact on strength and the pozzolanic characteristics with curing and showed significant improvement in the presence of 3% cement.
- CBR is also found to increase with SCBA content, and unsoaked sample showed higher CBR when compared to soaked sample.
- The split tensile test result indicates that the tensile strength of the soil is greatly improved. Since subgrade soil is more prone to failure due to tensile stresses than compressive loads, this result is very vital in terms of its application in subgrade stabilization.

- The XRD analysis confirms the formation of cementitious compounds such as CSH and CASH, while SEM images show a very dense microstructure of the stabilized sample.
- The test results follow that BA in combination with SCBA has the potential to be used as stabilizing agent in the construction of subgrade.

References

1. Batayneh M, Marie I, Asi I (2007) Use of selected waste materials in concrete mixes. Waste Manage 27(12):1870–1876
2. Raut S, Ralegaonkar R, Mandavgane S (2011) Development of sustainable construction material using industrial and agricultural solid waste: a review of waste-create bricks. Constr Build Mater 25(10):4037–4042
3. Cokca E, Yazici V, Ozaydin V (2009) Stabilization of expansive clays using granulated blast furnace slag (GBFS) and GBFS-cement. Geotech Geol Eng 27(4):489
4. Kim B, Prezzi M, Salgado R (2005) Geotechnical properties of fly and bottom ash mixtures for use in highway embankments. J Geotech Geoenviron Eng 131(7):914–924
5. Pandian N (2013) Fly ash characterization with reference to geotechnical applications. J Indian Inst Sci 84(6):189
6. Prabakar J, Dendorkar N, Morchhale R (2004) Influence of fly ash on strength behavior of typical soils. Constr Build Mater 18(4):263–267
7. Alrubaye AJ, Hasan M, Fattah MY (2017) Stabilization of soft kaolin clay with silica fume and lime. Int J Geotech Eng 11(1):90–96
8. Basha E, Hashim R, Mahmud H, Muntohar A (2005) Stabilization of residual soil with rice husk ash and cement. Constr Build Mater 19(6):448–453
9. Hughes PN, Glendinning S, Manning DA, White ML (2011) Use of red gypsum in soil mixing engineering applications. Proc Inst Civ Eng-Geotech Eng 164(3):223–234
10. Sharma AK, Sivapullaiah P (2012) Improvement of strength of expansive soil with waste granulated blast furnace slag. In: GeoCongress 2012: state of the art and practice in geotechnical engineering, pp 3920–3928
11. Sharma AK, Sivapullaiah P (2016) Ground granulated blast furnace slag amended fly ash as an expansive soil stabilizer. Soils Found 56(2)
12. Winnefeld F, Leemann A, Lucuk M, Svoboda P, Neuroth M (2010) Assessment of phase formation in alkali activated low and high calcium fly ashes in building materials. Constr Build Mater 24(6):1086–1093
13. Payá J, Monzó J, Borrachero MV, Díaz-Pinzón L, Ordóñez LM (2002) Sugar-cane bagasse ash (SCBA): studies on its properties for reusing in concrete production. J Chem Technol Biotechnol 77(3):321–325
14. Sridharan A, Sivapullaiah PV (2005) Mini compaction test apparatus for fine grained soils

Comparative Analysis of Utilization of Salts for Stabilization of Black Cotton Soil

Mamidala Prudhvi and Gajendran Chellaiah

Abstract Owing to its origin from chemical disintegration of basalt rock, black cotton soils often exhibit swelling and shrinkage characteristics due to change in water content. Black cotton soils spread about 20% of India's land territory. They are significantly located in the Deccan trap covering the circumferences of Maharashtra, Gujarat, Madhya Pradesh, Karnataka Andhra Pradesh, Tamil Nadu, Uttar Pradesh, and Rajasthan. It is also called regur. Generally, Indian black cotton soils are rich in montmorillonite, a sort of mud mineral. This mineral is responsible for swell–shrink performance of the soil. The soil changes in shading from black to profound black. Initially, it was viewed as that black color, because of the closeness of humidified natural problem. However, work has confirmed that the shade is because of the little grouping of titanium oxide. In this juncture, an effort has been made to study the effect of NaCl and $CaCl_2$ on physical properties of black cotton soil. A different measure of salt (2, 4, 6, and 8%) was added to the soil to consider salts' impact on the compaction behavior, stability limits, and compressive strength. The main discovery of this study was that the extension in every one of the chloride mixes increases the most remarkable dry thickness and declines the perfect dampness content. The liquid limit, plastic limit, and plasticity index values were decreased slightly with the increase in salt content. The strength increases as the salt content increases.

Keywords Soil stabilization · Black cotton soil · Atterberg limits · Strength

1 Introduction

The black cotton soil is a sort of far-reaching soil with high versatility and can hold dampness all through the dry season, which is why they are significant for developing harvests. It shows low bearing limit, low porousness, and high volume change due to the nearness of montmorillonite in its mineralogical content, and these properties make it ill-suited for the development of bank and other building structures. Dark cotton soil is far-reaching soil with the potential for shrinkage or

M. Prudhvi · G. Chellaiah (✉)
Karunya Institute of Technology and Sciences, Coimbatore 641114, India

T. G. Sitharam et al. (eds.), *Earthquake Geotechnics*, Lecture Notes in Civil Engineering 187, https://doi.org/10.1007/978-981-16-5669-9_16

growing under dampness change. They normally show high psychologists swell attributes with surface breaks, opening during the dry seasons, 50 mm or wider and a few mm profound. These breaks close during the wet season, and a lopsided soil surface is delivered by sporadic expansion. Such soils are particularly problematic as asphalt sub-grades. In India, about 51.8 million hectares of the land region are secured with dark cotton soil. The dark cotton soils are hard when dry yet lose their quality totally when in wet conditions. Dark cotton soils are an overall issue that represents a few challenges for structural architects. These dark cotton soils in India are helpful to ranchers yet risky to structural designers. Common building structures experience an enormous scope of harm because of progress in the properties of soil. Dark cotton soils consistently present tests to geotechnical engineers. This paper portrays an examination concerning the impact of expansion chloride mixes like NaCl and $CaCl_2$.

2 Experimental Program

The soil utilized in this paper was silty mud soil, which was brought from Vijayawada found east of the proposed Andhra Pradesh capital city. This kind of soils speaks to a broadly spread run of the mill soil in the center and south of Vijayawada. The soil examples were taken at a profundity of around 1 m beneath the ground surface. The properties of the soil and Atterberg limits are given in Table 1, while the characterization and the grain size conveyance of the soil appear in Figs. 1 and 2, respectively; the soil falsehoods over the A-line Fig. 1, along these lines the soil is delegated high versatility earth soil (CH) as per the bound together grouping framework.

Table 1 Properties of the soil used in this research

Property	Value
Liquid limit (%)	58
Plastic limit (%)	25
Plasticity index (%)	49
Specific gravity (kg m^{-3})	2.0
Clay fraction (%)	39
Silt fraction (%)	42
Sand fraction (%)	42

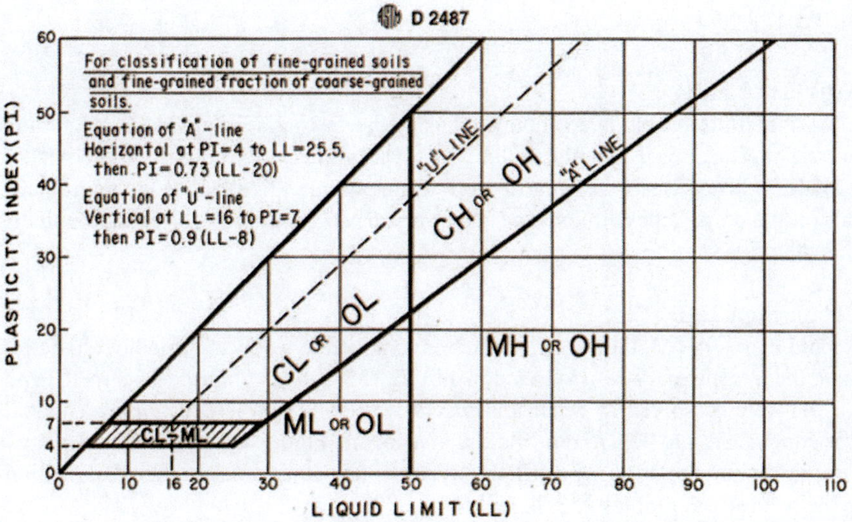

Fig. 1 Plasticity chart [1]

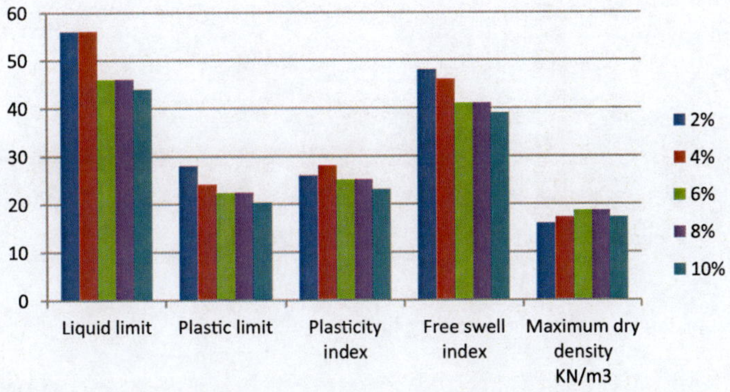

Fig. 2 Soil grain size conveyance

3 Soaking Solutions

Distilled Water: Distilled water was used in all tests

Chlorides Solutions

The concentration of every chloride salt (NaCl and $CaCl_2$) was acquired by dissolving in refined water to focus on each kind of salt in ppm and afterward blended in with soil. The dirt examples were set up by altered Procter test techniques as indicated by American Society for testing and materials (ASTM) (D 1557). Three unique focuses for each chloride salt (2, 4, 6, and 8%) were readied.

4 Soil Tests

Atterberg Limits

The liquid limit test has been conducted using the Cassagrande apparatus, according to ASTM (D423-66). The plastic limit test was conducted according to the ASTM (D424-59). These tests were carried out to investigate the effect of the addition of salt on the consistency limits. The effect of salt content on the Atterberg limits is shown in Fig. 3.

Compaction Test

According to ASTM, the modified proctor compaction test was carried out to determine the moisture-dry density relationship (D 1557). Each chloride salts (NaCl and $CaCl_2$) were dissolved in water and mixed with soil, then left for 1 day. The soil was compacted into 929.75 cm^3 mold in five layers. Figure 4 shows the dry density–moisture content relation for a different type and percentage of salts than that relation in the natural soil (Tables 2, 3, 4, and 5).

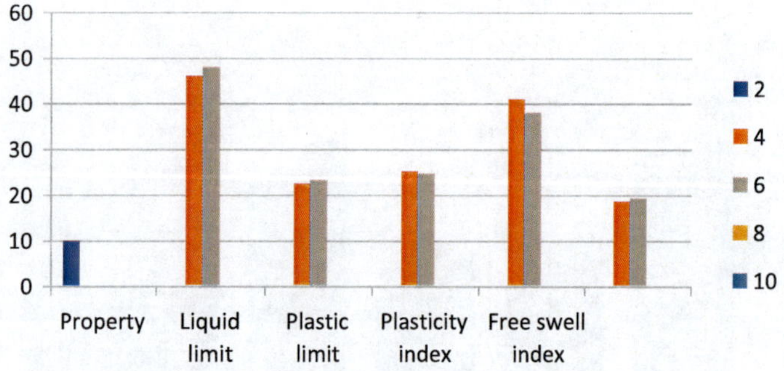

Fig. 3 Effect of salt content on the Atterberg limits

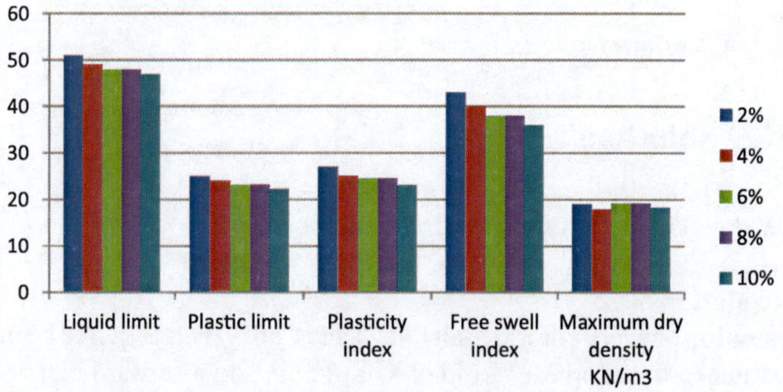

Fig. 4 Dry density–moisture content relation

Table 2 Index and engineering properties of untreated soil

S. No.	Property	For B.C soil
1	Liquid limit	62%
2	Plastic limit	29%
3	Plasticity index	48%
4	Free swell index	72%
5	Maximum dry density	17 (kN/m^3)

Table 3 Index and engineering properties of treated soil (NaCl)

S. No.	Property	2%	4%	6%	8%	10%
1	Liquid limit	56	56	46	46	44
2	Plastic limit	28	24	22.36	22.36	20.312
3	Plasticity index	26	28	25.12	25.12	23.12
4	Free swell index	48	46	41	41	39
5	Maximum dry density kN/m^3	16	17.2	18.59	18.59	17.24

Table 4 Index and engineering properties of treated soil ($CaCl_2$)

S. No.	Property	2%	4%	6%	8%	10%
1	Liquid limit	51	49	48	48	47
2	Plastic limit	25	24	23.22	23.22	22.3
3	Plasticity index	27	25	24.56	24.56	23.12
4	Free swell index	43	40	38	38	36
5	Maximum dry density kN/m^3	19	17.89	19.23	19.23	18.33

Table 5 Comparison between index and engineering properties of untreated soil and treated soil (NaCl and $CaCl_2$)

S. No.	Property	Untreated soil (%)	Treated soil (NaCl)	Treated soil ($CaCl_2$)
1	Liquid limit	62	46	48
2	Plastic limit	29	22.36	23.22
3	Plasticity index	48	25.12	24.56
4	Free swell index	72	41	38
5	Maximum dry density KN/m^3	17	18.59	19.23

5 Conclusion

Swelling characteristic is a main characteristic of black cotton soil, and it is observed that with the addition of salt, swelling pressure and swelling index have been reduced by a significant amount.

It is observed with the significant improvement in the test results with the addition of calcium chloride ($CaCl_2$) than sodium chloride (NaCl). The black cotton soil properties, i.e., liquid limit, plastic limit, and plasticity index, are substantially improved with the addition of sodium chloride and calcium chloride at an optimum dosage of 6%. Owing to properties of low cost and high abundance, NaCl has been recommended.

Reference

1. Bowles JE (1982) Engineering properties of soil, vol 2. McGraw-Hill, New York

Use of Waste Tyre Fibres as a Reinforcing Material for Road Subgrades: A Comparative Laboratory Analysis

Mohit K. Mistry, Shruti J. Shukla, and Chandresh H. Solanki

Abstract The utilization of waste products as a reinforcement material in different civil and geotechnical engineering projects is gaining momentum. In support of this rising approach, the present study proposes the use of waste tyre products for the improvement of expansive subgrade soil as a fibre reinforcement technique. The influence of various tyre products California bearing ratio (CBR) in soaked and unsoaked conditions was investigated. The test results indicated that the addition of waste tyre fibres increased the compaction characteristics, depending upon the fibre aspect ratio and dosage. The tyre fibres (both coated and uncoated) significantly enhanced the load penetration resistance capacity of expansive soil. The energy absorption capacity of tyre rubber plays an important role in influencing geotechnical properties. The cement coated fibres were found to be more beneficial as compared to the uncoated fibres. The improved properties of waste tyre fibre treated expansive soil can reduce the pavement thickness.

Keywords Expansive clay · Waste tyre fibres · Cement coating · Compaction test · California bearing ratio test

1 Introduction

Expansive clay soils are more prone to get swelled when it comes in contact with water. The excess water generates tremendous pore pressure in such soils, and hence, the existing structures on such expansive soil experiences high upward pressure in monsoon season. On the other side, such type of clayey soils get shrink when water gets evaporated or water gets expelled or groundwater table gets lowered in the summer season. Such fluctuation of the groundwater table causes severe damage to the lightweight structures including pavements and low rise buildings as well as multistorey buildings as well as retaining walls [1]. The significant land of the Gujarat state is covered with expansive soils [2]. The rapid growth of urbanization has increased the demand for road networks to connect the remote industry sites. The

M. K. Mistry (✉) · S. J. Shukla · C. H. Solanki
S V National Institute of Technology, Surat, Gujarat, India

T. G. Sitharam et al. (eds.), *Earthquake Geotechnics*, Lecture Notes in Civil Engineering 187, https://doi.org/10.1007/978-981-16-5669-9_17

obtainability of proper subgrade soil for the road network is always a major concern. The various ground improvement techniques like the use of geosynthetic products as continuous reinforcement, use of fly ash, lime, etc., as chemical admixtures are being used in practice. However, few of these techniques are uneconomical and ineffective after certain wetting–drying cycles. In the past few decades, the use of discrete reinforcement, in the form of various types of fibres (natural, synthetic, and waste) have become popular in research as well as practice [2–8]. In the present study, the waste tyre fibres have been used to improve the geotechnical property of expansive subgrade soil. The fibres possess the low specific gravity as compared to clayey soil, and hence, the resultant specific gravity of fibre-clay matrix reduces with the addition of fibres [9–11]. To overcome this, Mistry et al. [11] applied the cement coating on fibres and reported a significant improvement in specific gravity. The same concept has been used by the author for the present study. Both, cement coated and uncoated fibres were mixed in expansive soil and their CBR values were tested under the soaked and unsoaked conditions.

2 Experimental Materials and Methods

2.1 Materials

The soil used in the present study was sampled from the campus of S V National Institute of Technology, Surat and Gujarat. As per USCS classification system, the soil was found as high plasticity soil. The particle distribution and basic properties of soil are presented in Table 1.

To enhance the soil property, the waste tyre fibres with an average aspect ratio ($A_r = L/D$) as 4.5 were used as a reinforcing material. Where the L is the length of fibre and D is the diameter of the fibre. As the tyre fibres possess low specific gravity than

Table 1 Basic properties of soil

Properties/description	Values/category
Specific gravity of soil solids, G_s	2.56
% Sand (0.425–0.075 mm)	12
% Silt (0.075–0.002 mm)	58
% Clay (< 0.002 mm)	30
Liquid limit, LL (%)	65
Plasticity index, I_p (%)	34
Shrinkage limit, SL (%)	10
USCS classification	*CH*
Free swell index (%)	70
UCS value	114 kPa

Fig. 1 Waste tyre fibres: **a** uncoated; **b** cement coated

soil, it results in lower specific gravity of fibre-soil matrix. To enhance the specific gravity of fibre-soil matrix, fibres were coated with OPC 53 grade cement. Mistry et al. [11] in their study suggested a procedure to coat the fibres with cement. In the present study, the authors have followed the same procedure to coat tyre fibres. The cement slurry was prepared with 30% water content as the liquid limit of OPC 53 cement varies between 27 and 33, and hence, 30% water was used to prepare the cement slurry. Then, fibres were cured in that slurry only for 24 h to maintain the flexibility of fibres. The uncoated and cement coated fibres were mixed in four contents as 0.25, 0.50, 0.75 and 1.00% in high plasticity clay soil by dry mass of soil. Figure 1 represents the waste tyre fibres without cement coating and with cement coating.

2.2 Specimen Preparation and Experimental Programme

In the present study, modified compaction tests and California bearing ratio (CBR) tests have been carried out to investigate the influence of uncoated waste tyre fibres and cement coated waste tyre fibres on expansive subgrade soil. The CBR values depend highly upon the water (moisture) content and density of subgrade soil. Hence, each CBR mould was prepared using maximum dry unit weight (MDU) and optimum moisture content (OMC) obtained from modified compaction test. Each compaction test was performed as per IS 2720, part 8 [12]. To simulate the performance of both types of fibres in field conditions like normal rainfall and heavy rainfall, the representative laboratory unsoaked and soaked CBR tests were conducted. The CBR tests were performed by following IS 2720-part 16 [13]. The summary of CBR tests is illustrated in Table 2.

To simulate the field conditions, it is necessary to place the surcharge weights equivalent to the estimated weight of pavement on top of the specimen before testing [14]. The standard practice suggests that each 2.5 kg weight is equal to the weight of 6.5 mm construction height on the field. Accordingly, 5 kg weight was maintained as

Table 2 Summary of CBR tests

Test condition	Fibre type	Fibre content, f_c (%)
Unsoaked condition	Uncoated fibres	0%, 0.25%, 0.50%, 0.75% and 1.00%
	Cement coated fibres	0%, 0.25%, 0.50%, 0.75% and 1.00%
Soaked condition	Uncoated fibres	0%, 0.25%, 0.50%, 0.75% and 1.00%
	Cement coated fibres	0%, 0.25%, 0.50%, 0.75% and 1.00%

Note 0% represents the unreinforced soil

a surcharge load in each CBR test. For the design of flexible pavement, the CBR value matching with the 2.5 mm or 5 mm penetration can be used. Hazirbaba and Gullu [15] reported that in the case of fibre reinforcement, these penetration values may not be considered as true representative values of full mobilization of reinforcement. As the fibre requires higher distortion to reach its maximum tensile capacity. However, in the present study, the maximum penetration resistance capacity of the fibre-clay matrix was evaluated through the load vs penetration curve.

3 Results and Discussions

3.1 Compaction

A total 17 nos. of modified compaction tests were performed considering 1 test on untreated soil, 8 no. of tests on soil treated with uncoated fibres and 8 no. of tests on soil treated with cement coated fibres The values of MDU and OMC are presented in Table 3. The fluctuation in MDU is found with an increase in fibre content. However, in comparison with untreated expansive soil, the MDU and OMC are marginally increased with fibre content as well as with cement coating. Due to the high compaction energy, the tyre fibres get stretched and their rough surface

Table 3 Modified compaction test results for uncoated and cement coated waste tyre fibres

Fibre content, f_c (%)	Uncoated waste tyre fibres		Cement coated waste tyre fibres	
	MDU (kN/m^3)	OMC (%)	MDU (kN/m^3)	OMC (%)
0	15.80	18.00	15.80	18.00
0.25	16.79	20.50	17.10	18.20
0.50	16.39	23.50	15.84	22.20
0.75	16.30	18.50	16.80	19.00
1.00	13.64	29.00	15.90	20.10

interlocks the clay particles which do not allow particles to roll over each other. Hence, the MDU of fibre-clay matrix gets enhanced with the addition of uncoated fibres. On the other hand, the hydration product of cement binds the clay particles and fibres which results in the greater MDU of fibre-clay matrix with the addition of cement coated fibres. Past several studies [8, 11, 16, 17] reported that MDU of clayey soils gets reduced with tyre fibres when compacted at standard compaction energy.

From Table 3, it can be seen that in general the addition of fibres (uncoated as well as cement coated) has increased the OMC of expansive clay. The probable reason for the increased OMC is attributed to the rough surface and the larger size (> clay particles) which increases the voids in the fibre-clay matrix. From Table 3, it is also visible that with an increase in the percentage of fibre, the OMC of fibre-clay matrix gets fluctuated. This is probably due to the variation in the surficial gaps of fibre which led to the variation in void ratio. It is also observed that in comparison with uncoated fibres, the soil reinforced with cement coated fibres achieved the less OMC. It may be attributed to the applied cement coating on fibres which clogs the surficial gaps and restricts fibres to absorb water.

3.2 *California Bearing Ratio*

As showed in Table 2, a total 13 no of CBR tests were conducted under unsoaked and soaked conditions. The influence of uncoated and cement coated fibres on the CBR value of expansive subgrade soil in unsoaked and soaked conditions is illustrated in the following sections.

Effects of Uncoated and Cement Coated Fibres on *CBR* value of Cohesive Subgrade Soil in Unsoaked Condition

Figure 2 represents the penetration resistance capacity of expansive soil treated with uncoated waste tyre fibres and cement coated waste tyre fibres. It can be visible that with the addition of fibres, the penetration resistance capacity of expansive soil is enhanced significantly. The load penetration curve of virgin soil becomes flat after certain depth, whereas in case of fibre reinforced samples load penetration resistance capacity increases gradually with an increased depth which shows the ductile nature of the fibre-soil matrix. A similar trend of resistance was reported by Jha et al. [5].

In the present study, the slight fluctuation in penetration resistance has been observed with an increase in fibre percentage. The maximum penetration resistance was observed with 0.50 and 1.00% fibres in the case of uncoated fibres. Whereas in the case of cement coated fibres the maximum penetration resistance was reported only with 0.50% fibres. The influence of uncoated fibres and cement coated fibres on the unsoaked CBR value for 2.5 and 5 mm penetration depth is illustrated in Fig. 3. The increase in uncoated and cement coated fibres significantly enhances the CBR values for 2.5 mm and 5 mm penetration up to 0.75% fibre addition. Further with the 1% addition of fibres the CBR values get reduces. The probable reason for the

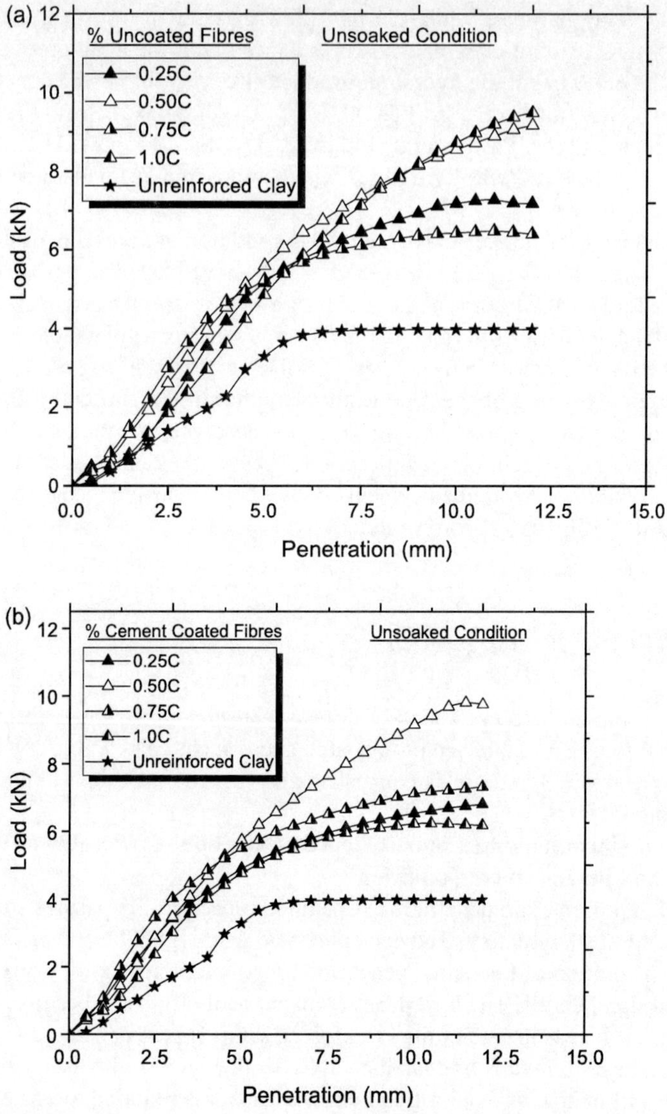

Fig. 2 Load versus Penetration curve of cohesive subgrade soil for a varying percentage of fibres in unsoaked condition: **a** uncoated fibres, **b** cement coated fibres

reduction might be attributed to the excessive interaction between fibre–fibre than that of fibre-soil. A similar observation was reported by Patel and Singh [18].

Effects of Uncoated and Cement Coated Fibres on *CBR* value of Cohesive Subgrade Soil in Soaked Condition

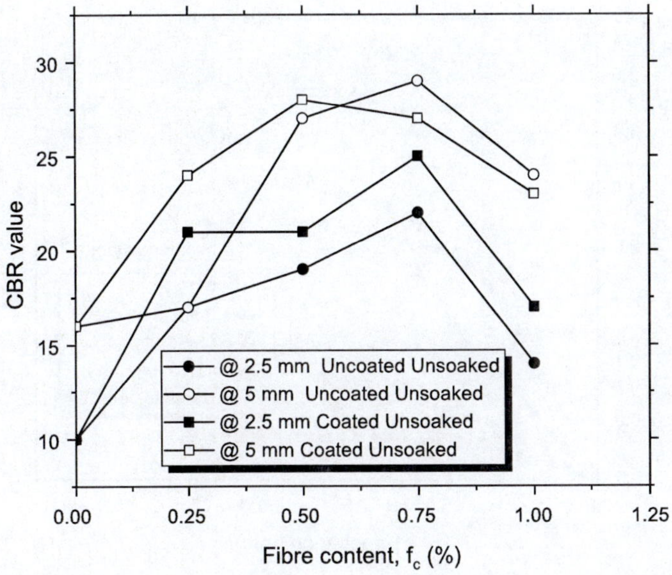

Fig. 3 Variation in CBR value of cohesive subgrade soil at 2.5 mm and 5 mm penetration depth for a varying percentage of fibres in unsoaked condition: **a** uncoated fibres, **b** cement coated fibres

Under the soaked condition, the load penetration resistance of expansive soil extensively increases with the addition of uncoated as well as cement coated fibres. It is visible from Figs. 2 and 4 that with the addition of fibres as compared to unsoaked condition, in soaked condition, the comparative improvement is more in expansive soil. In the present case, the load penetration curve of unreinforced soil achieves peak at 5 mm penetration depth with 0.25 kN load and then becomes flat. Whereas, with the addition of uncoated fibres, the peak resistance capacity increases up to 3.5 kN at 6.25 mm penetration depth (Fig. 4a).

It is observed in Fig. 4b that in the case of cement coated fibres, the penetration resistance capacity increases gradually till maximum recorded penetration depth (12.5 mm) and achieves maximum peak resistance load as 5.2 kN with the addition of 0.50% cement coated fibres. The influence of uncoated fibres and cement coated fibres on the soaked CBR value for 2.5 and 5 mm penetration depth is illustrated in Fig. 5. Similar to unsoaked condition, in soaked condition, the CBR values for 5 mm penetration were observed higher than the 2.5 mm penetration depth. It can be seen that with the addition of uncoated fibres the soaked CBR value increases from 2 to 13 for 0.50%, whereas, in the case of cement coated fibres, it increases from 2 to 14.

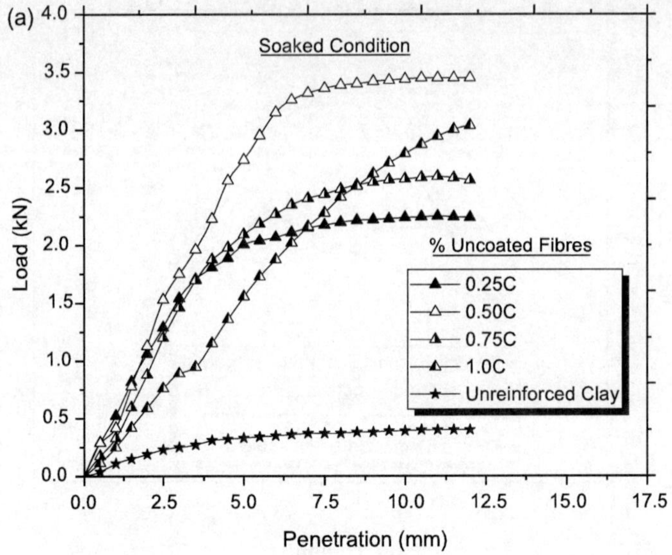

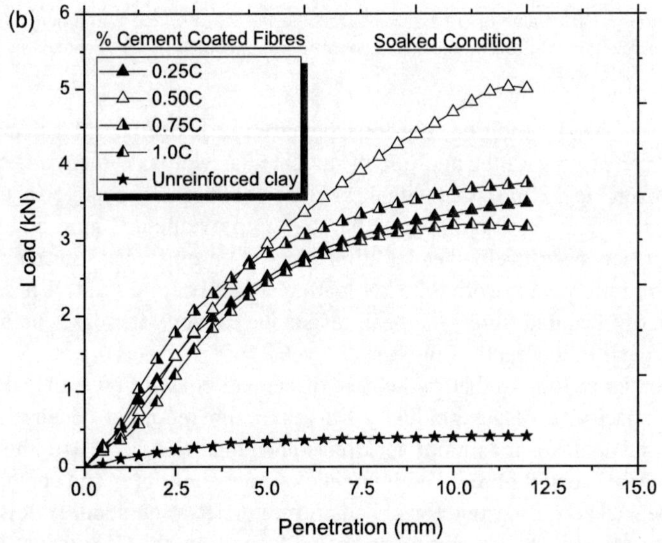

Fig. 4 Load versus Penetration curve of cohesive subgrade soil for a varying percentage of fibres in soaked condition: **a** uncoated fibres, **b** cement coated fibres

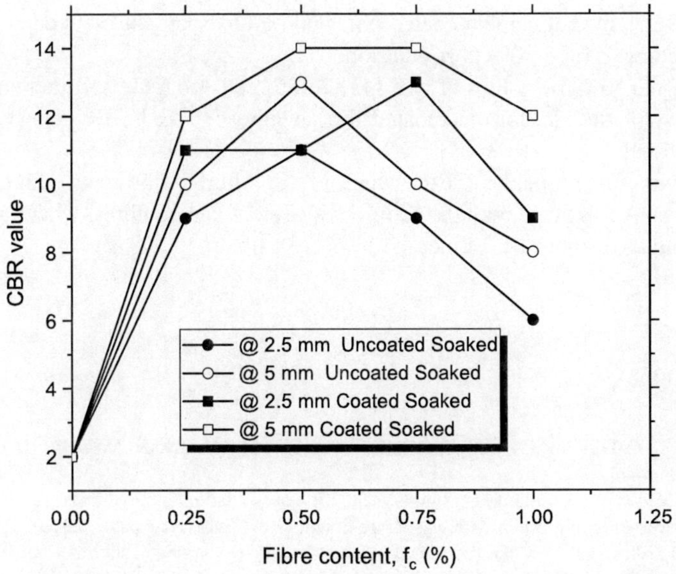

Fig. 5 Variation in CBR value of cohesive subgrade soil at 2.5 mm and 5 mm penetration depth for a varying percentage of fibres in soaked condition: **a** uncoated fibres, **b** cement coated fibres

4 Conclusions

The comparative laboratory investigation was conducted to check the influence of cement coated and uncoated waste tyre fibres on subgrade expansive soil. The following points were drawn:

- The inclusion of waste tyre fibres led to an increase in the service life of the pavement due to its ability to absorb the more energy and the better interfacial bonding between fibres, clay and hydration product of cement.
- The compaction behaviour of the cohesive subgrade soil is marginally affected by uncoated and cement coated waste tyre fibres. In comparison with uncoated fibres, the application of cement coating has marginally improved the MDU and OMC of cohesive subgrade soil. The slight fluctuation in MDU and OMC of subgrade soil has also been reported with the increase in fibre content.
- In general, as compared to unreinforced cohesive subgrade soil the CBR values enhance extensively with the addition of uncoated and cement coated fibres under both, unsoaked and soaked conditions.
- Under the unsoaked condition, CBR of cohesive subgrade soil reinforced with uncoated fibres increases gradually with penetration depth and achieves peak values when reinforced with 0.50 and 1% fibres, whereas under the soaked condition, it achieves peak values nearly at 6 mm penetration and then becomes constant.

- The CBR values of cement coated fibres under unsoaked and soaked conditions were achieved for 0.50% fibre addition.
- The unsoaked CBR values of tyre fibre reinforced subgrade soil increases by 3 times in with the addition of uncoated fibres, whereas soaked CBR values enhances by 6.5 times.
- Moreover, the unsoaked CBR values of tyre fibre reinforced subgrade soil increases by 2.8 times with the addition of cement coated fibres, whereas soaked CBR values enhance by 7 times.

References

1. Bowles JE (1979) Physical and geotechnical properties of soils. McGRAW-HILL International Book Company
2. Mistry M, Shukla S, Solanki C, Shukla S (2020) Consolidation and hydraulic conductivity of high plastic clay reinforced with waste tyre fibres. J Test Eval (Accepted, (2020)
3. Edil TB (2008) A review of environmental impacts and environmental applications of shredded scrap tires. Proc Int Work Scrap Tire Deriv Geomaterials - Oppor Challenges, IW-TDGM 2007. 3–18
4. Yasuhra K (2008) Recent Japanese experiences on scrapped tires for geotechnical applications. In: Yasuhara H (ed) Scrap tire derived geomaterials—opportunities and challeneges,. Taylor & Francis Group, London, pp 19–42
5. Jha JN, Choudhary AK, Gill KS, Shukla SK (2014) Behavior of plastic waste fiber-reinforced industrial wastes in pavement applications. Int J Geotech Eng. https://doi.org/10.1179/1939787914Y.0000000044
6. Cabalar AF, Karabash Z, Cabalar R, Karabash AF (2015) Journal of testing and evaluation california bearing ratio of a sub-base material modified with tire buffings and cement addition california bearing ratio of a sub-base material modified with tire buffings and cement addition California bearing ratio of a sub-base material modified with tire. J Test Eval 43:1–9. https://doi.org/10.1520/JTE20130070
7. Moghal AAB, Chittoori BCS, Basha BM (2018) Effect of fibre reinforcement on CBR behaviour of lime-blended expansive soils: reliability approach. Road Mater Pavement Des 19:690–709. https://doi.org/10.1080/14680629.2016.1272479
8. Mistry M, Venkateshwarlu P, Shukla S, Solanki C, Shukla SK (2020). Effect of placement of waste tyre fibres on unconfined compressive strength of clayey soil. https://doi.org/10.1007/978-981-13-7480-7_11
9. Al-Tabbaa A, Aravinthan T (1998) Natural clay-shredded tire mixtures as landfill barrier materials Al-Tabbaa A, Aravinthan T. Waste Management 18, 9–16 1998. 18, 1998
10. Subramaniam RM, Jeyapriya SP (2009) Study on effect of waste tyres in flexible pavement system. Indian Geotech Soc Chennai Chapter 20–24
11. Mistry M, Shukla T, Venkateswalu P, Shukla S, Solanki C, Shukla SK (2019). A new mixing technique for randomly distributed fibre-reinforced expansive soil. https://doi.org/10.1007/978-981-13-7010-6_15
12. IS 2720 Part 8 (1998) Determination of water content-dry density relation using heavy compaction)
13. IS 2720 Part 16 (1997) Laboratory determination of CBR
14. Mistry M, Vasoya A, Shukla S, Solanki C, Shukla SK (2020) Sustainable utilization of waste tyre products for pavement applications. In: Innovative civil engineering materials edited book volume as a part of special issue of journal for second ASCE India conference on 'CRSIDE 2020.' Cambridge Scholar Publishing

15. Hazirbaba K, Gullu H (2010) California bearing ratio improvement and freeze-thaw performance of fine-grained soils treated with geofiber and synthetic fluid. Cold Reg Sci Technol 63:50–60. https://doi.org/10.1016/j.coldregions.2010.05.006
16. Kalkan E (2013) Preparation of scrap tire rubber fiber-silica fume mixtures for modification of clayey soils. Appl Clay Sci. https://doi.org/10.1016/j.clay.2013.06.014
17. Priyadarshee A, Gupta D, Kumar V, Sharma V (2015) Comparative study on performance of tire crumbles with fly ash and Kaolin Clay. Int J Geosynth Gr Eng 1. https://doi.org/10.1007/s40891-015-0033-3
18. Patel SK, Singh B (2017) Experimental investigation on the behaviour of glass fibre-reinforced cohesive soil for application as pavement subgrade material. Int J Geosynth Gr Eng 3:0. https://doi.org/10.1007/s40891-017-0090-x

Liquefaction Mapping of Kolkata City Using FOSM Method

Srijani Sett and Kalyan Kumar Chattopadhyay

Abstract Determination of liquefaction potential is getting importance for prevention of the devastating effect of an earthquake. Several deterministic methods have been proposed for predicting liquefaction in the past but studies have shown that all these methods do not account for the different types of uncertainties associated with determination of liquefaction potential. Reliability analysis is a procedure which takes in account of several uncertainties and thereby produces results with reduced error. The city of Kolkata is actively seismic and is placed in seismic zones of III and IV. The soil is mostly alluvial with presence of sand layers along with soft silt and clay. Owing to its huge population, densely spaced buildings and increasing high-rise constructions, liquefaction potential mapping is essential to prevent future disasters. In this study, the liquefaction potential evaluation is conducted on 114 boreholes by the procedures of (Youd et al. in J Geotech Geoenviron 127:817–833, 2001) and Boulanger and Idriss (Boulanger, RW., Idriss, IM.: CPT and SPT-based liquifaction triggering procedures. Technical Report No. UCD/CGM-14/01, University of California, Davis, California, Centre for Geotechnical Modelling, 2014). An earthquake magnitude of 7 is considered and the analysis is done at depths of 7 m and 13 m. Based on these models, First Order Second Moment (FOSM) method is performed and the results are compared. It is concluded that reliability analysis conducted along with deterministic method gives more accurate results and shows that the city is most vulnerable to liquefaction at 7 m depth.

Keywords Reliability analysis · Liquefaction potential · FOSM method

S. Sett (✉) · K. K. Chattopadhyay
Department of Civil Engineering, Indian Institute of Engineering Science and Technology (IIEST) Shibpur, Howrah 711103, India

K. K. Chattopadhyay
e-mail: kkc@civil.iiests.ac.in

T. G. Sitharam et al. (eds.), *Earthquake Geotechnics*, Lecture Notes in Civil Engineering 187, https://doi.org/10.1007/978-981-16-5669-9_18

1 Introduction

During an earthquake, the seismic waves can generate cyclic loading conditions which may lead to a sudden increase in the pore water pressure of the soil. This makes the soil lose its shear strength completely and it starts behaving like a liquid. Any type of structures, like roadways, bridges, buildings, dams, etc., constructed on such liquefiable area are likely to get destroyed or at least damaged. An effective way to prevent such loss is determining the liquefaction potential of earthquake prone areas and thereby incorporating correct factor of safety in designs.

Seed and Idriss [1] developed a method, namely the "Simplified method", to determine the liquefaction potential of soil by evaluating the stresses induced in the soil due to an earthquake, termed as cyclic stress ratio (CSR). The relative density and triaxial test parameters were used to relate to the stresses causing liquefaction. This method afterwards was updated and improved by various researchers over the years [2–7] where SPT and CPT value of soil are used to evaluate the cyclic resistance ratio (CRR) of soil. A factor of safety, which is the ratio of CRR to CSR, determines the susceptibility of soil against liquefaction. Probabilistic SPT and CPT-based correlations for determination of liquefaction initiation in soil were proposed by Seed et al. [8] and Cetin et al. [9].

Deterministic models establish a limit for liquefaction susceptibility empirically, based on previous case histories and field observations. Empirical methods do not consider the uncertainties and variability in the load and resistance parameters in calculation of factor of safety [10]. Reliability analysis provides an opportunity to combine the various uncertainties in its calculation and expresses the liquefaction potential in terms of reliability index [11]. This reliability index can be used to evaluate probability of liquefaction which is a more consistent measure of liquefaction potential and is required in probabilistic-based design [12]. Probabilistic approach to evaluate liquefaction potential in terms of probability of liquefaction was proposed by various researchers using reliability methods of First Order Second Moment (FOSM), Advanced First Order Second Moment (AFOSM), First Order Reliability Method (FORM) where model and parameter uncertainties were considered [12–15]. Mapping functions between probability of liquefaction and its corresponding factor of safety values were established using Bayes theorem [10]. Jha et al. [11] proposed a combined method of FOSM and Point Estimate Method (PEM) which shows to be reliable for liquefaction potential analysis. Signar and Sil [16] evaluated the liquefaction potential of Guwahati city using deterministic methods and reliability analysis and used the results to construct liquefaction contour maps of the city.

The objective of this paper is to determine the liquefaction potential of Kolkata city by the deterministic methods proposed by Youd et al. [2] and Boulanger and Idriss [7]. An earthquake magnitude of 7 is used and the liquefaction potential is evaluated for depths of 7 m and 13 m. Based on this, reliability analysis is performed by FOSM method and liquefaction potential mapping is obtained for the region. The factor of safety and reliability index maps are compared and the most vulnerable regions of the study area are recognised.

2 Area of Study

The state of West Bengal lies in Eastern India surrounded by the Himalayas in the North and the Bay of Bengal in the South. The world's largest delta, Ganges Delta, covers the southern part of Bengal where the river Ganga empties into the Bay of Bengal. According to IS 1893-Part I: 2016, West Bengal has seismic zones of II to V with corresponding peak ground acceleration (PGA) of 0.1 g–0.36 g. Though it is not prone to earthquakes, in the last 100 years, earthquakes of magnitudes ranging from magnitude 5 to 8 have hit the state and its neighbouring regions (1857 Great Shillong Earthquake ($M = 8.1$), 1885 Bengal Earthquake ($M = 6.8$), 1934 Bihar-Nepal Earthquake ($M = 8.1$), 1950 Assam earthquake ($M = 8.7$), 1964 Sagar Island Earthquake ($M = 5.4$), 2011 Sikkim Earthquake ($M = 6.9$), 2015 Nepal Earthquake ($M = 7.8$)) [17]. This makes the region hazardous in nature.

The capital of West Bengal, Kolkata, is surrounded by the river Ganga on its western side. It has a huge commercial, industrial hub as well as high-rise residential complex and is well connected with the rest of the country by railway, roadway, air and water. As of the year 2020, the city has a population of nearly 14.9 million. The city is underlain with the major faults of Eocene Hinge Zone, Dhubri fault, Dauki fault, Garhmoyna-Khandaghosh fault, Jangipur-Gaibandha fault, Pingla fault, Debagram-Bogra fault, Rajmahal fault, Sylhet fault and Sainthia-Bahmani fault [17] shown in Fig. 1. The city falls in both the seismic zones of III and IV with a zone factor of 0.16 g and expected earthquake magnitude of 6 to 6.5 [18]. The nearest source of earthquake in 1964 and the farthest one in 1934 puts the maximum intensity in Kolkata city in the order of VII Modified Mercilli (MM) [17]. Therefore, an earthquake magnitude of 7 is chosen to study the liquefaction effects in Kolkata city.

The coordinates for the study area lies between 22° 33′ 15″ N to 22° 28′ 48″ N and 88° 19′ 06″ E to 88° 26′ 27″ E. A total of 26 sites distributed over Kolkata city are selected for this study. Four to five boreholes located in a common area are grouped together to form six different regions to assess the liquefaction susceptibility of the city in a better manner. Figure 2 shows the location of these sites and the six regions marked as Maidan, Tangra, Santoshpur, Tollygunge, Kalighat and Alipore. The soil profile of each borehole of the regions was studied and an average soil profile is presented in Table 1. The depths of each layer in a region are taken from the given borelogs and their corresponding descriptions are given. The upper and lower bound of the N values are selected from all the boreholes in a particular region to show the range of variability of each layer in a region.

3 Procedure

The liquefaction potential for all 114 boreholes in the study area is evaluated by both the procedures of Youd et al. [2] and Boulanger and Idriss [7] for an earthquake of magnitude 7. The reliability analysis using FOSM method is performed on the

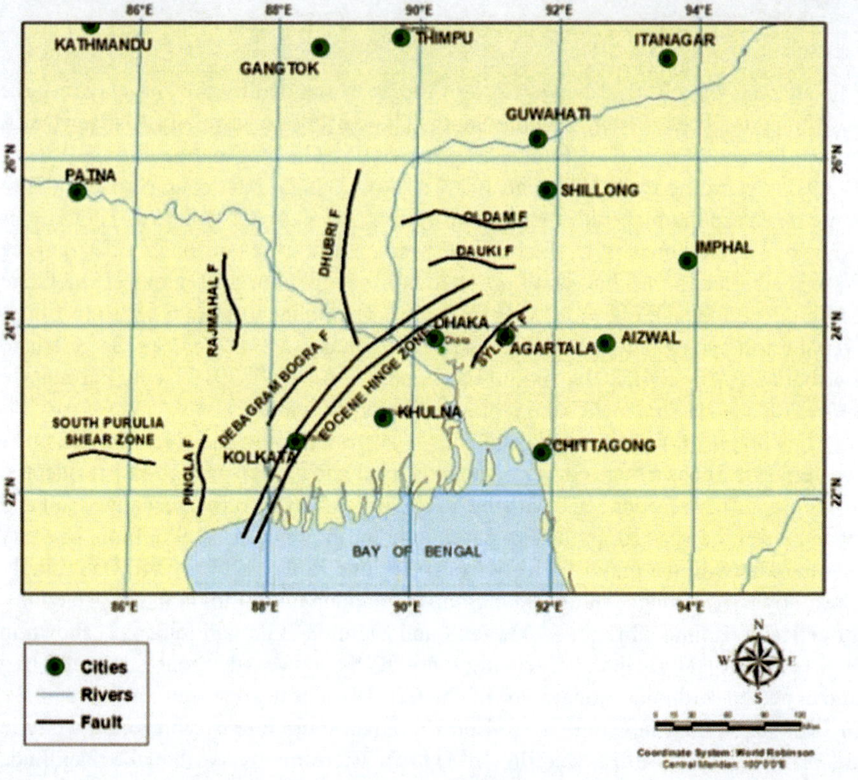

Fig. 1 Map of Eastern India and its surrounding countries showing the major faults planes

deterministic model of Boulanger and Idriss [7] and liquefaction contour maps are constructed for the study area.

3.1 Deterministic Approach

This approach utilizes a factor of safety to determine the liquefaction susceptibility of a soil. The factor of safety is a ratio of the resistance of the soil against liquefaction known as cyclic resistance ratio (CRR) to the forces experienced by the soil due to liquefaction known as cyclic stress ratio (CSR). When factor of safety (FS) is less than 1, the soil is termed as unsafe against liquefaction, FS is greater than 1, it is considered safe against liquefaction and FS is equal to 1, it is in limit state.

Cyclic Stress Ratio. Seed et al. [1] first drafted the equation to estimate CSR by using soil parameters like total vertical stress (σ_v), effective overburden stress $\left(\sigma_v'\right)$ along with earthquake parameters like maximum horizontal ground acceleration (a_{max}) and earthquake magnitude (M). The equation is given as

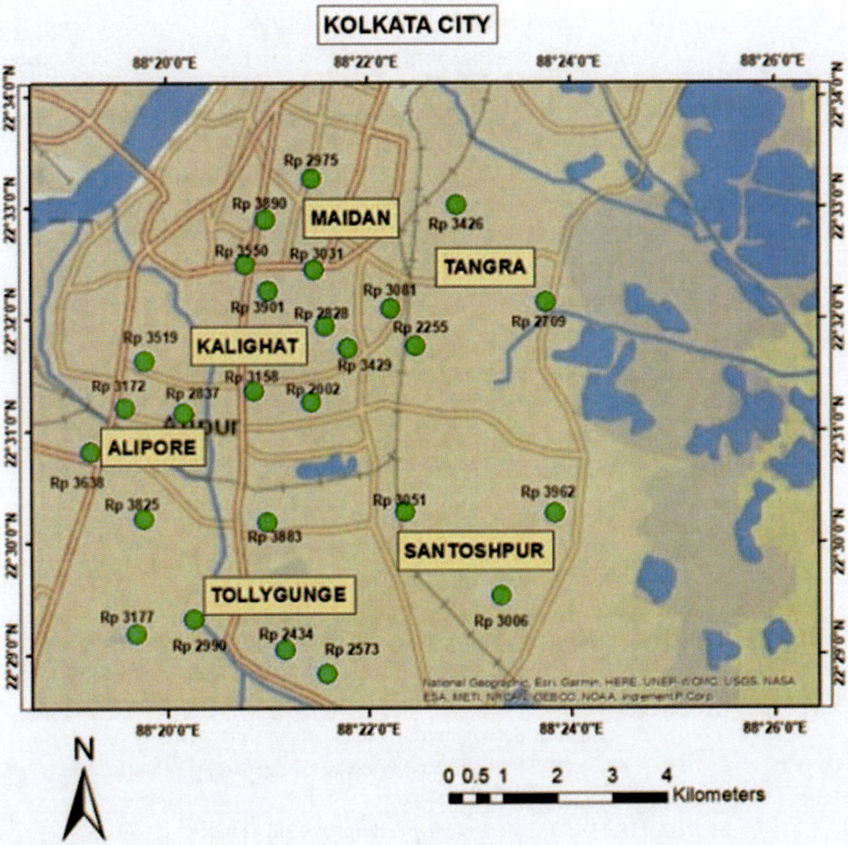

Fig. 2 Location of 26 sites (indicated by green dots) selected for this study in Kolkata city

$$\text{CSR} = 0.65 \times \frac{\tau_{\max}}{\sigma_v'} = 0.65 \times \frac{a_{\max}}{g} \times \frac{\sigma_v}{\sigma_v'} \times r_d \tag{1}$$

where $g =$ acceleration due to gravity, $r_d =$ shear stress reduction factor taking into account the dynamic response of soil profile.

Cyclic Resistance Ratio. It is determined by considering the various in situ soil parameters like CPT resistance, SPT blow count or Shear wave velocity V_s. In this study, the SPT of soil is considered to determine the value of CRR. Table 2 gives the equations proposed by Youd et al. [2] and Boulanger and Idriss [7] for the factors used to calculate values of CSR and CRR. The CRR equation is further modified for different earthquake magnitude and overburden pressure by using two correction factors, (i) Magnitude scaling factor (MSF), which accounts for the earthquake magnitude M under consideration and (ii) Overburden correction factor (K_σ), to account for overburden stresses at depth of interest. The corrected CRR value is expressed as

Table 1 Average soil profile of the different regions in the study area

Region	Stratum	Depth (m)	Characteristics of soil	Range of N value
Maidan	I	0–14.0	Soft to medium, brownish silty clay with fine sand and organic matter	4–16
	II	14.0–22.0	Medium dense, deep grey silty sand with clay binders	4–11
	III	22.0–35.0	Very stiff, yellowish grey clayey silt with presence of mica	30–39
Tangra	I	0.0–10.0	Very soft to soft, brownish grey silty clay with decomposed wood and rusty spots	2–8
	II	10.0–22.0	Soft to medium, deep grey clayey silt with sand mixture and kankars and presence of sand pockets	5–11
	III	22.0–35.0	Dense to very dense, yellowish grey, silty fine sand with mica	22–33
Santoshpur	I	0–10.0	Soft, brownish grey clayey silt with organic matter	3–9
	II	10.0–17.0	Very stiff, deep grey silty clay with presence of fine sand mixture	14–16
	III	17.0–30.0	Dense, yellowish grey silty sand with traces of mica	22–26
Tollygunge	I	0.0–6.0	Soft, brownish grey silty clay with decomposed wood and organic matter	3–12
	II	6.0–18.0	Loose to moderately dense, steel grey silty fine sand	15–21
	III	18.0–35.0	Medium dense, deep grey silty sand with mica	21–30
Kalighat	I	0–6.0	Soft to medium, brownish grey clayey silt with decomposed wood	4–5
	II	6.0–14.0	Soft, deep grey silty clay with presence of decomposed wood and organic matter	3–13
	III	14.0–25.0	Very stiff, deep grey clayey silt with sand mixture and mica	10–17
	IV	25.0–35.0	Hard, dense yellowish grey silty sand with mica	24–28
Alipore	I	0.0–10.0	Soft to medium, deep brownish grey clayey silt with presence of decomposed wood	4–7
	II	10.0–23.0	Soft to medium, brownish grey silty clay with sand mixture and organic matter	11–30
	III	23.0–35.0	Dense, steel grey fine sand with mica	19–28

Table 2 Comparison of equation for factors for the method of Youd et al. [2] and Boulanger and Idriss [7]

Factors	Youd et al. [2]	Boulanger and Idriss [7]
r_d	$r_d = 1.0 - 0.00765z$ $(z \leq 9.15)$ $r_d = 1.174 - 0.0267z$ $(9.15 \leq z \leq 23)$ Where z = Depth of soil strata (in metres)	$r_d = \exp[\alpha(z) + \beta(z)]$ $\alpha(z) = 1.0121 - 1.125\sin(\frac{z}{11.23} + 5.133)M$ $\beta(z) = 0.106 + 0.118\sin(\frac{z}{11.28} + 5.142)$
$CRR_{7.5}$	$CRR_{7.5,1} = \frac{1}{34-(N_1)_{60CS}} + \frac{(N_1)_{60CS}}{135} + \frac{50}{(45+(10\times(N_1)_{60CS})^2} - \frac{1}{200}$	$CRR_{7.5,1} = \exp\{\frac{(N_1)_{60CS}}{14.1} + \left(\frac{(N_1)_{60CS}}{126}\right)^2 - \left(\frac{(N_1)_{60CS}}{23.6}\right)^3 + \left(\frac{(N_1)_{60CS}}{25.4}\right)^4 - 2.8\}$
$(N_1)_{60CS}$	$(N_1)_{60CS} = \alpha + \beta\ (N_1)_{60}$ Where $\alpha = 0 \text{for} FC \leq 5\%$ $\alpha = \exp\left[1.76 - \left({}^{190}/_{FC^2}\right)\right]$ for $5\% < FC < 35\%$ $\alpha = 5.0$ for FC $\geq 35\%$ $\beta = 1.0$ for FC $\leq 5\%$ $\beta = \left[0.99 + \left(FC^{1.5}/1000\right)\right]$ for $5\% <$ FC $<$ 35% $\beta = 1.2$ for FC $\geq 35\%$	$(N_1)_{60CS} = (N_1)_{60} + \Delta(N_1)_{60}$ $\Delta(N_1)_{60} = \exp\{1..63 + \frac{9.7}{FC+0.01} - \left(\frac{15.7}{FC+0.01}\right)^2\}$ $(N_1)_{60}$= Field SPT N value after corrections where FC = Fines Content
MSF	$MSF = \frac{10^{2.24}}{Mw^{2.56}}$ where M_w = Earthquake magnitude	MSF= 1 + (MSF_{max}-1) {8.64 exp $(\frac{-M}{4}) - 1.325)\}$ MSF_{max}= 1.8 for sand MSF_{max}= 1.09 for clay & plastic silt
K_σ	$\left(K_\sigma = (\sigma'_{vo}/P_a)^{(f-1)}\right)$ where P_a = atmospheric pressure	$K_\sigma = 1 - C_\sigma \ln\left(\frac{\sigma'_{vo}}{P_a}\right) \leq 1.1$ $C_\sigma = \frac{1}{18.9-2.55\sqrt{(N_1)_{60CS}}} \leq 0.3$
C_N	$C_N = \sqrt{\frac{P_a}{\sigma'_{vo}}} \leq 1.7$	$C_N = \left(\frac{P_a}{\sigma'_{vo}}\right)^a \leq 1.7$ $a = 0.784 - 0.0768\sqrt{(N_1)_{60CS}}$

$$CRR_{M,K_\sigma} = CRR_{7.5,1} \times MSF \times K_\sigma \tag{2}$$

3.2 Reliability Analysis

Reliability analysis is defined as a process that implies limit state probabilities of a system under adverse loading conditions. In this study, two deterministic models are considered (Youd et al. [2] and Boulanger and Idriss [7]) and compared. The performance function is the factor of safety (FS) determined by the deterministic models. The limit state is taken at FS = 0 and its function is designated by Z, which takes the form $Z = R - S$, where R is the resisting load (CRR) and S is the load imposed on the soil (CSR). When $Z > 0$, liquefaction doesn't occur, $Z < 0$, liquefaction occurs and $Z = 0$ is considered as a limit state condition. The method used in this study to define the mean and standard deviations of the limit state is First Order Second Moment (FOSM) method.

First Order Second Moment (FOSM) Method. This is a commonly used method to determine the reliability index. This method requires only the linear (first order) terms of Taylor series and the first two moments (mean and standard deviation). The limit state function (Z) is formed on the basis of the random input variables CRR and CSR.

The mean and standard deviation of R and S are taken as μ_R, μ_S and σ_R, σ_S. The mean (μ_Z), standard deviation (σ_Z) and coefficient of variation (δ_z) of Z expressed as

$$\mu_z = \mu_R - \mu_S \tag{3}$$

$$\sigma_Z = \sqrt{\sigma_R^2 + \sigma_S^2} \tag{4}$$

$$\delta_z = \frac{\sigma_z}{\mu_z} \tag{5}$$

The mean (μ_{CSR}) and coefficient of variation (δ_{CSR}) of CSR is determined as

$$\mu_{\text{CSR}} = 0.65\left(\frac{\mu_{\max}}{g}\right)\left(\frac{\mu_{\sigma_v}}{\mu_{\sigma'_v}}\right)\left(\frac{\mu_{r_d}}{\mu_{\text{MSF}}}\right)\left(\frac{1}{\mu_{K_\sigma}}\right) \tag{6}$$

$$\delta_{\text{CSR}}^2 = \delta_{a\max}^2 + \delta_{\sigma_v}^2 + \delta_{\sigma'_v}^2 + \delta_{r_d}^2 + \delta_{\text{MSF}}^2 + \delta_{K_\sigma}^2 - 2\rho_{\sigma_v \sigma'_v}\sigma_v \sigma'_v \tag{7}$$

where $\rho_{\sigma_v \sigma'_v}$ = correlation coefficient between σ_v and σ'_v. The coefficient of variation of CRR is determined as

$$\delta_{\text{CRR}} = \frac{\Delta\text{CRR}}{2\mu_{\text{CRR}}} \tag{8}$$

Reliability index is a ratio of the mean and standard deviation of the limit state function; it helps to compare the reliability of different systems without calculating the absolute probability values, and represented as

$$\beta = \frac{\mu_Z}{\sigma_Z} = \frac{\mu_R - \mu_S}{\sqrt{\sigma_R^2 + \sigma_S^2}} \quad (9)$$

4 Results and Discussion

A comparative study on liquefaction susceptibility has been conducted for Kolkata city based on the deterministic methods and the reliability analysis. The soil properties used in this study are the bulk density, SPT value, depth of groundwater table and fines content varying along the depth of soil. An earthquake magnitude (M) = 7 and its corresponding peak ground acceleration ($a_{\max}$) of 0.17 g [19] have been used.

A total 114 number of boreholes, distributed over the study area, have been considered in the present study. The average depth of the boreholes range from 30 to 40 m and ground water table is observed at a depth 1–3 m below the ground surface in the study area. A typical borehole, located at 35, Panditya Road, Kolkata, highlighting the detailed soil properties along the depth has been shown in Table 3.

Table 3 Typical borelog located at 35, Panditya Road, Kolkata showing detailed soil properties

Depth (m)	Soil description	Bulk density (kN/m^3)	SPT N value	Fines content (%)
3.68	Brownish grey, clayey silt with rusty spots	18.64	5	95
5.23		18.64	6	95
7.18	Brownish grey, silty clay with organic matter and decomposed wood	14.62	4	86
8.73		14.62	3	86
10.68		14.32	4	92
12.23		14.32	3	92
14.18	Deep grey silty clay	18.05	8	91
17.68	Bluish grey silty clay	19.33	11	94
19.23		19.33	17	94
20.73		19.33	21	94
22.23		19.33	9	94
23.73		19.33	29	94
25.68	Brownish grey, clayey silt with fine sand and rusty spots	19.13	34	91
27.23		19.13	25	91
28.73		19.13	23	91
30.68	Deep grey, silty clay with rusty spots	18.44	21	92

The factor of safety and reliability index values are plotted against depth for two typical boreholes located at 35, Panditya Road, Kolkata (Fig. 3a and b) and Kolkata Port area, Taratola (Fig. 4a and b) and scatter plots are constructed for the collective 114 boreholes (Figs. 5 and 6) for $M = 7$ by the procedures of Youd et al. [2] and Boulanger and Idriss [7]. The factor of safety plots are divided into three regions, for the range of factor of safety 0.0 to 0.5 is considered as critical, 0.5 to 1.0 is moderately critical and >1.0 is considered as safe. Similarly, in the reliability index vs depth plots, the regions are marked as for reliability index <−2 is critical, −2 to 1 is moderately critical and > 1 is considered safe. The regions are indicated by dot and dash lines. Figures 3 and 4 indicate the top 15–20 m of the soil to be most susceptible to liquefaction beyond which the curves enter into the safe region. From Fig. 5, it is observed that maximum points fall in the moderately critical zone and

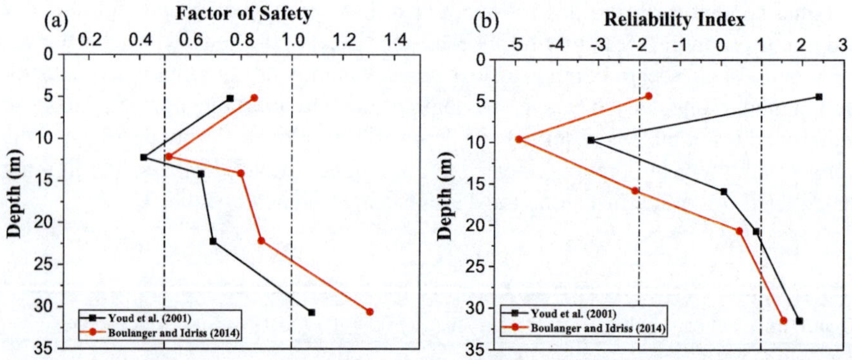

Fig. 3 The **a** Factor of safety versus Depth and **b** Reliability index versus depth shown for a typical borehole at 35, Panditya road, Kolkata, according to Youd et al. [2] and Boulanger and Idriss [7] for M = 7

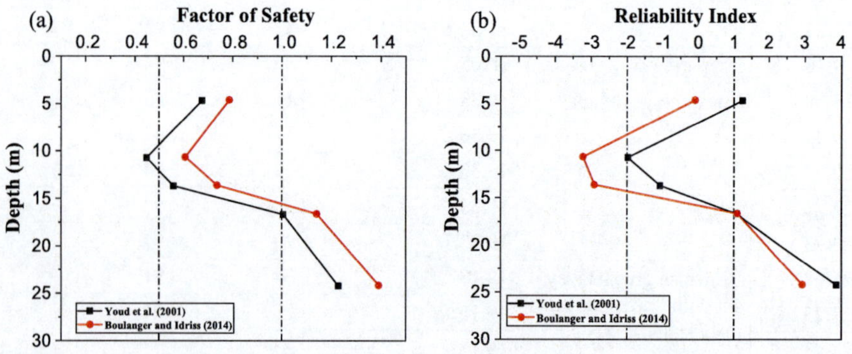

Fig. 4 The **a** Factor of safety versus depth and **b** Reliability index versus depth shown for a typical borehole at Kolkata Port area, Taratola, according to Youd et al. [2] and Boulanger and Idriss [7] for $M = 7$

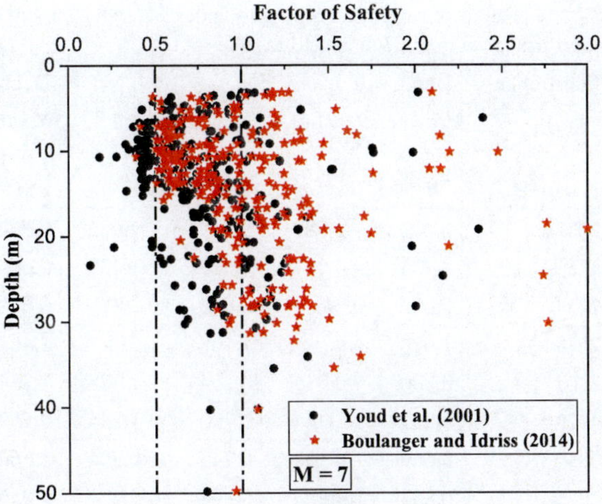

Fig. 5 Factor of safety versus depth according to Youd et al. [2] and Boulanger and Idriss [7]

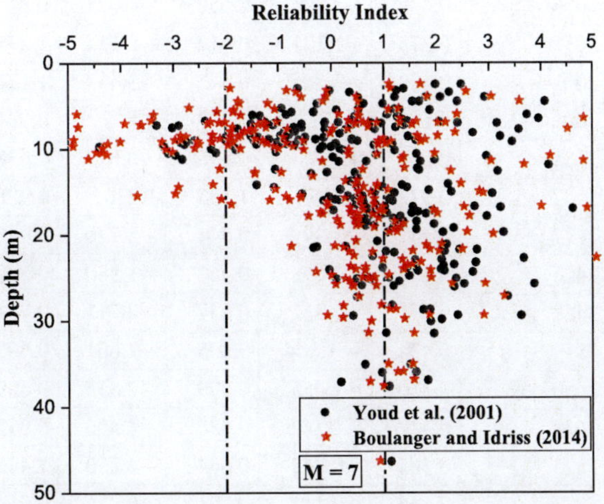

Fig. 6 Reliability index versus depth according to Youd et al. [2] and Boulanger and Idriss [7]

safe zone within a depth of 20 m from ground surface. However, a few points fall within the critical zone. The reliability index vs depth plot (Fig. 6) shows a handful of points in the critical region up to 10 m depth, though majority of the points fall in the moderately critical and safe zone till 30 m of depth. Therefore, the first and second layers of the soil profile are chosen for this study whose average depths are calculated at 7 m and 13 m, respectively. Factor of safety and reliability index have been calculated by the methods of Youd et al. [2] and Boulanger and Idriss [7] at these average depths for all the boreholes for an earthquake magnitude 7 in the study area and values of the most vulnerable boreholes of each site is displayed in Table 4.

Contour maps for factor of safety, prepared by the methods of Youd et al. [2] and Boulanger and Idriss [7], for depths 7 m and 13 m are presented in Figs. 7, 8, 9

Table 4 Factor of safety and reliability index value for methods of Youd et al. [2] and Boulanger and Idriss [7] at depth 7 m and 13 m

Location no	Depth = 7 m				Depth = 13 m			
	Youd et al. [2]		Boulanger and Idriss [7]		Youd et al. [2]		Boulanger and Idriss [7]	
	FS	β	FS	β	FS	β	FS	β
2002	0.402	−2.589	0.549	−2.982	0.714	3.202	1.007	2.836
2255	0.425	−0.287	0.529	−1.439	0.669	2.940	1.074	0.925
2434	0.602	0.741	0.928	0.770	0.658	2.192	1.223	1.921
2573	1.028	2.061	1.168	1.614	1.462	0.827	2.579	0.802
2709	0.330	−0.836	0.504	−1.228	1.234	2.126	1.883	5.113
2828	0.557	– 0.996	0.790	−1.245	0.559	1.273	0.948	0.395
2837	0.569	−0.138	0.745	−0.904	1.660	2.047	6.152	1.669
2975	0.802	1.460	0.864	0.318	1.056	0.501	1.496	1.959
2990	0.504	−0.158	0.706	−1.088	0.761	2.598	1.152	1.675
3006	0.462	– 2.853	0.656	−4.027	0.817	1.443	1.019	0.860
3031	0.375	−1.171	0.412	−2.125	1.213	2.129	1.846	2.264
3051	0.429	−2.386	0.547	−3.430	1.120	1.355	1.581	3.795
3081	0.655	0.451	0.977	0.515	0.934	3.150	1.738	4.069
3158	0.548	0.354	0.696	−0.672	0.638	2.230	1.043	1.034
3172	0.972	1.216	1.333	1.276	1.988	0.036	2.472	1.374
3177	0.447	−1.991	0.544	−2.770	0.943	3.017	1.353	2.440
3426	0.458	−0.698	0.562	−1.880	0.589	1.801	0.958	0.245
3429	0.575	−0.932	0.637	−2.711	0.741	1.818	1.024	1.135
3519	0.422	– 1.804	0.605	−2.601	0.603	2.105	1.042	2.234
3550	0.350	−1.455	0.499	−2.855	0.609	1.954	0.976	0.877
3638	0.487	– 3.095	0.630	−4.486	0.652	−2.946	1.019	0.614
3825	0.585	−0.842	0.644	−4.850	1.430	1.315	2.124	1.307
3883	0.646	−0.111	0.791	−0.836	0.814	1.893	1.549	2.067
3890	0.993	1.386	1.173	1.122	1.759	3.571	2.194	4.513
3901	0.472	−3.016	0.610	−4.824	0.791	3.712	2.544	1.160
3962	0.531	0.956	0.802	0.862	–	–	–	–

and 10. By comparing the contour maps, it is observed that Youd et al. [2] method is conservative than Boulanger and Idriss [7] method for predicting liquefaction susceptibility. This can be attributed to the fact that certain factors stress reduction factor (r_d), Magnitude Scaling Factor (MSF), Overburden correction factor (K_σ), corrected SPT N value ($(N_1)_{60CS}$), Correction for Overburden Pressure (C_N) are modified from the Youd et al. [2] procedure and the corrections are updated in the Boulanger and Idriss [7] procedure. Reliability index contour maps were also

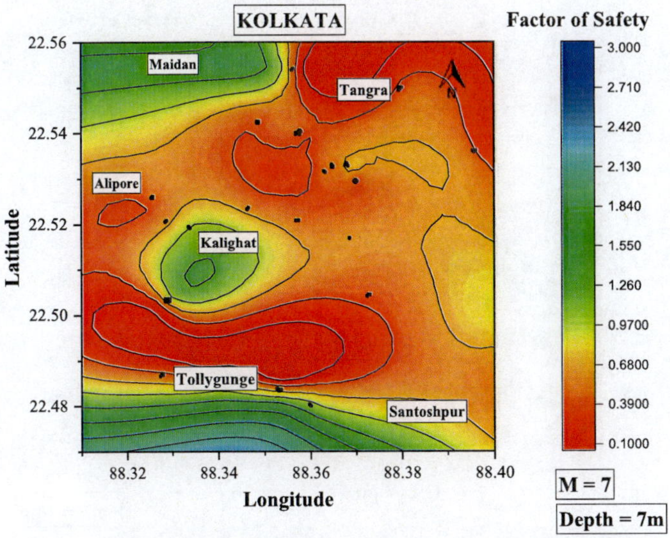

Fig. 7 Factor of safety contour map by Youd et al. [2] method at 7 m depth

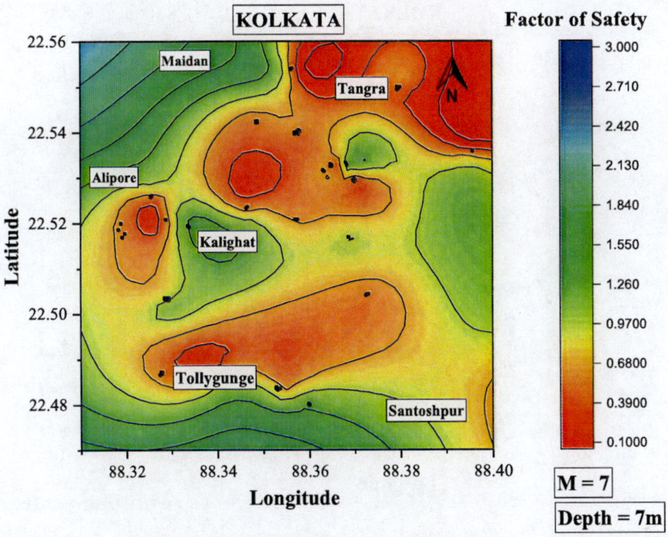

Fig. 8 Factor of safety contour map by Boulanger and Idriss [7] method at 7 m depth

prepared using the Boulanger and Idriss [7] method for the depth 7 and 13 m and presented in Figs. 11 and 12, respectively.

From the results, it can be observed that the factor of safety and reliability index values depend on a number of factors like the soil properties, presence of ground water table, earthquake magnitude, depth of soil, seismicity of the site, etc. Kolkata

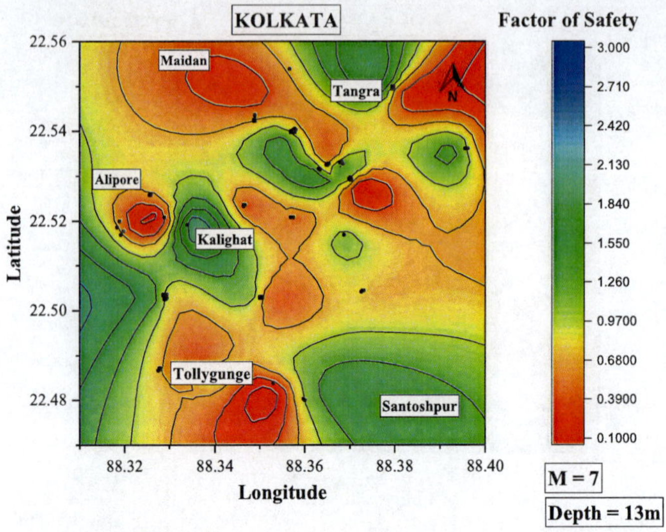

Fig. 9 Factor of safety contour map by Youd et al. [2] method at 13 m depth

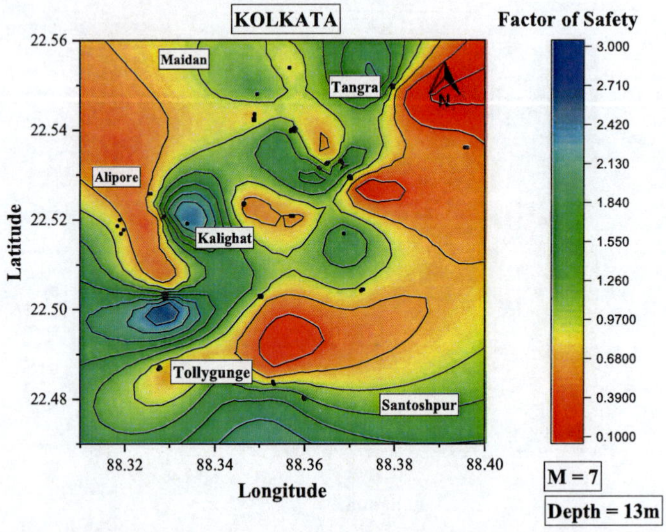

Fig. 10 Factor of safety contour map by Boulanger and Idriss [7] method at 13 m depth

is located on the banks of river Ganga and placed in the border of seismic zones III and IV is very much prone to earthquakes in the range of 6 to 7. The subsurface deposit of the city consists of silt and sand mixed with clay up to 20 m that is found highly susceptible to liquefaction, underlain by non-liquefiable dense sand and stiff clay.

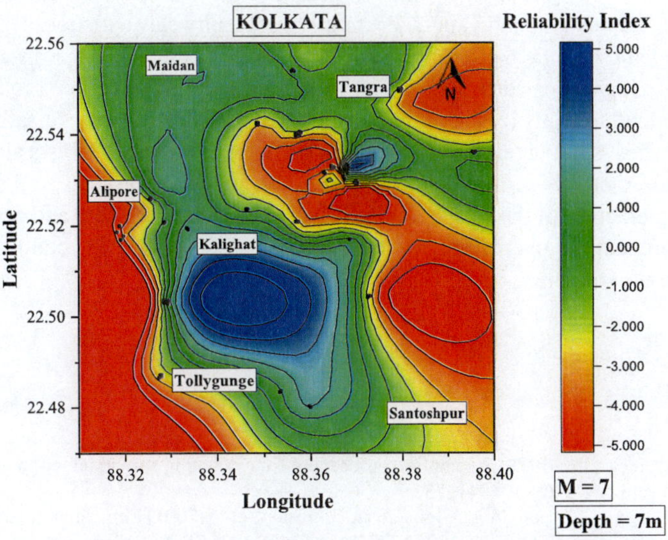

Fig. 11 Reliability index contour map by Boulanger and Idriss [7] method at 7 m depth

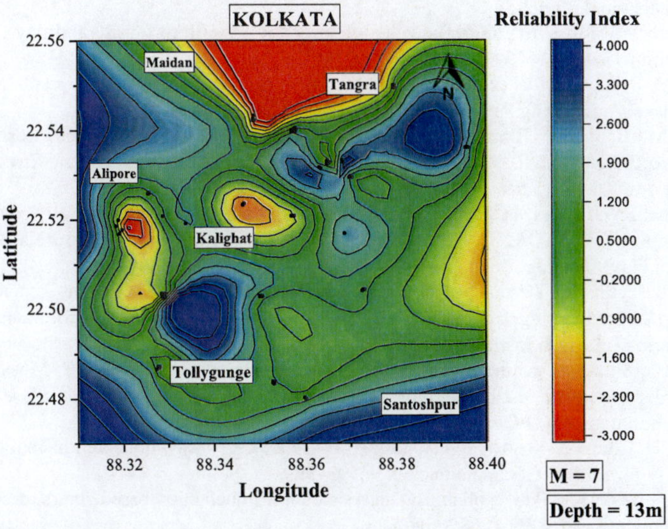

Fig. 12 Reliability index contour map by Boulanger and Idriss [7] method at 13 m depth

5 Conclusion

In the present study, the factor of safety and reliability index contour maps show Kolkata city to be more prone to liquefaction at depth 7 m in the regions of Alipore,

Tollygunge, Santoshpur and Tangra. Presence of soft, silty clay with traces of decomposed wood and organic matter in the top 10 m stratum and high ground water table level of these regions can attribute to their susceptibility to liquefaction. Liquefaction potential calculated only by deterministic methods gives lower factor of safety value as uncertainties is not taken into account. By utilizing reliability analysis along with deterministic methods in microzonation and seismic hazard projects, the susceptibility to liquefaction of the region can be determined more precisely and accurately. This will prevent the overestimation of factor of safety leading to economical and safe design of structures.

References

1. Seed HB, Idriss IM (1971) A simplified procedure for evaluating soil liquefaction potential. J Geotech Eng Div 97(9):1249–1273
2. Youd TL, Idriss IM, Ronald AD, Ignacio A, Gonzalo C et al (2001) Liquefaction resistance of soils: Summary report from the 1996 NCEER and 1998 NCEER/NSF workshops on evaluation of liquefaction resistance of soils. J Geotech Geoenviron 127(10):817–833
3. Boulanger RW, Idriss IM (2004) Evaluating the liquefaction or cyclic failure of silts and clays. Technical Report No. UCD/CGM-04/01, University of California, Davis, California, Center for Geotechnical Modeling
4. Boulanger RW, Idriss IM (2006) Liquefaction susceptibility criteria for silts and clays. J Geotech Geoenviron 132(11):1413–1424
5. Idriss IM, Boulanger RW (2008) Soil liquefaction during earthquakes. Monograph (MNO)-12, Earthquake Engineering Research Institute, Oakland, California
6. Idriss IM, Boulanger RW (2010) SPT-based liquefaction triggering procedures, technical report No.UCD/CGM-10/02, University of California, Department of Civil and Environmental Engineering, Davis, California (2010)
7. Boulanger RW, Idriss IM (2014) CPT and SPT based liquifaction triggering procedures. Technical Report No. UCD/CGM-14/01, University of California, Davis, California, Center for Geotechnical Modeling
8. Seed RB, Cetin KO, Moss RS, Kammerer AM, Wu JM et al (2003) Recent advances in soil liquefaction engineering: a unified and consistent framework. Berkeley, California: Earthquake Engineering Research Institute
9. Cetin KO, Seed RB, Kiureghain AD, Tokimatsu K, Harder LF, Kayen RE, Moss R (2004) Standard penetration test-based probabilistic and deterministic assesment of seismic soil liquefaction potential. J Geotech Geoenviron 130(12):14–1340
10. Juang CH, Chen CJ, Rosowsky DV, Tang WH (2000) CPT-based liquefaction analysis Part 2: reliability for design. Geotechnique 50(5):593–599
11. Jha SK, Suzuki K (2008) Reliability analysis of soil liquefaction based standard penetration test. Comput Geotech 36(4):589–596
12. Juang CH, Fang SY, Khor EH (2006) First order reliability method for probabilistic liquefaction triggering analysis using CPT. J Geotech Geoenviron 132(3):337–350
13. Low BK, Tang WH (1997) Efficient reliability evaluation using spreadsheets. J Eng Mech 123(7)
14. Juang CH, Rosowsky DV, Tang WH (1999) Reliability-based method for assessing liquefaction potential of soils. J Geotech Geoenviron 125(8):684–689
15. Hwang JH, Yang CW, Juang DS (2004) A practical reliability-based method for assessing soil liquefaction potential. Soil Dyn Earthq Eng 24(9):761–770
16. Signar L, Sil A (2018) Liquefaction potential assessment of Guwahati city using first-order second-moment method. Innov Infrastruct Solut 3:36

17. Nath SK, Adhikari MD, Maiti SK, Devaraj N, Srivastava N et al (2014) Earthquake scenario in West Bengal with emphasis on seismic hazard microzonation of the city of Kolkata India. Nat Hazards Earth Syst Sci 14:2549–2575
18. Govindaraju L, Bhattacharya S (2012) Site-specific earthquake response study for hazard assessment in Kolkata city India. Nat Hazards 61:943–965
19. Chakraborty P, Pandey AD, Mukherjee S, Bhargava A (2004) Liquefaction assessment for microzonation of Kolkata city. In: 13th world conference on earthquake engineering: conference proceedings, vol August 1–6, Paper no. 42, 13 WCEE Secretariat, Vancouver, BC., Canada
20. Juang CH, Chen CJ, Rosowsky DV, Tang WH (2000) CPT-based liquefaction analysis Part 1: determination of limit state function. Geotechnique 50(5):583–592

To Study the Impact on Bearing Ratio of Fine-Grained Subgrade Using Waste EPS

Shiladitya Mandal and Debangsu Mistry

Abstract With an objective to enhance bearing strength of soil, numerous studies on soil stabilization have been done over the past several decades through introducing new additive materials and new methodologies for different types of soil. Being a material of remarkable versatility, the use of expanded polystyrene (EPS) products has increased worldwide since its invention, simultaneously increasing the problem related to its proper disposal and accumulation as waste. Diverse studies on the properties of EPS have shown its viability as a geotechnical inclusion for soil improvement. Having good compressibility, high rupture and flexural strength, low permeability and high strength to density value, EPS can be utilized as a lightweight fill material. In this experimental study, soft clayey silt from Kolkata, West Bengal was selected as it has very low bearing capacity. Unsoaked CBR test was conducted on this soil with various percentages of waste EPS by weight of soil. The CBR values corresponding to 2.5 and 5 mm penetration for inclusion of varied amount of waste EPS has been calculated. A comparative study between the various percentages of waste EPS mixed with soil has been conducted by graphical representation. It is concluded that a mix of 0.25% waste EPS with soil showed a significantly higher value of CBR in comparison with that of untreated virgin soil. Therefore, the application of waste EPS for subgrade soil could prove as an innovative way of ground improvement as well as efficient waste management.

Keywords Waste expanded polystyrene (EPS) · CBR · Fine-grained soil

1 Introduction

The long-term performance of any road reflects the approach taken towards its design and construction. A pavement is an arrangement of vertically stacked layers such as surface course, base course, subbase course and subgrade, out of which the subgrade, being the lower most one, is the most valued component.

S. Mandal (✉) · D. Mistry
Department of Civil Engineering, Narula Institute of Technology, Kolkata 700109, India

T. G. Sitharam et al. (eds.), *Earthquake Geotechnics*, Lecture Notes in Civil Engineering 187, https://doi.org/10.1007/978-981-16-5669-9_19

The subgrade soil type greatly affects the strength and performance of the roadway. Poor performances or failures in majority of flexible pavements can be resisted by properly designing an efficient subgrade [1]. The concept of modern-day soil stabilization and ground improvement techniques came into picture in the late twentieth century. Numerous methods of soil stabilization have been widely studied over the last several years using various industrial wastes like cement kiln dust, saw dust, fly ash, lime, industrial waste sand, etc. [2–10], as well as domestic wastes like eggshell, waste plastic, etc. [11–15] which has resulted in achieving its objective. Stabilization of soil basically involves achievement of desired strength and reduction in consolidation. Properties like shear strength, compressibility and permeability have improved significantly by stabilizing the soil using proper additives following correct methodologies [16–20]. Soil stabilization using additives for obtaining greater bearing strength has also produced results in which there has been different feedback regarding variation in strength with time for the soil and additive mixture [12, 21].

Over the last few years, environmental issues have become highly alarming and are posing a huge threat to the present and future generations. The use of plastic has exponentially increased and its usage leads to severe disposal problems [22]. In countries like India, the process of waste disposal does not occur under a controlled environment. Waste products, instead of being properly treated and disposed, end up in dumps and garbage vats. They are then burned in huge piles in the open, causing massive air pollution and water pollution by leaching of hazardous chemicals into groundwater table [23]. Expanded polystyrene (EPS) products became popular during Second World War, and very soon its enormous production and public use in versatile fields made it a major contributor to the non-biodegradable waste problem [24]. EPS used in plates and cups for eating purposes and various other applications such as casing for products and packaging works is dumped in vats and put to fire [25] or left in the open where it eventually gets buried into the soil. Therefore, an effective method of its disposal is required which does not pollute the environment. The use of EPS with coarse grained soil such as sand has shown desirable results in improved bearing strength [26, 27], and from previous literature study, it has been observed that a very few research has been conducted to assess the impact on the CBR value of fine-grained soil with an inclusion of EPS. An urgency to deal with wastes and their disposal has required researchers to work and come up with solutions which could reduce pollution significantly and also strengthen civil engineering entities just the same way as an attempt has been taken for this particular work.

The objective of this paper is to incorporate waste EPS into fine-grained soil and study its impact on the California bearing ratio (CBR) value of the soil. The virgin soil is tested for its index properties, and its unsoaked CBR value has been determined. Waste EPS is mixed in soil in varying percentages of 0.05, 0.10, 0.15, 0.20, 0.25 and 0.30% of the weight of soil in an interval of 0.05%, and its corresponding CBR values are determined for unsoaked conditions. The results are compared, and an optimum mix of soil and waste EPS has been obtained.

2 Test Material

Kolkata is situated on the East bank of river Hooghly and due to the close proximity of Bay of Bengal, a prominent tidal activity is observed in the river. Situated at the flood plain of Hooghly River, soil under the city of Kolkata is mainly fine alluvial made of river deposits [28]. Fine-grained soil has been used for the present study, and the waste material that is utilized here as an additive is waste EPS. Their details are presented below.

2.1 Soil

The soil for this project was collected from Barrackpore Ferry Ghat. The first 1–2 m of the top soil was rejected as it does not reflect the true characteristics of the area, and the soil beneath the top layer was collected for this experiment.

The soil is yellowish grey in colour, soft to very soft by texture. After collection, it was thoroughly cleaned and large chunks of soil were broken down. It was then dried, sieved and stored in air tight containers. As per ISSCS, the soil was classified as silt of intermediate plasticity (MI). The virgin soil was tested for properties like specific gravity, grain size distribution, plastic limit, liquid limit, optimum moisture content and maximum dry density in accordance with IS 2720 codal provisions and is enlisted in Table 1. Modified proctor compaction test was performed on the soil as per IS 2720–1983.

Table 1 Properties of virgin fine-grained soil

Properties of soil	Values
Colour	Yellowish grey
Texture	Very soft to soft yellowish grey clayey silt
Moisture content	3%
Clay content	46%
Silt content	48%
Sand content	6%
Specific gravity	2.52
Liquid limit	41%
Plastic limit	26%
Bulk density	2.143 gm/cc
Optimum moisture content	15.18%
Maximum dry density	1.860 gm/cc

2.2 *Expanded Polystyrene (EPS)*

The chemical name of expanded polystyrene is poly(1-phenylethene-1,2-diyl) and is denoted by $(C_8H_8)_n$. Its closed-cell structure prevents the intrusion of water, thereby making it hydrophobic in nature. It does not react with soil; hence, it is non-biodegradable as well. The density of EPS plays a critical role in its compressive strength and other strength parameters. With the increase in density of the EPS, its compressive strength increases. At lower strain level, the material exhibits a linear stress–strain relation but that becomes irregular at a higher strain level. The creep or deformation effect due to long-term constant stress also increases with the increase in the density. EPS has a density of approximately 1% of soil [29]. It shows good compressibility and has high rupture and flexural strength, and its high strength to density value makes it suitable for light weight fill material [30]. The CBR value of EPS is almost similar to that of clayey soil but increases with the density [31]. It has also been observed that with the increase in density, the bearing carrying capacity and failure at settlement decrease [32]. When compressed, EPS deforms by bending, buckling; besides its excellent mechanical properties, low density and low permeability make it a suitable material for geotechnical application [33]. EPS is white in colour and is foamy in texture. To incorporate EPS evenly into soil, it has been shredded into fine fragments using mixer grinder and passed through 1.18 mm IS sieve [33]. The shredded and sieved waste EPS was preserved in plastic containers and used later in varying percentages by dry weight of soil (Fig. 1).

Fig. 1 Waste EPS shredded and sieved

3 Results and Discussion

An experimental study has been conducted on fine-grained soil mixed with varied percentages of waste EPS to understand its effect on the CBR value of the soil. Waste EPS was mixed in percentages of 0.05, 0.1, 0.15, 0.2, 0.25 and 0.3, respectively, with fine-grained soil. CBR test was performed for unsoaked conditions on the soil–waste EPS mix, which was prepared for sampling using the value of optimum moisture content obtained from a modified proctor compaction test of the virgin soil in accordance with IS-2720, and the results are displayed in Table 2.

The CBR value is calculated according to the equation mentioned in IS 2720-987. The equation is given below.

$$\mathrm{CBR} = \frac{P_T}{P_S} \times 100 \tag{1}$$

where P_T is corrected unit (or total) test load corresponding to the chosen penetration from the load penetration curve and P_s is nit (or total) standard load for the same depth of penetration as for P_T taken from the table given in Fig. 2 of IS:2720 (part 16)- 1987. In all cases of CBR performed, bearing ratio at 5 mm penetration was found to be greater than that corresponding to 2.5 mm penetration.

The load versus penetration plots for virgin soil and soil mixed with different percentages of waste EPS are represented graphically in Fig. 2. The CBR values obtained for both 2.5 and 5.0 mm penetration are plotted for the virgin soil and soil mixed with percentages of additives which is shown in Fig. 3, and the corresponding

Table 2 Load values corresponding to penetration values for different amount of waste EPS inclusion to the soil

Soil + waste EPS by weight of soil	Penetration (mm)	Load (kg)
Soil + 0% waste EPS	2.5	74.455
	5.0	145.525
Soil + 0.05% waste EPS	2.5	137.060
	5.0	252.130
Soil + 0.10% waste EPS	2.5	137.064
	5.0	262.283
Soil + 0.15% waste EPS	2.5	138.756
	5.0	275.820
Soil + 0.20% waste EPS	2.5	145.525
	5.0	284.281
Soil + 0.25% waste EPS	2.5	184.444
	5.0	323.200
Soil + 0.30% waste EPS	2.5	137.064
	5.0	267.360

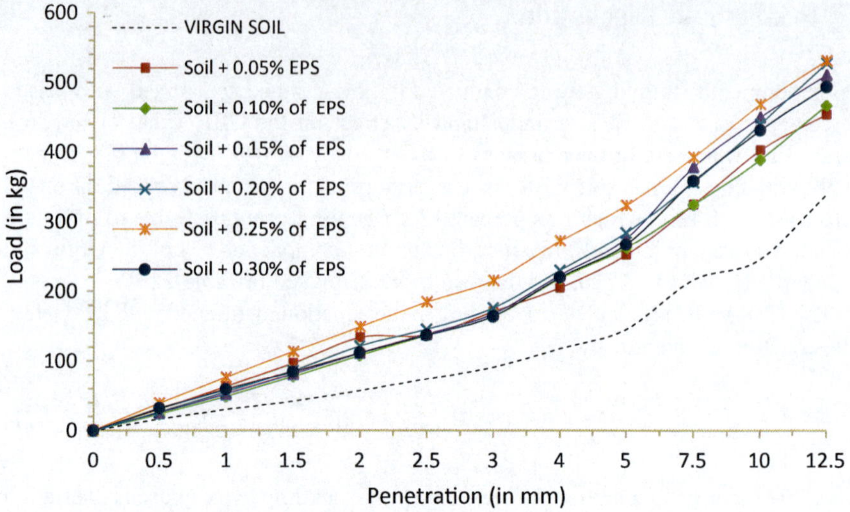

Fig. 2 CBR values for different soil samples with varying percentage of waste EPS mix

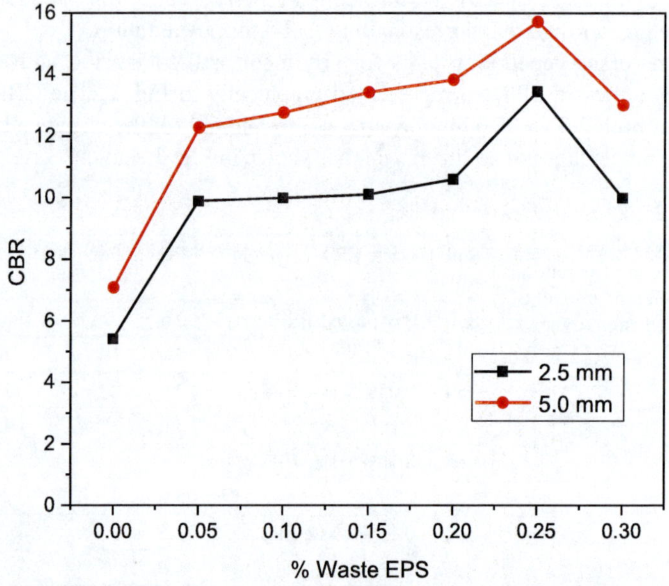

Fig. 3 CBR values corresponding 2.5 and 5 mm penetration for different soil samples with varying percentage of waste EPS mix

Table 3 CBR values for 2.5 and 5 mm penetration for different amount of waste EPS inclusion to the soil

Soil + waste EPS by weight of soil	CBR (%) corresponding to 2.5 mm penetration	CBR (%) corresponding to 5.0 mm penetration
Soil + 0% waste EPS	5.419	7.062
Soil + 0.05% waste EPS	9.980	12.251
Soil + 0.10% waste EPS	9.990	12.744
Soil + 0.15% waste EPS	10.110	13.400
Soil + 0.20% waste EPS	10.607	13.813
Soil + 0.25% waste EPS	13.440	15.700
Soil + 0.30% waste EPS	9.990	12.990

data are given in Table 3. It can be observed that mixing 0.05% of waste EPS increases the CBR value of virgin soil to 12%, and thereafter, with increasing percentages of additives in the soil, the CBR value shows a rising trend up to 0.25%. The maximum CBR value, 15.7%, is attained when soil is mixed with 0.25% of waste EPS after which the value decreases to 12.99% at 0.30% addition of waste EPS.

4 Conclusion

After successfully conducting the experimental work, it was found that the test sample prepared by mixing fine-grained soil with 0.30% of waste EPS by weight of soil showed greater difficulty while being placed in the CBR mould but the test sample with 0.25% of waste EPS showed better packing and significant increase in the CBR value compared to the former. The experimental data and the graphs clearly show that the optimum value of CBR was obtained for soil mixed with 0.25% waste EPS by weight of soil. From the experimental study, it was observed that for the optimum mix of waste EPS and soil, the design CBR value increased by 2.22 times compared to the CBR of virgin soil. This work could be beneficial in strengthening subgrades especially under the zone of heavy load-bearing pavements by making usage of discarded and waste EPS materials as a process of ground improvement coupled with effective waste management. EPS materials, due to their non-biodegradable nature and excellent polymeric properties, when used would definitely show higher durability and enhanced service characteristics in areas of applications. Unsoaked CBR values obtained from the experiment definitely concludes that soaked CBR values, when performed, would rather be more critical in determining the conditions of a pavement subgrade if simulated for the worst loading and rainfall events. Determination of CBR at values higher or lower than the OMC- MDD of the mixture of soil and various percentages of waste EPS remains as a limitation for the experiment performed. Future scope regarding performance of a triaxial test, unconfined compressive strength test, and various other engineering properties like specific gravity, liquid limit, plastic limit, etc., can be performed on the mixture of soil and 0.25% waste EPS by weight of soil.

References

1. Saxena RK (1991) Can failures be minimised and pavement performance improved by adequately designing and constructing road subgrades? J Ind Roads Congress 52(2):263–311
2. Aal-Swaidani A, Hammoud I, Meziab A (2016) Effect of adding natural pozzolana on geotechnical properties of lime-stabilized clayey soil J 8(5):714–72. https://doi.org/10.1016/j.jrmge.2016.04.002
3. Ameta NK, Wayal AS, Hiranandani P (2013) Stabilization of dune sand with ceramic tile waste as admixture. Am J Eng Res 2(9):133–139
4. Athanasopoulou A (2014) Addition of lime and fly ash to improve Highway separate soils. J Mater Civ Eng 26(4):773–775
5. Butt WA, Gupta K, Jha JN (2016) Strength behaviour of clay soil stabilized with sawdust Ash. J Geoengin 7
6. Consoli NC, da Silva Lopes L, Foppa D, Heineck KS (2019) Key parameters dictating strength of lime/cement-treated soils. P I Civ Eng Geotec 162(2):111—118
7. Edil TB, Acosta HA, Benson CH (2006) Stabilizing soft fine-grained soil with fly ash. J Mater Civ Eng 18(2). https://doi.org/10.1061/(ASCE)0899-1561(2006)18:2(283)
8. Kavak A, Akyarl A (2007) A field application for lime stabilization. Environ Geol 51(6):987–997. https://doi.org/10.1007/s00254-006-0368-0
9. Obuzor GN, Kinuthia JM, Robinson RB (2011) Enhancing the durability of flooded low-capacity soils by utilizing lime-activated ground granulated blast furnace slag (GGBS). Eng Geol 123(3):179–186
10. Sengupta A, Mukherjee S, Ghosh A (2017) Improvement of bearing ratio of subgrade using compacted fly ash layer. Geotech Geol Eng 35(2):1885–1894
11. Kumar M, Tamilarasa V (2014) Effect of egg shell powder in the index and Engineering properties of soil. IJETT 11(7)
12. Okagbue CO (2007) Stabilization of clay using Wood Ash. J Mater Civ Eng 19(1):14–18
13. Sameer S, Giridhar K, Krishna YM (2016) Stabilization of subgrade with waste plastic as soil stabilizer in flexible pavement. IJIRSET 5(10)
14. Vijaya HM, Paul W, Vardhan J (2018) Improvement of subgrade properties by using eggshell powder. IJSR 17(2):479–503
15. Yadav AK, Gaurav K, Kishor R, Suman SK (2017) Stabilization of alluvial soil for subgrade using Rice Husk Ash, Sugarcane Bagasse Ash and Cow Dung Ash for rural roads. Int J Pavement Res Technol 10(3):254–261
16. Choudhary AK, Krishna AM (2016) Experimental investigation of interface behaviour of different types of granular soil/geosynthetics. Int J Geos Ground Eng 2(4)., (2016) https://doi.org/10.1007/s40891-016-0044-8
17. Joe MA, Rajesh AM (2015) Soil stabilization using Industrial waste and lime. IJSRET 4(7)
18. Mishra NK (2012) Strength characteristics of clayey sub-grade soil stabilized with fly ash and lime for road works. Indian Geotech J 42(3):206–211
19. Sharma NK, Swain SK, Sahoo UC (2012) Stabilization of a clayey soil with fly ash and lime: a micro level investigation. Geotech Geol Eng 30(5):1197–1205. https://doi.org/10.1007/s10706-012-9532-3
20. Wild S, Kinuthia JM, Jones GI, Higgins DD (1998) Effects of partial substitution of lime with ground granulated blast furnace slag (GGBS) on the strength properties of lime-stabilised sulphate-bearing clay soils. Eng Geol 51(1):37–53
21. Park DW, Vo HV (2015) Evaluation of air foam stabilized soil of dressed soil waste as a pavement subgrade layer KSCE J Civ Eng 19(7):2091–2097. http://worldcat.org/issn/12267988

22. Verma R, Vinoda KS, Papireddy M, Gowda ANS (2016) Toxic pollutants from plastic waste-a review. Procedia Environ Sci 35:701–708
23. Graca B, Bełdowska M, Wrzesień P et al (2014) Styrofoam debris as a potential carrier of mercury within ecosystems. Environ Sci Pollut R 21:2263–2271
24. Boyer RF (1981) Anecdotal history of styrene and polystyrene. J Macromol Sci A 15(7):1411–1434
25. Wiedinmyer C, Yokelson RJ, Gullett BK (2014) Global emissions of trace gases, particulate matter, and hazardous air pollutants from open burning of domestic waste. Environ Sci Technol 48(16):9523–9530
26. Chenari RJ, Fard MK, Maghfarati SP, Pishgar F, Machado SL (2016) An investigation on the geotechnical properties of sand–EPS mixture using large oedometer apparatus. Constr Build Mater 113:773–782
27. Miao L, Wang F, Han J et al (2013) Properties and applications of cement-treated sand-expanded polystyrene bead lightweight fill. J Mater Civ Eng 25(1)
28. Das D, Chattopadhyay BC (2009) Characterization of soil over Kolkata municipal area. IGC-2009
29. Horvath JS (1994) Expanded polystyrene (EPS) geofoam: an introduction to material behaviour. Geotext Geomembranes 13(4):263–280. https://doi.org/10.1016/0266-1144(94)90048-5
30. Beju YZ, Mandal JN (2017) Expanded polystyrene (EPS) geofoam: preliminary characteristic evaluation. Procedia Eng 189:239–246
31. Beju YZ, Mandal JN (2019) California bearing ratio (CBR) behaviors of eps geofoam: experimental and numerical studies. https://doi.org/10.1007/978-3-319-78981-1_15
32. Mandal JN, Ram Rathan Lal B, Padade AH, Dutta S (2014) Bearing capacity improvement on soft soil using expanded polysterine (EPS) geofoam
33. Ossa A, Romo MP (2009) Micro- and macro-mechanical study of compressive behavior of expanded polystyrene geofoam. Geosynth Int 16(5):327–338. https://doi.org/10.1680/gein.2009.16.5.327
34. Gao H, Liu J, Liu H (2011) Geotechnical properties of EPS composite soil. J Geotech Eng 5(1):69–77. https://doi.org/10.3328/IJGE.2011.05.01.69-77

Stochastic Response Surface-Based Probabilistic Analysis of Geosynthetic Reinforced Soil Slope Under Pseudo-Static Conditions

Ekansh Agarwal, Anindya Pain, and Shantanu Sarkar

Abstract A probabilistic analysis of geosynthetic reinforced slope (GRS) is performed in the present study. The pseudo-static approach under the framework of limit equilibrium method of horizontal slices is employed to assess the internal stability of GRS. Simple ($2N + 1$) formulation is used for the same, where N stands for number of slices/reinforcements. The probabilistic analysis is carried out using collocation-based stochastic response surface (CSRS) method, given its accuracy and efficiency. A third order Hermite polynomial chaos expansion (HPCE) is employed for the same. The internal friction angle of soil (ϕ), unit weight (γ) and reinforcement tensile strength (T_u) are selected as random variables. The results are examined for two angles of slope inclination viz. $\beta = 50°$ & $60°$ to show the versatility of CSRS method with different β. The results from the contemporary study are validated with Monte-Carlo simulation for same set of parameters to check for the accuracy and efficiency of CSRS method. The effect of coefficient of variation (CoV) of ϕ on the probability of failure of GRS is examined in detail. The study concludes that CSRS method incorporating an accurate deterministic model is highly efficient and powerful approach to conduct the probabilistic analysis of GRS.

Keywords Horizontal slice method · Collocation · Reliability · Pseudo-static · CSRS · Geosynthetic reinforced slope

1 Introduction

Geosynthetic reinforced slopes (GRS) are used extensively as stability measure in construction due to their simple design and ease of construction. Earlier, unreinforced slopes were in practice which were less stable and hence, to compensate for their lack

E. Agarwal (✉) · A. Pain · S. Sarkar
Academy of Scientific and Innovative Research (AcSIR), Ghaziabad 201002, India

A. Pain
e-mail: pain_anindya@cbri.res.in

Geotechnical Engineering Group, CSIR-Central Building Research Institute, Roorkee 247667, India

T. G. Sitharam et al. (eds.), *Earthquake Geotechnics*, Lecture Notes in Civil Engineering 187, https://doi.org/10.1007/978-981-16-5669-9_20

of stability they used more area. However, this is not feasible in today's era and as a result GRS started getting attention as a stabilization method for unstable slopes. The deterministic analysis of these slopes is usually performed using the factor of safety approach which practices increasing the factor of safety to an uneconomical extent. But it has been proven many a times that two naturally identical slopes may have the same value of factor of safety but different values of probability of failure (P_f), putting a question mark on the effectiveness of deterministic analysis. This arose the need to develop the probabilistic analysis which takes in to account the uncertainty of the input parameters.

The design of GRS is usually done with the help of limit equilibrium methods as they show a sensible agreement with the laboratory data. The limit equilibrium method used in the present study is Horizontal Slice method (HSM) originally developed by Low and Xu [1]. Various researchers [2–7] had used HSM and showcased its different formulations based on the equilibrium equations. The simple $(2\ N + 1)$ formulation of HSM is used in the present study owing to its simplicity and efficiency in analyzing the internal stability of GRS.

Many researchers had analyzed the unreinforced soil and rock slopes probabilistically [8–13], however, there is a paucity of literature on reliability analysis of reinforced GRS. The analysis of GRS had been attempted only by a few [14, 15], but unfortunately, the methodology adopted by them is comparatively less efficient.

A traditional way of estimating the probability of failure is calculating the limit state function for a certain number of realizations which is known as Monte-Carlo simulation (MCS). However, it is highly computationally inefficient. Another approach is constructing a dummy model of the original deterministic model using Kriging, support vector machines, multivariate adaptive regression splines (MARS), neural network, polynomial chaos expansion (PCE), etc. However, a highly common approach is the Response Surface Method (RSM) [16–18], but it may not always lead to a convergent solution.

As a result, collocation-based stochastic response surface (CSRS) method is used in the present study to perform the reliability analysis of GRS, as it is highly efficient and acceptably accurate as compared to other methods of analysis. Hermite polynomial chaos expansion (HPCE) is performed to expand the unknown random output vector. CSRS method replaces the original deterministic model by a meta model and the analysis is performed on the same. To the best of the author's knowledge, the algorithms used in the study are original and no such study combining an accurate deterministic model with the efficient CSRS method for the probabilistic analysis of GRS is conducted till date.

2 Deterministic Model

To analyze the GRS deterministically, $(2\ N + 1)$ formulation of HSM is used in the pseudo-static framework. Soil is considered as dry, homogenous and linearly elastic in nature. No effect of pore water pressure is considered and the backfill is chosen to be cohesionless. The failure surface is selected as log-spiral. The (*2 N + 1*) formulation satisfies the horizontal equilibrium for the whole wedge and vertical equilibrium for individual slices together which reduces the number of equations and unknowns to *2 N + 1*. The equations of equilibrium reported in Shahgholi et al. [2] are used in a similar manner to calculate the total required tensile strength of the reinforcement (T_j) from Eq. (1).

$$T_j = \sum_{i=1}^{N} N_i \sin\alpha + \sum_{i=1}^{N} W_i k_h - \sum_{i=1}^{N} S_i \cos\alpha \tag{1}$$

where S_i = shear force upon base of slice, N_i = normal force upon base of slice, W_i = weight of slice, k_h = coefficient of horizontal seismic acceleration, and N = number of slices/reinforcements.

The non-dimensional form of T_j may be expressed according to Eq. (2):

$$K = T_j / 0.5\gamma H^2 \tag{2}$$

where γ = unit weight of the soil and H = height of the slope.

K is optimized to identify the critical failure surface by formulating an algorithm in *MATLAB*. θ_o and θ_h are allowed to vary from 0° to 120° where $\theta_h - \theta_o > 0$ (Fig. 1).

The results from the present study are verified by comparing them with Nouri et al. [4] (Fig. 2) for $H = 5$ m, $\beta = 60°$, $\gamma = 18$ kN/m^3, $\phi = 20° - 45°$, $k_h = 0$ and 0.15, and $k_v = 0$. The results are in accordance with the above mentioned literature which prove the compatibility of the proposed algorithm to optimize K.

3 Probabilistic Analysis

As discussed earlier, probabilistic analysis gives us a detailed information regarding the stability of GRS. In this section, probabilistic analysis is performed using the robust technique of collocation-based stochastic response surface besides validating the results with the published literature.

3.1 Limit State Function

A limit state function is the foundation of any probabilistic analysis technique and mainly used to differentiate between the acceptable and unacceptable performance. In the present study, limit state function [$g(X)$] is defined by:

$$g(X) = \frac{T_u}{T_i} - 1 \tag{3}$$

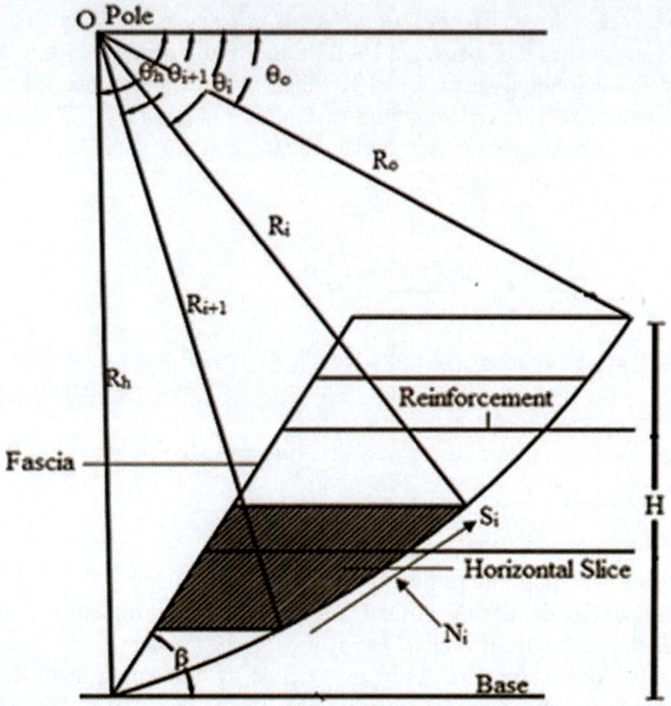

Fig. 1 Schematic diagram of GRS

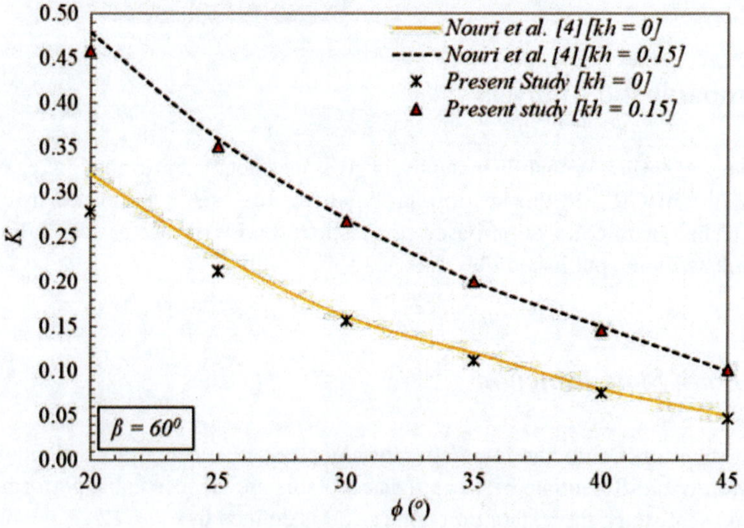

Fig. 2 Comparison of K from the deterministic analysis

where T_u/T_i = Factor of safety (*FoS*), T_u is the ultimate tensile strength of reinforcement, and T_i is the maximum load in a particular layer of reinforcement which is given by:

$$T_i = \left(Y_{r,j} \times \gamma\right) K (S_v \times S_h) \tag{4}$$

where $Y_{r,j}$ is the depth of the *j*th reinforcement layer from the top of the slope, S_v is the vertical spacing between the two reinforcements given by *H/N*, and S_h is the horizontal spacing which is chosen as 1 m.

A trial's safety is dependent upon the limit state function as follows:

$$\begin{aligned} &\text{if } g(X) \le 0, \text{ trial is marked as unsafe} \\ &\text{if } g(X) > 0, \text{ trial is marked as safe} \end{aligned}$$

Accordingly, the associated probability of failure (P_f) is calculated as:

$$P_f = \sum_{j=1}^{\text{NOR}} (\text{count}_M)/\text{NOR} \tag{5}$$

where count_M is the number of times the value of $g(X) \le 0$ and *NOR* is number of realizations.

3.2 Collocation-Based Stochastic Response Surface Method

The CSRS method is basically an extension of the traditional response surface method but the only difference between them being the type of input accepted by each of these. The former one takes inputs in the form of random variables unlike the latter one which uses deterministic quantities as inputs [19]. The steps to be followed to calculate the probability of failure of GRS using CSRS method are illustrated in flowchart shown in Fig. 3.

The prime objective of CSRS method is to form a meta-model of the deterministic algorithm and then performing the analysis on this Meta model. Hermite polynomial chaos expansion (HPCE) is used in the present study to approximate the original model by expanding the basis function. Hermite polynomials are of one-dimensional as well as multi-dimensional nature. The latter one are more complicated but more accurate than the former. Therefore, in this study, multi-dimensional Hermite polynomials are chosen for expansion owing to the accuracy of the method.

Mollon et al. [20] had stated that there is a minuscule difference between the PCE's of 3rd, 4th and 5th order. Moreover, as the order increases, the values of new coefficients in the increased order are negligible besides the values of common coefficients remaining the same. As the random variables considered are less in number (3), therefore, a PCE of 3rd order is selected for analysis in the present

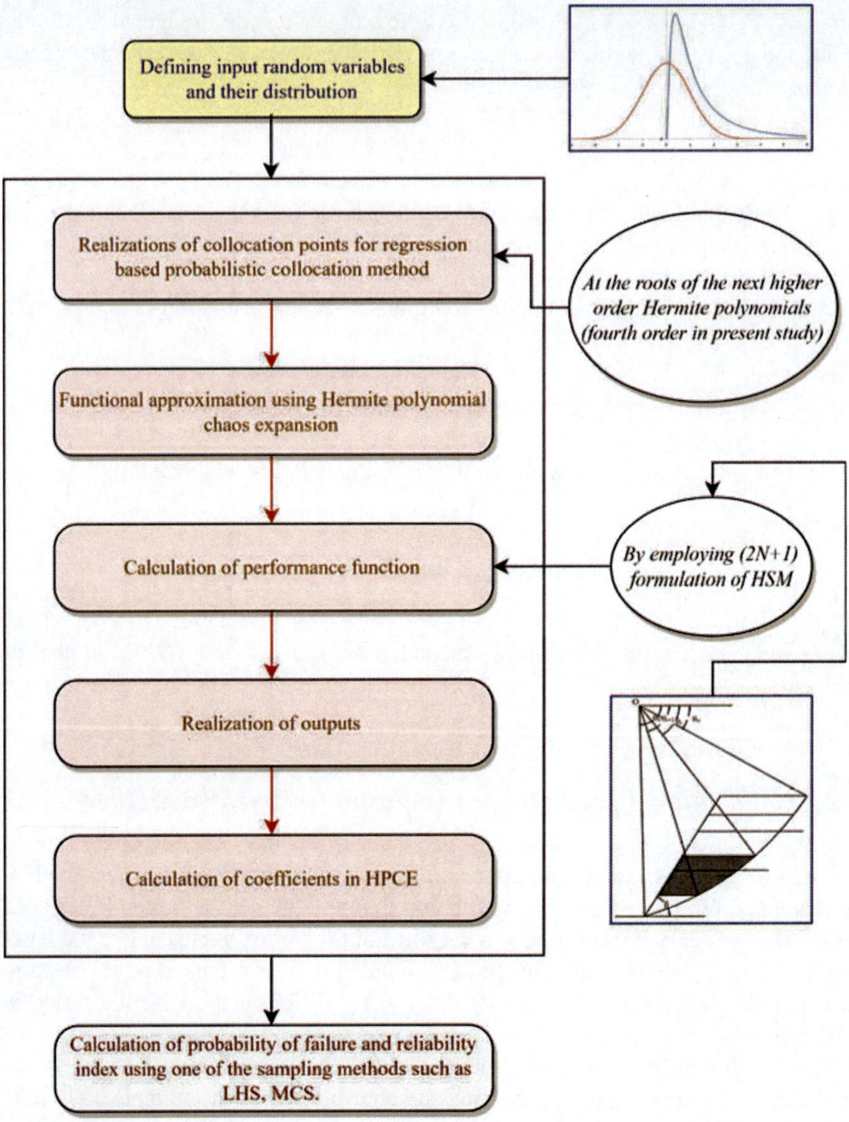

Fig. 3 Flowchart of CSRS

study as it is acceptably accurate in producing results for P_f in the order of 10^{-3} or above. Also, the benefits of higher order are less whereas the time taken by them for simulation is quite high. The number of terms (NT) in the expanded form of third order multi-dimensional Hermite polynomial with n number of random variables is given by:

$$\mathrm{NT} = 1 + 3n + \frac{3n(n-1)}{2} + \frac{n(n-1)(n-2)}{6} \quad (6)$$

Generally, the equations produced after the expansion of Hermite polynomials are very cumbersome and not easy to solve. To ease out the process of solving, regression-based stochastic collocation method is used which is explained in the following section.

Regression-based Stochastic Collocation Method. This method is used to get a random solution in stochastic space unlike its conventional counterpart, which produced the deterministic results. A crucial step in implementing this method is the selection of collocation points for the response surface. This is done by selecting the collocation points at the roots of next higher order polynomial for every considered random variable. In addition, zero is also considered as one of the roots as it is the region of the highest probability for a normal random variable [21]. After the selection of collocation points, unknown coefficients in the PCE are determined. In the present study, a PCE of third order with 3 random variables is considered whose expansion is given in Eq. (7)

$$\begin{aligned} X \approx\ & c_o + c_1 y_{a1} + c_2 y_{a2} + c_3 y_{a3} + c_4(y_{a1}^2 - 1) + c_5(y_{a2}^2 - 1) + c_6(y_{a3}^2 - 1) \\ & + c_7(y_{a1}^3 - 3y_{a1}) + c_8(y_{a2}^3 - 3y_{a2}) + c_9(y_{a3}^3 - 3y_{a3}) + c_{10} y_{a1} y_{a2} + c_{11} y_{a1} y_{a3} \\ & + c_{12} y_{a2} y_{a3} + c_{13}(y_{a1} y_{a2}^2 - y_{a1}) + c_{14}(y_{a1} y_{a3}^2 - y_{a1}) + c_{15}(y_{a2} y_{a1}^2 - y_{a2}) \\ & + c_{16}(y_{a2} y_{a3}^2 - y_{a2}) + c_{17}(y_{a3} y_{a1}^2 - y_{a3}) + c_{18}(y_{a3} y_{a2}^2 - y_{a3}) + c_{19} y_{a1} y_{a2} y_{a3} \end{aligned} \quad (7)$$

where X is a random output vector expressed as a function of random input variables, $f\ (Y_1, Y_2, Y_3)$ and (y_{a1}, y_{a2}, y_{a3}) are the viable realizations from a standard normal vector (Y_1, Y_2, Y_3).

Further solving Eq. (7) gives Eq. (8).

$$X = f(y_{a1}, y_{a2}, y_{a3}) = c_o + c_1 b_{a1} + c_2 b_{a2} + \cdots + c_{19} b_{a19} \quad (8)$$

where $(1, b_{a1}, b_{a2}, ..., b_{a19})$ denote the basis functions and $(c_0, c_1, c_2, ..., c_{19})$ are the unknown coefficients.

The unknown coefficients are determined using Eq. (9):

$$M^T M c = M^T f \quad (9)$$

where M is a m x NT matrix whose a_{th} row is illustrated as $(1, b_{a1}, b_{a2}, ..., b_{a19})$, m is the number of realizations; c is $NT \times 1$ vector containing unknown coefficients and f is a $NT \times 1$ vector whose a_{th} element is given by $f\ (y_{a1}, y_{a2}, y_{a3})$.

The number of unknown coefficients in a HPCE is given by Ghanem and Spanos [22]. In the present study, the number of unknown coefficients calculated by following the above said literature is 20. Moreover, the rank of the matrix M is 20, which is equal to the number of coefficients and the size of matrix, confirming that M has a full rank.

After the unknown coefficients are calculated, the P_f of GRS may be calculated using any of the sampling-based methods.

3.3 Comparison of Probabilistic Model

To verify the formulation, the results from the present study are compared with Basha and Babu [23] for $H = 5.5$ m, $\phi = 37°$, $\beta = 78.7°$,γ = 18 kN/m^3, $k_h = 0.216$, $N = 10$, $\delta/\phi = 0.8$, $T_u = 20$ kN/m, and $k_v = 0$. The results are in accordance with the said literature with minor differences noted due to difference in methodologies used (Table 1). The present study uses an efficient and robust CSRS method in comparison to the conventional First Order Reliability Method (FORM) used by Basha and Babu [23]. Moreover, Basha and Babu [23] had assumed the value of radial reaction as constant along the failure surface which contributes to a higher value of K and ultimately, a lower value of $\beta_{\min}$. The value of reliability index ($\beta_{\min}$) for tension mode of failure from the present study is 2.75 in comparison to 2.230 from Basha and Babu [23].

4 Results and Discussions

The random variables used in the present study along with their distribution are tabulated below (Table 2). The influence of CoV(ϕ) on the reliability index of GRS

Table 1 Comparison of $\beta_{\min}$ from CSRS method with Basha and Babu (2012) for Seiken wall for $H = 5.5$ m, $\beta = 78.7°$, $\gamma = 18$ kN/m^3, $\phi = 37°$, $k_h = 0.216$, $N = 10$, $\delta/\phi = 0.8$, $T_u = 20$ kN/m, $CoV = 5\%$, and $k_v = 0$

Seiken wall (Toho-Oki earthquake)	Stability against tension failure	
	$\beta_{\min}$	$P_f(\%)$
Basha and Babu [23]	2.23	1.3
Present study	2.75	0.3

Table 2 Attributes of the random variables used in the present study

Random variable	Properties		
	mean (μ)	(CoV) %	Distribution
γ (soil unit weight)	18 kN/m^3	5	Normal
ϕ (internal friction angle of the soil)	35°	5–15	Log-normal
T_u (ultimate tensile strength)	20 kN/m	5	Normal

for two angles of slope inclination with the horizontal (β) viz. 50° and 60° is presented and plotted.

The prime reason behind choosing two different values of β is just to showcase the versatility of the proposed algorithm in correctly estimating the P_f for any given value of β. The values are chosen as 50° and 60° because the inclination angle with the horizontal in case of a reinforced slope is less than 70°. However, the algorithm gives the reader a flexibility to choose any angle of slope inclination as per their convenience. Moreover, the comparison of CSRS method with crude MCS is illustrated by performing a regression analysis for same set of parameters.

4.1 *Regression Analysis*

Regression analysis is a powerful tool to check the dependency of independent variables on dependent variables. Stepping upon this advantage, regression analysis is performed in the present study to check the competency of CSRS method and it's fitting with crude MCS. The analysis is run for 100 testing samples and the results are plotted in Fig. 4, which show an excellent fitting with a high value of $R^2 = 0.99$.

4.2 *Influence of CoV(Φ) on P_f*

In reliability analysis of reinforced slopes, CoV (ϕ) is a parameter which greatly influences the P_f and reliability index ($\beta_{\min}$) of these slopes. Figure 5 shows the influence of CoV(ϕ) on the probabilistic values of GRS for $\beta = 50°$ and 60°. The CoV (γ, T_u) are kept constant while the CoV(ϕ) is varied from 5 to 15% considering the actual field conditions. The results reveal that as the CoV (ϕ) increases, the P_f of GRS also increases and consequently, the reliability index decreases. This is due to the reason that as the value of CoV(ϕ) increases, the values of standard deviation and ultimately the variance also increase which depict a large deviation in the mean values of random variables considered, eventually leading to an increased P_f.

In Fig. 5 for $\beta = 50°$, $H = 5$ m, $k_v = 0$, $T_u = 20$ kN/m, $\gamma = 18$ kN/m^3, $\phi = 35°$, $k_h = 0.2$, and $N = 5$, the value of P_f for *CoV* (ϕ) = 5% is less than CoV (ϕ) = 10% and 15% by 99.62% and 99.81%, respectively.

A similar trend is followed in case of $\beta = 60°$, $H = 5$ m, $k_v = 0$, $T_u = 20$ kN/m, $\gamma = 18$ kN/m^3, $\phi = 35°$, $k_h = 0.2$, and $N = 5$. The value of P_f for *CoV* (ϕ) = 5% is less than CoV (ϕ) = 10% and 15% by 91.98% and 96.10%, respectively. These results prove the competency of CSRS method to accurately predict the P_f with increasing CoV (ϕ).

A point to be noted is that for the same set of parameters, the P_f for $\beta = 60° > P_f$ for $\beta = 50°$, which justifies the correctness of the algorithm and gives the confidence to conduct this study.

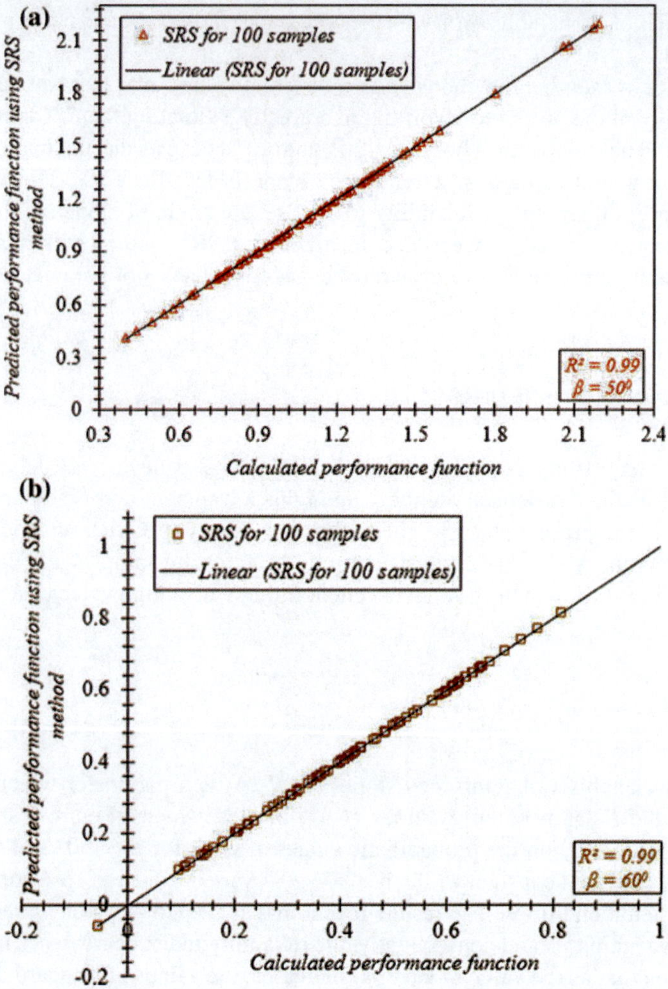

Fig. 4 Comparison between the values of limit state function determined by SRS method and MCS for: **a** $\beta = 50°$, $H = 5$ m, $k_v = 0$, $T_u = 20$ kN/m, $\gamma = 18$ kN/m^3, $\phi = 35°$, $k_h = 0.2$, and $N = 5$; **b** $\beta = 60°$ for $H = 5$ m, $k_v = 0$, $T_u = 20$ kN/m, $\gamma = 18$ kN/m^3, $\phi = 35°$, $k_h = 0.2$, and $N = 5$

A detailed description of the values obtained for $\beta = 50°$ and 60°, for different values of CoV (ϕ) is given in Table 3.

5 Summary and Conclusions

In the present study, a robust and accurate method of collocation-based stochastic response surface (CSRS) to probabilistically analyze the geosynthetic reinforced

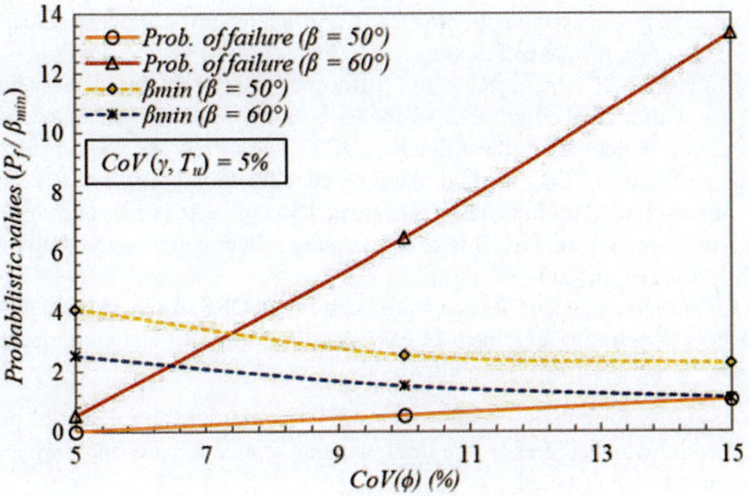

Fig. 5 Influence of CoV (ϕ) on probabilistic values (P_f and $\beta_{\min}$) of GRS

Table 3 Probabilistic values for $\beta = 50°$ and 60° for different values of CoV (ϕ)

	$\beta = 50°$			$\beta = 60°$		
	CoV(ϕ) (%)					
	5	10	15	5	10	15
	$\beta_{\min}$			$\beta_{\min}$		
P_f(%)	0.002	0.53	1.07	0.52	6.48	13.35
$\beta_{\min}$	4.1	2.56	2.30	2.56	1.52	1.11

slope, is described in detail. A simple yet efficient $(2N + 1)$ formulation of horizontal slice method is used for deterministic analysis. The results are in very well accordance with the published literature. To the best of the author's comprehension, no such study to probabilistically analyze the GRS incorporating a simple and efficient formulation of HSM is performed till date. The internal friction angle of soil (ϕ), unit weight (γ) and tensile strength of the reinforcement (T_u) are chosen as random variables and the analysis is done for $\beta = 50°$ & $60°$. The probabilistic model is executed using the CSRS method. Hermite polynomial chaos expansion (HPCE) is used for the same and the collocation points are obtained followed by the calculation of unknown coefficients. The efficiency and accuracy of CSRS method is complemented by performing a regression analysis with MCS (100 testing samples) which gives a high value of R^2 equal to 0.99. As an integral part of this study, the influence of *CoV* (ϕ) on the P_f and $\beta_{\min}$ of GRS is demonstrated by plotting the results besides tabulating them in a table for both, $\beta = 50°$ & $60°$. On the basis of the results, the following conclusions are drawn:

- A simple (*2 N + 1*) formulation of HSM is efficient and accurate in calculating the values of performance functions.
- A 3rd order PCE is acceptable for performing the CSRS-based probabilistic analysis with a high efficiency. An increase in the order of PCE increases the complexity and decreases the efficiency.
- While performing CSRS method, the limit state function is evaluated 125 times in comparison to MCS in which it is evaluated 50,000 or 10^5 times or may be 10^6 times for similar accuracy. This confirms an excellent computational efficiency of the proposed method.
- The CoV(ϕ) has a great influence on the stability of GRS. An increase in CoV(ϕ) leads to a substantial increment in the value of P_f of GRS and consequently, a decrement in the value of reliability index.

The proposed formulation presents an efficient, accurate, and powerful probabilistic method which addresses the limitations of its counterparts and may lead to an economic design of GRS.

References

1. Lo SR, Xu DW (1992) A strain-based design method for the collapse limit state of reinforced soil walls or slopes. Can Geotech J 29(5):832–842
2. Shahgholi M, Fakher A, Jones CJFP (2001) Horizontal slice method of analysis. Geotechnique 51(10):881–885
3. Fakher A, Nouri H, Shahgholi M (2002) Limit equilibrium in reinforced soil walls subjected to seismic loads. In: Proceedings of the 3rd iranian international conference on geotechnical engineering and soil mechanics, Tehran, pp 281–286
4. Nouri H, Fakher A, Jones CJFP (2006) Development of horizontal slice method for seismic stability analysis of reinforced slopes and walls. Geotext Geomembr 24(3):175–187
5. Nouri H, Fakher A, Jones CJFP (2008) Evaluating the effects of the magnitude and amplification of pseudo-static acceleration on reinforced soil slopes and walls using the limit equilibrium horizontal slices method. Geotext Geomembr 26(3):263–278
6. Khosravizadeh M, Dehestani M, Kalantary F (2016) On the seismic stability and critical slip surface of reinforced slopes. Soil Dyn Earthq Eng 85:179–190
7. Farshidfar N, Keshavarz A, Mirhosseini SM (2020) Yield acceleration of reinforced soil slopes using the modified pseudo-dynamic method. SN Appl Sci 2:1–14
8. Li KS, Lumb P (1987) Probabilistic design of slopes. Can Geotech J 24(4):520–535
9. Bhattacharya G, Jana D, Ojha S, Chakraborty S (2003) Direct search for minimum reliability index of earth slopes. Comput Geotech 30(6):455–462
10. Low BK, Tang WH (2007) Efficient spreadsheet algorithm for first-order reliability method. J Eng Mech 133(12):1378–1387
11. Lü Q, Low BK (2011) Probabilistic analysis of underground rock excavations using response surface method and SORM. Comput Geotech 38(8):1008–1021
12. Metya S, Bhattacharya G (2014) Probabilistic critical slip surface for earth slopes based on the first order reliability method. Ind Geotech J 44(3):329–340
13. Metya S, Mukhopadhyay T, Adhikari S, Bhattacharya G (2017) System reliability analysis of soil slopes with general slip surfaces using multivariate adaptive regression splines. Comput Geotech 87:212–228

14. Luo N, Bathurst RJ (2018) Probabilistic analysis of reinforced slopes using RFEM and considering spatial variability of frictional soil properties due to compaction. Georisk: Assess Manage Risk Eng Syst Geohazards 12(2):87–108
15. Hamrouni A, Daniel DIAS, Sbartai B (2020) Soil spatial variability impact on the behavior of a reinforced earth wall. Front Struct Civ Eng 15(6)
16. Wong FS (1985) Slope reliability and response surface method. J Geotech Eng 111(1):32–53
17. Faravelli L (1989) Response-surface approach for reliability analysis. J Eng Mech 115(12):2763–2781
18. Rajashekhar MR, Ellingwood BR (1993) A new look at the response surface approach for reliability analysis. Struct Saf 12(3):205–220
19. Isukapalli SS, Roy A, Georgopoulos PG (1998) Stochastic response surface methods (SRSMs) for uncertainty propagation: application to environmental and biological systems. Risk Anal 18(3):351–363
20. Mollon G, Dias D, Soubra AH (2011) Probabilistic analysis of pressurized tunnels against face stability using collocation-based stochastic response surface method. J Geotech Geoenviron Eng 137(4):385–397
21. Li D, Chen Y, Lu W, Zhou C (2011) Stochastic response surface method for reliability analysis of rock slopes involving correlated non-normal variables. Comput Geotech 38(1):58–68
22. Ghanem RG, Spanos PD (2003) Stochastic finite elements: a spectral approach. Dover Publication Inc., Mineola, New York
23. Basha BM, Babu GS (2012) Target reliability-based optimisation for internal seismic stability of reinforced soil structures. Geotechnique 62(1):55–68

Impact of Residual Strength of Soil in Reactivated Landslide: Case Studies of a Few Landslides in Different Regions of the Globe

Pankaj Kundu and Neha Gupta

Abstract The residual shear strength is an essential parameter for designing mitigation techniques in reactivated landslides. The post-failure movement characteristics of reactivated landslides are generally affected by the residual strength behavior of the sliding surface. The factors which contribute to the residual strength such as displacement rate, mineralogical composition of the soil, salinity of the pore fluid, and soil aging are discovered by various researchers and are summarized, and their correlation with the residual shear strength has been discussed. Also, some case studies are presented from different countries where reactivated landslides have been observed ("Marappalam" landslide in India and "Ogoto" and "Kuchi-Otani" landslide in Japan and landslide in Three Gorge Reservoir in China), and behavior of these failure considering the residual shear strength are presented here.

Keywords Residual shear strength · Clay fraction · Soil aging · Marappalam landslide · Three gorge reservoir

1 Introduction

The reactivated landslide is a residual slide that moves slowly at the slip surface mainly due to an increase in the groundwater table. Sliding is a failure phenomenon that occurs mostly in clayey soil, and when it reaches equilibrium, it is said to be in the residual state due to the earlier movements. The minimum drained strength of a soil mass gives the direct estimation of the residual shear strength of the soil. The residual shear strength of a soil is associated with reactivated landslide and is used for many geotechnical applications such as slope stability and landslide simulation.

Residual shear strength is the minimum drained shear strength attained in the soil for large displacements. The strength of soil at the slip surface is reduced to its residual value when a fault or slip occurs, so the residual strength must be known for this preexisting shear surface for designing mitigation techniques for this reactivated

P. Kundu (✉) · N. Gupta
Indian Institute of Technology (IIT), Roorkee 247667, India
e-mail: pkundu@eq.iitr.ac.in

T. G. Sitharam et al. (eds.), *Earthquake Geotechnics*, Lecture Notes in Civil Engineering 187, https://doi.org/10.1007/978-981-16-5669-9_21

landslide [1]. In this study, we will discuss the factors affecting the residual shear strength and its impact on some reactivated landslides by considering different case studies.

2 Factors Affecting Residual Shear Strength

Residual shear strength of soil depends upon multiple factors such as displacement rate [1–3] mineralogical composition of soil [4, 5], the salinity of the pore fluid [6], and soil aging [7, 8].

2.1 *Displacement Rate*

Residual strength is less affected by the variation of displacement rate, but significant strength gain has been observed when the displacement rate is faster than 100 mm/min and concludes that variation in strength is negligible when the rate of displacement are in the range of 0.002–0.001 mm/min [1].

Three types of variation in residual strength are found with changing of displacement rate: (1) increase, (2) decrease, and (3) does not change. During slow shearing, soil grains align parallel to the shearing direction, but as the rate increases, their alignment changes from sliding to turbulent sharing, and as a result, shear strength increases [2].

The effect of displacement rate for residual strength is increasing with a decrease in effective normal stress. With higher effective stress, residual strength is achieved at lower displacement rates [1, 3].

2.2 *Mineralogical Composition of Soil*

If the clay fraction in the soil is less than 25%, the residual shearing resistance is greater than 20°, and when it is more than 50%, then the residual strength is completely controlled by the sliding of clay minerals. When clay fraction is in the range between 25 and 50%, then soil shows transitional behavior. Higher than 50% of the clay fraction has little impact on residual strength [1].

The residual friction angle is controlled by both clay fraction and the mineralogical composition of the clay [4]. The drained residual shear strength of a soil is influenced by the type of clay mineral and quantity of the clay particles, which is reduced by increasing clay fraction and the liquid limit of the soil. The shear strength shows linear behavior for clay fraction up to 45% but nonlinearity induced if clay fraction is greater than 50% then liquid limit ranges between 60 and 220 [5].

2.3 *Pore Water Chemistry of Soil*

The residual strength of the soil is affected by the nature of its pore fluid components. The behavior of clayey soil with distilled water and sodium chloride solution is studied by Tiwari et al. 2005 [6] and found that the shear strength of kaolin is not affected by the sodium chloride solution, but the bentonite shows significant improvement in residual shear strength up to approximately 200%, due to the inward salt diffusion.

The residual friction angle of the soil sample prepared with seawater is generally 3°–5° higher than the soil with distilled water. But also swelling can be observed in those soils when NaCl starts leaching, and that results in a decrease in shear strength [7].

2.4 *Soil Aging*

Soil with low plasticity can produce a strength gain of about 8% after one day of aging and 13% after 230 days. And for the soils with high plasticity, the strength gain is about 6% after one day and 43 after 230 days [8].

The strength gain in residual soil has been observed with aging shows the strength gain in fine samples is less compared to those samples with the same fine sizes with wide grain size distribution. The test infers strength gain of 20% and 30% for Rosazzo flysch and Montona flysch over a period of one month [8].

3 Laboratory Experiments for Residual Strength Test

A new ring shear apparatus has been developed by Bishop et al. 1971 [9] to measure the residual strength of the soil. The main advantage of using a ring shear apparatus apart from its long operation time and complicated construction is that it can be sheared in one direction for any amount of displacement. This allows the soil particle to orient in the same direction of shearing and to develop the condition of residual strength. The another advantage of using a ring shear apparatus is that the cross-sectional area of the sample does not change during shear.

Stark and Eid 1994 [5] conducted a torsional ring shear test on 32 clay and shale specimen and shows that the drained residual failure envelops nonlinear and also there is a decrease in secant residual friction angle with an increase in effective normal stress.

Gibo et al. 2002 [10] observed that the strength recovery of the soil containing a large number of clay particles is negligible whereas, there is a large recovery of shear strength of soil containing huge amounts of silt and sand particles at effective stress less than 100 kPa. The rate of strength recovery can be improved by decreasing effective stress.

Residual condition on a slip surface can be achieved when clay particles become parallel to the direction of shearing under a certain effective normal stress. A higher degree of particle alignment is possible at the residual condition by increasing the effective normal stress, thus reducing the residual friction angle. By increasing or decreasing the effective normal stress at the residual condition, swelling or compression may be observed, followed by a peak before reaching to a new residual condition [11].

4 Case Study

4.1 Kuchi-Otani and Ogoto Landslide, Japan

Gratchev et al. [12] investigated the shear strength of soil from "Kuchi-Otani" landslide and "Ogoto" landslide. "Kuchi-Otani" landslide located in Hyogo prefecture, Japan, was reactivated in 2003 (see Fig. 1) [12]. "Ogoto" landslide is located in the vicinity of Biwa Lake, situated in Otso city, Japan. Due to their location, it may impose threat to local people and cause several damages so residual shear strength of these reactivated landslides which is studied.

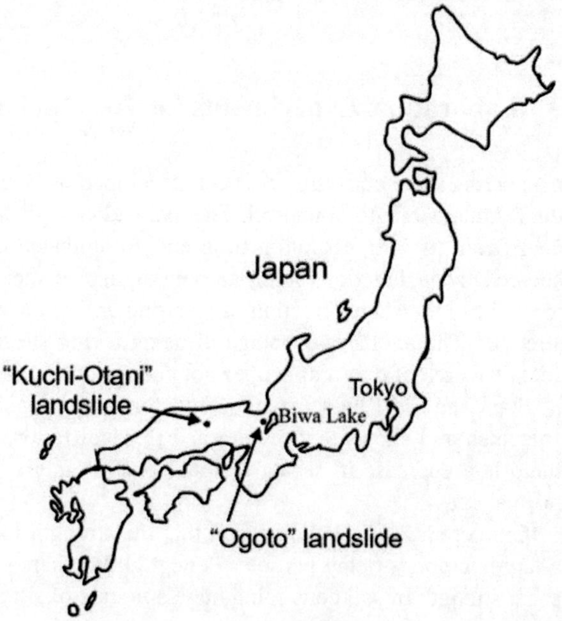

Fig. 1 Location of studied landslide [12]

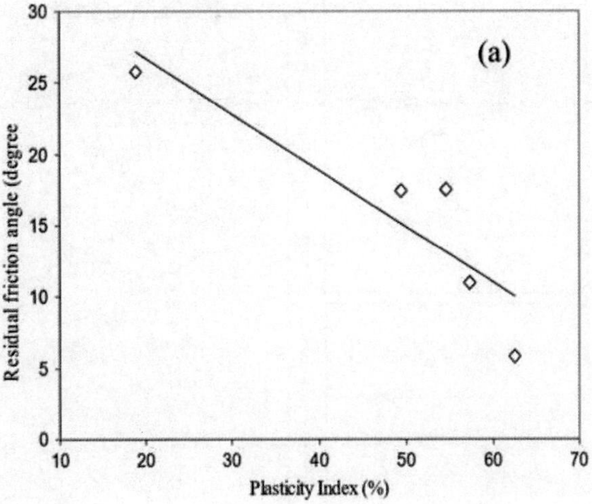

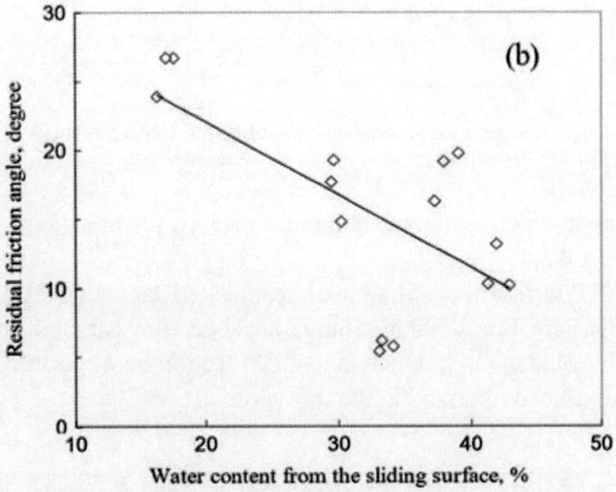

Fig. 2 Relations between residual shear strength and **a** plasticity index, and **b** water content [12]

The shear strength of the clayey soil is decreased significantly by the presence of a high amount of clay content in those soil. Also, a higher amount of plasticity and higher water content tend to decrease the shear strength in soil (see Fig. 2) [12].

4.2 Marappalam Landslide, India

Frequently occurred landslide is the major cause of damaged road network and local infrastructure in Marappalam area (see Fig. 3) [13] of Nilgiris district, and among them, 1993 and 2009 landslide are the major landslides in the history of Nilgiris landslide. The study area is located along the national highway (NH181) and railway

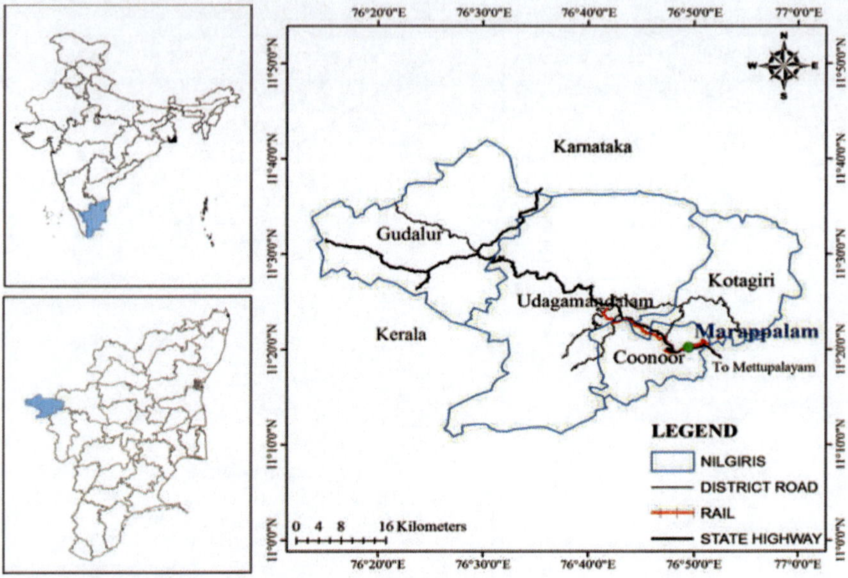

Fig. 3 Location map: **a** India, **b** Tamil Nadu, **c** Nilgiris and Marappalam [13]

track which is damaged due to the landslide at Marappalam in November 2009 (see Fig. 4) [13].

The residual strength of reactivated landslide of Marappalam area of Nilgiris district, Tamil Nadu, India, has been investigated by Senthilkumar and Chandrasekaran [14], where landslides are frequently occurred due to rainfall and lead to significant damage to the infrastructure.

Fig. 4 View of Marappalam 2009 landslide. **a** View of damaged road network **b** Damaged railway track and debris materials along the run-out distance [13]

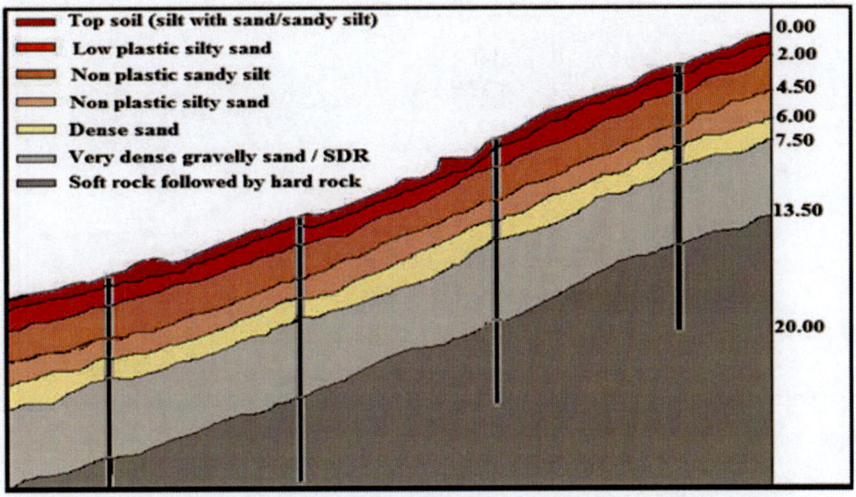

Fig. 5 Weathering profile of Marappalam slope [13]

The weathering effect of the Marappalam slope is investigated by Senthil Kumar et al. [13], and they classify the soil slope into six layers (see Fig. 5). The top surface soil layer of grade V and grade VI is completely weathered to silty sand layer up to a depth of 4.5 m, followed by a highly weathered rock layer of grade IV up to 9.0 m.

It has been observed that topsoil layer contains a higher clay fraction due to a high degree of weathering and the amount of clay fraction decreases as the depth increases as an effect of weathering negligible at a higher depth.

As the residual strength of the soil is a function of a clay fraction present in a clay mineral, at the topsoil, the residual strength decreases, and as we increase the depth of the soil the residual strength increases.

4.3 Three Gorge Reservoir, China

More than 4200 landslides have been observed in the bank of the Three Gorges Reservoir (TGR), China [15]; the rapid rate of infiltration can cause several shallow landslides. Changes in the water level play a key role in the instability of slopes in the reservoir area and may cause a devastating landslide. Also, the movement of landslide and the recovery of residual shear strength are studied by Miao et al. (2014) for understanding the mechanism of a slow-moving landslide in the Jurassic Red Strata in TGR [17]. Many reactivated landslides have been discovered in the Wanzhou area which is located in the middle of Three Gorges Reservoir (Fig. 6) [16]. Reactivation of Qianjiangping landslide and Anlesi landslide in the Jurassic strata has been observed many times, which comprises layers of mudstone, siltstone, and sandstone [15] and therefore studies the residual shear strength of such area which

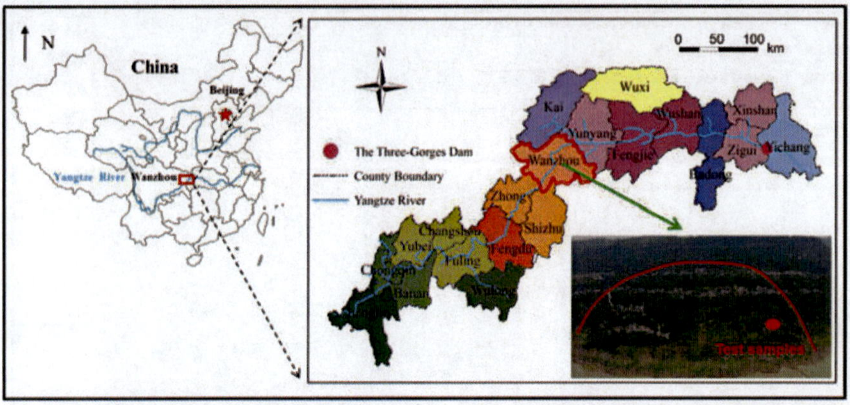

Fig. 6 Location of Wanzhou [16]

is of utmost importance for analyzing the failure mechanism of those reactivated landslides.

The effect of over-consolidation ratio and rate of shearing influencing the residual strength are examined by a series of multistage ring shear tests under drained conditions. The soil samples collected for testing were SM type (silty sand) and show that the fast shearing involves larger local dilation that may increase the void ratio and water content in the shear zone, which can reduce the shear strength [16]. The study also reveals that residual strengths of landslide soils of silty sand texture are not significantly different whether the specimen is overconsolidated or normally consolidated, so the normally consolidated sample can be used to measure the drained residual shear strength of a landslide soil [16].

Miao et al. [17] examine the effect of shear rate and the flocculation of water level in the reservoir for the recovery of residual shear strength; the residual shear strength of soil can be recovered in a short time in case of slow-moving landslides. Deceleration appears when residual shear strength is greater than the sliding force, which stops the movement any further. Also, the response of landslide can be delayed due to the flocculation of water level in the reservoir [17].

5 Conclusions

The residual shear strength of the soil is less affected by the displacement rates when the rate of displacement is in the range of 0.002–0.001 mm/min which is much below the highest rate of movement found in landslides that are 0.005 mm/min [1].

The residual friction angle is highly influenced by the type of clay minerals and the quantity of clay particles. It behaves linearly up to clay fraction of 45% for liquid limit ranges between 60 and 220; after nonlinearity arises [5], more studies are needed to study the behavior of residual shear strength with different soil types.

The presence of NaCl in the pore water can increase the shear strength of soil up to 200% [6]; more research is needed to further investigate the presence of another chemical compound in the pore water and its contribution to the shear strength. The regain of strength in residual soil is observed up to 30% in Montona flysch after 30 days which can significantly help in designing the countermeasures for landslide mitigation [8]. More number of case studies are required for better understanding and to assign suitable mitigation technique that will lead to a more economical and feasible solution for that particular reactivated landslide.

References

1. Skempton AW (1985) Residual strength of clays in landslides, folded strata and the laboratory. Géotechnique 35(1):3–18
2. Wang G, Suemine A, Schulz WH (2010) Shear-rate-dependent strength control on the dynamics of rainfall-triggered landslides, Tokushima Prefecture, Japan, Earth Surf. Proc Landforms 35:407–416
3. Kimura H, Nakamura S, Vithana SB, Sakai K (2014) Shearing rate effect on residual strength of landslide soils in the slow rate range. Landslides 11(6):969–979
4. Collotta T, Cantoni R, Pavesi U, Ruberl E, Moretti PC (1989) A correlation between residual friction angle, gradation and the index properties of cohesive soils. Geotechnique 392:343–346
5. Stark TD, Eid HT (1994) Drained residual shear strength of cohesive soils. J Geotech Geoenviron Eng 1201:269–273
6. Tiwari B, Tuladhar GR, Marui H (2005) Variation in residual shear strength of the soil with pore water chemistry. J Geotech Geoenviron Eng, ASCE 131(12):1445–1456
7. Stark TD, Choi H, McCone S (2005) Drained shear strength parameters for analysis of landslides. J Geotech Geoenviron Eng 1315:575–588
8. Carrubba P, Del Fabbro M (2008) Laboratory investigation on reactivated residual strength. J Geotech Geoenviron Eng 134:302–315. https://doi.org/10.1061/(ASCE)1090-0241(2008)134:3(302)
9. Bishop AW, Green GE, Garga VK, Andresen A, Brown JD (1971) A new ring shear apparatus and its application to the measurement of residual strength. Géotechnique 21(4):273–328
10. Gibo S, Egashira K, Ohtsubo M, Nakamura S (2002) Strength recovery from residual state in reactivated landslides. Géotechnique 52(9):683–686
11. Mesri G, Huvaj-Sarihan N (2012) Residual shear strength measured by laboratory tests and mobilized in Landslides. J Geotech Geoenviron Eng 138:585–593. https://doi.org/10.1061/(ASCE)GT.1943-5606.0000624
12. Gratchev I, Sassa K, Fukuoka H (2005) The shear strength of clayey soils from reactivated landslides 9
13. Senthilkumar V, Chandrasekaran SS, Maji VB (2017) Geotechnical characterization and analysis of rainfall—induced 2009 landslide at Marappalam area of Nilgiris district, Tamil Nadu state, India. Landslides 14(5):1803–1814. https://doi.org/10.1007/s10346-017-0839-2
14. Senthilkumar V, Chandrasekaran SS (2018) Evaluation of residual shear strength of landslide reactivated soil. In: Wu W, Yu H-S (Eds) Proceedings of china-Europe conference on geotechnical engineering, Springer Series in Geomechanics and Geoengineering. Springer International Publishing, Cham, pp 1516–1520. https://doi.org/10.1007/978-3-319-97115-5_136
15. Yin Y, Wang H, Gao Y, Li X: Real-time monitoring and early warning of landslides at relocated Wushan Town, the Three Gorges Reservoir, China, p 11

16. Li D, Yin K, Glade T, Leo C (2017) Effect of over-consolidation and shear rate on the residual strength of soils of silty sand in the Three Gorges Reservoir. Sci Rep 7(1):5503. https://doi.org/10.1038/s41598-017-05749-4
17. Miao H, Wang G, Yin K, Kamai T, Li Y (2014) Mechanism of the slow-moving landslides in Jurassic red-strata in the Three Gorges Reservoir, China. Eng Geol 171:59–69. https://doi.org/10.1016/j.enggeo.2013.12.017

Comparative Study of Seismic Slope Displacement Analyses Based on Probabilistic Modeling of Soil Parameters

Mao-Xin Wang, Wenqi Du, and Dian-Qing Li

Abstract The Newmark-type slope displacement analysis and the computed displacement of a sliding mass have been increasingly used to evaluate the co-seismic slope performance. This study develops a probabilistic procedure to assess earthquake-induced displacements of flexible slopes incorporating the depth-dependent variability of undrained shear strength (c_u) and shear wave velocity (V_s). This procedure is used to compare the non-stationary random modeling of soil parameters (with increasing trends of mean μ and standard deviation σ) with the stationary random modeling (with spatially uniform μ and σ) in terms of several aspects. Using a suite of acceleration–time histories selected, probabilistic analyses are conducted in a generic slope example based on Monte Carlo simulations. The results indicate that using a deterministic analysis by best-estimate soil parameters generally leads to a lower displacement hazard level than the probabilistic modeling. Comparative analyses illustrate that considering varying yet homogeneous c_u and V_s generally results in a larger variance of yield acceleration (k_y) and larger probability of exceedance for displacement than those of the spatially variable soil case. This observation is more evident for the stationary modeling involving smaller k_y values. Also, ignoring the increasing trends of c_u and V_s generally yields deeper failure surfaces and smaller average accelerations. Furthermore, a negative correlation is observed between k_y and the fundamental period of sliding mass ($T_{s,\text{mass}}$). This study could be useful to better quantify the soil property uncertainty within the framework of performance-based seismic slope displacement assessment.

Keywords Seismic slope displacement · Decoupled analysis · Soil spatial variability · Random field

M.-X. Wang (✉) · W. Du · D.-Q. Li
State Key Laboratory of Water Resources and Hydropower Engineering Science, Institute of Engineering Risk and Disaster Prevention, Wuhan University, Wuhan 430072, PR China
e-mail: wangmx@whu.edu.cn

T. G. Sitharam et al. (eds.), *Earthquake Geotechnics*, Lecture Notes in Civil Engineering 187, https://doi.org/10.1007/978-981-16-5669-9_22

1 Introduction

Assessing co-seismic slope stability is an important topic in geotechnical earthquake engineering. The Newmark-type analysis has been increasingly utilized to compute the earthquake-induced permanent displacement of a sliding mass. This type of index is useful in practice when the serviceability of structures after earthquakes is mainly controlled by the deformation (e.g., Kramer [8]; Kare et al. [7]). Initiated by Newmark's pioneering work [13], the rigid-block model (e.g., Rathje and Saygili [16]; Wang et al. [20]) is most appropriate for shallow landslides with stiff materials because the internal deformation of sliding mass can be neglected (i.e., assuming a rigid mass). When deeper slope failures or softer materials are involved, however, the dynamic response of the sliding mass is significant and can be accounted for using a so-called decoupled sliding block approach [11]. In short, the average acceleration within a flexible mass is derived from a dynamic response analysis, and the acceleration is then taken as the input of a conventional sliding block analysis. The one-dimensional (1D) dynamic response is commonly used due to its effectiveness and practicality (e.g., Rathje and Antonakos, [15]).

To account for different sources of uncertainty (e.g., seismic loading and slope properties) associated with the prediction of expected displacement levels, the probabilistic seismic slope displacement analysis has attracted increasing attention in recent years (e.g., Rathje and Saygili [16]; Du and Wang [2]). The representation of seismic loading can be well achieved by selecting/simulating one or multiple suits of ground motions that properly match the specified target distribution of ground motion parameters (e.g., response spectra) or alternatively by conducting the fully probabilistic displacement hazard analysis. In contrast, the treatment of the soil property uncertainty is relatively simplified, although this uncertainty is well recognized and some studies have revealed that the resultant influence on the displacement results is considerable (e.g., Du et al. [4]). Furthermore, the slope materials are commonly assumed to be homogeneous, which is, however, unrealistic due to the depositional and post-depositional processes of geo-materials in slopes. Recently, Wang and Rathje [22] conducted a probabilistic displacement assessment of a slope in non-homogeneous soils that have increasing trends of undrained shear strength (c_u) and shear wave velocity (V_s) with depth. The uncertainty of c_u and V_s in that study was approximately quantified by logic tree, so the inherent spatial variability of soil properties (e.g., Phoon and Kulhawy [14]; Li et al. [10]) is neglected.

To address this issue, this paper aims to develop a procedure for probabilistic analysis of flexible slope displacements incorporating spatially variables c_u and V_s, and further compare different probabilistic modeling types based on the analysis results. This procedure specifically consists of four modules: (1) probabilistic modeling of spatially variables c_u and V_s; (2) estimation of the slope's yield acceleration (k_y) considering non-homogeneous c_u; (3) dynamic response of the slope system considering non-homogeneous V_s; and (4) computation of the sliding displacement. The remaining part of this paper starts with a description of the procedure, followed by a slope example for illustration and comparative studies. This procedure could be applied to the performance-based probabilistic seismic design for heterogeneous soil slopes.

2 Methodology

2.1 Probabilistic Modeling of c_u and V_s

The random field theory [18] is adopted to account for the spatial variability of soil properties, and the covariance matrix decomposition method (e.g., Wang et al. [21]) is used to generate random field (RF) realizations of c_u. A RF consists of n_e interdependent random variables, each corresponding to a spatial location of the slope profile. To consider the increasing trend of c_u with depth, the non-stationary RF simulation method described by Griffiths et al. [6] is adopted. The general idea is to model the strength c_{u0} at the crest level as a lognormal stationary RF with mean of μ_0 and standard deviation of σ_0. The ratio of the randomized strength c_{u0i} to μ_0 is used to scale the best-estimate value of c_u at the i-th field element (or spatial location). A realization of the c_u field can thus be obtained as

$$c_{ui} = \frac{c_{u0i}}{\mu_0}(\mu_0 + r_{cu}z_i),\ i = 1, 2, \ldots, n_e \tag{1}$$

where z_i is the depth of the i-th element; r_{cu} is the gradient of strength changing with z, and $r_{cu} = 0$ corresponds to the stationary RF modeling; c_{ui} is the generated realization of c_u at that element. For illustration purpose, the correlation between each two stationary field variables of $\ln c_{u0}$ is quantified by the theoretical single exponential auto-correlation function (e.g., Tang et al. [17]):

$$\rho(\tau_x, \tau_y) = \exp\left[-2\left(\tau_x/\delta_h + \tau_y/\delta_v\right)\right] \tag{2}$$

where τ_x and τ_y are the distances in horizontal and vertical directions between the two filed elements, respectively; δ_h and δ_v are the horizontal and vertical scales of fluctuation, respectively. The parameter δ_h is typically more than one order of magnitude larger than δ_v [14], which means a more significant variation of soil properties in the vertical direction. This study focuses on describing the vertical trend of soil parameters, so $\delta_h = \infty$ is assumed throughout the paper. Detailed steps for generating the stationary lognormal RF realization (i.e., c_{u0i} in this study) are given elsewhere [10]. Similar to Wang and Rathje [22], V_s is obtained using a generic empirical model [12]:

$$V_{si} = \left(c_{ui}/0.09\right)^{1/1.25} \tag{3}$$

where V_{si} is the shear wave velocity realization at the i-th element.

2.2 Estimation of the Slope Yield Acceleration

The mass sliding is initiated once the input acceleration–time history exceeds the yield acceleration k_y (in units of g, gravitational acceleration), which is typically obtained as the pseudo-static acceleration (i.e., the product of g and the pseudo-static coefficient k_h) that results in factor of safety (FS) of 1. To allow for efficient probabilistic calculations, Bishop's simplified method is incorporated. Including the external horizontal seismic force and considering the undrained slope case, FS can be obtained as (e.g., Duncan et al. [5])

$$FS = \frac{\sum_{j=1}^{n_s} \frac{c_{uj}\Delta l_j \cos\alpha_j + W_j \tan\phi_{uj}}{\cos\alpha_j + (\sin\alpha_j \tan\phi_{uj})/FS}}{\sum_{j=1}^{n_s} \left(W_j \sin\alpha_j + k_h W_j d_v / R\right)} \tag{4}$$

where n_s is the number of slices of a sliding mass; α_j is the inclination of the base of the jth slice; W_j is the weight of that slice; R is the radius of the slip circle; d_{vj} is the vertical distance between the center of gravity of the j-th slice and the center of the slip circle; c_{uj} and ϕ_{uj} are the undrained shear strength and friction angle at the field element for that slice's base, respectively. By substituting FS = 1 and $\phi_{uj} = 0$ into Eq. (4), k_y can be calculated as

$$k_y = \frac{\sum_{j=1}^{n_s} \left[c_{uj}\Delta l_j - W_j \sin\alpha_j\right]}{R \sum_{j=1}^{n_s} W_j d_{vj}} g \tag{5}$$

Note that the base of each slice passes through one RF element only (i.e., c_{uj} has a unique value). One should specify a set of potential slip surfaces (PSSs) and then perform Eq. (5) for each PSS. The PSS resulting in the smallest k_y is identified as the critical slip surface (CSS), and the sliding block above the CSS is used in the permanent displacement analysis [13].

2.3 Calculation of the Average Acceleration–Time History

Within the decoupled approach, the dynamic response of the identified sliding block is considered using the 1D equivalent-linear seismic response analysis (e.g., Du and Pan [1]). The sum of the thickness (h_{mass}) of the sliding block and the distance from the block base to bedrock is taken as the total thickness H of a 1D column. Multiple horizontal soil layers are included such that the generated V_{si} (see Sect. 2.1) is assigned onto the corresponding layer. The destabilizing force–time history is considered equal to the product of the average acceleration $k\ (t)$ and the mass of the sliding block [15], such that $k\ (t)$ can be calculated as

$$k(t) = \tau_h(t)/\sigma_v \cdot g \tag{6}$$

where $\tau_h(t)$ is the horizontal shear stress–time history at the depth h_{mass}; σ_v is the total vertical stress at that depth.

2.4 *Computation of the Sliding Displacement*

Given the seismic loading $k\ (t)$ and the slope's dynamic resistance k_y, the permanent displacement D can be computed as

$$D = \int I\big(\dot{A}(t)\big)\dot{A}(t)dt \tag{7}$$

where the sliding acceleration $A\ (t)$ is determined as $k\ (t) - k_y$; $\dot{A}(t)$ is the sliding velocity obtained by numerically integrating $A\ (t)$ once; $I\ (\cdot)$ is an indicator function that is equal to one if $\dot{A}(t) > 0$ and zero otherwise.

3 Illustrative Example

The presented procedure is applied to an undrained clay slope example for illustrative and comparative study. Figure 1 displays the slope geometry, which has a height of 15 m and a slope angle of 26.6°. Other deterministic parameters include r_{cu} of 3 kPa/m and soil's unit weight of 18.5 kN/m^3, as well as the modulus reduction and damping ratio curves of Vucetic and Dobry [19] for modeling the nonlinear soil properties.

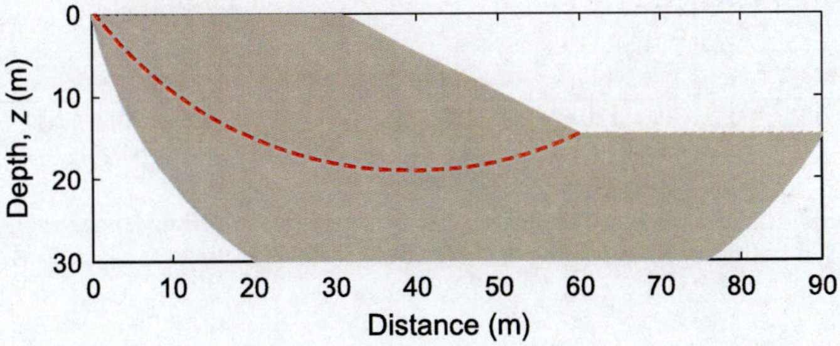

Fig. 1 Slope geometry. The critical slip surface and other potential slip surfaces are marked in a dashed red line and solid gray lines, respectively

The mean and standard deviation of the log-normally distributed c_{u0} are assumed to be $\mu_0 = 35$ kPa and $\sigma_0 = 10.5$ kPa, respectively, corresponding to a coefficient of variation (COV) of 0.3 (e.g., Phoon and Kulhawy [14]). A large number of PSSs is specified as plotted in Fig. 1, and the CSS with $k_y = 0.26$ g is identified by using the best-estimate c_u (i.e., $c_{u0} = \mu_0$).

Besides the non-stationary RF modeling using $\delta_v = 4$ m for c_{u0}, we also consider the non-stationary random variable (RV) modeling with $\delta_v = \infty$ (i.e., randomized c_u profile without fluctuation), which means that c_{u0} is homogeneous in each random realization. Three typical c_u realizations for the two models from slope crest to base are shown in Fig. 2a and b. The former case is, to some extent, more realistic as revealed by borehole data (e.g., Li et al. [9]). Also, Fig. 2c and d presents the c_u realizations for the more often used stationary RF modeling and stationary RV modeling, respectively, where $\mu_0 = 80$ kPa (associated with the mid-point depth value of the best-estimate c_u trend), COV = 0.3, and $r_{cu} = 0$ are specified. All of the four models will be considered in the following analysis for comparison.

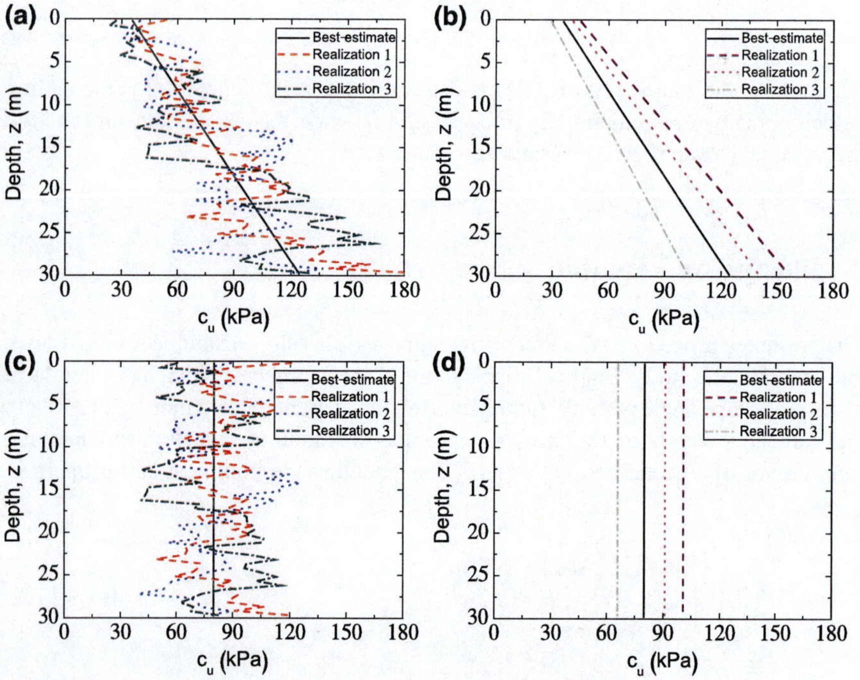

Fig. 2 Mean and three randomized profiles of c_u for **a** non-stationary RF modeling, **b** non-stationary RV modeling, **c** stationary RF modeling, and **d** stationary RV modeling

3.1 Comparison of Yield Acceleration and Fundamental Period

After performing the Monte Carlo simulation, the k_y distributions for the four random models are obtained from Eq. (4) and are displayed in Fig. 3 using the boxplot. The negative k_y values are set to be 0.01 g to represent highly vulnerable slopes (Du et al. [4]). Several observations can be made. First, the median k_y value is generally smaller than the associated best-estimate k_y value, especially for the non-stationary RF modeling. Second, the two RV models tend to produce larger variance of k_y than the two RF models, which would further yield larger uncertainty of slope displacements. Third, the stationary models generally result in smaller k_y than the non-stationary models. The possible reason is that the former models generally yield deeper failure surfaces (larger h_{mass}), corresponding to larger weights of sliding masses.

As another important parameter in quantifying the earthquake-induced deformation of flexible slopes, the fundamental period of sliding mass ($T_{s,\text{mass}}$) for each random model is derived from the average thickness of the sliding mass (h_{mass}) and the average shear wave velocity over that thickness ($\overline{V}_s$) using $T_{s,\text{mass}} = 4h_{\text{mass}}/\overline{V}_s$ [22]. Some previous studies directly model k_y and $T_{s,\text{ mass}}$ as independent random variables to represent the soil property uncertainty for simplicity. For practical interest, this study makes a preliminary investigation of the correlation between k_y and $T_{s,\text{ mass}}$ using the analysis results. Figure 4 displays the distribution of $T_{s,\text{ mass}}$ with respect to k_y for the two RF models. It is seen that the $T_{s,\text{ mass}}$ values are generally within the range of 0.2–0.5 s that is judged to be representative enough.

It is observed that k_y and $T_{s,\text{ mass}}$ are negatively correlated, which is qualitatively consistent with the finding of Wang and Rathje [22] that the high strength slope case produces lower best-estimate $T_{s,\text{ mass}}$ than the low strength case. Since Fig. 5 shows that the distribution of h_{mass} is insensitive to k_y, the negative correlation between k_y

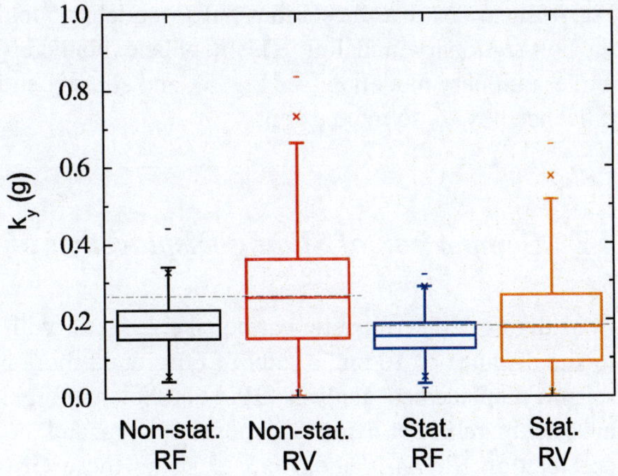

Fig. 3 Distributions of k_y based on realizations of soil parameters using four random models. In these boxplots, the dash-dotted line represents the best-estimate value of k_y for each model

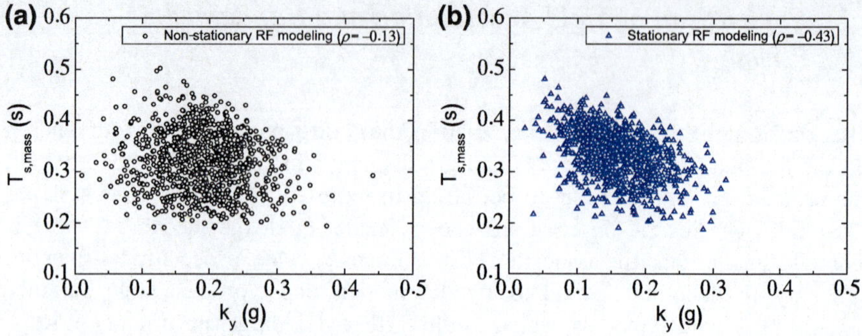

Fig. 4 Distribution of $T_{s,\,\mathrm{mass}}$ versus k_y for different random models

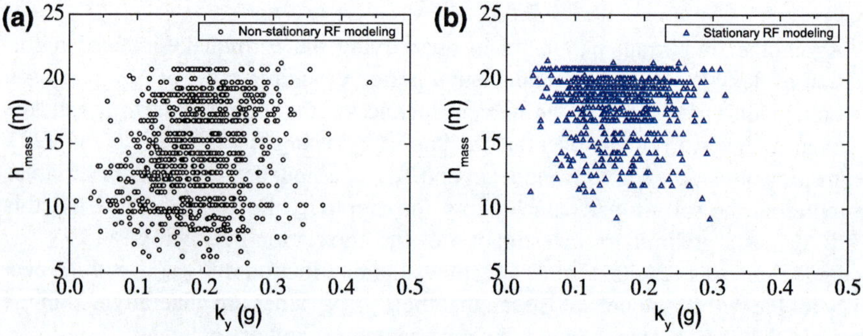

Fig. 5 Distribution of h_{mass} versus k_y for different random models

and $T_{s,\,\mathrm{mass}}$ should be mainly resulted from the variable $\overline{V}_s$; larger soil stiffness (i.e., $\overline{V}_s$) generally corresponds to larger k_y due to the used relationship between c_u and V_s and vice versa. Thus, a smaller k_y is expected as $T_{s,\,\mathrm{mass}}$ becomes larger. Furthermore, Fig. 4 illustrates that the stationary RF modeling yields a stronger correlation than the non-stationary modeling. This may be explained by the narrower range of h_{mass} in the stationary modeling (see Fig. 5), and consequently, the dispersion of $T_{s,\,\mathrm{mass}}$ is influenced by $\overline{V}_s$ to more extent.

3.2 Comparison of Sliding Displacement Results

Considering a deterministic earthquake scenario with magnitude of 7 and source-to-site distance of 10 km, a total of 60 ground motions (GMs) are selected for the seismic displacement analysis [3]. As is shown in Fig. 6, the selected GMs generally have comparable 5% damped response spectra, and the median value of peak ground acceleration (PGA) is about 0.3 g. Using these GMs as inputs for 1D response

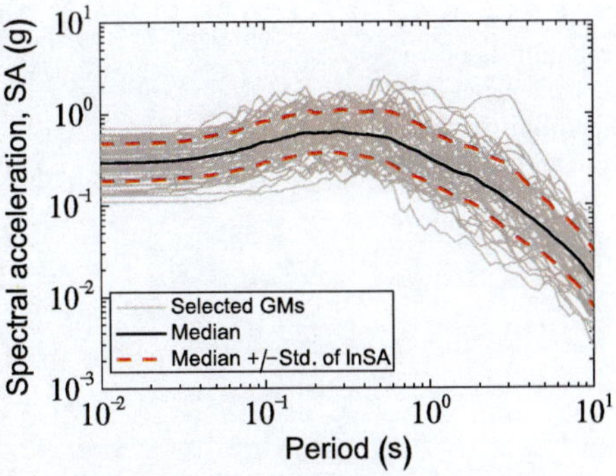

Fig. 6 Spectral distribution of selected 60 ground motions

analyses of individual randomized V_s profiles, the average acceleration $k(t)$ for each realization can be obtained (see Sect. 2.3). The peak value of $k(t)$ is called $k_{\max}$, whose ratio to PGA of the inputted GM (i.e., $k_{\max}$/PGA) is computed. Noted that $k_{\max}$/PGA is an important parameter in most existing empirical-based decoupled Newmark models.

The distribution of $k_{\max}$/PGA versus the ratio of $T_{s,\,\text{mass}}$ to the mean period of inputted GM (i.e., $T_{s,\,\text{mass}}/T_m$), corresponding to each random model, is shown in Fig. 7. Also plotted in this figure is the trend produced by an existing predictive model [15] with PGA = 0.3 g. The main observations are summarized as follows:

(1) The data distribution of the random modeling is generally comparable with the empirical model trend, though resulting in substantial variance of $k_{\max}$/PGA. Specifically, the variance is larger in the RF modeling, which is not surprising, because the associated slope failure mode varies more considerably than the RV modeling (larger variance of h_{mass}).

(2) The difference of data patterns between Fig. 7a and b illustrates that the non-stationary modeling cases produce larger $k_{\max}$/PGA than the stationary

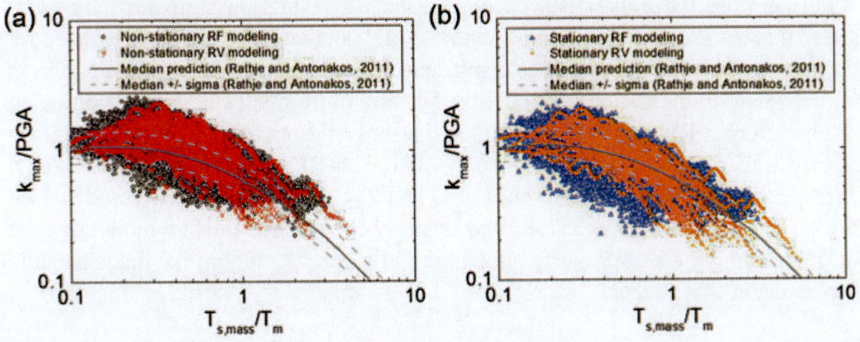

Fig. 7 Distribution of $k_{\max}$/PGA versus $T_{s,\,\text{mass}}/\text{T}_\text{m}$ for different random models. The Rathje and Antonakos [15] model trend with PGA = 0.3 g is also shown

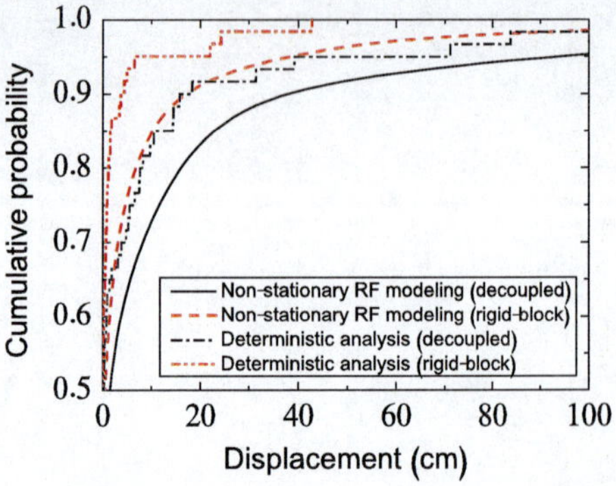

Fig. 8 Cumulative probabilities of displacements for deterministic and probabilistic cases

modeling cases. This is because the former ones generally correspond to shallower failure masses (smaller h_{mass} in Fig. 5). As a result, it is seen from Fig. 7b that k_{max}/PGA for the Rathje and Antonakos [15] model is slightly underestimated when $T_{s,\,mass}/T_m$ is within the range of 0.5 to 1. In general, the empirical model developed using homogeneous soil profiles can yield reasonable prediction of k_{max}/PGA even in cases with non-homogeneous c_u and V_s. Yet, a larger uncertainty may be included when the soil spatial variability is considered (corresponding to the RF modeling).

The displacement values are then obtained from Eq. (5) for different models. Figure 8 compares the results for decoupled and rigid-block models regarding both the probabilistic modeling and the deterministic analysis (using only the best-estimate soil parameters and 60 GMs). It is seen that the decoupled model produces larger displacement hazard level (i.e., smaller cumulative probability) than the rigid-block model, which is reasonable considering the range of the T_s/T_m ratio in this case. In addition, the displacement hazard level for the random model is generally higher, implying that the displacement estimation ignoring the soil uncertainty is non-conservative.

Furthermore, the probabilistic decoupled analysis results are compared in Fig. 9. It is seen that the RV modeling (ignoring the spatial fluctuation) generally yields larger displacements at moderate-to-high displacement levels than the RF modeling. This is more evident for the stationary modeling, because more small k_y values are obtained by the stationary RV modeling (see Fig. 3). The non-stationary RF modeling results in slightly higher displacement level than the stationary RF modeling in this case study, corresponding to the slight difference of k_y and $T_{s,\,mass}$ between the two models (see Fig. 4). These results could be useful to properly choose probabilistic modeling type for quantifying the soil property uncertainty in slope-specific probabilistic seismic displacement assessment.

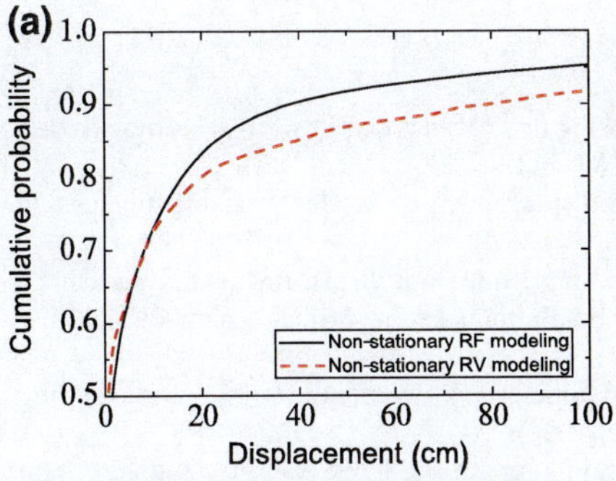

Fig. 9 Cumulative probabilities of various displacements for different random models

4 Summary and conclusions

This study has introduced the random field (RF) theory to account for the spatial variability of c_u and V_s in the seismic slope displacement analysis. The non-stationary RF modeling (with the spatial fluctuation) and non-stationary random variable (RV) modeling (without the spatial fluctuation) that include the depth-dependent soil variability were compared with the stationary RF modeling and stationary RV modeling (ignoring the varying trends of c_u and V_s). Based on a slope example, comparative analyses were conducted in terms of the distributions of two important slope property parameters (i.e., k_y and $T_{s,\,mass}$), as well as the distributions of k_{max}/PGA and sliding displacement obtained using a suit of ground motions. Comparative results lead to the following major conclusions:

1. Ignoring the spatial variability of c_u and V_s (i.e., RV modeling) results in larger variance of k_y compared with the RF modeling. The stationary RF/RV modeling using the mid-depth value of the c_u trend as the statistical mean generally leads to lower k_y level than the non-stationary RF/RV modeling, and therefore, it may overestimate the earthquake-induced susceptibility of slope.
2. The parameters k_y and $T_{s,\,mass}$ are found to be negatively correlated; a stronger correlation is observed when the depth-dependent trends of c_u and V_s are neglected (i.e., the stationary random modeling). Additionally, the k_{max}/PGA ratio is generally larger for the non-stationary modeling in which shallower failure masses and higher average motions within the masses are more likely to occur.
3. The displacement hazard might be underestimated if the soil property uncertainty is not considered. The RV modeling generally yields higher displacement hazard (i.e., larger probability of exceedance for a displacement threshold) than the RF modeling, while this difference is more evident for the stationary modeling. The crudest stationary RV modeling may lead to the largest displacement hazard.

References

1. Du W, Pan TC (2016) Site response analyses using downhole arrays at various seismic hazard levels of Singapore. Soil Dyn Earthq Eng 90:169–182
2. Du W, Wang G (2016) A one-step Newmark displacement model for probabilistic seismic slope displacement hazard analysis. Eng Geol 205:12–23
3. Du W, Wang G (2018) Ground motion selection for seismic slope displacement analysis using a generalized intensity measure distribution method. Earthquake Eng Struct Dynam 47(5):1352–1359
4. Du W, Wang G, Huang D (2018) Influence of slope property variabilities on seismic sliding displacement analysis. Eng Geol 242:121–129
5. Duncan JM, Wright SG, Brandon TL (2014) Soil strength and slope stability, 2nd edn. John Wiley & Sons, New Jersey
6. Griffiths DV, Huang J, Fenton GA (2015) Probabilistic slope stability analysis using RFEM with non-stationary random fields. In: Schweckendiek T, van Tol AF, Pereboom D et al (eds) Geotechnical safety and risk V. IOS Press, Amsterdam, pp 704–709
7. Kare V, Nayak C, Jagadale U, Deulkar W (2020) Earthquake response of 3D frames with strap footıng considering soil structure interactıon. Techno-societal 2018. Springer, Cham, pp 895–904
8. Kramer SL (1996) Geotechnical earthquake engineering. Practice Hall, New Jersey
9. Li DQ, Qi XH, Phoon KK, Zhang LM, Zhou CB (2014) Effect of spatially variable shear strength parameters with linearly increasing mean trend on reliability of infinite slopes. Struct Saf 49:45–55
10. Li DQ, Wang MX, Du W (2020) Influence of spatial variability of soil strength parameters on probabilistic seismic slope displacement hazard analysis. In: Engineering Geology, pp 105744
11. Makdisi FI, Seed HB (1978) Simplified procedure for estimating dam and embankment earthquake-induced deformations. J Geotech Eng Div 104(7):849–867
12. Moon SW, Ku T (2018) Undrained shear strength in cohesive soils estimated by directional modes of in-situ shear wave velocity. Geotech Geol Eng 36(5):2851–2868
13. Newmark NM (1965) Effects of earthquakes on dams and embankments. Geotechnique 15(2):139–160
14. Phoon KK, Kulhawy FH (1999) Characterization of geotechnical variability. Can Geotech J 36(4):612–624
15. Rathje EM, Antonakos G (2011) A unified model for predicting earthquake-induced sliding displacements of rigid and flexible slopes. Eng Geol 122(1–2):51–60
16. Rathje EM, Saygili G (2008) Probabilistic seismic hazard analysis for the sliding displacement of slopes: scalar and vector approaches. J Geotech Geoenviron Eng 134(6):804–814
17. Tang XS, Wang MX, Li DQ (2020) Modeling multivariate cross-correlated geotechnical random fields using vine copulas for slope reliability analysis. Comput Geotech 127:103784
18. Vanmarcke EH (2010) Random fields: analysis and synthesis. 2nd edn. World Scientific Publishing Co. Pte. Ltd., Singapore
19. Vucetic M, Dobry R (1991) Effect of soil plasticity on cyclic response. J Geotech Eng 117(1):89–107
20. Wang MX, Huang D, Wang G, Li DQ (2020) SS-XGBoost: A machine learning framework for predicting Newmark sliding displacements of slopes. J Geotech Geoenviron Eng 146(9):04020074
21. Wang MX, Tang XS, Li DQ, Qi XH (2020) Subset simulation for efficient slope reliability analysis involving copula-based cross-correlated random fields. Comput Geotech 118:103326
22. Wang Y, Rathje EM (2018) Application of a probabilistic assessment of the permanent seismic displacement of a slope. J Geotech Geoenviron Eng 144(6):04018034

Comparison of Small-Strain Shear Modulus of Beach and Volcanic Sands

Shaurya Sood, Gabriele Chiaro, Thomas Wilson, and Mark Stringer

Abstract The small-strain shear modulus (G_{max}) of soils is an essential dynamic parameter for any seismic design. This paper aims to investigate the G_{max} of two reconstituted sands—beach and volcanic sands of variable shape and grain size distribution. A series of bender elements tests were carried out on dry samples at different frequencies and confining pressures. The interpretation of results is carried out through first-time arrival method. The results indicate lower G_{max} values of volcanic sand in comparison with beach sand, which can be attributed to lower particle density and higher void ratios for volcanic sand. Using the available relationships in the literature, the G_{max} values were estimated, and a brief comparison was made between measured and estimated values.

Keywords Small-strain shear modulus · Beach sand · Volcanic sand

1 Introduction

The tectonic location of New Zealand urges an understanding of the soil properties under seismic loading. In addition to the typical hard-grained soils such as beach sand, the geologic activities in the past have led to the formation of volcanic soils. One such volcanic soil known as pumice is abundantly found around the Rotorua region of North Island. Pumiceous soils can be either pyroclastic (airfall tephra and close to origin) or epiclastic (eroded deposits such as in river basins). The distinguishing feature of pumice from hard-grained soils is the presence of internal voids, which makes them lightweight and highly crushable. The voids in volcanic soils are created due to rapid liberation of high-pressure entrained gases upon the ejection of molten lava from the Earth's crust. Therefore, their engineering properties require a

---

S. Sood (✉) · G. Chiaro · M. Stringer
Department of Civil and Natural Resources Engineering, University of Canterbury, Christchurch, New Zealand
e-mail: shaurya.sood@pg.canterbury.ac.nz

T. Wilson
Department of Earth and Environment, University of Canterbury, Christchurch, New Zealand

T. G. Sitharam et al. (eds.), *Earthquake Geotechnics*, Lecture Notes in Civil Engineering 187, https://doi.org/10.1007/978-981-16-5669-9_23

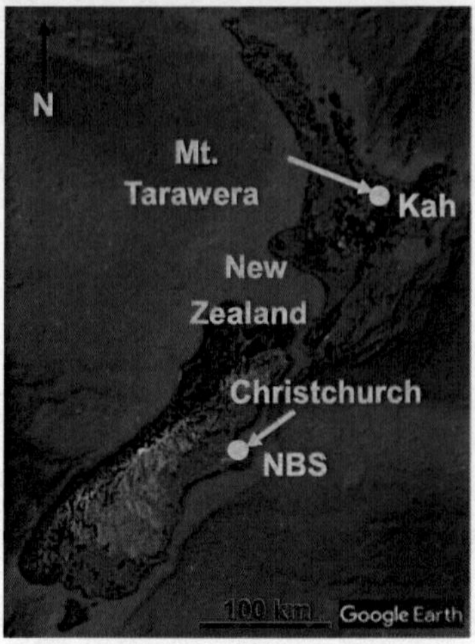

Fig. 1 Sampling locations

deeper investigation and are expected to be different from hard-grained soils. Both these kinds of sands have potential application as geomaterials in civil engineering applications due to high shear strength when prepared at high relative density. The abundance, lightweight nature and high shear strength of pumiceous sand make it a promising material to be used for geotechnical purposes, for example, as a filler in land reclamation sites.

The small-strain shear modulus (G_{max}) is an essential parameter for predicting deformations and seismic design of geostructures. In this paper, we present the bender element results on two reconstituted dry sands, one of them being hard-grained beach sand while the other being volcanic in origin. The volcanic pumiceous soil has been sampled from Rotorua region around Mt. Tarawera. The hard-grained sand chosen for comparison is taken from New Brighton beach in Christchurch region (Fig. 1).

2 Materials and Experimental Procedure

2.1 Materials Used

Two granular soils were used in this study—one hard grained while the other being vesiculated volcanic in nature. The hard-grained sandy soil from New Brighton Sand (NBS) is fairly uniform, while the airfall pyroclastic pumice contained particles up to 37.50 mm. In this study, the pumice particles with grain sizes between 0.075–2 mm were used (Fig. 2). These have been referred to as "Kaharoa" ashes (Kah), the term Kaharoa being used from the 1305 AD Mt. Tarawera eruption which was predominantly rhyolitic [1]. The index properties such as grain size characteristics,

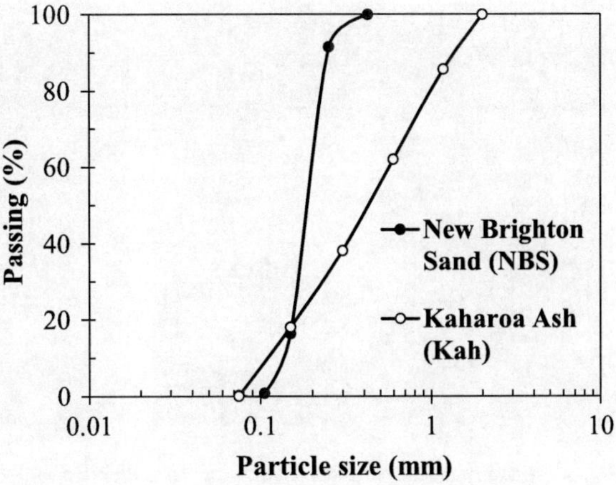

Fig. 2 Grain size distributions of materials used

Table 1 Index properties of materials used

Test Soil	Material	Grain size characteristics					
		D_{50} (mm)	C_u*	C_c*	G_s	e_{max}	e_{min}
New Brighton Sand (NBS)	Beach sand	0.19	1.58	1.02	2.68	0.93	0.58
Kaharoa ash (Kah)	Pumiceous sand	0.45	4.92	0.85	2.37	1.23	0.78

* $C_u = D_{60}/D_{10}$, $C_c = D_{30}^2/(D_{60} x D_{10})$

specific gravity (G_s), and void ratios (e_{max} and e_{min}) so determined are summarized in Table 1.

Both soils can be classified as poorly graded sands according to the specification given by ASTM D [2]. The G_s tests were determined according to ASTM D [3] by water pycnometry and vacuum application. It is worthwhile to mention here that the G_s of pumice deposits was determined by crushing individual particle to eliminate the effect of internal air voids [4].

The maximum (e_{max}) and minimum (e_{min}) void ratios were determined by JIS A [5]. The G_s of Kah is lower than that of NBS, indicating its lightweight nature, and the void ratios (e_{max} and e_{min}) of Kah ashes are higher than that of NBS. On observing the representative particles through a microscope, it was seen that the NBS consisted of subrounded to rounded particles while the Kah contained angular particles (Fig. 3). The presence of voids in Kah ashes is also evident as compared to hard-grained NBS particles in the same figure.

Fig. 3 SEM images of particles (natural beach sand NBS and volcanic sand Kah)

2.2 *Experimental Procedure*

All tests were performed on dry specimens 61 mm in diameter and 130 mm in height in a triaxial cell type fitted with bender elements (BEs)—transmitter on the top cap and receiver on the pedestal. The specimens were prepared in dense state at similar initial relative density of 70%. The sample preparation was carried out by dividing the total mass into three layers of equal dry mass, and then, compaction was carried out by horizontally tapping the side of the mold with a wooden hammer until required densification was achieved. After the setup, the BE tests were carried out at isotropic confining pressures (*p*') of 25, 50, and 92 kPa, respectively. Sine waves with frequencies of 5–15 kHz were employed as input signals using function generator. The function generator provided excitation, which led to the vibration of transmitter BE, and shear wave was produced though the soil matrix from the top cap BE. The receiver BE converted this wave into an electrical pulse. Both these transmitted and received signals were stacked and viewed graphically through the oscilloscope and then stored on a computer.

3 Results and Discussion

The shear wave velocity measurements (V_s) in this study were carried out using a pair of bender elements (BEs). The V_s was calculated by measuring the travel distance (*d*) and travel time (*t*) as

$$V_s = d/t \tag{1}$$

The travel distance was taken as the distance between the tips of the two BEs. The shear waves were propagated vertically through the samples producing particle motion in the horizontal direction. Different methods have been suggested in the literature for the estimation of first time arrival of shear waves in the analysis of BE tests [6]. The first time arrival method is used in this study for the calculation of V_s. The BE tests for both sands were carried out for initial relative density of 70% in dry states under variable frequencies (5, 7, 10, and 15 kHz) and isotropic confining pressures of 20, 50, and 92 kPa. A typical example of the input and output shear waves from the BE tests on NBS under 92 kPa isotropic confining pressure at frequencies of 5, 7, 10, and 15 kHz is shown in Fig. 4. The arrows in this figure indicate the selection of the first arrival times of the shear waves in these tests. A reference frequency of 10 kHz was used in all the tests to evaluate the first arrival time of shear waves.

Using the tip to tip distance between the BEs and the travel time between the input and output signals, the V_s (from Eq. 1) and ultimately the small-strain shear modulus (G_{max}) of both soils were measured as

$$G_{max} = \rho \cdot V_s^2 \tag{2}$$

where ρ is the mass density of the soil. A similar procedure was followed for determining t, V_s, and G_{max} for Kah ashes ($e_0 = 0.92$, $D_r = 70\%$) under similar variable frequencies and confining pressures. It was found that the frequency had little or no

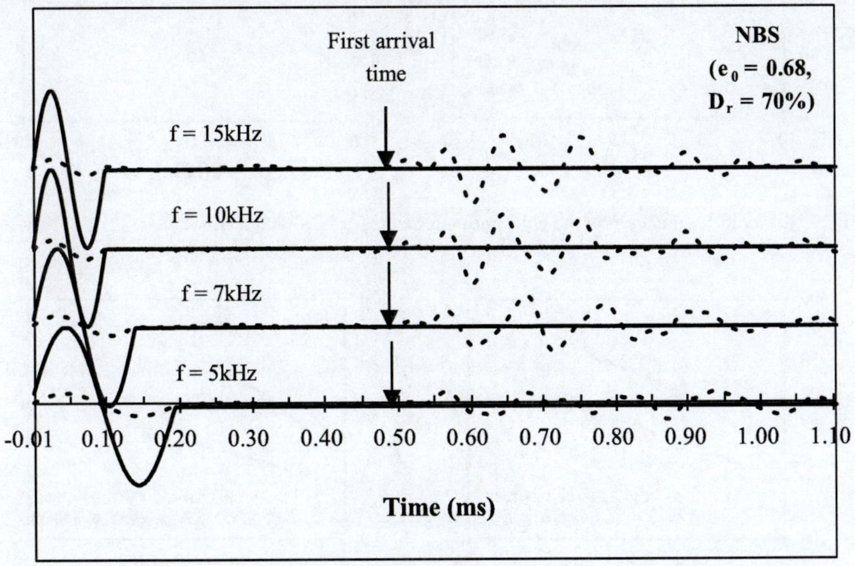

Fig. 4 Analysis of BE test results based on the first arrival time for NBS—$e_0 = 0.68$, $D_r = 70\%$, $p' = 92$ kPa, and $q = 0$ kPa (amplitude = 5 V, frequencies, and $f = 5, 7, 10$, and 15 kHz)

effect on the V_s and G_{max} measurements for both NBS and Kah ashes (Fig. 5) at a certain confining pressure. In the same figure, the increase of G_{max} with the increase in p' is evident for both sands though the magnitude of G_{max} for Kah is significantly lower than NBS. The effect of confining pressure p' on the V_s and G_{max} has been presented separately for both sands in Fig. 6. The lower V_s and G_{max} for Kah can be attributed to factors such as internal voids, lower particle density, and higher void ratios in comparison with the hard-grained nature of NBS. The difference in void ratios for the two studied soils can be attributed to their different grain sizes and shapes [7]. In comparison with subrounded NBS, which consists of fairly identical shaped particles, the volcanic Kah is angular in shape and has a wider range of particle sizes (as viewed through electron microscope), which results in larger void ratio values.

The G_{max} of granular soils has been found to be highly dependent on the mean effective confining pressure p' and void ratio e. To inculcate the effects of both p' and e, a more general form G_{max} can be written as

$$G_{max} = A \cdot f(e) \cdot \left(\frac{p\prime}{p_a}\right)^n \tag{3}$$

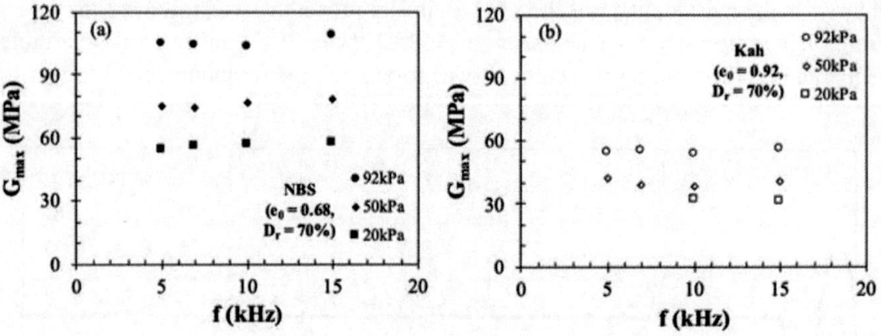

Fig. 5 Effect of frequency f on the G_{max} measurements for NBS and Kah sands

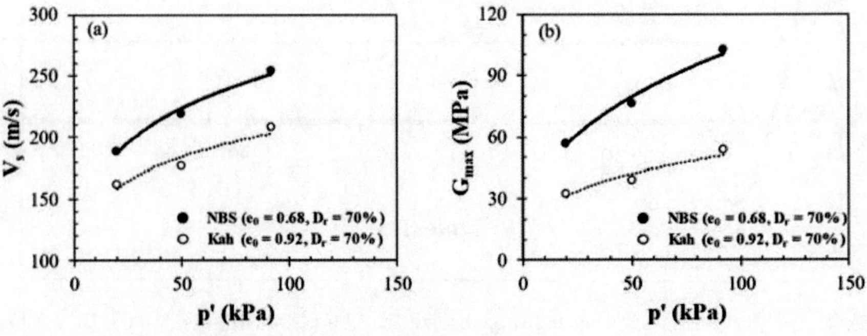

Fig. 6 Effect of confining pressure p' on the V_s and G_{max} measurements for NBS and Kah sands

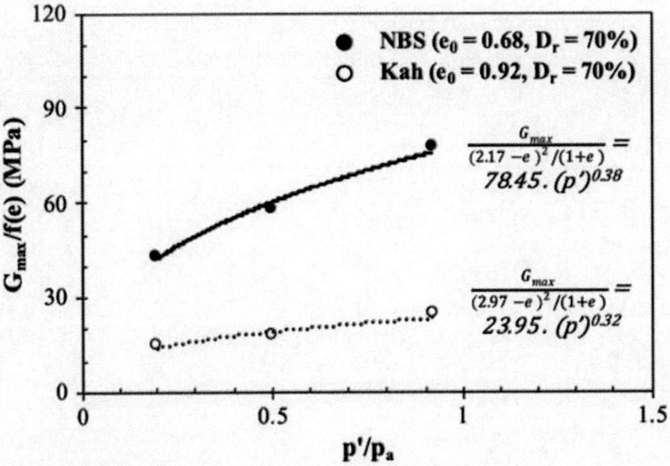

Fig. 7 Effect of confining pressure p' on the G_{max} for NBS and Kah sands

where f (e) is the void ratio function, p' is the mean effective confining pressure, p_a (1 atm = 100 kPa) is the atmospheric pressure used for normalization purposes, and A and n are experimentally determined model parameters. Various expressions for f (e) have been derived in the literature in the form of ratio or power relationships. Hardin and Richart [8] proposed empirical relationship f (e) for rounded and angular sands as

$$f(e) = \frac{(c-e)^2}{(1+e)} \tag{4}$$

where $C = 2.17$ and 2.97 for rounded and angular sands, respectively. Using Eq. 4 for studied materials—NBS (subrounded to rounded particles) and Kah ashes (angular particles), the normalized G_{max} was plotted against p'/p_a (shown in Fig. 7).

Although the data points are less, the normalized G_{max} increases with p' for both sands, but relatively higher G_{max} values are shown by hard-grained NBS in comparison with vesiculated Kah ashes. The increase of G_{max} with the increase of confining pressure p' can be explained due to the development of higher magnitude forces at contact points [9]. The values of the pressure-dependent n factor of both sands are less than 0.5, the n value of Kah ($n = 0.32$) being lesser than that of NBS ($n = 0.38$). A-factor values of approximately 78 and 24 are determined for subrounded NBS and angular Kah, which are closer to literature values proposed by Hardin and Richart [8] (70 for rounded and 33 for angular particles). The lesser A and n values for Kah suggest lesser dependence of G_{max} on p' as compared to NBS. Some additional tests at higher confining pressures are needed to justify this trend.

It is important here to mention that the two soils differ significantly in terms of particle shape, particle size distribution, and void ratios. The particle shape of

Table 2 Components and proposed expressions for G_{max} of sands

Study	Expression	*A* (MPa)	*x*	*n*
Saxena and Reddy [10]	$A. (1 / (0.3 + 0.7e^2)) \cdot {P_a}^{1-n} . p'^n$	428.2	–	0.574
Menq [11]	$A. e^x. (p'/p_a)^n$	$67.1{C_u}^{-0.2}$	$-1-(D_{50}/20)^{0.75}$	$0.48{C_u}^{0.09}$
Wichtmann and Triantafyllidis [12]	$A. ((x - e)^2/(1 + e)) {P_a}^{1-n} . p'^n$	$1563 + 3.13{C_u}^{2.98}$	$1.94*\exp(-0.066C_u)$	$0.40{C_u}^{0.18}$
Senetakis et al. [13]	$A. e^x. (p'/p_a)^n$	$57.01 - 5.88C_u$ (Beach) $52.02 - 3.04C_u$ (Volcanic)	$-0.28C_u - 0.98$	0.47 0.55

the two sands was considered to be in general categories through image viewing—subrounded to rounded for NBS and angular for Kah. To account for the effect of gradation characteristics on the G_{max} for both sands, some past studies [10–13] co-relating the parameters *A* and *n* in the Eq. 3 to the coefficient of uniformity, C_u, and the mean grain size D_{50} were considered. A summary of the proposed component parameters and expressions for the G_{max} of sands using the four mentioned studies is provided in Table 2. Using the derived expressions in this table, the G_{max} values of NBS and Kah ashes were then estimated and compared to the measured values.

The plots of measured against estimated G_{max} values for NBS and Kah are presented in Figs. 8 and 9, respectively, with maximum error involved from all four studies. With the amount of dataset available in this study, most relationships considered seem to underestimate the G_{max} values for NBS and Kah sands. Out of the considered models from the literature, the relationship proposed by Witchmann and Triantafyllidis [12] seems to produce most accurate results with ± 10% error for subrounded NBS followed by Menq [11] (±33% error) and Senetakis et al. [13] (±38% error). For angular Kah, Menq [11] seems to produce best predictions (± 38% error) followed closely by Saxena and Reddy [10] and Senetakis et al. [13], which seem to underestimate G_{max} values to nearly similar extents (±45% error). The maximum error obtained in the Saxena and Reddy [10], Menq [11], Wichtmann and Triantafyllidis [12], and Senetakis et al. [13] model estimations for the two soils was 58, 38, 75, and 45%, respectively. Alternatively, the relationship proposed by Saxena and Reddy [10] seems to produce least accurate results for NBS with an error of around ± 58%. While for Kah, the relationship by Wichtmann and Triantafyllidis [12] seems to be producing least accurate prediction with ± 75% error. It is worth-while to note that all these estimations were made qualitatively through general shape consideration for the two soils and also the two soils differed in gradations. To verify the effect of C_u on G_{max}, additional tests by adjusting similar gradation for both sands should be carried out.

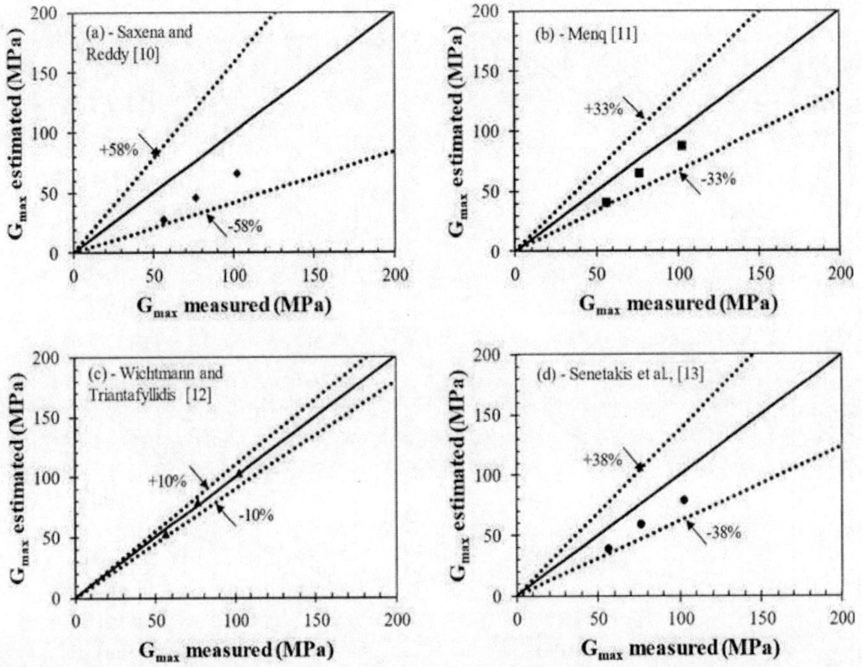

Fig. 8 Measured $G_{\max}$ values of this study verses estimated values for NBS from considered empirical relationships

4 Conclusions

The small-strain shear modulus ($G_{\max}$) was determined under same initial relative density and dry isotropic confining pressure conditions for two different types of sand—beach NBS and volcanic Kah. The following conclusions can be drawn as follows:

- There seems to be no substantial effect of frequency on the $G_{\max}$ measurements for both sands at a certain confining pressure.
- The volcanic Kah shows significantly lower $G_{\max}$ values in comparison with beach NBS. This is mainly attributed to different gradation, lower particle density, and higher void ratios for vesiculated Kah in comparison with hard-grained NBS.
- The effect of mean effective confining pressure p' on $G_{\max}$ is similar for both types of sands. $G_{\max}$ increases with the increase in p', though this relationship seems to be less significant for Kah in comparison with NBS due to intra-particle voids.
- Using general considerations of shape, the considered empirical relationships seem to underestimate the G_{max} values for both sands.
- Some additional tests at higher confining pressures need to be carried out to justify the values of parameter n (<0.5) in the $G_{\max}$ expression. In addition, by considering

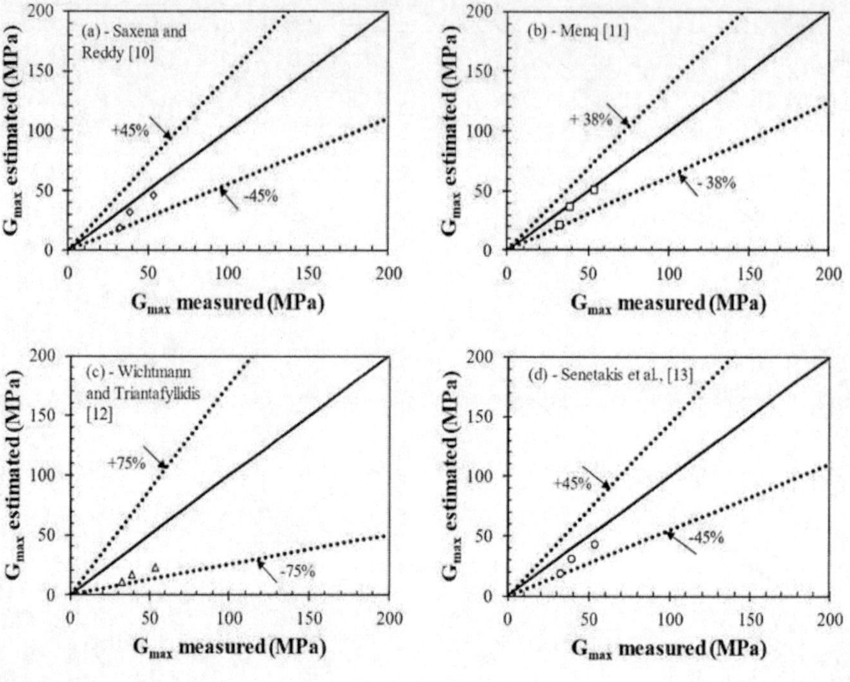

Fig. 9 Measured G_{max} values of this study verses estimated values for Kah from considered empirical relationships

similar grain size distribution and ultimately C_u for both soils, the effect of similar gradation on G_{max} for these two different sands can also be verified.

Acknowledgements All the tests were conducted at the Geomechanics Laboratory of University of Canterbury, Christchurch, New Zealand. We would like to thank Mr. Saile Faitotonu for his assistance. We would also like to thank Dr. Sean Rees, who provided valuable suggestions into bender element testing. We also wish to acknowledge the financial support provided by Department of Civil and Natural Resources Engineering and Department of Earth and Environment at the University of Canterbury, DEVORA and QuakeCoRE.

References

1. Lowe DJ, Balks MR (2019) Introduction to tephra-derived soils and farming, Waikato-Bay of Plenty, North Island, New Zealand. University of Wisconsin-Platteville Education Abroad Program, Winterim Field Trip (15-17 January, 2019). School of Science (Earth Sciences), University of Waikato, Hamilton, p 82
2. ASTM D 2487—17 (2017) Standard practice for classification of soils for engineering purposes (Unified Soil Classification System). ASTM International, West Conshohocken, PA

3. ASTM D 854—14 (2014) Standard test methods for specific gravity of soil solids by water pycnometer. ASTM International, West Conshohocken, PA
4. Wesley LD (2001) Determination of specific gravity and void ratio of pumice materials. Geotech Test J 24(4):418–422
5. JIS A 1224 (2009) Test method for minimum and maximum densities of sands
6. Yamashita S, Kawaguchi T, Nakata Y, Mikami T, Fujiwara T, Shibuya S (2009) Interpretation of international parallel test on the measurement of G_{max} using bender elements. Soils Found 49(4):631650
7. Liu X, Yang J, Wang G, Chen L (2016) Small-strain modulus of volcanic granular soil: an experimental investigation. Soil Dyn Earthq Eng 86:15–24
8. Hardin BO, Richart FE Jr (1963) Elastic wave velocities in granular soils. J Soil Mech Found Div 89(SM1):33–65
9. Santamarina JC, in collaboration with Klein K, Fam M (2001) Soils and Waves. J. Wiley and Sons, Chichester, UK, P 488
10. Saxena S, Reddy K (1989) Dynamic Moduli and Damping Ratios for Monterey No. 0 Sand by Resonant Column Tests. Soils Found 29(2):37–51
11. Menq F-Y (2003) Dynamic properties of sandy and gravelly soils. Ph.D. Dissertation, University of Texas at Austin, Austin, TX
12. Wichtmann T, Triantafyllidis T (2009) Influence of the grain-size distribution curve of quartz sand on the small strain shear modulus G_{max}. J. Geotech. Geoenviron. Eng. 135(10):1404–1418
13. Senetakis K, Anastasiadis A, Pitilakis K (2012) The small-strain shear modulus and damping ratio of quartz and volcanic sands. Geotech Test J 35(6):964–980

Toward an Early Warning System in Guwahati City for Landslides Triggered by Rainfall and Earthquake

Jai Prakash Kumar, Amarjeet Paswan, Sudhanshu Awasthi, Olympa Baro, and Risha Mal

Abstract Guwahati city located in the northeastern state of Assam is prone to landslides. Each year, there are reports of loss of life and property due to landslides. Conventional methods of slope stability are yet to find a practical application in many of the human inhabited hill slopes of the city. Instead, the practice of cutting of slopes at steep angles to build houses is a rampant practice in Guwahati city. Hence, this study suggests the setting up of landslide early warning system (LEWS), where the people residing in landslide prone areas would be warned ahead of a fatal slope failure. To achieve this aim, this study proposes a two component LEWS consisting of a landslide susceptibility database and real-time monitoring of the site conditions. Based on this framework, the landslide prone hill slopes in Guwahati city have been identified. In situ properties of the hill slopes such as the soil type, permeability, slope angle, and elevation are proposed to be used as preparatory input parameters for the landslide susceptibility database component of the LEWS. The second component would be real-time monitoring of the changes in the pore water and slope angle of the hill slopes. Based on the preset threshold limits, an alarm would be issued to warn the residents of the site.

Keywords Landslide early warning system · Rainfall · Earthquake · Guwahati city

1 Introduction

Landslides are a form of natural disaster common to the hilly regions across the world including the Indian subcontinent. In addition to the Himalayan region of the country, landslides are a common occurrence in the northeastern region of India. The northeastern state of Assam which is composed of an array of small hillocks is particularly

J. P. Kumar · A. Paswan · S. Awasthi · O. Baro (✉)
Department of Civil Engineering, National Institute of Technology Silchar, Assam 788010, India
e-mail: olympa@civil.nits.ac.in

R. Mal
Department of Electrical Engineering, National Institute of Technology Silchar, Assam 788010, India

T. G. Sitharam et al. (eds.), *Earthquake Geotechnics*, Lecture Notes in Civil Engineering 187, https://doi.org/10.1007/978-981-16-5669-9_24

susceptible to landslides. The presence of steep hillocks and an average annual rainfall of 200 cm [1] makes landslides an annual phenomenon in the state. In addition to the naturally occurring rain, haphazard cutting of otherwise stable slopes at steep angles to accommodate more houses results in landslides. The annual occurrence of landslides in the capital city of Guwahati is an apt example of such man-made disasters. Although conventional methods of retaining structures and internal slope reinforcements are available, very few practical applications of these are seen in Guwahati city. Past reports show that places such as Kahilipara, Kharghuli, Chandmari, and Fatasil Ambari within the city experience landslides on an annual basis. Each year, at least one casualty is reported in Guwahati city due to landslides during the monsoon season.

In addition to the heavy annual rainfall in the region, the seismicity levels in Guwahati city is also very high. As per [2], the city falls under seismic zone V which is the highest level of seismicity in the country. Taking into account this high seismicity, several studies have been conducted to understand the earthquake (EQ) hazard levels, liquefaction potential, and seismic site conditions of the city. These studies have revealed that Guwahati city is surrounded by innumerable tectonic faults, and the movement along these faults has resulted in the high seismicity of the region. Some of the noteworthy faults located in the vicinity of Guwahati city are Kopili fault, Oldham fault, and Dhubri fault which were responsible for the 1869 Cachar EQ (M_W 7.5), 1897 Assam EQ (M_W 8.1), and 1930 Dhubri EQ (Mw 7.1), respectively. Each of these above-mentioned EQs has contributed toward damage to infrastructure as well as change in topography of Assam in various forms including landslides [3]. Studies suggest the possibility of recurrence of such EQs in the near future, thus, indicating the occurrence of the future EQ-induced landslides.

The above discussion clearly shows that the high annual rainfall, high levels of seismicity, and ongoing man-made changes to the hills of Guwahati city could eventually lead to the next fatal landslide. Thus, this study is aimed at preventing future casualties due to landslide events in Guwahati city with the help of a landslide early warning system (LEWS). This LEWS is proposed keeping in mind the possibility of issuing an alarm with sufficient time window for the residents of a vulnerable region to escape to a safer area. The LEWS is designed based on pore pressure, ground movement, soil type, permeability of the soil, slope angle, elevation, and land cover of hill slopes in Guwahati city which have recently experienced landslides. Using this past data, threshold values of the above-mentioned parameters are set which when exceeded would issue an alarm. Thus, this LEWS would be equipped with sensors to perform real-time monitoring of the various parameters that could trigger a future rainfall-induced landslide. It has to be mentioned here that there is limited information of EQ-induced landslides in Guwahati city. One incident was recorded in 1982 in Birubari, Guwahati, where four lives were lost after a landslide had occurred during heavy rains followed by an EQ of M_W 4.7 [4]. Though in recent times, no major ($7.0 \leq M_W \leq 7.9$) or great ($M_W \geq 8.0$) EQs have occurred near Guwahati city; based on historical records, the probability of such events leading to EQ-induced landslide during the monsoon season cannot be ruled out. Thus, taking into account the high levels of seismicity, this proposed LEWS is also equipped with accelerometers which would detect any changes in the slope angle of the hill and also measure the Richter scale EQ.

Landslides occur when the slope undergoes some processes and that changes its condition from stable to unstable. This is essentially due to decrease in shear strength of slope materials, to an increase in shear stress. A change in the stability of slope can be caused by number factors acting together such as saturation by continual rainfall which results in increase of pore water pressure that causes reduction in shear strength of slope materials. Increase of hydrostatic pressure in cracks and fractures, erosion, anthropogenic activities, and ground shaking due to EQ are also some of triggering factors responsible for slope instability. Based on reports by Geological Survey of India (GSI) and newspaper articles different types of landslides failure occurred in Guwahati.

On June 13th, 2017, at Anandpur, Guwahati, the types of landslides failure were shallow translational debris slide caused due to slope excavation and triggered by increased rainfall. On August 24th, 2018, Jorabat and Kharghuli areas of Guwahati and on June 10th, 2019, at Bezbaruah Nagar, Guwahati, landslides were occurred due to heavy rainfall, failure mechanism was shallow planar failure, types of material were debris, and types of movements were translational slide.

2 Study Area

Guwahati is the capital city of Assam located on the banks of the Brahmaputra River. To the south of the city is the state of Meghalaya, Lokpriya Gopinath Bordoloi International Airport in the west, and Chandrapur township in the east. Guwahati city covers an area of about 600 km^2 and is located between the coordinates 25.59 °N and 91.33 °E and 26.15°N and 91. 10°E. The city is adorned by several hills which gives the city a unique and uneven topography. The plain areas of the city are mostly alluvial soils transported by the Brahmaputra River. Contrary to this, the hill slopes of Guwahati city are composed of residual soils such as reddish silty clay and yellowish silty sand [5]. Past studies and reports have shown that the soils present in the hill slopes are prone to erosion by the heavy annual rainfall experienced by the city. The average annual rainfall recorded in Guwahati city is 169.8 cm. The heavy rains in combination with the soil type and rampant deforestation of the slopes result in landslides. As per the Geological Survey of India, most of the hills in the city are at high risk for landslides. In recent times, the city has witnessed exponential growth in population, and a large chunk of which have settled in the landslide prone hilly areas. Thus, leading to repetitive reports of loss of life and property from these areas during each monsoon season. This study is an attempt toward counteracting the losses induced by landslides with the help of a LEWS.

3 Proposed Framework

The LEWS proposed in this study is intended to issue alarms for rainfall and/or EQ-induced landslides. Past records show that most of the rainfall-induced landslides in Guwahati city had occurred between the months of May and September. This is in accordance with the rainfall data obtained from the Indian Meteorological Department (IMD). As per the IMD, the maximum amount of rainfall in Guwahati city occurs between May and September every year. Figure 1 shows the rainfall distribution in months from 2014 to 2018 in Guwahati city.

Past studies such as [6] found that in addition to the rainfall, landslides are triggered due to other factors such as soil type, permeability, slope angle, elevation, and land cover. Other parameters which could trigger landslides are slope angle and elevation of the hill slopes. Ayalew and Yamagishi [7] observed that slope angle effects the pore water, pore pressure levels, and hydraulic conductivity of the soil. Relatively, elevation of the hill had lesser effect on the initiation of landslides; however, with increasing slope angle, hills with higher elevation were found to be more prone to landslides. In another study, Beguería [8] studied the effect of landcover on landslides. It was observed that hills which showed signs of human activity had lesser resistance to landslides even after being reclaimed by nature. On the contrary, pristine hills covered with forests showed greater resilience to landslides.

From the above discussion, it can be observed that the occurrence of landslides is a combination of various factors such as pore water of soil, rainfall intensity, soil type, permeability, slope angle, elevation, and land cover. To issue an early warning, it is essential that in addition to real-time monitoring, information regarding in situ

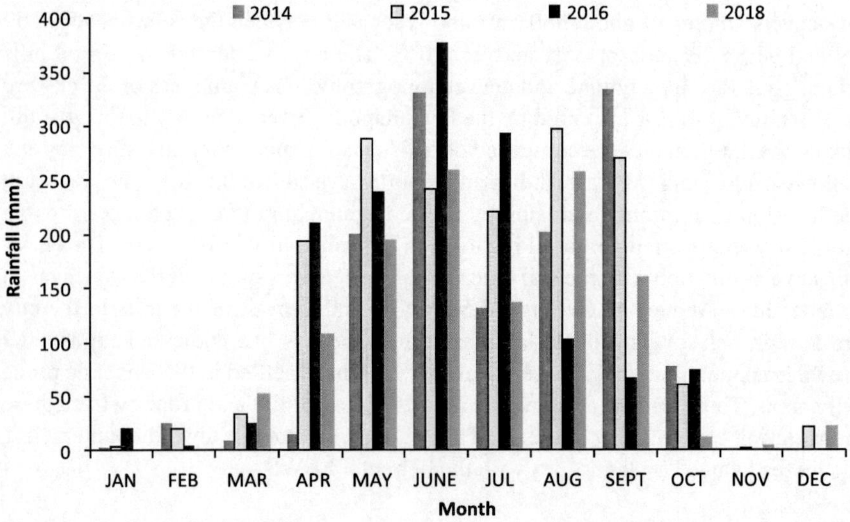

Fig. 1 Rainfall distribution in months

parameters is available in the LEWS. Thus, this study proposes a framework for the LEWS wherein the functionality of the system is divided into two components: (1) landslide susceptibility database and (2) real-time monitoring.

3.1 Landslide Susceptibility Database

To develop the landslide susceptibility database in situ pore water, soil type, permeability, slope angle, elevation, and land cover are used as preparatory input parameters. A detailed database of the preparatory input parameters would be programmed into the Arduino microcontroller of the LEWS. This database would, thus, act as a measure of in situ soil parameters as well as a standard for stable condition of hill slopes in Guwahati city.

3.2 Real-time Monitoring

During heavy rainfall, pore water of the soil is increased which in turn increases the pore water pressure, thus, making the hill slope susceptible to landslides. As rainfall intensifies, it leads to loosening of the soil particles resulting in erosion which in turn increases the slope angle [9]. This increase in slope angle intensifies the possibility of both rainfall and EQ-induced landslides. Thus, monitoring the pore water and slope angle of hill slopes in real time are essential steps toward the functionality of the LEWS. To achieve the above-stated aims, this study proposes to install soil moisture sensors at hill slopes identified as prone to landslides. Further, to detect the changes in slope angle, accelerometers would also be installed at the vulnerable hill slopes. The data collected from the soil moisture sensors and the accelerometers would be sent to Arduino microcontroller. The data collected in real time would be then checked for crossing of pre-determined threshold limits.

The conceptual framework of the LEWS proposed in this study is shown in Fig. 2.

4 Results and Discussion

Several studies have been conducted to decipher the soil conditions in Guwahati city. One such study was performed by Das and Saikia [5] where the soil types available in hills slopes of Guwahati city were identified. Das and Saikia [5] concluded that the hill slopes of Guwahati city are mostly composed of two types of residual soils reddish silty clay and yellowish silty sand. Triaxial tests performed by Das and Saikia [5] revealed that the permeability of the reddish silty clay and yellowish silty sand is 1.864×10^{-7} m/sec and 1.208×10^{-6} m/sec, respectively. Based on the study

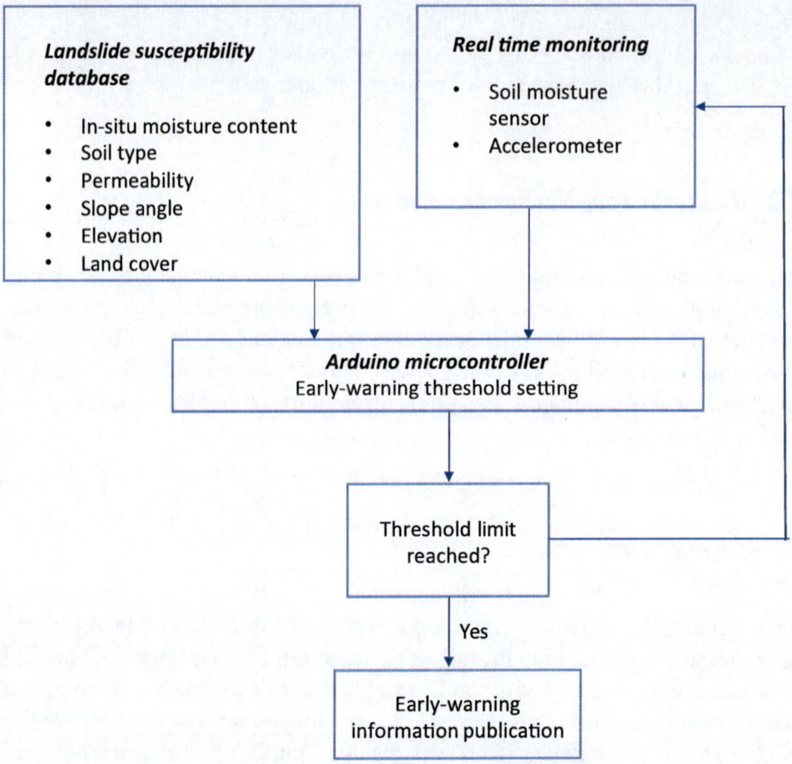

Fig. 2 Conceptual framework of LEWS proposed for Guwahati city

performed by Das and Saikia [5], the two soil types and the respective permeabilities are used as preparatory input parameters for the proposed LEWS.

For the slope angle and elevation, database has been developed based on hill slopes which have undergone landslide failure in recent times. Based on past reports, it is observed that hill slopes at locations such as Kahilipara and Fatasil Ambari have undergone multiple landslides in the recent past. Using geographic information system (GIS)-based remote sensing data, the average slope angles and elevation of these hill slopes are found. Table 1 shows the locations of hill slopes which have undergone landslide failure between 2012 to 2020 with the average slope angles, elevations, and year of occurrence. The locations of hill slopes shown in Table 1 are obtained based on reports by Geological Survey of India (GSI) and newspaper articles. Future field investigations at the same sites would reveal the land cover and in situ moisture content. In addition to the landslide susceptibility database, real-time monitoring of the soil moisture content and slope angle is proposed in this study. For the monitoring of soil moisture content, a moisture sensor would be used in the landslide prone hill slopes.

Table 1 Locations of landslides with the average slope angles and elevations of the hill slopes

S. No.	Location	Elevation (m)	Average slope angle (degrees)	Year of occurrence
1	Dhirenpara	178	17	June 2012
2	Sonaighuli	116	16	September 22nd, 2014
3	Hengrabari	109	12	September 23rd, 2014
4	Narengi	172	19	September 23rd, 2014
5	Maligaon	149	19	August 2018
6	Sunsali	170	13	August 2018
7	Noonmati	123	11	August 2018
8	Fatasil Ambari	166	20	August 2018
9	Gandhi mandap, South Sarania hills	136	17	May 2019
10	Kahilipara	137	15	June 25th, 2019
11	Amiya nagar, Chandmari	193	14	July 10th, 2019
12	Jalukhari	121	14	July 2019
13	Amingaon	135	17	July 2019
14	Matiapahar, Chakardeo village	168	18	July 2019
15	Panjabari	126	13	May 26th, 2020
16	Nillachal hills	235	19	June 2020
17	Fatasil Ambari	169	21	June 19th, 2020
18	Kahilipara	122	12	June 26th, 2020
19	Kharghuli	85	9	June 29th, 2020

Similarly, an accelerometer would also be installed at site to constantly monitor the changes in slope angle.

A prototype of the LEWS for Guwahati city is developed in this study which will be installed at the hill slope sites mentioned in Table 1 in the near future. In this prototype, a moisture sensor and an accelerometer are connected to an Arduino microcontroller (see Fig. 2). This Arduino microcontroller is programmed to convert the analog data collected by both the sensors to digital form. Based on the future field investigations, the Arduino microcontroller would be programmed to set a threshold value for moisture content. Presently, this prototype is tested by setting an arbitrary threshold value of moisture content. For slope angle, the GIS data obtained as shown in Table 1 are used as separate threshold values for each hill slope and programmed into the Arduino microcontroller. Further, in addition to the real-time data, the Arduino microcontroller is also programmed with the landslide susceptibility database. As per the framework shown in Fig. 2, the both landslide susceptibility database and real-time data with preset threshold values are programmed into the

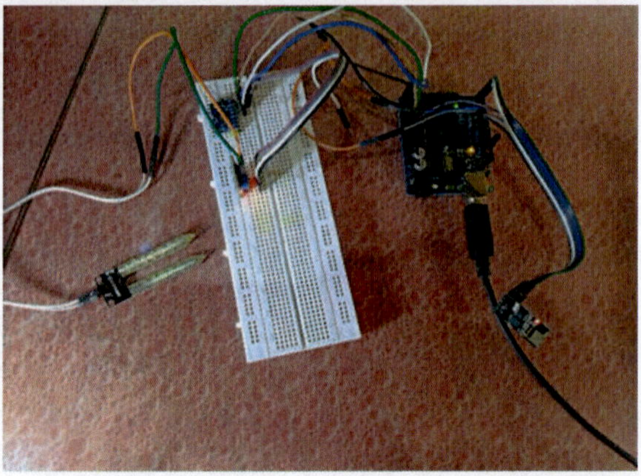

Fig. 3 LEWS prototype developed in this study

Arduino microcontroller. Once the threshold values are exceeded, the Arduino microcontroller sends a signal to the buzzer connected to the prototype which sounds an alarm of a probable landslide failure. Figure 3 shows the LEWS prototype developed in this study.

5 Future Work

The prototype of the LEWS proposed in this study is currently equipped with moisture sensor; however, a threshold value is yet to be set for the moisture content from field investigations. The field investigations would be performed to detect the in situ pore water of the landslide prone hill slopes identified in this study. Based on the data collected, a threshold value would be set for the pore water which when exceeded would trigger the alarm. Further, the accelerometer equipped in the LEWS is currently used for detection in changes of slope angle only. This study intends to use the same accelerometers for detection of arrival of P-waves at the identified sites as a future prospect. Since the seismic activities could also turn to be major cause for inducing landslides, proper detection of any ground shakes should be taken care of. This sensor easily detects ground movements in any co-ordinate axis. Since least destructive P-waves travel relatively faster than more destructive S-waves, it is feasible for accelerometer to detect early signs of occurrence.

Therefore, all types of systems would get short moment to act. MEMS-based accelerometers are used for this purpose as they have less power requirement for functioning. These accelerometers even have built-in digitizer to banish any unnecessary noise. The real-time monitoring of the arrival of P-waves, change in slope angle, and increase in moisture content of soil in combination with the landslide susceptibility database would help to develop a robust LEWS for Guwahati city.

References

1. Rakhecha PR (2016) Assessment of water resources and seasonal prediction of rainfall in India. In: Proceedings of the international association of hydrological sciences, vol 374, pp 151–157
2. IS 1893 (Part 1) (2016) Criteria for earthquake resistant design of structures. Fifth revision. Bureau of Indian Standards, New Delhi 110002
3. Baro O, Kumar A (2015) A review on the tectonic setting and seismic activity of the Shillong plateau in the light of past studies. Disaster Adv 8(7):34–45
4. Prakash S (2011) Historical records of social-economically significant landslides in India. J South Asia Disaster Stud 4(2):177–204
5. Das UK, Saikia BD (2010) Shear strength of unsaturated residual soils of the hills in Guwahati. In: Proceedings of the Indian geotechnical conference, GEOtrendz, pp. 680–682
6. Hong Y, Adler RF (2008) Predicting global landslide spatiotemporal distribution: integrating landslide susceptibility zoning techniques and real-time satellite rainfall estimates. Int J Sedim Res 23(3):249–257
7. Ayalew L, Yamagishi H (2005) The application of GIS-based logistic regression for landslide susceptibility mapping in the Kakuda-Yahiko Mountains, Central Japan. Geomorphology 65:15–31
8. Beguería S (2006) Changes in land cover and shallow landslide activity: a case study in the Spanish Pyrenees. Geomorphology 74:196–206
9. Jing X, Chen Y, Pan C, Yin T, Wang W, Fan X (2019) Erosion failure of a soil slope by heavy rain: laboratory investigation and modified GA model of soil slope failure. Int J Environ Res Public Health 16:1075–1086

Effect of Frequency Content of Earthquake Ground Motions on the Dynamic Behavior of Tiered Geo-Synthetic Reinforced Soil Retaining Wall

Anindita Gogoi and Arup Bhattacharjee

Abstract The seismic behavior of any structure depends on its dynamic properties and the characteristics of induced earthquake. The dynamic properties of structures are primarily defined by fundamental frequency at which it vibrates naturally. On the other hand, ground motion contains different predominant frequencies which are identified as one of the most influencing ground motion parameter to explain the structural response during earthquake. If the frequency of earthquake seismic wave gets closer to the natural frequency of structure, then the structure will oscillate with its highest amplitude and may undergo severe damages. This study is basically concerned with the dynamic behavior of reinforced retaining wall with tiered configuration. The objective is to study the effect of the frequency content of earthquake ground motion on the dynamic behavior of tiered geo-synthetic reinforced soil wall. For this purpose, single and multi-tiered retaining walls with different offset lengths are numerically developed using PLAXIS 2D. The accuracy of numerical modeling has been confirmed by a validation analysis. A set of three earthquake ground motions with different frequency content are considered as input dynamic excitations, and dynamic behavior of the walls are evaluated in terms of acceleration amplification and horizontal displacement.

Keywords Frequency content · Tiered geo-synthetic reinforced soil retaining wall · Dynamic behavior

1 Introduction

Earthquake is a random phenomenon reckoned to be nature's one of the most destructive force whose wrath brings great destruction in engineering structures. In reality, earthquakes generate seismic waves with different predominant frequencies. In addition, every structures have natural frequencies at which it takes few second to vibrate

A. Gogoi (✉)
Jorhat Engineering College, Jorhat 785007, India

A. Bhattacharjee
Civil Engineering. Jorhat Engineering College, Jorhat 785007, India

T. G. Sitharam et al. (eds.), *Earthquake Geotechnics*, Lecture Notes in Civil Engineering 187, https://doi.org/10.1007/978-981-16-5669-9_25

naturally. If the frequency of earthquake becomes closer to the fundamental frequency of the structure. Then, the structure will vibrate with higher amplitude even at low frequency content and undergo tremendous damages. Therefore, understanding the dynamic behavior of structure is very much essential to mitigate the structural damages induced by earthquake. Geo-synthetic reinforced soil (GRS) retaining wall is the most common application of reinforced earth system and used as an economic alternative of earth retaining wall. It is a composite body composed of layers of polymeric reinforcing elements and compacted soil fill. In seismically active areas, tall GRS retaining walls are often constructed in multi-tiered configuration as it exhibits good seismic performance during an earthquake. Liu et al. [8] suggested that such type of configuration increase the fundamental frequency of the GRS retaining wall to some measure.

The main objective of the present analysis is to study effect of frequency content of earthquake ground motion on the dynamic behavior of tiered reinforced soil retaining wall. The study is carried out numerically. Finite element software PLAXIS 2D is used for numerical modeling of single-tiered and multi-tiered wall under dynamic loading condition. The three different earthquakes with different frequencies are considered as input dynamic excitation. The dynamic responses of the walls are studied in terms of acceleration amplification factor and horizontal displacement.

2 Validation of Finite Element Procedure

An experimental modular block reinforced soil wall reported by Ling et al. [5] is adopted for validation analysis. The physical model wall is 2.8 m high, 5 m long and 2 m wide constructed on 20 cm thick foundation and comprises of concrete modular facing blocks of dimension 24 × 30 × 45 cm. The wall is reinforced with five layer of polyester geogrid of length 205 cm which are placed at a vertical distance of 60 cm. The wall is subjected to a record of 0.8 g Kobe earthquake. To prevent seismic waves reflecting from the steel walls during shaking, 10 cm thick expanded polystyrene (EPS) boards are used at the front and back ends of the steel container.

The physical model wall is simulated using finite element program PLAXIS 2D. In the numerical model, plane strain 15 nodded triangular element is used as it provides higher degree of accuracy in the calculation of stresses and displacement. The stage construction procedure is implemented to simulate the construction procedure of GRS retaining wall. The properties of backfill soil and foundation soil used in finite element model are taken from the experimental model Ling et al. [5]. The geogrid element, facing block and EPS board properties are adapted from Ren et al. [10] and Burke et al. [2].

2.1 Modeling of Backfill Soil and Foundation Soil

The backfill soil and foundation soil is modeled using linear elastic perfectly plastic Mohr–Coulomb model. The input parameters for Mohr–Coulomb model are elastic modulus (E), cohesion (c), frictional angle (ϕ), Poisson's ratio (ν), and dilatancy angle (ψ).

2.2 Modeling of Facing, Geogrid and EPS Board

The geo-synthetic reinforcements are modeled using the five node geogrid element having modulus of axial stiffness (EA). The facing blocks are modeled using linear elastic model, and input parameters consist of elastic modulus (E), Poisson's ratio (ν), and unit weight (γ). The EPS boards are modeled using linear elastic model, and input comprises of modulus of elasticity (E), Poisson's ratio (ν) and unit weight (γ).

The model property used for numerical simulation is tabulated in Table 1.

Table 1 Material properties of finite element model

Backfill soil properties	
Elastic modulus (kN/m^2)	159×10^3
Cohesion (kPa)	1
Mass density (kN/m^3)	14.30
Poisson's ratio	0.33
Angle of friction (°)	38
Dilatancy angle (°)	8
Facing wall	
Elastic modulus (kN/m^2)	2×10^6
Mass density (kN/m^3)	23
Poisson's ratio	0.2
EPS board	
Elastic modulus (kN/m^2)	2×10^6
Mass density (kN/m^3)	1
Poisson's ratio	0.2
Geogrid	
Axial stiffness (kN/m)	680

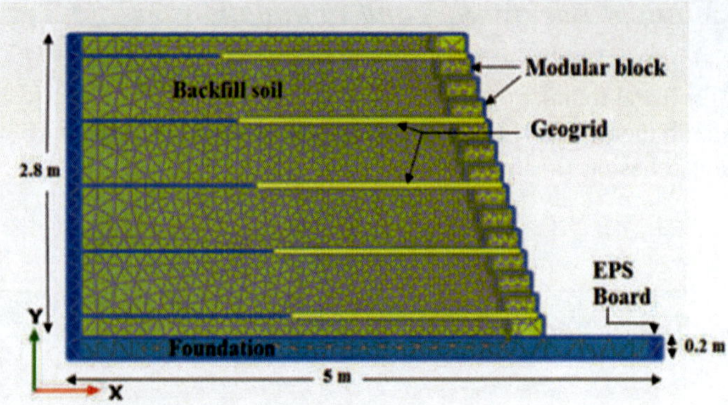

Fig. 1 Numerical model (PLAXIS 2D)

2.3 *Modeling of Interface Element*

To simulate the soil-structure interaction criteria, interface elements are used in between two dissimilar materials. The degree of the interaction is controlled by strength reduction factor R_{inter}. The R_{inter} are chosen as 0.7, 0.65 and 0.5 for interface between soil–geogrid, soil–concrete and concrete–geogrid.

2.4 *Boundary Condition*

In the numerical model, standard fixities are applied at the model boundaries. In PLAXIS, standard fixities restrained the base of the model against horizontal and vertical displacement, and model sides are restrained against horizontal displacement. During dynamic excitation, model base is fixed in vertical direction, and in the horizontal direction, earthquake load is applied as prescribed displacement. Adsorbent boundaries are provided to the far sides of the model boundaries. Because it ensures that seismic waves are absorbed at model boundaries without reflecting into the soil medium. The peak ground acceleration of 0.8 g Kobe Earthquake (1995) is assigned as earthquake load. Damping ratio of 5% is provided to simulate the soil damping. The geometry of numerical model obtained in PLAXIS 2D is shown in Fig. 1.

2.5 *Horizontal Displacement of Facing*

Figure 2 shows the comparisons of the simulated result and experimental result conducted by Ling et al. [5]. The maximum displacements at the top are found to

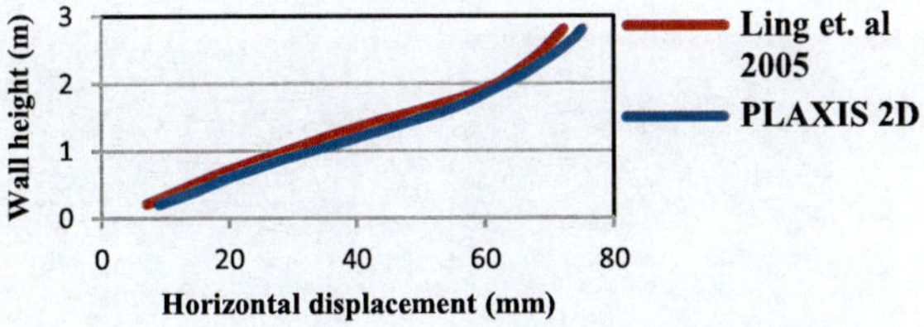

Fig. 2 Horizontal displacement measured by Ling et al. [5] and finite element model

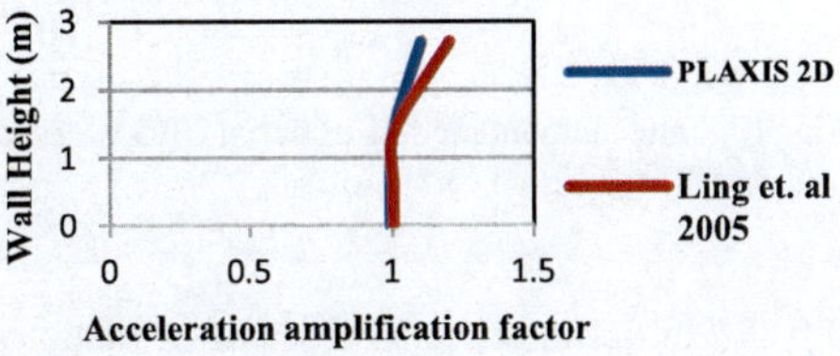

Fig. 3 Horizontal acceleration amplification measured by Ling et al. [5] and finite element model

be 75 and 72 mm for finite element and physical model, respectively. Therefore, it can be concluded that finite element model satisfactorily represents the deformation pattern of the experimental model wall.

2.6 *Acceleration Amplification*

The horizontal acceleration amplification at the end of dynamic excitation obtained from finite element analysis and physical test conducted by Ling et al. [5] is shown in Fig. 3. The amplification factor is given for the ratio of maximum acceleration in the backfill, typically at the top of backfill, to the acceleration at the foundation level [5]. The maximum acceleration at the top of the backfill is found to be 1.2 and 1.1 for finite element model and experimental model, respectively. It is observed from the figure that finite element model is satisfactorily able to describe amplified acceleration developed in experimental model.

3 Model Subjected to Earthquake Ground Motions

3.1 *Numerical Modeling of Single-Tiered and Multi-Tiered Reinforced Soil Retaining Wall*

Implementing the validated finite element procedure, dynamic study is conducted on tiered geo-synthetic reinforced soil (GRS) retaining wall at a height of 9 m. For this study, single-tiered, two-tiered and three-tiered walls with three offset distances of

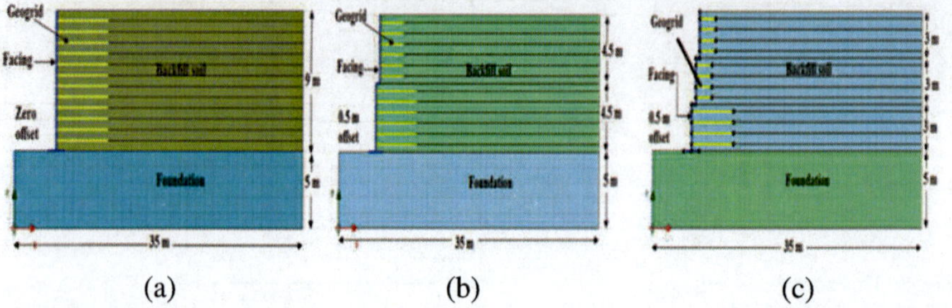

Fig. 4 Finite element models of tiered GRS wall. **a** Single-tiered (Zero offset). **b** Two-tiered (0.5 m offset). **c** Three-tiered (0.5 m offset)

0.5, 1.5 and 2 m are selected. Offset distances and reinforcement length are considered as per FHWA [3] guideline. The walls are assumed to be constructed on rigid foundations of size 35 × 5 m, and concrete modular panels of thickness 0.15 m are used as facing element. The unit weight of modular panel block is considered 3.6 kN/m^2 and that of poisons ratio is 0.1. Modular blocks are modeled as plate element which has axial stiffness (EA) and bending stiffness (EI) as 4.74 × 10^6 kN/m and 8893 kN/m^2, respectively. The Young's modulus (E) value is calculated using 5000 $\sqrt{f_{ck}}$. Here, compressive strength of concrete is taken as 40 kN/mm^2. The other material model properties and dynamic boundary conditions are considered same as that of the validated model (Fig. 1).

3.2 *Fundamental Frequency Calculation*

The fundamental frequency of the GRS reinforced walls are estimated theoretically and numerically. In theoretical method, fundamental frequencies are estimated using Wu [12] equations reported by Hatami and Bathurst [4] and Krishna and Bhattacharjee [1]. In Wu [12], equation fundamental frequencies are estimated based on the backfill height and shear modulus of the backfill soil.

Wu equation [12]

$$\mathrm{GF} = \sqrt{1 + \left(\frac{1}{2-\upsilon}\right) * \frac{H}{B}} \tag{1}$$

$$f_{1} = \frac{1}{4H}\sqrt{\frac{G}{\rho}}$$

$$f_{11} = \mathrm{GF} * f_{1}$$

Here, GF is the shape factor, 'f_1'is the FF (fundamental frequency) of an equivalent one dimensional (1D) linear elastic model, 'H' is the height, 'G' is shear modulus, 'ρ' is the density of soil, 'ν' is the Poisson's ratio, and 'B' is the width of foundation of GRS wall.

By using this formula, the fundamental frequencies are calculated for all of the wall models. In this analysis, shear modulus of soil (G), density of soil (ρ), and Poisson's ratio are taken as 60,000 kPa, 14.30 kN/m^3 and 0.33, respectively. The fundamental frequencies of the walls are tabulated in Table 2.

Numerically, fundamental frequencies are determined from the free vibration response obtained after a small dynamic excitation. Table 3 represents fundamental frequencies estimated using numerical method. As can be seen from Table 3, the fundamental frequency of GRS retaining wall is increased to some extent with increase in tier offsets. The fundamental frequency determined by using Wu [12] is different from that obtained from the numerical methods. An insignificant variation in fundamental frequencies of wall with zero offset and other tiered offset is obtained by Wu [12] equation. But considerable variations in fundamental frequencies are obtained after application of free vibration to the numerical models of different offsets.

Table 2 Fundamental frequencies of the walls using Wu [12] equation

No. of tiers	Offset distance(m)	Wall height (m)	Average width of wall (m)	Fundamental frequency (Hz)
Single tier	0.0	9.0	30.00	2.15
Two tire	0.5	9.0	29.75	2.17
	1.5	9.0	29.25	2.17
	2.0	9.0	29.00	2.18
Three tier	0.5	9.0	29.50	2.17
	1.5	9.0	28.50	2.18
	2.0	9.0	28.00	2.19

Table 3 Fundamental frequencies of the walls using numerical method

No. of tier	Offset distances (m)	Fundamental frequency (Hz)
Single tier	0.0	2.10
Two tier	0.5	2.43
	1.5	2.44
	2.0	2.48
Three tier	0.5	2.47
	1.5	2.52
	2.0	2.55

3.3 Earthquake Ground Motions

The geo-synthetic reinforced soil retaining wall models are subjected to a set of three earthquake ground motions. The earthquake (EQ) recorded during the 1989 Loma Prieta EQ, 1995 Kobe EQ, and 2001 Bhuj EQ are adopted for the study. The actual ground motion time history of all earthquakes and the frequency content of the earthquake acceleration histories by means of fast Fourier transforms (FFT) are presented in Fig. 5. It can be seen that all ground motions contain different peak ground acceleration (PGA) levels, durations, and frequency contents. The predominant frequencies of earthquakes are 0.673, 1.27, and 1.47 Hz for the Loma Prieta EQ, Bhuj EQ, and Kobe EQ, respectively.

To study the effects of real earthquake ground motions on the seismic behavior of GRS walls, the PGAs of all three ground motions are scaled to one PGA of 0.2 g, and

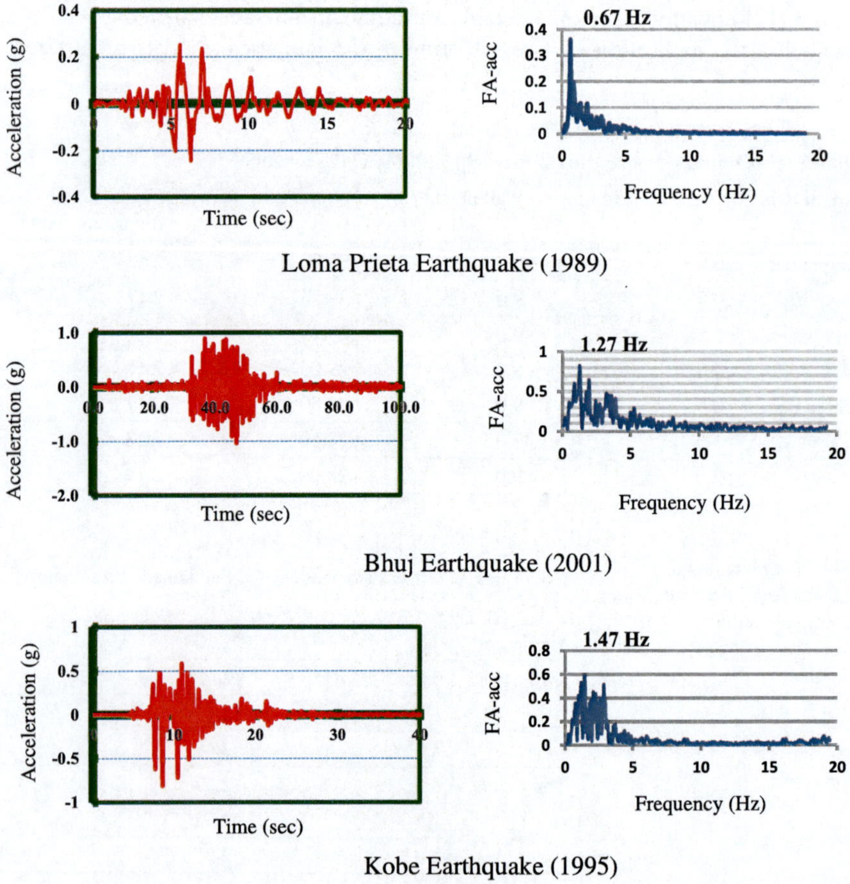

Fig. 5 Acceleration time history and FFTs of Loma Prieta, Bhuj, and Kobe Earthquake

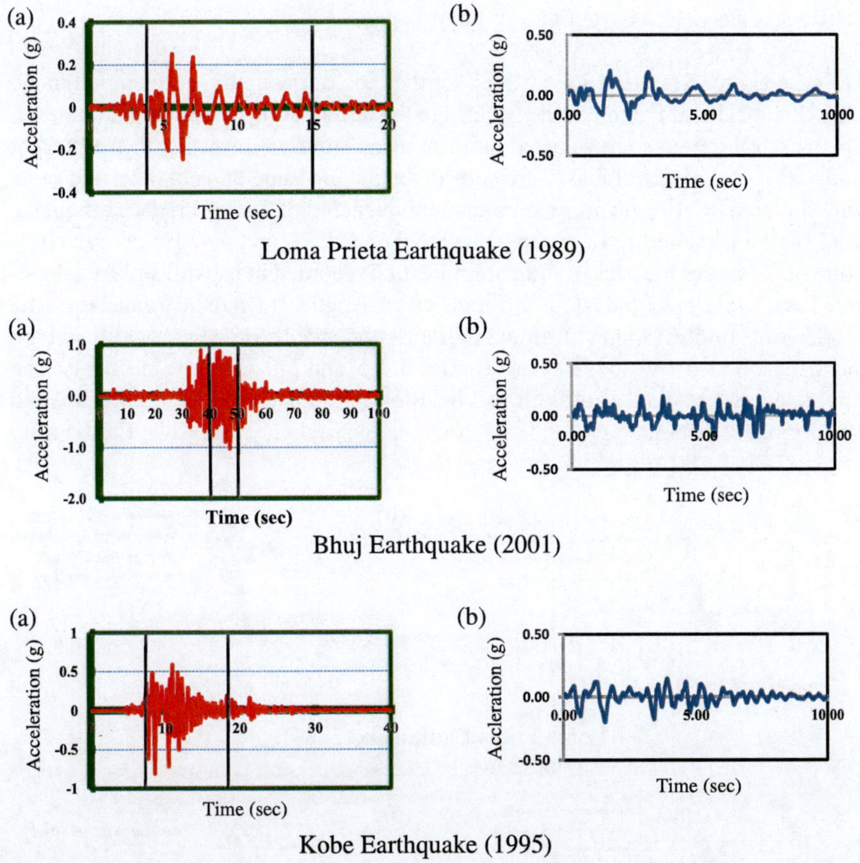

Fig. 6 Earthquake ground motion records for different earthquakes. **a** Actual acceleration histories. **b** Scaled histories for 10 s duration and 0.2 g PGA

duration of 10 s is used in numerical simulation. Scaled ground motions are shown in Fig. 6. Earthquake ground motions are applied in terms of prescribed displacement at the base of walls.

3.4 Model Responses Subjected to Three Different Earthquake of Excitation

Scaled earthquake motions are applied at the bottom of the walls as the input excitation. The wall responses are observed in terms of acceleration amplification and horizontal displacement.

3.4.1 Acceleration Amplification Factor

The acceleration amplification is given for the ratio of maximum acceleration in the backfill typically at the top of the backfill to the acceleration at the foundation level. The measuring points are selected at 10 m away from toe of walls. In the present study, a set of three earthquakes are considered having same PGA of 0.2 g and same time duration of 10 s. The predominant frequencies of Loma Prieta earthquake, Bhuj Earthquake, and Kobe Earthquake are 0.667, 1.27, and 1.47 Hz, respectively. The Fig. 7a shows the acceleration amplification recorded at backfill soil for single-tiered and two-tiered wall with different offset lengths for three earthquakes. The acceleration amplifications for Loma Preita Earthquake recorded at backfill soil are found to be 1.33, 1.38, 1.52, and 1.61 for 0, 0.5, 1.5, and 2 m offsets, respectively. The amplifications recorded at backfill soil for Bhuj Earthquake are 2.2, 2.38, 2.45, and 2.57 for offset distances of 0, 0.5, 1.5, and 2 m, respectively. For Kobe Earthquake,

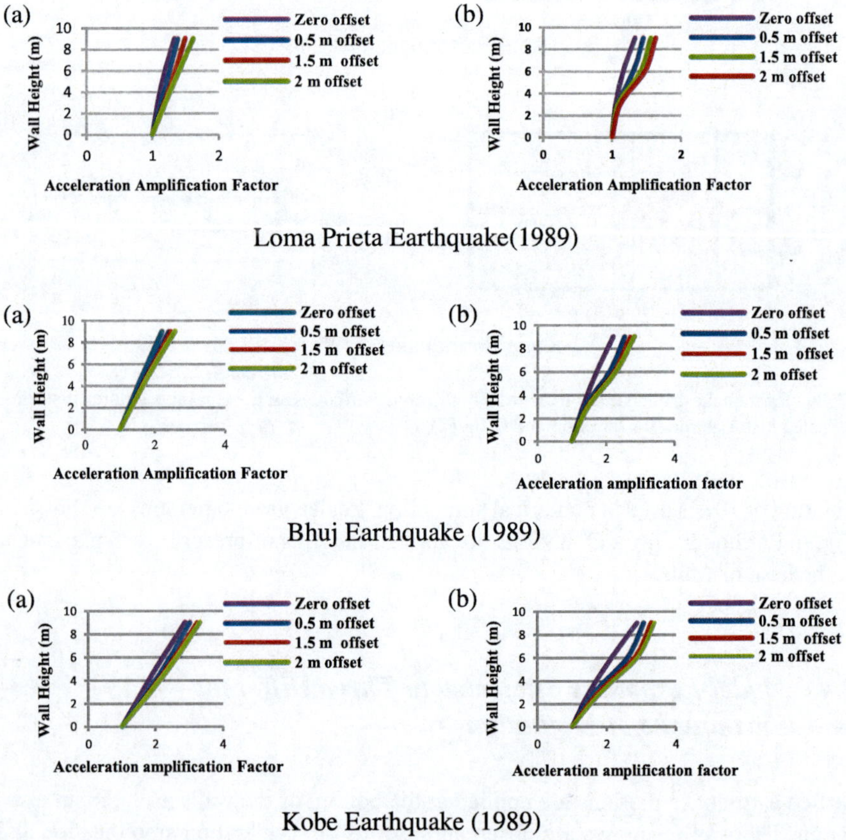

Fig. 7 Acceleration amplifications at backfill of **a** Two-tiered wall, **b** Three-tiered wall subjected to different earthquakes

acceleration amplifications are found to be 2.87, 3.01, 3.22, and 3.31 for offset distances of 0, 0.5, 1.5, and 2 m, respectively.

Figure 7b also shows the comparison of single- and three-tiered wall response in terms of acceleration amplification for three earthquake ground motions. The acceleration amplifications recorded at backfill soil for Loma Preita Earthquake are found to be 1.45, 1.56, and 1.63 for offset distances of 0.5, 1.5, and 2 m, respectively. The acceleration amplifications for Bhuj Earthquake are obtained as 2.53, 2.69, and 2.82 for offset distances of 0.5, 1.5, and 2 m, respectively. For Kobe Earthquake, acceleration amplifications are found to be 3.06, 3.28, and 3.37 for offset lengths of 0.5, 1.5, and 2 m, respectively. The acceleration amplifications are moderately higher in multi-tiered wall due to the effect of facing.

It can be seen from figures that single-tiered and multi-tiered GRS walls shows maximum responses in terms of acceleration amplifications when walls are allowed to vibrate under Kobe Earthquake. The minimum responses are observed for Loma Prieta Earthquake. The reason for getting the maximum response for Kobe Earthquake can be justified with its predominant frequency which is 1.47 Hz. This is close to the fundamental frequency of the walls considered. As can be seen from Table 2, the fundamental frequency of single-tiered wall is 2.1 Hz and values increase with increase in tier offset. The predominant frequencies of Bhuj Earthquake and Loma Prieta Earthquake are less than the Kobe Earthquake. As a result, walls are less amplified under Bhuj and Loma Preita Earthquake.

3.4.2 Horizontal Displacement

Figure 8a presents the comparison of horizontal displacement of the single-tiered and two-tiered wall for different offset distances for three different earthquakes. The horizontal displacements observed for Loma Prieta Earthquake are 147, 116, 82, and 70 mm for 0, 0.5, 1.5, and 2 m offsets, respectively. The displacements observed for Bhuj Earthquake are 346, 295, 242, and 192 mm for the offset distances of 0, 0.5, 1.5, and 2 m offsets, respectively. For Kobe EQ displacements are found to be 576, 545, 509, and 475 mm, respectively, for offset distances of 0, 0.5,1.5, and 2 m, respectively.

The comparison of horizontal displacements of single-tiered and three-tiered wall for three earthquakes are presented in Fig. 8b. For Loma Prieta Earthquake, the displacement are 74.03, 53, and 32 mm for three-tiered wall with offset distances of 0.5, 1.5, and 2 m, respectively. The displacements for Bhuj Earthquake are 288, 217, and 187 mm for three-tiered wall with offset distances of 0.5, 1.5, and 2 m, respectively. For Kobe Earthquake, the displacements are obtained as 494, 447, and 366 mm for offset distances of 0.5, 1.5, and 2 m, respectively.

The study shows that wide range of displacement occurs in retaining wall under different earthquake ground motions. It can be seen that with increase in tier offset, wall displacement significantly decreases. The overall wall responses are maximum for Kobe Earthquake, while wall responses are minimum for Loma Prieta Earthquake. The maximum response of walls for Kobe Earthquake is due to higher frequency

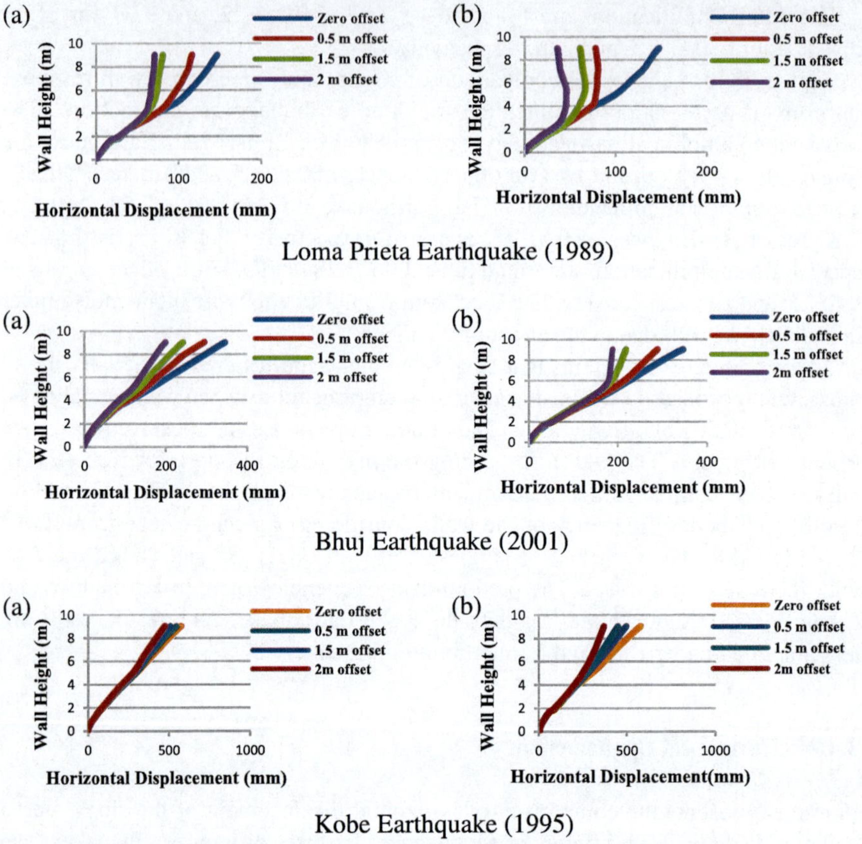

Fig. 8 Horizontal displacement of **a** Two-tiered wall, **b** Three-tiered wall subjected to different earthquakes

content which is 1.47 Hz. This is closer to the fundamental frequency of walls. It can also be seen that with increase in tier offset, horizontal displacement of wall decreases. The smaller displacement in tiered wall is due to its higher frequency content which increases with increase in tier offsets. The fundamental frequency of three-tiered wall is highest among all the walls. Therefore, difference between the fundamental frequencies of three-tiered wall and predominant frequencies of the earthquake are also highest among all. Thus, three-tiered walls show least horizontal displacement despite of higher acceleration amplification factor. The fundamental frequency is inversely proportional to height of the wall as show in Eq. 1. As the multi-tiered wall is constructed with reduced height, therefore, fundamental frequency of multi-tiered wall is higher than vertical wall. The smaller displacement in multi-tiered wall is also due to smaller inertial force exerted by the smaller soil mass in upper tiers.

4 Conclusion

In the present analysis, a numerical model is adopted to simulate the GRS retaining wall. The numerical model is validated with results acquired from the physical model test reported in the literature. Implementing the validated finite element procedure singe-tiered and multi-tiered numerical models are developed. Parametric studies are conducted in terms of horizontal displacement and acceleration amplification factor. The study is focused on the effect of the frequency content of different earthquake ground motions on the dynamic behavior of tiered reinforced soil retaining wall. Following conclusions can be made from the present study.

1. In this study, when GRS walls are being subjected to earthquake ground motions with different frequency content, walls shows maximum response in terms of acceleration amplification and horizontal displacement when predominant frequency of earthquake gets closer to the fundamental frequency of walls. Therefore, it can be concluded that dynamic response of reinforced retaining wall depends on its fundamental frequency and frequency content of the earthquake.
2. The fundamental frequency of walls increases with increase in offset distance. Due to the higher fundamental frequency content, multi-tiered wall exhibits smaller deformation despite higher acceleration amplification factor. Hence, it can be suggested that GRS wall in multi-tiered configuration is a more effective solution to mitigate the structural damages of tall GRS wall during earthquake.

This study is covered only for the frequency effect of earthquake ground motions on the dynamic behavior of the tiered wall. The effect of different PGA and time of duration of earthquake is not considered in this study.

References

1. Bhattacharjee A, Krishna AM (2016) Behavior of rigid-faced reinforced soil-retaining walls subjected to different earthquake ground motions. Int J Geo-Mech © ASCE, ISSN 1532-3641
2. Burke C, Ling HI, Liu H (2004) Seismic response analysis of a full-scale reinforced soil retaining wall. In: 17th ASCE engineering mechanics conference
3. FHWA Publication (2010) Mechanically stabilized earth walls and reinforced soil slopes design and construction guidelines, vol I & II. Publication No. FHWA-NHI-10–024, US Department of Federal Highway Administration (FHWA), US
4. Hatami K, Bathurst RJ (2000) Effect of structural design on funda-mental frequency of reinforced-soil retaining walls. Soil Dyn Earthquake Eng 19:137–157
5. Ling HI, Mohri Y, Leshchinsky D, Burke C, Matsushima K, Liu H (2005) Large-scale shaking table tests on modular-block reinforced soil retaining walls. J Geotech Geoenviron Eng 131(4):465–476. ISSN 1090-0241/2005/4
6. Ling HI, Yang S, Leshchinsky D, Liu H, Burke C (2010) Finite-element simulations of full scale modular-block reinforced soil retaining walls under earthquake loading. J Eng Mech ASCE 135(5):653–661
7. Liu H (2011) Comparing the seismic responses of Single and multi-tiered Geosynthetic reinforced soil walls. Geo-Frontiers, ASCE

8. Liu H, Yang G, Ling HI (2014) Seismic response of multi-tiered reinforced soil retaining walls. Soil Dyn Earthquake Eng 61–62:1–12
9. Reference Manual (2016) PLAXIS 2D. In: R.B.J. Brinkgreve (ed) Delft University of Technology & PLAXIS, The Netherlands
10. Ren F, Zhang F, Xu C, Wang G (2016) Seismic evaluation of reinforced-soil segmental retaining walls. Geotext Geomembr 44:604–614
11. Tutorial Manual (2016) PLAXIS 2D. In: Edited Brinkfreve RBJ (eds), Deft University of Technology & PLAXIS, Netherlands
12. Wu G (1994) Dynamic soil-structure interaction, pile foundations and retaining structures PhD thesis. University of British Columbia, Vancouver, Canada

A Review on Behavior of Piled Raft Foundations Under Various Loads

P. A. Amalu and B. R. Jayalekshmi

Abstract Piled raft foundations are a combined system of deep foundation and shallow foundation. In this type of foundation systems, the pile footings are provided beneath a raft or mat footing. Thus, the piled raft foundations can be constructed either by providing raft foundation connected to the pile footings or by providing a cushion layer between raft and pile footings. The first type of piled raft foundation is generally termed as connected piled raft foundations, and the latter piled raft foundations. Piled raft foundations is termed as disconnected can be subjected to different types of loadings, like vertical and inclined loading depending on the type of superstructure, lateral loading induced by earth pressure as well as seismic loads and moments induced by the eccentricity in loadings. Numerous studies have been conducted on the behavior of piled raft foundations under independent and combined effects of these loadings. This article aims to summarize these literature works to review the behavior of piled raft foundations under static and dynamic loadings acting in different directions.

Keywords Piled raft foundations · Static loading · Dynamic loading

1 Introduction

When the applied load from a superstructure to ground exceeds the load carrying capacity of a designed raft foundation, the method of incorporating pile foundation beneath the raft foundation can be adopted. These foundation systems are generally referred to as combined piled raft foundations. These systems can be designed in two types; the first type includes raft foundation and pile footings connected. Here, the raft acts as a pile cap, but with considerable load bearing capacity, whereas the second type of foundation systems, including raft foundation and pile foundation, is constructed without any connection. In these types of foundations, a gap will be provided between the bottom of raft foundation and top of pile footings. The conventional design of piled raft footings has considered either the load bearing by

P. A. Amalu (✉) · B. R. Jayalekshmi
Department of Civil Engineering, National Institute of Technology Karnataka, Surathkal, India

T. G. Sitharam et al. (eds.), *Earthquake Geotechnics*, Lecture Notes in Civil Engineering 187, https://doi.org/10.1007/978-981-16-5669-9_26

raft and settlement by pile foundation or the other way around. These designs resulted in uneconomical designs [1]. Therefore, developing a method that incorporates the combined load bearing effect of pile foundation and raft footing was focused on many literatures. Different methods of analysis were proposed by researchers on the design of piled raft foundation systems. The pioneer research work includes the book on pile foundation analysis and design [2]. With the development in computational techniques, rigorous analysis was carried on the piled raft foundations under various loading conditions. These analyses were done with piled raft systems subjected to different types of loading. The behaviors of piled raft foundations under static and dynamic loadings are not similar [3]. There was a marginal variation in the results of deformation and load proportions in these foundations under dynamic loading conditions, when compared to static loading conditions. This was due to the incapability of incorporating all the parameters involved in the soil–structure interfaces and the load distribution pattern between raft and pile. It was pointed out in many literatures that the proportion of load shared between pile and raft foundation is not a constant value under the influence of lateral loads [3–5] and [6]. However, the load proportions carried by both these foundations were not varying considerably under static vertical loads.

The response of piled raft foundations also varies depending on the type of connection between pile and raft foundations. In the studies conducted by Horikoshi et al. [6] and Matsumoto et al. [7], different types of joints within a connected piled raft system were used. The results of these analysis highlighted that depending on whether the connection was rigid, hinged, or semi hinged, the behavior of piled raft foundations changed under lateral loadings. These studies also pointed out that these changes were not considerable under vertical loading. Therefore, behavior of piled raft foundations under different types of loading is a vast area of research in which extensive numerical studies and experimental works are actively in progress for over a decade.

The piled raft foundations are advantageous over conventional raft or pile foundations when the cost of construction increases considerably for constraining the settlement of the supported structure within desirable limits. The combined action of pile and raft foundation imparts larger horizontal stiffness to the entire foundation system which reduces the lateral deformations under horizontal loading conditions. Further, the pile footings in a piled raft foundation act as a settlement reducer, whereas the raft foundations share the extra load acting more than the capacity of the pile footings. Thus, these piled raft foundations are a feasible solution where the depth of hard strata to provides an end bearing pile larger. Also, these types of foundation systems are more economical in the case of larger vertical loads, especially when designed to support high rise buildings [8].

The aim of this article is to summarize some of the pioneer and innovative literature works conducted on both the connected and the disconnected piled raft foundation systems. This article categorizes these findings based on the type of loadings applied on piled raft foundation.

2 Bearing Capacity of Piled Raft Foundations

The bearing capacity of piled raft foundation could be estimated as the cumulative capacities of raft and pile foundations separately [9]. The bearing capacity of raft foundation (Q_R) can be determined using the following equation [10].

$$Q_R = \left(F_q N_q q + F_c N_c c + F_\gamma N_\gamma \gamma B\right) A \tag{1}$$

where N_q, N_c, and N_γ are the bearing capacity factors, γ is the unit weight of soil, c is the cohesion, q is the vertical stress at the foundation level depending on the soil friction angle, B is the width, and A is the area of the raft. The factors F_q, F_c, and F_γ account for the shape and inclination factors.

In case of clayey soils, (1) can be modified to get the bearing capacity (Q_{RC}).

$$Q_{\text{RC}} = F_c N_c c_u A \tag{2}$$

The bearing capacity of pile groups (Q_{GP}) in clayey soils can be estimated using the following equation.

$$Q_{\text{GP}} = \eta n Q_{\text{SP}} \tag{3}$$

where η is the efficiency factor for the pile group, n is the number of piles, and Q_{SPC} is the capacity of a single pile which can be determined using the following equation [11].

$$Q_{\text{SP}} = A_P N_c c_p + \sum_{i=1}^{n} \alpha_i c_i A_{si} \tag{4}$$

where A_p is area of pile, c_p is the cohesion, α_i is the adhesion factor, c_i is the average cohesion for the ith layer, and A_{si} is the surface area of pile shaft.

From Eqs. (1)–(4), the bearing capacity of piled raft foundation (Q_{PR}) can be estimated as the summation of bearing capacities of pile and raft foundation as detailed in the following equation [12].

$$Q_{\text{PR}} = Q_{\text{RC}} + Q_{\text{GP}} \tag{5}$$

In case of sandy soils, (1) can be modified to get the bearing capacity (Q_{RS}) [13].

$$Q_{\text{RS}} = 0.5 \gamma B N_\gamma S_\gamma \tag{6}$$

where N_γ and S_γ are the factors representing the bearing capacity and shape of the foundation. The bearing capacity of the pile group can be estimated using Eq. (3)

where the value of Q_{SP} is determined using Eq. (7).

$$Q_{SP} = A_p\left[0.5D\gamma N_\gamma + P_D N_q\right] + \sum_{i=1}^{n} k_i P_{Di} \tan\delta A_{si} \tag{7}$$

where D is the diameter of pile, γ is the effective unit weight of the soil at the end of pile, P_D is the effective overburden pressure, $k_{i,}$ P_{Di} is the coefficient of earth pressure and overburden pressure corresponds to ith layer, and δ is the friction angle.

The bearing capacity of the piled raft foundation was then determined by using Eq. (5).

3 Performance of Piled Raft Foundations Under Static Loading Conditions

3.1 Foundations Subjected to Vertical Loads

The proportion of vertical load shared between pile foundation and raft foundation in a piled raft foundation was investigated using experimental and analytical methods by Matsumoto et al. [7]. The study evidently pointed out that the settlement of a piled raft footing system was lesser when compared to the settlement of a pile group in cohesionless soils. The percentage of load carried by the raft was found to vary between 30 and 60% of the vertical load which was validating the results of Liang et al. [14] despite the type of connection between the pile and raft foundation. However, a higher load proportion was shared by raft foundation in the case of piled raft connected rigidly. Since the study incorporated the different types of connection between the raft and pile, the effect of these connections on the behavior of the footing system was found to be insignificant.

An early study on the performance of disconnected pile and raft footing was done by Jose et al. [15]. A layer of sand gravel was provided as a cushion between these footings. The literature suggests a novel system of piled raft footing with short and long pile combinations below the cushion layer. After rigorous finite element analysis and a field study, it was suggested by the authors that a provision of cushion layer could effectively redistribute the vertical loading between short and long piles. However, the optimum thickness of this layer and the required modulus of elasticity of the material must be determined for each case. Therefore, in projects where shallow depths of soft soils are available, short piles can be used to improve the shear strength and thereby modify the shallow layers to effectively contribute to stress distribution from vertical loading.

Regardless of the loose to dense state of sandy soils, the longer piles tend to withstand higher vertical loads in a piled raft system [16]. The literature extends this observation to point out that even with fewer numbers of piles; increase in length can improve the vertical load carrying capacity of a piled raft system in cohesionless soils. An increase in area of raft foundation had also resulted in the corresponding improvement in bearing strength of piled raft systems. It was noted that there was an optimum number of piles above which further increase in the number of piles did not contribute to the improvement in bearing strength of the piled raft system [16].

3.2 Foundations Subjected to Horizontal Loads

The type of connection between raft and pile foundations has considerable effect on the response of the system toward horizontal load as stated by Mattsson et al. [17]. In their elaborative study, it was found that the hinge jointed type was least tilted compared to rigid jointed pile raft foundations under cyclic horizontal loading. The cyclic horizontal loading in this literature is analogous to a simplified form of earthquake loading, whereas the horizontal displacement of rigid jointed piled raft foundation was comparatively less. The findings of the study also highlighted that even though the type of connection between pile and raft footing had less effect in the load proportion carried by pile, the effect was considerable in raft which had a rigid connection. Thus, when a piled raft footing system is subjected to an earthquake loading, better resistance to lateral displacement and developed shear forces is offered by rigid jointed systems.

3.3 Foundations Subjected to Combined Vertical and Horizontal Loads

The detailed case study of disconnected piled raft foundation was detailed with field observations and numerical analysis of the observed results by Tarenia and Patra [18]. The collected parameters of soil and piled raft system by Tarenia and Patra [18] were used to numerically model the same system under the combined effect of both vertical and horizontal loads by Chanda et al. [19]. The study suggested certain obvious results in the reduced settlement and improved load carrying capacities of pile groups compared with a single pile in a piled raft system. The important conclusion drawn from the analysis by Chanda et al. [19] suggested that a nine-pile square group system in disconnected piled raft foundation can safely withstand a lateral load around 5% of vertical load when considered under simultaneous loading, whereas connected piled raft foundations can withstand a lateral load around 10% of the vertical load due to its improved horizontal stiffness. Therefore, it can be inferred that the connected piled raft foundations offer better resistance to combined effect of vertical and lateral loadings. It was observed from the studies of Badelow and Poulos [20] that under the combined action of these forces along with moments, the capacity to withstand lateral loads improved in the piled raft foundation. Further, the proportion of lateral load taken by raft foundation was found to be reduced at higher strain levels.

3.4 Foundations Subjected to Eccentric Loading

The effect of piled raft systems under eccentric loading was experimentally evaluated by Matsumoto et al. [7]. The effect of eccentric loading on square-, circular-, and strip-type raft foundations with connected and disconnected piles was extensively studied. The results highlighted that under eccentric loading connected piled raft foundations were advantageous over disconnected foundation systems in resisting the rotational moments developed. However, the disconnected piled raft showed higher bearing capacity. This was because of the effect of gravel cushion layer provided in between the pile and raft footing. The effect of length of pile was almost identical in both types of piled raft foundations, which evidently suggested that an increase in length of pile improved the load bearing capacity. A noticeable result that must be pointed out from this study is regarding the spacing of piles beneath the raft foundation. When the spacing was increased in the connected piled raft system, the load bearing capacity increased as well. On contrary, a reduction in spacing resulted in more load carrying capacity in the case of disconnected footing systems. Considering the eccentric loading, closer spaced piles provided better resistance to the developed moments.

4 Performance of Piled Raft Foundations Under Dynamic Loading Conditions

4.1 Disconnected Piled Raft Foundations

Throughout the literature on the study of behavior of piled rafts under static horizontal loading, it is commonly seen that the connected piled raft systems offer better resistance to horizontal displacement. However, the load transfer from superstructure to pile foundation will obviously be higher for rigid jointed piled raft when compared with disconnected piled raft. In this point of view, Saadatinezhad et al. [4] suggested that non-connected piled raft foundations appeared to be economical and efficient under dynamic seismic loading. The results from experimental study and analysis with finite element modeling highlighted that the shear forces and moments induced by the earthquake loadings were lesser for non-connected piled raft systems. The literature suggested that the lateral displacement in non-connected piled rafts can be reduced by improving the horizontal stiffness of the cushion material. The effect of this cushion was studied in detail in the numerical investigation conducted by Xiaolei et al. [5]. While considering dynamic loads, the presence of a cushion layer had obviously damped the vibrations and in effect reduced the acceleration of seismic motion up to 15% [5]. The literature brought up an important finding about the influence of elastic modulus of cushion. It was concluded by the authors that changing the elastic modulus of the cushion layer had not contributed much to the seismic absorption. This conclusion can be vital while selecting the material for cushion as well as improving its properties. As expected, the literature also inferred that the higher the thickness of the cushion layer the more will be its damping efficiency. Saadatinezhad et al. [4] discussed the load transferring mechanisms of the cushion layer by considering the effects due to the propagation of seismic waves from ground to superstructure. The findings of Xiaolei et al. [5] suggested that cushion between raft and pile transfer almost entire seismic loads to the superstructure under frequent ground motions even though it may vary a little in case of severe earthquake.

The cushion layer in the unconnected piled raft foundation can be selected depending on various criteria depending upon the type of loading conditions. When the loading conditions are dynamic, the provision of a cushion layer with higher thickness can damp the seismic waves. Therefore, it is evident that the thickness and stiffness of the cushion layer are the two major parameters that influence the performance of cushion layers under different loading conditions, whereas in eccentric loading conditions, Matsumoto et al. [7], the connected piled raft foundations offer better resistance to additional moments. Therefore, in such cases, lesser thickness of cushion layer are preferred. The studies of Saadatinezhad et al. [4] have also highlighted that the horizontal stiffness of the cushion layer is also an important parameter to determine the performance of cushion layer.

4.2 Combined Piled Raft Foundation

The response of the rigid connected piled raft foundation system was different to that of disconnected foundations. These effects were verified over a broad spectrum of research conducted on the rigid piled raft foundations under dynamic loading. Among the early stages of these studies, Horikoshi et al. [6] put forwarded the experimental results using shake table tests on combined piled raft foundations. It was observed that the dynamic loading conditions had not altered the proportion of load taken by pile foundations alone, whereas any lateral displacement had contributed to an increase in the load taken by pile foundations. Therefore, under dynamic loading conditions, the lateral displacement time history must be evaluated to understand the load sharing between the pile and raft foundations. In the recent studies by Kumar et al. [3], more load proportion was taken by pile foundations under pseudo-static loads. However, the literature also highlighted that there was a marginal variation among the results obtained under pseudo-static and dynamic loading analysis. These inferences clearly state that the dynamic analysis of a piled raft system is necessary to understand the real responses of the piled raft system with respect to the surrounding soil. Such rigorous analysis will not be an easy task to compute. From many of the research works on connected piled raft foundations, it can be generalized that the displacements and moments attain peak values at the top of pile foundations.

5 Numerical Evaluation to Analyze the Performance of Piled Raft Foundations Under Dynamic Loading Conditions

The behavior of piled raft foundation under seismic loading was numerically evaluated using the ANSYS software based on finite element method. Three cases were studied where the first case studied the behavior of raft foundation alone in clayey soil, the second case studied behavior of connected piled raft foundation, and third case focused on the behavior of unconnected piled raft foundation with cushion layer. A group of sixteen square pile footings of dimension 0.4 × 0.4 m and depth of 12 m was modeled with a square raft foundation of dimension 8 × 8 × 1 m and cushion of depth 0.5 m with an elastic modulus of 40 MPa. The foundation was modeled in a clayey soil domain of 16 × 16 × 16m as shown in Fig. 1. The acceleration time history data of El-Centro earthquake 1940 of peak ground acceleration 0.318 g and duration of 31.18 s were applied to study the response of the raft and piled raft foundations. A superstructure load of 0.2 MPa was applied on top of the raft foundation which represented the load from a multi-story superstructure. The lateral

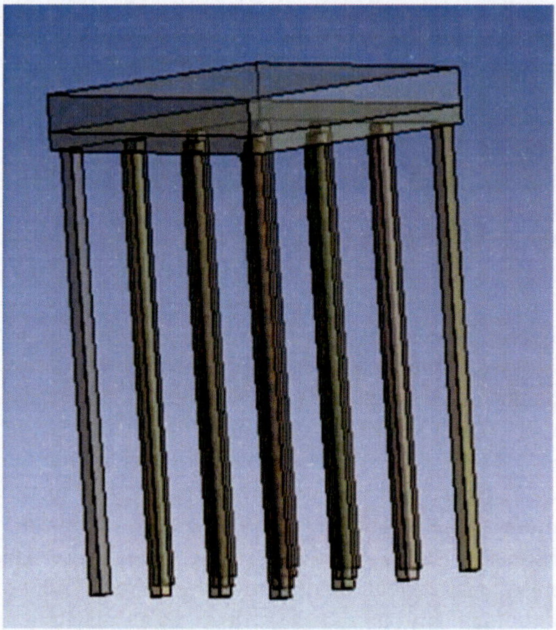

Fig. 1 Unconnected piled raft model

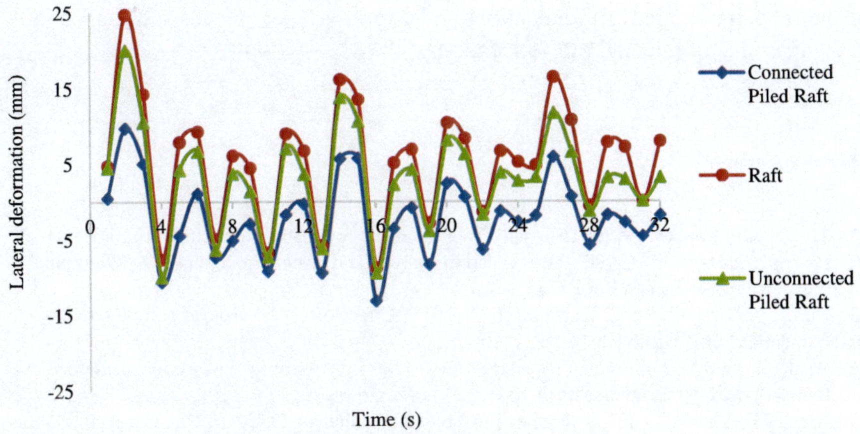

Fig. 2 Lateral deformation time history plot at the top of the raft

deformation time history plot at the top of the raft was obtained after conducting a transient structural analysis in the ANSYS software. The curves thus plotted in Fig. 2 highlight that the piled raft foundations offer better resistance to lateral deformation when compared to structures supported by only raft foundation. The unconnected piled raft foundation reduced the lateral deformation by almost 19% when compared with raft foundation, whereas the connected piled raft foundation reduced the lateral deformation further by 51% due to the rigid connection between the pile head and raft foundation.

6 Conclusion

The actual response of piled raft foundations under different loading conditions was simulated and numerically evaluated by researchers. Even though the actual load transferring mechanisms of a piled raft foundation under dynamic loads cannot be evaluated with higher accuracy, the results from various studies by incorporating appropriate assumptions were reasonable with field studies. However, it was evident that the design of a piled raft foundation system under vertical load will be conservative, and thereby, the effect of lateral loadings must be accounted for seismic loading was considered as a major type of lateral loading in the piled raft systems. Since the nature of seismic loading is of dynamic type, the deformation analysis based on pseudo-static approaches could give marginal difference in the results. Further, the selection of type of connection between the pile and raft foundation is also an important criterion in designing the piled raft foundation in seismic zones. The non-connected piled raft foundation could reduce the structural response under seismic loads to be transferred to the pile heads, whereas the lateral displacement could be more as compared to connected piled raft systems. This paper provides an overview on the behavior of piled raft foundations under various types of loads and connections at the interface of pile and raft foundations.

References

1. Jayarajan P, Kouzer KM (2015) Analysis of piled raft foundations. Indian J Sci 16(51):51–57
2. Poulos HG, Davis EH (1980) Pie foundation analysis and design. In: Lambe TW, Whitman RV (eds.), Rainbow-Bridge Book Co
3. Kumar A, Choudhury D, Katzenbach R (2016) Effect of earthquake on combined pile-raft foundation. Int J Geomech 16(15):1–16
4. Saadatinezhad M, Lakirouhani A, Asli SJ (2019) Seismic response of non-connected piled raft foundations. Int J Geotech Eng 1–15
5. Xiaolei H, Yaokun L, Jing J, Junhao Y, Weichen L, Baicheng D (2016) Numerical simulation on the seismic absorption effect of the cushion in rigid-pile composite foundation. Earthq Eng Eng Vib 15:369–378
6. Horikoshi K, Matsumoto T, Hashizume Y, Watanabe T (2003) Performance of piled raft foundations subjected to dynamic loading. Int J Phys Model Geotech 2:51–62
7. Matsumoto T, Nemoto H, Mikami H, Yaegashi K, Arai T, Kitiyodom P (2010) Load tests of piled raft models with different pile head connection conditions and their analyses. Soils Found 50(1):63–81
8. Badelow F, Poulos HG (2016) Geotechnical foundation design for some of the world's tallest buildings. Japan Geotech Soc Spec Publ 2(2):96–108
9. Sanctis LD, Mandolini A (2006) Bearing capacity of piled rafts on soft clay soils. J Geotech Geoenviron Eng 132(12):1600–1610
10. Terzaghi K (1943) Theoretical soil mechanics
11. Design and Construction of Pile Foundations—Code of Practice. Is Code 2911 (Part1/Sec 3) (2010)
12. Poulos HG (2000) Practical design procedures for piled raft foundations. Design applications of raft foundations, pp 425–467
13. Fattah MY, Yousif MA (2014) Bearing capacity of pile group and piled raft foundations on sandy soil. J Eng Dev 17(2):64–96
14. Liang FY, Chen LZ, Shi XG (2003) Numerical analysis of composite piled raft with cushion subjected to vertical load. Comput Geotech 30(6):443–453
15. Jose M, Divya Krishnan K, Ravichandran PT (2020) Behaviour of vertically loaded piled raft system on cohesionless soil. Materials today: proceedings, pp 3–7
16. Sawada K, Takemura J (2014) Centrifuge model tests on piled raft foundation in sand subjected to lateral and moment loads. Soils Found 54(2):126–140
17. Mattsson N, Menoret A, Simon C, Ray M (2013) Case study of a full-scale load test of a piled raft with an interposed layer for a nuclear storage facility. Geotechnique 63(11):965–976
18. Tarenia K, Patra N (2020) Behaviour of disconnected and connected piled-raft foundations subjected to vertical and lateral loads simultaneously. Geo-Congress 1:33–44
19. Chanda D, Saha R, Haldar S (2020) Behaviour of piled raft foundation in sand subjected to combined V-M-H loading. Ocean Eng 216:1–20
20. Azizkandi AS (2020) Experimental study on connected and non-connected piled raft foundations subjected to eccentric loading. Int J Civ Eng 18(7):743–761

Uncertainty Analysis in Liquefaction Potential Assessment

Nikhil Gupta, Ranjan Kumar, and Kapilesh Bhargava

Abstract Soil sites are prone to liquefaction during earthquake. For establishment of various important structures, e.g., nuclear power plant, research reactors, etc., potential site need to be evaluated with regard to liquefaction. There are deterministic methods namely empirical and analytical which are available for liquefaction assessment. But uncertainties associated with various input parameters are not considered by these deterministic methods. Near field and far field earthquake affect liquefaction significantly. In the present study, uncertainty analysis is carried out in liquefaction of a soil site for 9 near field and 7 far field earthquakes. Uncertainty parameters, namely mean, standard deviation, and type of distribution which this random variable follows are estimated from 38 borehole data of site. From these uncertainty parameters and observed data, probability of liquefaction for respective bore hole is evaluated by theory of probability. The results obtained from near field and far field earthquake were compared. Probability of liquefaction has been observed to be higher for far field earthquakes in comparison to near field earthquakes.

Keywords Liquefaction · Near field and far field earthquake · Random variable

1 Introduction

This soil site lies in seismic zone II near coastal area. The type of soil in this site is marine deposits along the coast and aeolian deposits of red sand teri and other types of sand near the coast. The formation of teri is made up mainly of red stained quartz with an admixture of fine red clayey dust and fine grains of iron ore. Drilling of 38 bore hole has been carried out up to 30 m depth. SPT data available for all these bore holes. Liquefaction is defined as the transformation of a granular material

N. Gupta (✉)
Chemical Technology Group, Bhabha Atomic Research Centre Mysore, Mysore 571130, India

R. Kumar
Civil Engineering Division, Bhabha Atomic Research Centre, Mumbai 400094, India

K. Bhargava
Nuclear Recycle Board, Bhabha Atomic Research Centre, Anushaktinagar, Mumbai 400094, India

T. G. Sitharam et al. (eds.), *Earthquake Geotechnics*, Lecture Notes in Civil Engineering 187, https://doi.org/10.1007/978-981-16-5669-9_27

from a solid to a liquefied state as a consequence of increase in pore water pressure and reduced effective stress. The change of state occurs most readily in loose to moderately dense uniformly graded saturated granular soil with poor drainage, such as silty sands or sands, and gravel. Shear strength of such type of soil is only due to particle-to-particle inter granular friction. When soil is subjected to cyclic loading, excess positive pore water pressure is generated. As a consequence of this, water trapped between the soil particle try to separate soil particle, i.e., strength which is gained by inter-granular contact between particle is reduced to a significant value. Due to drastic reduction of strength, soil loses its shape and behaves like liquid. This phenomenon of soil is called liquefaction. In case of loose soil, due to cyclic stress, if negative pore water pressure is generated, the soil is compacted which leads to ground settlement. Therefore, structure may sink or destabilize, if supported by such soil. Hence, for construction of important building like nuclear structures, liquefaction potential of site needs to be evaluated. Near field and far field earthquake affect liquefaction potential significantly, although some deterministic methods are available for assessment of liquefaction potential but input parameter which is used in deterministic method have some limitations. Uncertainty associated with this input parameter cannot be accounted by this deterministic method. Hence, there is a need for evaluation of liquefaction potential for a site taking into account for uncertainty associated with the input parameters. In the present study, coastal site is selected, and soil geotechnical test data of 38 bore hole is collected. Probability of liquefaction is calculated for near field and far field earthquake ground motion.

2 Uncertainty Analysis

Total stress, effective stress, max acceleration, % of fines, and SPT count are the uncertain input parameters. To account for uncertainty, there is a need to evaluate the distribution which these input parameters follow. Chi square test and K-S test are two commonly used tests for evaluating the distribution of this random variables [1]. The Chi square test is based on the error between the observed and assumed probability density function (PDF) of the distribution, and the K-S test is based on the error between the observed and assumed cumulative distribution function (CDF) of the distribution. The above tests need the data of mean and standard deviation of random variables for random variables total stress, effective stress, % of fines, and SPT count case; these uncertainty parameters are calculated by geotechnical data available for 38 bore holes and uncertainty parameter for max acceleration is obtained from output 38 data of deconvolution analysis of PGA using DEEPSOIL V7.0 software [2]. Mean and COV (coefficient of variation) of each random variable are given in Table 1.

Probability distribution test if the error obtained for a particular distribution is less than a significance level value than the distribution is fit for that random variable 5% significance level means 5 out of a total of 100 different samples, the assumed theoretical distribution cannot be an acceptable.

Table 1 Mean and coefficients of variations of random variable

Depth	Total stress		Effective stress		a_{max}/g		SPT value		% of fines	
	Mean	COV	Mean	COV	Mean	COV	Mean	COV	Mean	COV
1–2	24.21	0.038	9.78	0.094	0.106	0.151	16.16	0.441	44.3	0.583
2–3	40.22	0.038	16.16	0.093	0.103	0.136	35.38	0.229	30.4	0.312
3–4	56.29	0.037	22.61	0.092	0.099	0.121	19.89	0.314	38.9	0.421
4–5	73.36	0.026	30.05	0.064	0.098	0.122	23.09	0.409	42.7	0.313
5–6	88.93	0.027	36	0.068	0.087	0.115	28.35	0.470	47.2	0.009
6–7	106.2	0.024	43.68	0.057	0.093	0.075	28.16	0.487	30.0	0.288
7–8	120.2	0.032	48.06	0.080	0.076	0.118	30.09	0.465	35.4	0.369
8–9	137.5	0.031	55.71	0.077	0.072	0.097	57.57	0.116	30.9	0.587
9–10	153.4	0.029	61.94	0.073	0.067	0.060	34.5	0.557	25.3	1.197

In chi square test, it is observed that for total stress error obtained for depth 3–4 m total error obtains from normal and log normal distribution is 2.69 and 6.3, respectively, and for depth 7–8 m total error obtains from normal and log normal distribution is 2.89 and 6.48 which is less than 5% significance level value of chi square test which is 7.815 for 6 intervals.

In K-S test, it is observed that for total stress error obtained for 28 data of depth 3–4 m, total error obtains from normal and log normal distribution is 0.156 and 0.259, respectively, and for 20 data of depth 7–8 m, total error obtains from normal and log normal distribution is 0.163 and 0.261. 5% significance level value of K-S test is 0.242 for 28 data and 0.294 for 20 data.

Error obtained from normal distribution is less than the error obtained from log normal distribution in both chi square, and K-S test also error obtained from normal distribution is less than significance level value. Hence, normal distribution is fit for total stress case.

The result obtained for other random variables effective stress, a_{max}/g, and percentage of fines follow normal distribution curve, and SPT N value follows log normal distribution curve.

2.1 *Probability of Liquefaction for Entire Site*

For each random variable, mean and standard deviation at every one meter depth is taken from Table 1. For these mean and standard deviations, 40 random variables are generated for each parameter at every one meter depth. Corresponding to these random variables, 40 no's of CSR and CRR are calculated by empirical method [3]. From these 40 CRR and CSR values, mean, standard deviation, and coefficient of variance are obtained. Difference in CRR and CSR value is defined as liquefaction performance function Z. Since CRR and CSR are random variables, and hence, Z is also a random variable. If the mean value and standard deviations of CRR and CSR are μ_R, μ_S, and σ_R, σ_S, respectively, then according to first order and second moment method, the mean value μ_z, the standard deviation σ_z and covariance coefficient δ_z of Z can be defined as [3]

$$\mu_Z = \mu_R - \mu_S \tag{1}$$

$$\sqrt{\sigma_R^2 + \sigma_S^2} = \sigma_Z \tag{2}$$

$$\delta_Z = \frac{\sigma_Z}{\mu_Z} = \frac{\sqrt{\sigma_R^2 + \sigma_S^2}}{\mu_R - \mu_S} \tag{3}$$

Reliability index β is defined as inverse of coefficient of variation δ_z

$$\beta = \frac{1}{\delta_Z} = \frac{\mu_Z}{\sigma_Z} = \frac{\mu_R - \mu_S}{\sqrt{\sigma_R^2 + \sigma_S^2}} \tag{4}$$

From the above equations statics for the Z can be calculated. It is observed that the PDF of CRR and CSR are better described by lognormal distribution than normal distribution, then the reliability index β, and the liquefaction probability can be expressed as

$$\beta = \frac{\mu_Z}{\sigma_Z} = \frac{\mu_{LnR} - \mu_{LnS}}{\sqrt{\sigma_{LnR}^2 + \sigma_{LnS}^2}} = \frac{\mathrm{Ln}\left[\frac{\mu_R}{\mu_S}\left(\frac{\delta_S^2+1}{\delta_R^2+1}\right)^{0.5}\right]}{\left[\mathrm{Ln}\left(\delta_R^2+1\right)\left(\delta_S^2+1\right)\right]^{0.5}} \tag{5}$$

where

σ_{LnR} is logarithmic standard deviation of R
σ_{LnS} is logarithmic standard deviation of S

$$P_f = \emptyset(-\beta) = 1 - \emptyset(\beta) \tag{6}$$

From Eqs. (5) and (6), β and probability of liquefaction are calculated at every one-meter depth. Summary of these results shown in Table 2

Table 2 Summary of Probability of liquefaction for whole site

Depth (m)	CSR		CRR		β	PL
	Mean	COV	Mean	COV		
1–2	0.161	0.128	0.443	0.616	1.473	0.07038
2–3	0.160	0.131	0.986	0.089	11.564	0
3–4	0.158	0.146	0.610	0.496	2.544	0.00548
4–5	0.154	0.109	0.717	0.434	3.393	0.00035
5–6	0.154	0.131	0.783	0.392	3.906	4.70E-05
6–7	0.141	0.106	0.718	0.477	3.295	0.00049
7–8	0.156	0.147	0.839	0.326	4.700	1.30E-06
8–9	0.148	0.129	0.772	0.399	3.907	4.70E-05
9–10	0.150	0.129	1.000	0.000	14.778	0

Table 3 Summary of probability of liquefaction for BH 14

Depth (m)	CSR		CRR		β	P_L
	Mean	COV	Mean	COV		
1.5 m	0.158	0.149	0.265	0.576	0.688	0.246
4.5 m	0.160	0.117	0.565	0.540	2.194	0.014
6 m	0.159	0.113	0.793	0.390	3.930	0.000
9 m	0.156	0.137	0.539	0.632	1.819	0.034

2.2 *Uncertainty Analysis for Critical Boreholes*

Observed value of variable for 4 critical boreholes namely BH14, BH26, BH31, BH33 are taken as mean value and COV taken from Table 1. For these mean and standard deviations, 40 random variable are generated for each random variable. Corresponding to these random variables, 40 no's of CSR and CRR is calculated [4]. From this 40 CRR and CSR values mean, standard deviation and coefficient of variance of CRR and CSR are obtained and from Eqs. (5) and (6). β and probability of liquefaction are calculated [4]. Summary of obtain results is shown in Tables 3, 4, 5, and 6 and Fig. 1.

2.3 *Probability of Liquefaction Due to Near Field and Far Field Earthquake Ground Motions*

Nine near field and 7 far field [5] earthquake ground motions is used to evaluate probability of liquefaction. DEEPSOIL V7.0 software [2] is used which can perform 1D site specific ground response analysis by assuming soil stress strain behavior

Table 4 Summary of probability of liquefaction for BH 26

Depth (m)	CSR		CRR		β	P_L
	Mean	COV	Mean	COV		
1.5	0.177	0.172	0.237	0.377	0.593	0.277
4.5	0.173	0.131	0.406	0.667	1.089	0.138
6	0.172	0.131	0.458	0.701	1.216	0.112
9	0.168	0.162	0.409	0.774	0.951	0.171

Table 5 Summary of Probability of liquefaction for BH 31

Depth (m)	CSR		CRR		β	P_L
	mean	COV	mean	COV		
1.5	0.176	0.171	0.264	0.732	0.304	0.380
4.5	0.172	0.149	0.248	0.573	0.424	0.336
7.5	0.164	0.117	0.606	0.560	2.199	0.014

Table 6 Summary of probability of liquefaction for BH 33

Depth (m)	CSR		CRR		β	P_L
	Mean	COV	Mean	COV		
3	0.167	0.174	0.470	0.521	1.788	0.037
5	0.164	0.118	0.241	0.807	0.197	0.422
9	0.158	0.120	0.326	0.754	0.744	0.229

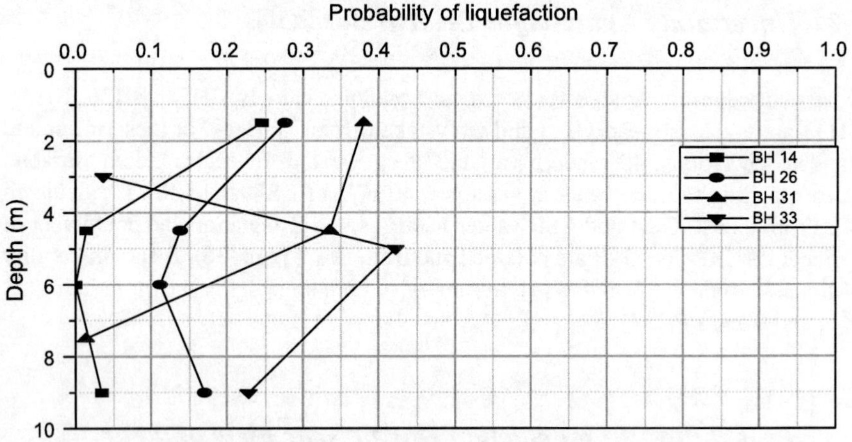

Fig. 1 Probability of liquefaction with depth for BH 14, BH 26, BH 31 and BH 33

nonlinear and equivalent linear. CSR is obtained in output of DEEPSOIL software for each earthquake at a particular depth using 9 near field earthquake. Nine CSRs are obtained for nonlinear and equivalent linear analysis. Mean value and COV of CSR are calculated for this 9 near field earthquake ground motions. In the similar manner, mean value and COV of CSR are calculated for 7 far field ground motions.

For CRR, mean value is taken as observed value and COV for a particular depth is taken from Table 2. For this mean and COV values of CRR and CSR, β and probability of liquefaction are calculated from Eqs. (5) and (6). Output results for BH 26 are shown in Tables 7, 8, 9, and 10. Graph between depth and probability of liquefaction is also drawn which is given in Fig. 2.

Table 7 Probability of liquefaction in nonlinear analysis of near field earthquake

Depth m	CSR		CRR		β	P_L
	Mean	COV	Mean	COV		
1.5	0.161	0.095	0.212	0.616	0.205	0.419
4.5	0.146	0.058	0.306	0.434	1.558	0.060
6	0.140	0.070	0.319	0.400	1.919	0.028
9	0.118	0.164	0.270	0.400	1.827	0.034
10.5	0.107	0.211	0.470	0.100	7.207	0.000

Table 8 Probability of liquefaction in eq-linear analysis of near field earthquake

Depth m	CSR		CRR		β	P_L
	Mean	COV	Mean	COV		
1.5	0.172	0.013	0.212	0.616	0.087	0.465
4.5	0.157	0.055	0.306	0.434	1.382	0.083
6	0.146	0.082	0.319	0.400	1.810	0.035
9	0.122	0.163	0.270	0.400	1.754	0.040
10.5	0.110	0.211	0.470	0.100	7.069	0.000

Table 9 Probability of liquefaction in non-linear analysis of far field earthquake

Depth m	CSR		CRR		β	P_L
	Mean	COV	Mean	COV		
1.5	0.173	0.116	0.212	0.616	0.090	0.464
4.5	0.154	0.231	0.306	0.434	1.316	0.094
6	0.147	0.286	0.319	0.400	1.559	0.060
9	0.131	0.383	0.270	0.400	1.344	0.090
10.5	0.126	0.390	0.470	0.000	3.679	0.000

Table 10 Probability of liquefaction in eq. linear analysis of far field earthquake

Depth m	CSR		CRR		β	P_L
	Mean	COV	Mean	COV		
1.5	0.171	0.048	0.212	0.616	0.103	0.459
4.5	0.156	0.202	0.306	0.434	1.31	0.095
6	0.148	0.264	0.319	0.4	1.571	0.058
9	0.132	0.363	0.27	0.4	1.345	0.089
10.5	0.128	0.37	0.47	0	3.81	0

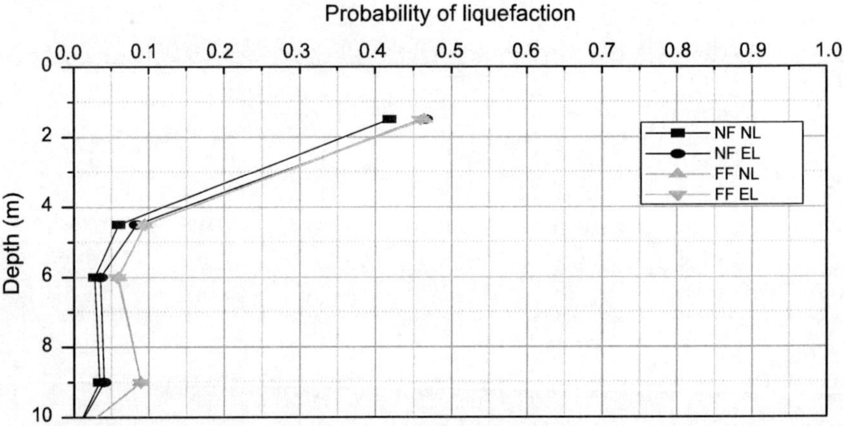

Fig. 2 Depth versus probability of liquefaction for near field and far field EQ by NL and EL analysis for BH 26. Where NF NL is nonlinear analysis of near field earthquake. NF EL is equivalent linear analysis of near field earthquake. FF NL is nonlinear analysis of far field earthquake. FF EL is equivalent linear analysis of far field earthquake

3 Results and Conclusion

Since deterministic methods of liquefaction assessment have some limitations, probabilistic methods are used for the assessment of the same. For uncertainty analysis, five random variables, namely total stress, effective stress, maximum acceleration, percentage fines, and SPT N value have been considered. After conducting statistical tests namely Chi square test and K-S test, normal distribution and log normal distribution are fitting for these random variables. But error obtained in normal distribution is less than that in log normal distribution. Hence, normal distribution is adopted in this study for 4 random variables except SPT N value. For SPT N value, log normal distribution is adopted.

Probability of liquefaction for whole site was evaluated. Probability of liquefaction was observed as 7% for 1–2 m depth, 0.5% at 3–4 m depth and at other depths, no liquefaction was observed. Liquefaction was observed in only four boreholes out of thirty-eight boreholes.

Then uncertainty analysis of the parameters for these four boreholes was done, and probability of liquefaction was estimated for non-linear and equivalent linear analysis separately. For borehole no. 14, liquefaction probability at 1.5 m and 4.5 m was found to be 24.5% and 1.4%, respectively. For borehole no. 26, liquefaction probability at 1.5 m, 4.5 m, 6 m and 9 m was found to be 27.65%, 13.8%, 11.1%, and 17.07%, respectively. For borehole no. 31, liquefaction probability at 1.5 m, 4.5 m, and 7.5 m was found to be 38%, 33%, and 1.3%, respectively. For borehole no. 33, liquefaction probability at 3 m, 5 m and 9 m was found to be 3.69%, 42.19%, and 22.8%, respectively.

Probability of liquefaction (PL) of BH 26 was evaluated for near field and far field motions. It is observed that probability of liquefaction obtained from far field earthquake is generally more than probability of liquefaction obtained from near field earthquake in both nonlinear and equivalent linear analysis. It is because for each frequency content of far field earthquake has more data, hence, for each frequency content sum of amplitude of acceleration is more in far field earthquakes.

References

1. Haldarand A, Mahadevan S ((2005)) Probability, reliability and statistical methods in engineering design. Wiley
2. Hashash YMA, Musgrove MI, Harmon JA, Ilhan O, Groholski DR, Phillips CA, Park D (2017) DEEPSOIL7.0, User Manual
3. Youd TL, Idriss IM (2001) Liquefaction resistance of soil: summary report from the 1996 NCEER and 1998 NCEER/NSF workshops on evaluation of liquefaction resistance of soil. Geotech Geoenviron Eng 817–833
4. Hwang JH, Yang CW, Juang DS (2008) A practical reliability-based method for assessing soil liquefaction potential. Soil Dyn Earthq Eng 24(2004):761–770
5. Heydari M, Mousavi M (2015) The Comparison of Seismic Effects of Near-field and Far-field Earthquakes on Relative Displacement of Seven-storey Concrete Building with Shear Wall. Curr World Environ 10(Special Issue 1):40–46
6. LNCS Homepage. http://www.strongmotioncentre.org. Last Accessed 15 Apr 2019

A Study on Seismic-Induced Liquefaction of Soil and Response of a Group of Piles

M. K. Pradhan, Praveen Kumar, V. S. Phanikanth, and K. Srinivas

Abstract In the present study, the response of the pile foundation subject to seismic load in both non-liquefied and liquefied soil is investigated by developing a mathematical code. Further, the response of pile foundation is compared for both the cases of soils. It is observed from the results that the responses of pile foundation are different in both cases. It is also observed from pile response that pile embedded in liquefiable soil warrants more attention. Therefore, the evaluation of the liquefaction potential of soil is very much essential before planning of construction of any important facility or structure. Hence, in the present study, a mathematical code is also developed to evaluate the liquefaction potential of soils. A parametric study is also carried out for pile foundation in liquefied soils using developed code. During an evaluation of liquefaction potential of soil at any depth using the methods available in the literature, the $a_{\max}$ (maximum acceleration) at ground level is used. In the present study, a method is also proposed to evaluate the liquefaction potential of soil considering the actual acceleration of soil at the depth of concern and including the soil properties, site parameters that are responsible for soil acceleration amplification.

Keywords Liquefaction of soil · Group of piles · Soil–pile–structure interaction · Seismic response

1 Introduction

The foundation system is the substructure that transfers the loads from the superstructure to a soil stratum beneath it. Pile foundations are more commonly used as a deep form of foundations system to transmit vertical, lateral, uplift, and seismic loads where soils at shallow depth possess a comparatively lower load-bearing capacity. Various responses of pile such as deflection, bending moment, shear force along the

M. K. Pradhan (✉) · P. Kumar
Homi Bhabha National Institute and BARC, Trombay, India
e-mail: mpradhan@barc.gov.in

M. K. Pradhan · P. Kumar · V. S. Phanikanth · K. Srinivas
BARC, Trombay, India

T. G. Sitharam et al. (eds.), *Earthquake Geotechnics*, Lecture Notes in Civil Engineering 187, https://doi.org/10.1007/978-981-16-5669-9_28

depth of the pile are very important parameters in the analysis and design of pile foundation. It is observed, severe damages, failures of lifelines, and structures supported on pile foundations occurred during major past earthquakes such as Niigata (1964), Kobe (1989), Chi-Chi (1995), and Bhuj (2001). All those events demonstrated the destroying effects of seismically induced liquefaction of soil. However, sometimes construction of important structures such as nuclear facilities, bridges, harbors, ports, and high-rise structures supported on pile foundations is inevitable/unavoidable on location which is susceptible to seismic-induced liquefaction.

1.1 Pile Foundation in Liquefied

Liquefaction is a phenomenon that occurs in loose, saturated, cohesionless soil wherein the strength and the stiffness of the soil is reduced due to repeated cyclic loading such as earthquake shaking or any other rapid load. Due to the cyclic loads, there is an increase of pore water pressure of soil under an un-drained condition followed by a reduction in effective vertical stress. After certain cycles of repeated loads, the pore pressure built up in loose saturated cohesionless soil. The contact between soil grains becomes minimal. The shear strength (which depends upon its effective stress) gets reduced up to such a level that the soil grains start to move readily with respect to each other. Hence, the soil tends to behave like a liquid mass; i.e., liquefaction is initiated. Soil loses its shear strength, stiffness and starts behaving like a liquid. This phenomenon is known as liquefaction of soils. Flow liquefaction can produce massive flow slides and contributes to the sinking or tilting of heavy structures, the floating of light buried structures, and to the failure of retaining structures. Similarly, cyclic mobility can cause slumping of slopes, settlements of buildings, lateral spreading, etc.

As stated by Kramer [1], liquefaction phenomena can affect buildings, bridges, buried pipelines, and other constructed facilities in different ways. The extensive devastation of pile foundations due to the liquefaction of soil during past earthquake events worldwide was very common. Damages of pile occurred in Alaska earthquake (19,645), Mexico City earthquake, Loma Prieta earthquake, Kobe earthquake, Kutch earthquake are a few of them [2–4]. Failures of pile foundation in liquefied soils and the importance of consideration of liquefaction criteria of soils in pile designs are presented by various researchers [5–9].

Madabhushi et al. [10] presented that the reason for the pile failure of Showa Bridge was on account of the bending moment for lateral spreading of soils. However, Bhattacharya [2, 3] concluded that the mechanism of pile failure of Showa Bridge as buckling failure due to the loss of lateral restraint of soil in liquefied zone. Failure of pile foundations in liquefiable areas during an earthquake was attributed to liquefaction-induced lateral spreading in most of the reported case histories by Hamada, Tokimastu et al., Ishihara and Finn et al. [6, 7, 9]. The failure of various structures supported on the pile foundation stated above shows the limitation of present methodologies that are available for the design of pile foundation in the liquefied soil layer under earthquake conditions. A flowchart is presented in Fig. 1

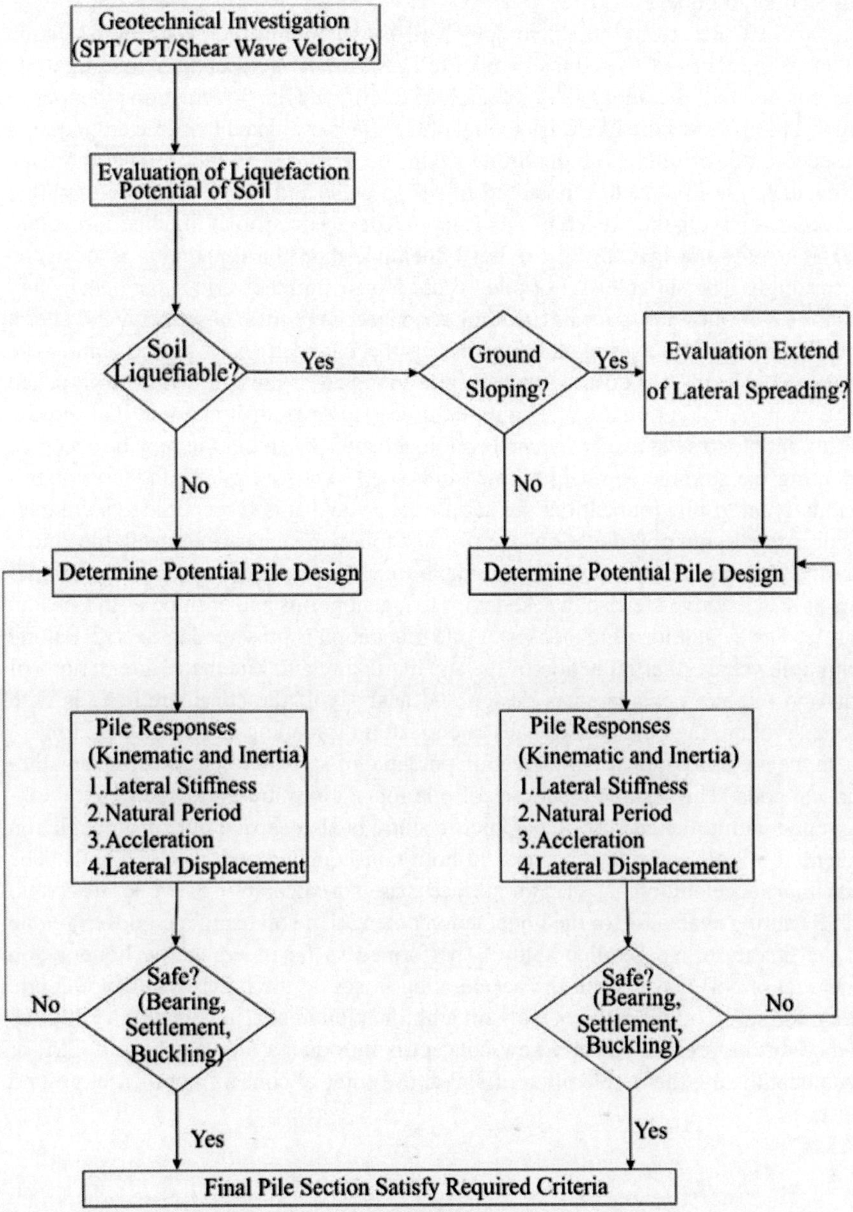

Fig. 1 Flow diagram of site identification and design of pile foundation in liquefiable soils

showing important steps along with various parameters to be considered in the analysis and design of safe and economic pile foundation particularly at the site with soil susceptible to liquefy.

It can be seen from the above flowchart that, before initiating the construction of any important projects, it is essential to carry out the geotechnical investigation. Various methods are adopted by researchers to evaluate the liquefaction potential of soils [11–13]. The liquefaction potential of the site is evaluated from the engineering properties of soil and site conditions using the available methods. The approach to evaluate the liquefaction potential of soil is presented in Sect. 2.2 along with a flowchart showing the procedures and steps. After evaluation of liquefaction potential, if it is found that the soil is not liquefiable then the foundation is designed accordingly. The suitable size of pile is determined, and the various parameters and responses of piles, i.e., stiffness, natural period, accelerations, base shear, and lateral displacements, are investigated considering the intended loads and assuming the proposed section. It is ensured that the pile foundations are safe with the proposed section, material geometry, and configurations. However, if it is found that the site is susceptible to seismically induced soil liquefaction, then the site may be accepted adopting the suitable engineering measures, i.e., improved pile foundation or soil stabilization. If pile foundations are adopted at a site that is susceptible to soil liquefaction, the design of foundation is performed following the various available guidelines, codes of pile foundation in liquefiable soil. The responses and various parameters as stated above are also checked with stringent norms and ensured as the system is safe. The evaluation of responses of pile foundation is presented in Sect. 2.1 along with a flowchart. Inertial action of the superstructure and kinematic interactions of soil and pile are considered in design and analysis. If the considered site is with sloped ground, then the lateral loads due to lateral spreading also accounted for.

In the present paper, the liquefaction potential of soil is evaluated using a mathematical code. The seismic responses of pile foundations are evaluated in both non-liquefied and liquefied soil. A parametric study is also carried out for liquefaction potential of soil and pile responses in both non-liquefied and liquefied soils. The maximum acceleration ($a_{\max}$) at the surface is used in calculation of cyclic stress ratio (CSR) during evaluation of the liquefaction potential of soil using methods available in the literature. No detailed study is performed so far in evaluating liquefaction potential of soil considering the acceleration in soil at the point of concern/depth. Also, the study on the influence of soil amplification in soil liquefaction is limited. Hence, in the present paper, a new concept is introduced in evaluation of CSR at each depth using the acceleration of soil at the point of concern not $a_{\max}$ at ground surface.

2 Methods

In the present study, code is developed to investigate the responses of pile foundations in non-liquefied soils as well as in liquefied soils. The brief about the code and flowchart is presented in Sect. 2.1. The results are compared for both the cases and presented in Sect. 3.1. A code is also developed to evaluate the liquefaction potential of soils. A parametric study is performed using the code and is presented in Sect. 3.2.

2.1 Mathematical Code for Pile Response in Non-Liquefied and Liquefied Soil

Responses of piles are important factors that are considered in the analysis and design of pile foundation in non-liquefied soils and liquefied soils. Madabhushi et al. [10] proposed the methods to evaluate various important parameters such as soil stiffness, natural frequency, effective length and flexibility of the pile, inertia loading on the pile, kinematic interaction in non-liquefied soils and liquefied soils (in the two-layer system and three-layer system). The code is developed considering all relevant parameters, i.e., as suggested in available literature (Madabhushi et al. [10]). However, fine details are not presented in flowcharts due to space restraint. The flowchart for the developed code to study the response of pile in liquefied soil is presented in Fig. 2.

The intended load considered by authors in the analysis was 9.4 MN. Hollow steel pipe was considered as pile foundation by authors with a diameter of 0.75 m and wall thickness of 12 mm. The soil profile was of 8.00 m thick silty layer overlaying a 22-m deep dense sand deposit. The water table was considered at ground level. The saturated unit weight of the silty sand layer was considered as 17 kN/m^3, and dense sand was 19 kN/m^3. The critical state of friction angle for both soils was assumed as 32°. Peak ground acceleration considered by authors in his study was of 0.2 g. Based on the proposed methods by Madabhushi et al. [10], a code is developed in the present study to investigate the responses of piles varying the various parameters. The results are presented in Sect. 3.1.

2.2 Code for Liquefaction Potential of Soils: Parametrical Study

Evaluation of the liquefaction potential of soils is very much essential before selecting, analyzing, and designing the types of foundation systems to be adopted for some important structures. Various methods of evaluations of the liquefaction potentials are available in the literatures [11–13]. Factor of safety (FS) against liquefaction of soil is the ratio of cyclic resistance ratio (CRR) to the cyclic stress ratio (CSR) and evaluated as Eqs. (1), (2), and (3); [1, 13, 14]

$$\text{FS} = \frac{\text{MSF} \times \text{CRR}_{7.5}}{\text{CSR}} \tag{1}$$

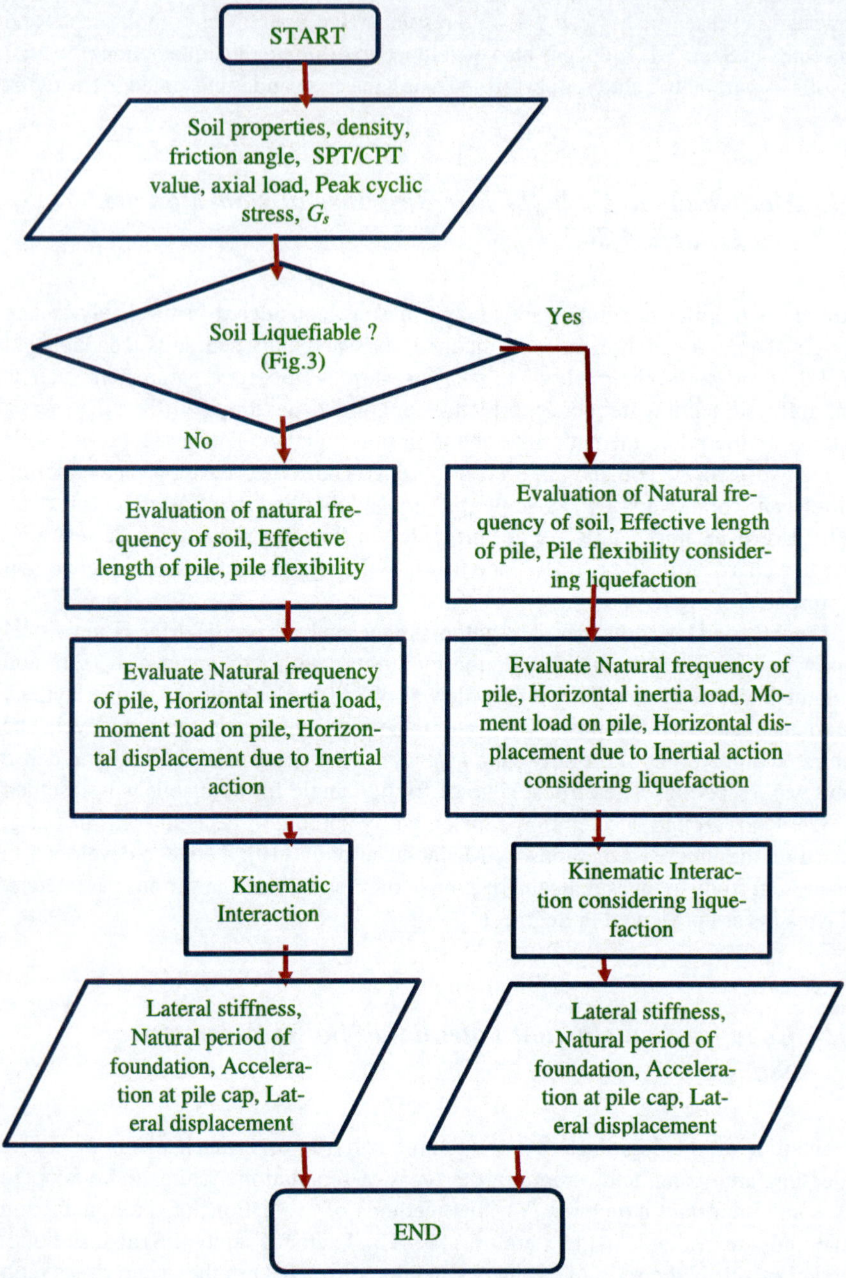

Fig. 2 Flowchart of the procedure of developed code to evaluate responses of pile in no liquefiable and liquefiable soil

where

$$\mathrm{CRR}_{7.5} = \frac{1}{34 - (N1)_{60\mathrm{cs}}} + \frac{(N1)_{60\mathrm{cs}}}{135} + \frac{50}{\left[10 \times (N1)_{60\mathrm{cs}} + 45\right]^2} - \frac{1}{200} \quad (2)$$

where

$\mathrm{CRR}_{7.5}$ is standard cyclic resistance ratio for a 7.5 magnitude earthquake obtained from SPT value.
$(N1)_{60\mathrm{cs}}$ is SPT value corrected for energy, borehole, rod, fine content, etc. [13]
MSF is magnitude scaling factor that depends on the magnitude of the earthquake (other than 7.5)

$$\mathrm{CSR} = 0.65 \times \left(\frac{a_{\max}}{g}\right) \times \left(\frac{\sigma_{v0}}{\sigma'_{v0}}\right) \times r_{\mathrm{d}} \quad (3)$$

where $a_{\max}$ = peak ground acceleration (PGA) in g,

r_{d} = stress reduction factor depends upon depth of consideration (function of depth).

If the factor of safety against liquefaction is less than 1.00 (FOS $\leq$ 1.2 as per IS 1893, 2016), then the soil is considered as susceptible to liquefy. In the present paper, mathematical code is developed based on the method proposed in available literatures to evaluate cyclic stress ratio (CSR), cyclic resistance ratio (CRR) and factor of safety (FOS) of soils against liquefaction. The flowchart of the procedure of developed code to evaluate liquefaction potential of soil is presented in Fig. 3. The code is developed considering all relevant parameters, i.e., corrected SPT value ($N_{1,60}$) cs (correction for energy, overburden, fine content), stress reduction factor (r_{d}), magnitude scaling factor correction, etc., as suggested in available literatures. However, fine/minutes details are not presented in flowcharts due to space restraint. Further, parametric studies are performed on various parameters, viz. standard penetration test (SPT) values, the density of soils, fine contents, relative density, and maximum amplitude of the earthquake. The results are presented in Sect. 3.2

2.3 *Proposed Method: Evaluation of Liquefaction Potential Considering Soil Amplification*

Maximum acceleration ($a_{\max}$) at ground level is considered during evaluation of liquefaction potential of soil as per the procedure available in literatures. During earthquakes, the bedrock stratum gets excited depending on various seismic characteristics, viz. magnitude of earthquakes. As the excitation moves from the bedrock

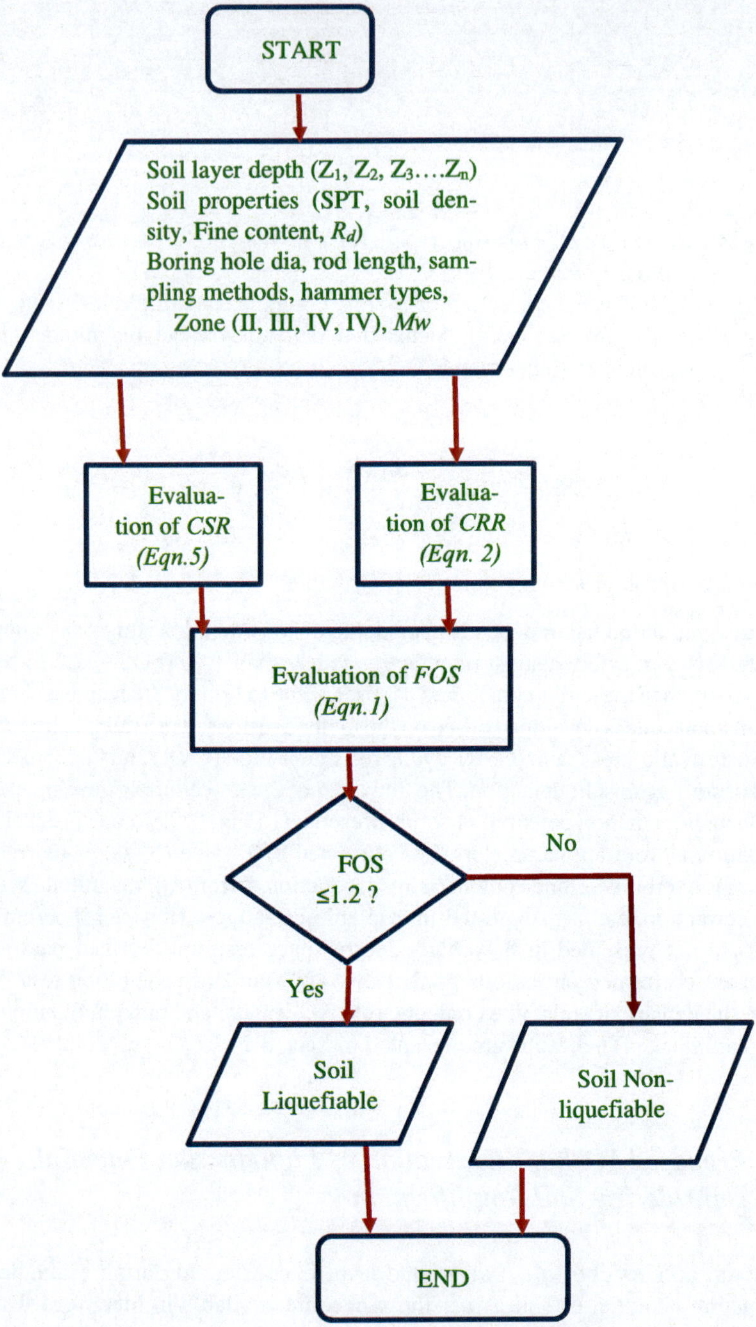

Fig. 3 Flowchart of the procedure of developed code to evaluate liquefaction potential of soil

stratum toward the foundation system, it propagates through the soil mass and its characteristics get changed. The local modification of a wave motion between bedrock and soil outcrop depends upon the geotechnical parameters of bedrock, nature, and profile of deposited soil above the bedrock. The schematic representation of the same is shown in Fig. 4.

The excitation behavior of earthquake motion at bedrock level may get amplified during its upward propagation through the soil layers. The various parameters affecting the amplifications are densities of soils, rigidity, soil thickness as well as the intensity of seismic motion. The amplification study in cohesionless soil will also be helpful and will trigger the research in the liquefaction phenomenon, which is not much considered so far by researchers.

Peak ground acceleration value (PGA) at bedrock and soil transform factor is required for calculation of dynamic amplification factor (DAF). The steps involved in evaluation of DAF and responses at soil surface are presented in the flowchart form in Fig. 5.

Considering uniform undamped soil on rigid rock and choosing two points both at the top and bottom of the soil layer, the transfer function is evaluated as Eq. (4) [1].

$$F_1(\omega) = \frac{1}{\mathrm{COS}(\omega H/V_s)} \tag{4}$$

where 'w' is circular frequency of ground shaking, 'H' is depth of soil layer, and 'V_s' is shear wave velocity in the soil medium.

The modulus of the transfer function is the amplification function. Considering uniform damped soil on rigid rock and choosing two points both at the top and bottom

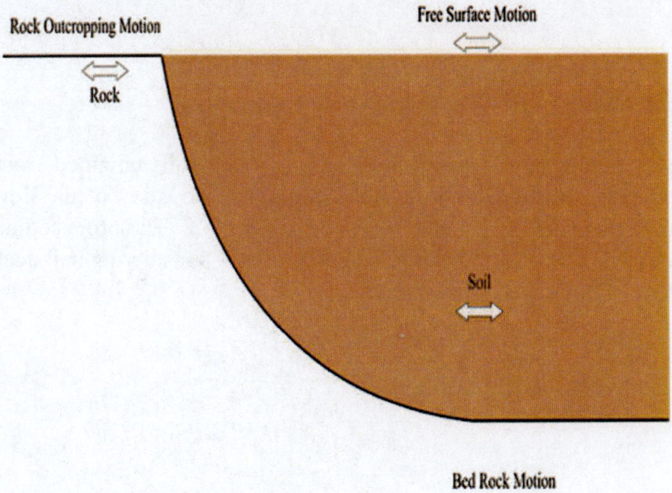

Fig. 4 Motion at bedrock, free surface, and rock outcropping

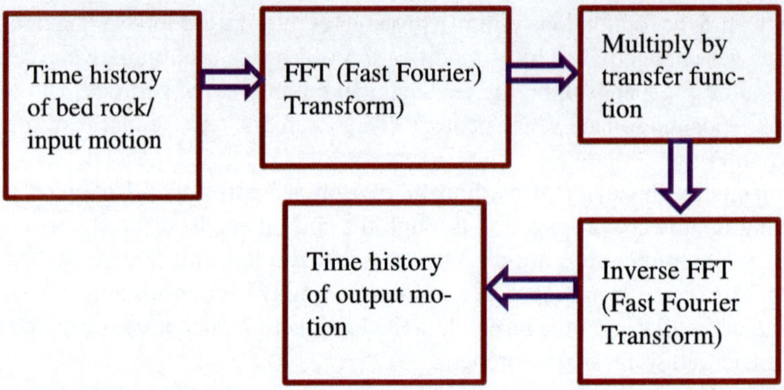

Fig. 5 Steps involved in calculation of ground responses from bedrock level to ground surface

of soil layer the transfer function is evaluated as Eq. (5).

$$F_2(\omega) = \frac{1}{\mathrm{COS}[\omega H/v_s(1 + i\xi)]} \quad (5)$$

Acceleration in each layer/depth of soil can be evaluated from various approaches, viz. from ground response analysis, as presented above or by using software like DEEPSOIL. In the present study, a new concept is introduced to evaluate the liquefaction potential of soil or CSR using the acceleration of soil observed in each layer/point despite $a_{\max}$ at ground surface. The various parameters affecting the soil acceleration amplifications, viz. densities of soils, rigidity, soil thickness, and intensity of seismic motion, also can be considered in liquefaction potential evaluation as briefed in Sect. 3.3

3 Results and Discussion

Responses of pile in non-liquefied soils and liquefied soils obtained from mathematical code are presented in Sect. 3.1. The parametric study of the liquefaction potential of soils is performed and presented in Sect. 3.2. The proposed method (a future study) to evaluate liquefaction potential of soil considering soil acceleration amplification is introduced in Sect. 3.3.

3.1 *Responses of Pile from Mathematical Code: Liquefied and non- Liquefied Soils*

Madabhushi et al. [10] proposed the methods to evaluate various important parameters such as soil stiffness, natural frequency, effective length and flexibility of the pile, inertia loading on the pile, kinematic interaction in non-liquefied soils and liquefied soils (in two-layer system and three-layer system). Based on the methods proposed by Madabhushi et al. [10], a code is developed in the present study to investigate various responses of piles by varying various parameters. The responses of group of pile foundations consisting of four piles, varying the pile diameter and length with input base excitation of 0.2 g in two layers' soil system, (both layers are non-liquefied soils with different soil properties) are presented in Table 1. Soil is considered as two layers soil system consisting of both top and bottom as non-liquefied layer as NL1 and NL2, respectively. Similarly, the responses of the group of pile in liquefied soils with horizontal ground are presented in Tables 2 and 3.

It is observed from Table 1 that by increasing the diameter of the pile in non-liquefied soil the stiffness of foundation increases, whereas the natural period of foundation and peak horizontal displacement decreases.

The responses of the group of pile foundation consisting of 4 piles with pile diameter of 0.75 m and pile length of 20.00 m embedded in top liquefied soil layer with input base excitation of 0.2 g are presented in Table 2 and for base excitation of 0.3 g are presented in Table 3. It is assumed that the liquefied soil layer lost its stiffness. The soil is considered as two layers soil system consisting of top liquefied layer (L1) and bottom non-liquefied layer (NL1). The stiffness of dense sand with an approximate relative density of 70% is assumed as 8500 kN/m^3.

It is observed from Tables 2 and 3 that in the two-layer soil system, with increasing the depth of top liquefied soil layer the stiffness of foundation decreases, whereas the natural period of foundation and peak horizontal displacement increases. It is also observed from Tables 2 and 3 for the top layer liquefied soil layers, that by increasing input base excitation from 0.2 to 0.3 g, the peak horizontal displacement of the pile increases.

The variation in stiffness, natural period, lateral displacement, and acceleration of group of pile for pile length of 20 m, pile diameter of 0.75 m, and input base excitation 0.2 g in non-liquefied soil and liquefied soil with horizontal ground are presented in Table 4.

It is observed from Table 4 that for a pile of 20-m long and 0.75 m diameter with 0.2 g input base excitation and 8 m top layer liquefied soil (L1), there is an approximate reduction in stiffness (93.77%), increase in natural period (300.88%), increase in lateral displacements (579.03%), and reduction in the acceleration of the pile cap (57.74%) in liquefied soils while comparing with non-liquefied soil. Similarly, for 3 m top layer liquefied soil, there is an approximate reduction in stiffness (67.99.77%), an increase in natural period (76.75%), an increase in lateral displacements (32.0%), and a reduction in the acceleration of the pile cap (57.74%) while comparing with non-liquefied soil.

Table 1 Responses of pile in non-liquefied soils (top non-liquefied layer NL1-8.00 m, bottom non-liquefied layer NL2 = 22.00 m, $a_{max} = 0.2$ g)

Pile Dia (*D*)	Pile length (*L*)	Effective length of Pile (L_{ad})	T_u (m)	Z_u (m)	Pile type	Stiffness in MN/m	Natural frequency of pile group (Hz)	Natural period (s)	Acceleration of pile group (m/s^2)	Peak horizontal displacement (m)
0.50	20	6.701	4.568	6.94	Flexible	29.30	1.777	0.562	4.179	0.033
0.75	20	8.840	4.569	6.94	Flexible	45.72	2.220	0.451	4.179	0.021
1.50	20	14.139	4.569	6.93	Flexible	97.31	3.283	0.308	4.179	0.010
0.50	15	6.7010	4.56	5.20	Flexible	29.30	1.777	0.562	4.179	0.033
1.00	15	10.740	4.560	5.20	Flexible	62.59	2.597	0.385	4.179	0.015
1.50	15	14.139	4.569	5.20	Flexible	97.31	3.238	0.308	4.179	0.0101

Note L_{ad} effective length of pile; T_u elastic length of pile or Depth of fixity of pile; Z_u L/T

Table 2 Responses of pile in liquefied soils with horizontal ground (pile length = 20.00 m, pile diameter = 0.75 m, a_{max} = 0.2 g)

Depth L1 (m)	Depth NL1 (m)	T_u (m)	Fixity depth L_f (m)	Stiffness MN/m	Natural frequency of pile group (Hz)	Natural period (s)	Acceleration of pile group (m/s^2)	Peak horizontal displacement (m)
12.0	22.0	2.158	3.884	1.191	0.358	2.790	1.765	0.348
8.0	22.0	2.158	3.884	2.845	0.553	1.805	1.765	0.145
5.0	22.0	2.158	3.884	6.809	0.856	1.167	1.765	0.060
3.0	22.0	2.158	3.884	14.635	1.256	0.796	1.765	0.028

Table 3 Responses of the pile in liquefied soils with the horizontal ground (a_{max} = 0.3 g, Pile length L = 20.00 m, Pile diameter = 0.75 m)

Depth L1 (m)	Depth NL1 (m)	Stiffness of pile foundation MN/m	Natural frequency of pile group (Hz)	Natural period of pile foundation (s)	Acceleration of pile group (m/s^2)	Peak horizontal displacement of pile foundation (m)
12.0	22.0	1.191	0.358	2.790	2.648	0.522
8.0	22.0	2.845	0.553	1.085	2.648	0.218
5.0	22.0	6.809	0.856	1.167	2.648	0.091
3.0	22.0	14.63	1.256	0.796	2.648	0.042

Table 4 Variation in stiffness, natural period, lateral displacement, and acceleration of pile foundation in non-liquefied and liquefied soil (pile length L = 20 m, $NL1$ = 22 m, pile diameter (D) = 0.75 m, a_{max} = 0.2 g)

Depth L1 (m)	Variation in stiffness of foundation in non-liquefied and liquefied soil	Variation in natural period of foundation in non-liquefied and liquefied soil	Variation in Lateral displacement of foundation in non-liquefied and liquefied soil	Variation in Acceleration of foundation in non-liquefied and liquefied soil
12.0	97.393	−519.440	−1521.302	57.746
8.00	93.777	−300.880	−579.034	57.746
5.00	85.106	−159.118	−183.699	57.746
3.00	67.990	−76.751	−32.004	57.746

In the sloped ground, the lateral spreading of the soil exerts additional pressures on the pile due to liquefaction of soils. Hence, the responses of soils may be only of residual responses considering lateral spreading of soil in the sloping ground or maybe a combination of lateral spreading and transient response due to earthquake.

Table 5 Responses of pile in liquefied soils with sloping ground considering lateral spreading and inertia effects (pile length (L) = 20 m, pile diameter (D) = 0.75 m)

Liquefied depth L1 (m)	Pile cap depth	Input base excitation S_a (g)	Soil Layer system	Horizontal displacement due to lateral spreading alone (m)	Horizontal displacement due to Lateral spreading and inertia effects (m)
8.00	1.5	0.1	2	0.042	0.120
8.00	0.5	0.1	2	0.030	0.109
8.00	1.5	0.2	2	0.042	0.198
8.00	0.5	0.2	2	0.030	0.187
8.00	1.5	0.3	2	0.042	0.277
8.00	0.5	0.3	2	0.030	0.265
8.00	1.5	0.1	3	0.219	0.297
8.00	0.5	0.1	3	0.089	0.168
8.00	1.5	0.2	3	0.219	0.375
8.00	0.5	0.2	3	0.089	0.246
8.00	1.5	0.3	3	0.219	0.454
8.00	0.5	0.3	3	0.089	0.324

The responses of the group of pile in sloping ground for two-layered (top liquefied and bottom non-liquefied) soil system and three-layer soil system with intermediate liquefied soils are presented in Table 5.

It is observed from Table 5 that in sloped liquefied ground with increased input base excitation there is an increase in displacement of the pile due to the combined effects of lateral spreading and inertia effects. With an increased depth of the crust layer in three-layered soil system (intermediate liquefied zone), it is observed that there is an increase in both the lateral displacements due to lateral spreading and also due to combining action with inertial effects.

After comparing the results obtained from the mathematical code which is presented in Tables 1, 2, 3, and 4, it is observed that the lateral displacements of the pile in liquefied soils increased considerably. Hence, it warrants the attention of designer during the design of pile foundation in liquefied soils. The liquefaction of soils reduces the stiffness of the soil, reduces the acceleration at the top of foundation, and lengthens the natural period of the foundation. All these factors should be considered during the design of a safe pile foundation in liquefiable soils.

3.2 Liquefaction Potential of Soils: Parametrical Study

There were several high magnitude earthquakes that occurred in past, viz. Tohoku earthquake, 2011 ($M_w = 9.1$, Alaska earthquake, 1964 ($M_w = 9.2$), Sumatra earthquake, 2004 ($M_w = 9.2$), etc. Hence, the influences of magnitude of the earthquake and a_{max} on earthquake-induced liquefaction need in depth understanding. In the present section, a code is developed based on the methods available in literatures and codes to evaluate the liquefaction potential of soils. The parametric study is performed varying various parameters to investigate the liquefaction potential of soils. The results of CSR, CRR, FOS, etc., with a maximum amplitude of acceleration of 0.1, 0.16, 0.24, and 0.36 g with $M_w = 6.75$ are presented in Table 6. The results of CSR, CRR, FOS with a maximum amplitude of acceleration of 0.1, 0.16, 0.24, and 0.36 g with $M_w = 9.25$ are presented in Table 7. After evaluation of liquefaction potential of soil, if it is found that the factor of safety against liquefaction of the soil is more than one, then it is considered that the soil is not to be susceptible to liquefy.

In each base excitation (0.1, 0.16, 0.24, and 0.36 g), the SPT values are changed along the depth for all cases, viz. case-1, case-2, case-3, and case-4. There are various research papers available for the relationship between SPT value of soils and their relative density [15–17]. In the present study, the approximate value of relative densities of cohesionless soils is assumed based on the observed site value and values for clean sand available in literatures.

It is observed in case-1 from Table 6 that with the magnitude of an earthquake of $M_w = 6.5$ and amplitude of maximum base excitations of 0.1 g the soil is not susceptible to liquefy. However, after increasing the maximum base excitation to 0.36 g (case-4) it is observed at some depths the soil is susceptible to liquefy.

It is observed in case-5 from of Table 7 that with the magnitude of earthquake $M_w = 9.25$ and amplitude of maximum base excitations 0.1 g the soil is susceptible to liquefy at an approximate depth of 15 m. However, after increasing the maximum base excitation to 0.36 g (case-8) it is observed that the soil is susceptible to liquefy at more depths. After comparing the results from Table 6 ($M_w = 6.5$) and Table 7 ($M_w = 9.25$), it is observed that with an increased magnitude of an earthquake the chances of occurrence of liquefaction of soils increased. Further, a parametric study performed in liquefaction potential of soil varying the SPT values at different depth with $M_w = 7.5$ and $a_{max} = 0.24$ g and is presented in Table 8.

Most designers use a value of 1.2–1.5 as a factor of safety against liquefaction (FOS) of soils depending on soil layers, the seismicity of the region, and the intended performance of the structure (importance of the structure). It is further clarified by code that when the design ground motion is conservative, earthquake-related permanent ground deformation is generally small if FS is greater than 1.2 [13]. As a thumb rule, there is no liquefaction at a depth of more than 15 m. However, it is reported by researchers that liquefaction also occurred at greater depth. It is possible for generating excessive pore pressure at even a depth of 30 m which is sufficient to overcome the stiffness created by overburden pressure and exceeds the threshold

Table 6 Parametric study in liquefaction potential of soils ($M_w = 6.75$)

Case	Depth (Z) m	SPT (N)	$(N_1)_{60cs}$	Density of soil (kN/m^3)	D_r (%)	FC %	CSR	CRR	FS	Remark
Case-1 0.1 g $M_w = 6.75$	0.50	10	23.722	17.25	45	65	0.1501	0.3781	2.51	NLIQ
	5.00	15	33.027	17.50	50	55	0.1423	1.7854	12.54	NLIQ
	10.0	20	33.184	18.25	55	45	0.1275	1.9349	15.17	NLIQ
	15.0	30	40.643	18.50	60	45	0.1070	0.1735	1.621	NLIQ
	20.0	40	44.423	19.25	65	45	0.0848	0.2364	2.787	NLIQ
	25.0	50	48.441	19.50	70	45	0.0663	0.2676	4.039	NLIQ
Case-2 0.16 g Mw = 6.75	0.50	10	23.722	17.25	45	65	0.2402	0.3781	1.573	NLIQ
	5.00	15	33.027	17.50	50	55	0.2276	1.7850	7.842	NLIQ
	10.0	20	33.184	18.25	55	45	0.2040	1.9349	9.486	NLIQ
	15.0	30	40.643	18.50	60	45	0.1713	0.1735	1.013	NLIQ
	20.0	40	44.423	19.25	65	45	0.1357	0.2364	1.741	NLIQ
	25.0	50	48.441	19.50	70	45	0.1060	0.2676	2.524	NLIQ
Case-3 0.24 g $M_w = 6.75$	0.50	10	23.722	17.50	60	65	0.3603	0.3781	1.049	NLIQ
	5.00	15	33.027	17.50	60	55	0.3414	1.7850	5.228	NLIQ
	10.0	20	33.184	17.50	60	45	0.3060	1.9349	6.324	NLIQ
	15.0	30	40.643	17.50	60	45	0.2569	0.1735	0.675	LIQ
	20.0	40	44.423	17.50	60	45	0.2036	0.2364	1.161	NLIQ
	25.0	50	48.441	17.50	60	45	0.1590	0.2676	1.683	NLIQ
Case-4 0.36 g $M_w = 6.75$	0.50	10	23.72	19.50	70	65	0.540	0.378	0.69	LIQ
	5.00	15	33.02	19.50	70	55	0.512	1.785	3.48	NLIQ
	10.0	20	33.18	19.50	70	45	0.458	1.934	4.21	NLIQ

(continued)

Table 6 (continued)

Case	Depth (Z) m	SPT (N)	$(N_1)_{60cs}$	Density of soil (kN/m^3)	D_r (%)	FC %	CSR	CRR	FS	Remark
	15.0	30	40.64	19.50	70	45	0.385	0.173	0.45	LIQ
	20.0	40	44.42	19.50	70	45	0.3054	0.236	0.77	LIQ
	25.0	50	48.44	19.50	70	45	0.238	0.267	1.12	NLIQ

Note SPT standard penetration test, D_r relative density of soil, *FC* fine content
CSR cyclic stress ratio, *CRR* cyclic resistance ratio, *FS* factor of safety against liquefaction

Table 7 Parametric study in liquefaction potential of soils ($M_w = 9.75$)

Case	Depth (Z) m	SPT (N)	$(N_1)_{60cs}$	Density of soil (kN/m^3)	D_r (%)	FC %	CSR	CRR	FS	Remark
Case-5 0.1 g $M_w = 9.25$	0.50	10	23.722	17.25	45	65	0.150	0.168	1.124	NLIQ
	5.00	15	33.027	17.50	50	55	0.142	0.796	5.600	NLIQ
	10.0	20	33.184	18.25	55	45	0.127	0.863	6.775	NLIQ
	15.0	30	40.643	18.50	60	45	0.107	0.077	0.723	LIQ
	20.0	40	44.423	19.25	65	45	0.084	0.105	1.244	NLIQ
	25.0	50	48.441	19.50	70	45	0.066	0.119	1.802	NLIQ
Case-6 0.16 g $M_w = 9.25$	0.50	10	23.722	17.25	45	65	0.240	0.168	0.702	LIQ
	5.00	15	33.027	17.50	50	55	0.227	0.796	3.500	NLIQ
	10.0	20	33.184	18.25	55	45	0.204	0.863	4.234	NLIQ
	15.0	30	40.643	18.50	60	45	0.171	0.077	0.452	LIQ
	20.0	40	44.423	19.25	65	45	0.135	0.105	0.777	LIQ
	25.0	50	48.441	19.50	70	45	0.106	0.119	1.126	NLIQ
Case-7 0.24 g $M_w = 9.25$	0.50	10	23.722	17.5	60	65	0.360	0.168	0.468	LIQ
	5.00	15	33.027	17.5	60	55	0.341	0.796	2.333	NLIQ
	10.0	20	33.184	17.5	60	45	0.306	0.863	2.822	NLIQ
	15.0	30	40.643	17.5	60	45	0.256	0.0775	0.301	LIQ
	20.0	40	44.423	17.5	60	45	0.203	0.105	0.518	LIQ
	25.0	50	48.441	17.5	60	45	0.159	0.119	0.751	LIQ
Case-8 0.36 g $M_w = 9.25$	0.50	10	23.722	19.5	70	65	0.540	0.168	0.312	LIQ
	5.00	15	33.027	19.5	70	55	0.512	0.796	1.555	NLIQ
	10.0	20	33.184	19.5	70	45	0.458	0.863	1.881	NLIQ

(continued)

Table 7 (continued)

Case	Depth (Z) m	SPT (N)	$(N_1)_{60cs}$	Density of soil (kN/m^3)	D_r (%)	FC %	CSR	CRR	FS	Remark
	15.0	30	40.643	19.5	70	45	0.385	0.077	0.201	LIQ
	20.0	40	44.423	19.5	70	45	0.305	0.105	0.345	LIQ
	25.0	50	48.441	19.5	70	45	0.238	0.119	0.500	LIQ
	0.50	50	48.441	19.5	70	45	0.2385	0.267	1.122	NLIQ

Table 8 Parametric variation in liquefaction potential of soils with $M_w = 7.50$ and $a_{max} = 0.24$ g

Case	Depth (Z) m	SPT (N)	(N1)60cs	Density (kN/m3)	Dr (%)	FC %	CSR	CRR	FS	Remark
Case-9	0.50	10	23.722	17.25	45	65	0.360	0.288	0.801	LIQ
	5.00	10	24.193	17.25	45	55	0.347	0.297	0.855	LIQ
	10.0	10	20.294	17.25	45	45	0.328	0.221	0.673	LIQ
	15.0	10	18.230	17.25	45	45	0.279	0.187	0.671	LIQ
	20.0	10	16.539	17.25	45	45	0.231	0.165	0.714	LIQ
	25.0	10	15.385	17.25	45	45	0.183	0.151	0.824	LIQ
Case-10	0.50	25	50.922	18.50	60	65	0.360	0.336	0.934	LIQ
	5.00	25	52.066	18.50	60	55	0.347	0.349	1.005	NLIQ
	10.0	25	42.321	18.50	60	45	0.328	0.200	0.609	LIQ
	15.0	25	37.162	18.50	60	45	0.279	0.043	0.155	LIQ
	20.0	25	32.934	18.50	60	45	0.231	1.068	4.616	NLIQ
	25.0	25	30.049	18.50	60	45	0.183	0.414	2.264	NLIQ
Case-11	0.50	50	96.256	19.50	70	65	0.312	0.295	0.945	LIQ
	5.00	50	87.022	19.50	70	55	0.301	0.499	1.654	NLIQ
	10.0	50	69.947	19.50	70	45	0.284	0.464	1.631	NLIQ
	15.0	50	60.905	19.50	70	45	0.242	0.439	1.810	NLIQ
	20.0	50	53.497	19.50	70	45	0.200	0.2663	1.325	NLIQ
	25.0	50	48.441	19.50	70	45	0.159	0.204	1.285	NLIQ

of liquefaction [18]. It is reported that a great earthquake can generate stresses to produce a condition of liquefaction in unconsolidated soils, even at a depth of 300 m [18].

If the soil at a site is found to be susceptible to liquefy as presented above, then the liquefaction consideration should be emphasized in the analysis and design of pile foundation. The frictional contribution of liquefied soil layer toward pile bearing capacity is neglected. The size, section, and materials of piles are selected suitably to ensure safe against buckling failure. Suitable engineering measures like ground improvements also can be adopted to mitigate the liquefaction potential of soil.

3.3 Proposed Method: Liquefaction Potential of Soil Considering Soil Acceleration Amplification

As presented in Sect. 2.2, for calculation of the factor of safety against liquefaction of soil, the computation of CRR and CSR is required. For calculating CSR at any depth based on methods available in literatures, the a_{max} (maximum acceleration) at ground level is used along with a stress reduction constant r_d (a function of depth

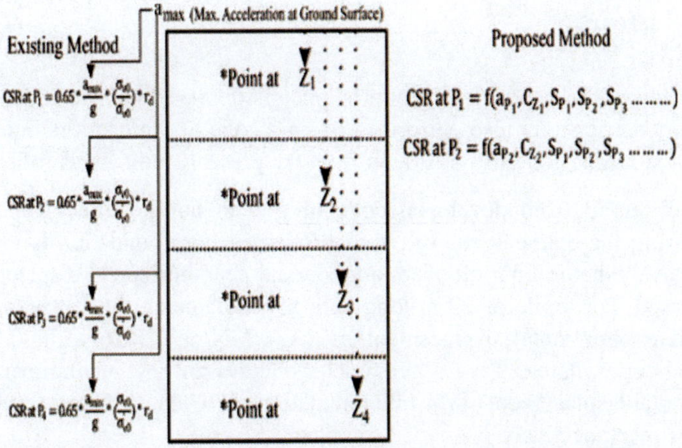

Fig. 6 Evaluation of cyclic stress ratio in existing methods and proposed method

only) as presented in Eq. 3. During an earthquake, the acceleration of soil along the height may amplify. The soil properties responsible for acceleration amplification along the height of soil may have effects on soil liquefaction. This criterion is not accounted for in the existing methods for evaluation of liquefaction potential of soil. This present paper ends with an introduction of a method to evaluate the liquefaction potential of soil at any depth considering the acceleration of soil particularly of that depth (a_{zi}) which is proposed as a future study. Also, envisaged considering the soil acceleration amplification as a function of soil engineering properties. The proposed method is presented schematically in Fig. 6.

In Fig. 6, the soil is shown as a four layers soil system. The left side of soil layers presents the existing methods to evaluate CSR at any depth (layers 1, 2, 3, and 4) using the $a_{\max}$ which is the acceleration at surface of the ground. However, in the method which is shown in right side of Fig. 4, it is proposed to consider the actual acceleration ($a_{\max,\,p1}$, $a_{\max,\,p2}$, etc.) of soil at that depth. The cyclic stress ratio of soil during the evaluation of liquefaction potential can be evaluated considering the observed acceleration in each point a_{p1}. The acceleration at any depth can be evaluated using DEEPSOIL, or using the technics presented in Sect. 2.3 and shown in Fig. 5. In the proposed method, site parameters C_{z1} (similar to r_{d}), the soil properties (S_{P1}, S_{P2}..) responsible for soil amplification may be considered. The liquefaction potential of soil at various depth evaluated based on the proposed method (considering the acceleration at each depth of soil and soil amplification) may differ from the liquefaction potential of soil evaluated using methods available in the literature. However, to derive the final conclusion, it requires a more extensive study.

4 Conclusions

In the present work, response of group of piles is observed using a mathematical code. A parametric study also performed using a code to evaluate the liquefaction potential of soil. The conclusions drawn from the present study are as follows.

1. It is observed from developed code of pile in non-liquefied soil that, by increasing the diameter the pile the stiffness of pile foundation is increased, whereas the natural period of foundation and peak horizontal displacement is decreased. For a pile of 20 m long and 0.5 m diameter, the stiffness, natural period and horizontal displacement are 29.30 MN/m, 0.5626 s, and 0.0335 m, respectively, whereas, for 1.5 m diameter pile the stiffness, natural period, and horizontal displacement of the pile foundation are 97.311 MN/m, 0.3088 s, and 0.0101 m, respectively.
2. Further, while comparing the results for a pile of 20-m length and 0.75 m diameter in non-liquefied soil and liquefied soil with 8 m thick top liquefied layer soil at input base excitation of 0.2 g, it is observed that there is reduction in stiffness (93.77%), an increase in natural period (300.88%), increase in lateral displacements (579.03%), and reduction in acceleration of the pile cap (57.74%).
3. In a two-layer soil system, with increased top liquefied soil layer thickness, it is observed that the stiffness of foundation is decreased, whereas the natural period of foundation and peak horizontal displacement is increased. For a pile of 20 m long and 0.5 m diameter at 0.2 g input base excitation with a top liquefied soil thickness of 3 m, the stiffness, natural period, and horizontal displacement are 14.635 MN/m, 0.7962 s, and 0.0284 m, whereas for 12 m top liquefied layer the stiffness, natural period and horizontal displacement are 1.191 MN/m, 2.79 s, and 0.3482 m.
4. For a pile of 20 m long and 0.5 m diameter at 0.3 g input base excitation with a top liquefied soil thickness of 3 m the stiffness, natural period and horizontal displacement are 14.635 MN/m, 0.7962 s and 0.0284 m, respectively, whereas for 12 m top liquefied layer the stiffness, natural period, and horizontal displacement are 1.191 MN/m, 2.79 s, and 0.3482 m.
5. It is also observed from a pile foundation in liquefied soil that by increasing the input base excitation from 0.2 to 0.3 g, the peak horizontal displacement of the pile also increases. In the soil layer system possessing top liquefied layer of a thickness of 12 m, 8 m, 5 m, and 3 m with the input base excitation of 0.2 g, the observed peak horizontal displacements of pile are 0.3482 m, 0.1459 m, 0.0609 m, and 0.0284 m, respectively. Similarly, with increased input base excitation of 0.3 g, the observed peak horizontal displacements of pile are 0.5224 m, 0.2188 m, 0.0914 m, and 0.0425 m, respectively.
6. It is observed in the sloped liquefied ground with increased input base excitation, that there is an increase in displacement of the pile due to the combined effects of lateral spreading and inertia effects. Pile of 20 m length, 0.75 m diameter, and pile cap having 0.5 m depth with input base excitation of 0.1 g the horizontal displacement of 0.0309 m due to lateral spreading and lateral displacement of

0.1092 m in the pile due to lateral spreading and inertia effects is observed. But, with increased input base excitation of 0.3 g the horizontal displacement of 0.0309 m due to lateral spreading and 0.2658 m due to combine action of lateral spreading and inertia effects is observed.

7. With the increased depth of the crust layer in the three-layered soil system (intermediate liquefied zone), it is observed that lateral displacement increases due to lateral spreading as well as due to combining action with inertial effects.
8. The factor of safety against liquefaction of soil and the susceptibility of soil to liquefy depends on the magnitude of earthquake (M_w), the maximum amplitude of base excitations (a_{max}), SPT values, depth of point of consideration, density of soil and fine content, etc.
9. A method is introduced as future study to evaluate the liquefaction potential of soil considering the acceleration of soil at point of concern and soil properties and site parameters responsible for soil amplification.

References

1. Kramer SL (2008) Geotechnical earthquake engineering. International series in civil engineering and engineering mechanics, 5th edn, Prentice-Hall
2. Bhattacharya S (2003) Pile instability during earthquake liquefaction. Ph.D. thesis, Cambridge University, U.K
3. Bhattacharya S, Madabhushi S, Bolton MD (2004) An alternative mechanism of pile failure in liquefiable deposits during earthquakes. Geotechnique 54(3):203–213
4. Bhattacharya S, Madabhushi S (2008) A critical review of methods for pile design in seismically liquefiable soils. Bull Earthq Eng. https://doi.org/10.1007/s10518-008-9068-3
5. Boulanger RW, Chang D, Brandenberg S, Armstrong RJ, Seismic design of pile foundations for liquefaction effects. In: Pitilakis KD (ed) Earthquake geotechnical engineering, pp 277–302
6. Finn WD, Fujita L (2002) Piles in liquefiable soils: seismic analysis and and design issue. Soil Dyn Erthq Eng 22(9):731–742
7. Hamada M, O'Rourke TD (1992) Case studies of liquefaction and lifeline performance during past earthquake. NCEER-92 Nat Centre Earthq Eng Res 1
8. Liyanapathirana DS, Poulos HG (2005) Seismic lateral response of piles in liquefying soil. J Geotech Geo-environ Eng ASCE 131(12):1466–1479
9. Tokimatsu K, Asaka Y (1998) Effects of liquefaction-induced ground displacements on pile performance in the 1995 Hyogoken-Nambu earthquake. Soils Found 2:163–177
10. Madabhushi G, Knappett J, Haigh S (2010) Design of pie foundation in liquefiable soil. Imperial College Press
11. Idriss IM, Seed HB (1974) Seismic response by variable damping finite element. J Geotech Eng Div ASCE 100(1):1–13
12. Seed HBS, Idriss IM (1970) A siple procedure for evaluating soil liquefaction. Earthquake Engineering Research Centre, Barkeley, California
13. Standards IIS (1893) (Part 1) Criteria for erthaquake resistannt design of structures, part-1, general provision of building. Bureau of Indian Standards, (2016) Sixt Revision
14. Youd TL, Idriss IM (2001) Liquefaction resistance of soil: Summary report from the 1996 NCEER and 1998 NCEER/NSF workshops on evaluation of liquefaction resistance of soils. J Geotech Geoenviron Eng 127(4)
15. Cubrinovski M, Ishihara K (1999) Emprirical corelation between SPT N-value and relative density of sandy soil. Soils Found 39(5):61–71

16. Meyerhof GG (1957) Discussion on research on determining the density of sand by penetration testing. In Proceedings of 4rth international conferences on soil mechanics and foundation engineering. vol 1:110
17. Skempton AW (1996) Standard penetration test procedures and the effects in sands of overburden pressure, relative density, particle size, ageing and overconsolidations. Geotechniques 36(3):425–447
18. Stewart D, Knox R (1995) What is the maximum depth liquefcation can occure? In: 3rd international conference on recent advances in geotechnical earthquake engineering and soil dynamics, Louis, Missouri

Effect of Static Load Inclination on Settlement of Strip Footing Under Cyclic Excitations

Suvendu Kumar Sasmal and Rabi Narayan Behera

Abstract One of the key problems in foundation engineering is to observe the settlement of foundation. This problem becomes more complex in the context of cyclic stress reversal. The behavior of shallow foundations was reported experimentally and statistically considering the load inclination, by a number of researchers. However, the studies were limited to only static case. In many cases, the direction of load application of static load may fluctuate while mechanically applied cyclic load remains constant. In order to simulate such a system, in this study, a modified approach has been followed to observe the settlement of foundation under combined influence of pulse load and inclined static load. The footing-soil system is modeled with the help of Beam on Nonlinear Winkler Foundation Model (BNWF) analysis using OpenSEES. The soil is taken as dense sand with relative density (Dr) = 69%. Initial results from the model are validated against existing analytical and numerical solutions. Then the settlement pattern under long-term cyclic load is analyzed, and the influence of increase in load inclination is discussed. It is observed that the total settlement is controlled by the static and cyclic load intensity which in turn are influenced by the angle of static load inclination (α).

Keywords Shallow foundation · Dense sand · Pulse load · Load inclination · Settlement

List of Symbols

Dr	Relative density of soil, (%)
ϕ	Angle of internal friction
γ	Unit weight of soil
E	Young's modulus
ν	Poisson's ratio

S. K. Sasmal · R. N. Behera (✉)
Department of Civil Engineering, National Institute of Technology Rourkela, Rourkela 769008, India

T. G. Sitharam et al. (eds.), *Earthquake Geotechnics*, Lecture Notes in Civil Engineering 187, https://doi.org/10.1007/978-981-16-5669-9_29

D_f/B	Embedment ratio
s_s	Static settlement of footing
s	Total settlement of footing (static + cyclic)
B	Width of footing
α	Angle of load inclination (in degrees)
$q_{d(max)}$	Intensity of cyclic load
q_u	Ultimate bearing capacity (UBC), centric vertical loading condition
q_{ui}	Ultimate bearing capacity (UBC), centric inclined loading condition

1 Introduction

The analysis of shallow foundation response is a cumbersome task which is attributed to the complex nature of the underlying soil. The analysis of settlement is one of the key research areas in the field of geotechnical engineering. The expressions for calculating the settlement of foundation were reported by Meyerhof [1] and Meyerhof [2]. However, with time, different new techniques also emerged. One major approach of studying the shallow foundation settlement response is to consider the load inclination. Hansen [3] studied the inclined load response using the method of characteristics. The field tests to observe the response of the shallow foundations were conducted by Muhs and Weiss [4]. Vesic [5] gave the expressions for the inclination factors. With the load inclination, the failure surface of the foundation alters, resulting in reduction of bearing capacity and the settlement.

The settlement of shallow foundation considering load inclination was reported in a number of researches. Patra et al. [6] observed the ultimate bearing capacity and settlement of strip footing resting on dense and medium dense sand under the influence of eccentrically inclined static load. They highlighted the role of load inclination in controlling the behaviour of the foundation. In the context of cyclic load, the response of plane strain footing was reported by Raymond and Komos [7]. The behavior of square foundation under cyclic load was given by Das et al. [8]. The cyclic response of circular foundation was discussed in Sawicki et al. [9]. The influence of incremental cyclic load was described in Tafreshi et al. [10].

From the review of the literature, it is clear that the study on settlement of the footing due to only static inclined load is well established. However, there may be certain circumstances where the footing is subjected to constant cyclic loads where the angle of load application of the static load varies, on which there are lack of research. Hence, this study is an approach to observe the settlement under such loading conditions. In this study, a strip footing resting on dense sand is numerically modeled to observe the settlement response. The suitability of the model is first established by comparing the initial outcomes with existing results and by observing the static load-settlement response. The study is then extended to find out the displacement of the footing under the influence of 10^6 load cycles. The settlement pattern is discussed in detailed. Then, the reduction in settlement of the foundation due to load inclination (α) is also discussed.

2 The BNWF Model

Beam on Nonlinear Winkler Foundation model (BNWF) is one of the major emerging techniques in numerical geotechnics. The nonlinear cyclic responses of shallow foundation were successfully reported using this model by Allotey and Naggar [11], Harden et al. [12], and Gajan et al. [13]. The settlement of strip footing due to vertical pulse load was reported by Sasmal and Behera [14, 15] using the BNWF model.

The soil-footing system and loading pattern used in this study are shown in Fig. 1. Figure 1a illustrates the loading condition used in this study. The soil-footing interface is modeled with the help of nonlinear springs as shown in Fig. 1b. The details of the

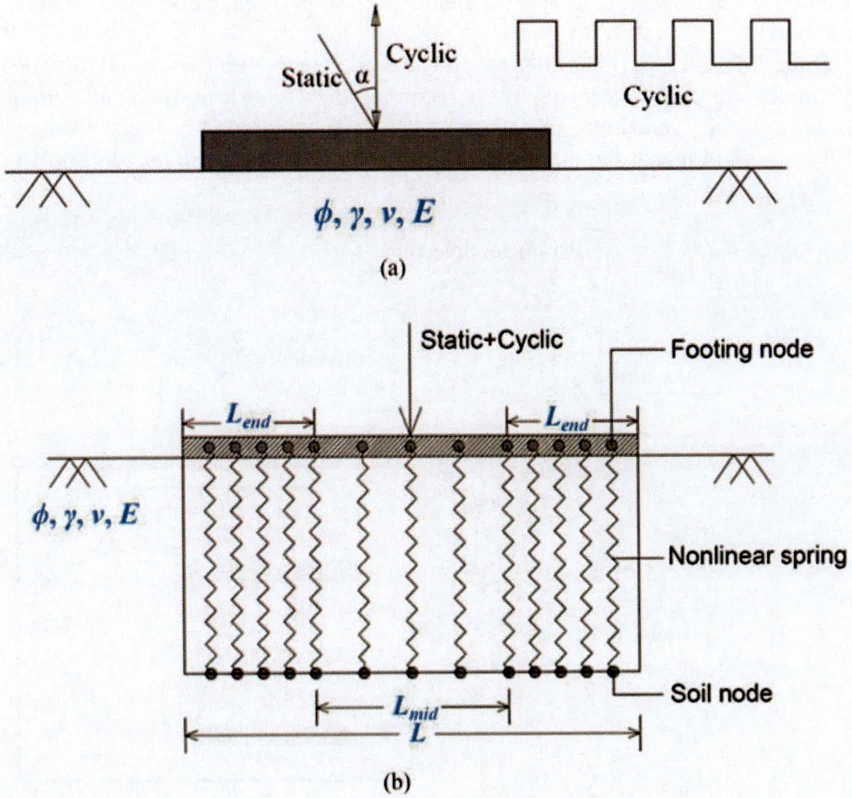

Fig. 1 Foundation system considered for the numerical model **a** loading system **b** BNWF model

BNWF model are discussed below. The present model is created using OpenSEES [16].

2.1 Parameter Selection and Model Details

The soil (Dense sand) parameters are taken as the values reported in Patra et al. [6]. The value of E and ν is taken from EPRI [17] using the range of Dr %(relative density) mentioned in Das [18]. The values of the soil parameters are given in Fig. 2. The soil-footing interface is modeled using nonlinear mechanistic springs. 101 number of nodes are used to create the footing. One-dimensional elastic beam column elements are used to join the footing nodes.

Each footing node (Degree of freedom = 3) is attached to a soil node (Degree of freedom = 0); situated vertically below, with the help of nonlinear mechanistic springs. Each spring has drag and closure components along with the elastic and plastic components. The stiffnesses of the springs are calculated using the expressions given by Gazetas [19]. In addition, for embedded foundation, embedment factor is multiplied.

Three different types of springs are used to create the model. These are QzSimple, PySimple, and TzSimple material for recording the vertical, passive, and sliding response of the foundation, respectively. The steps mentioned in Raychowdhury [20] are used to model the mechanistic springs. Typically, the end length (Fig. 1b)

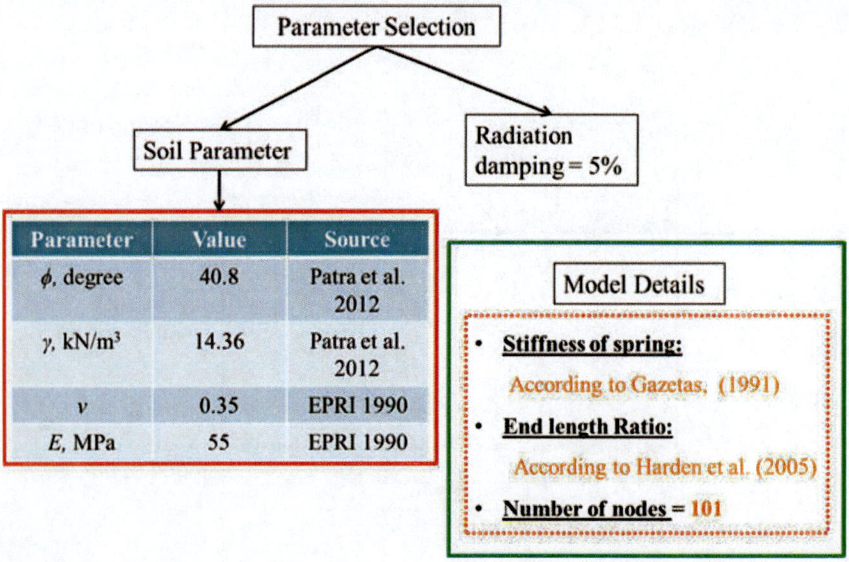

Fig. 2 Parameter selection and model details

influences the response of the footing, which is selected according to Harden et al. [12].

3 Results and Discussion

The model outcomes are explained in the following sections.

3.1 Validation of Present Results

The initial results in the case of inclined load condition (reduction factors), obtained from the present study are compared with the reduction factors obtained according to Meyerhof [2] and Patra et al. [6]. The comparison is shown in Fig. 3. It is inferred from Fig. 3 the present numerical model is capable of producing reasonable good results. The reduction factor (RF) is calculated as;

$$\mathrm{RF} = \frac{q_{ui}}{q_u} \tag{1}$$

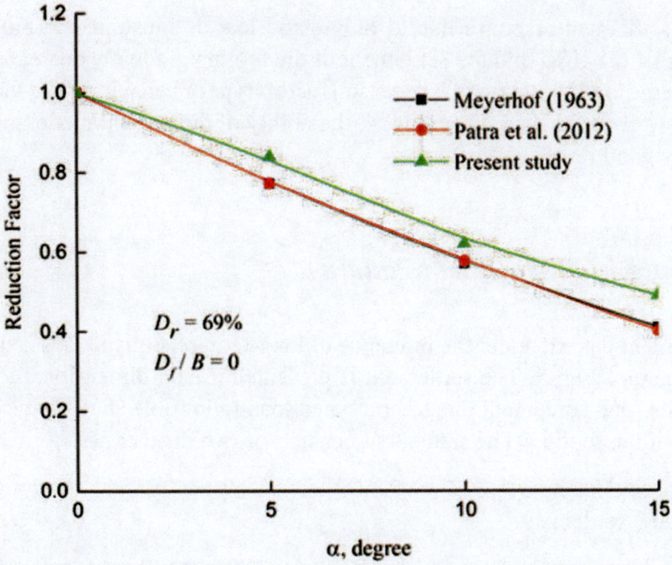

Fig. 3 Comparison of the present results with reported experimental and analytical solutions

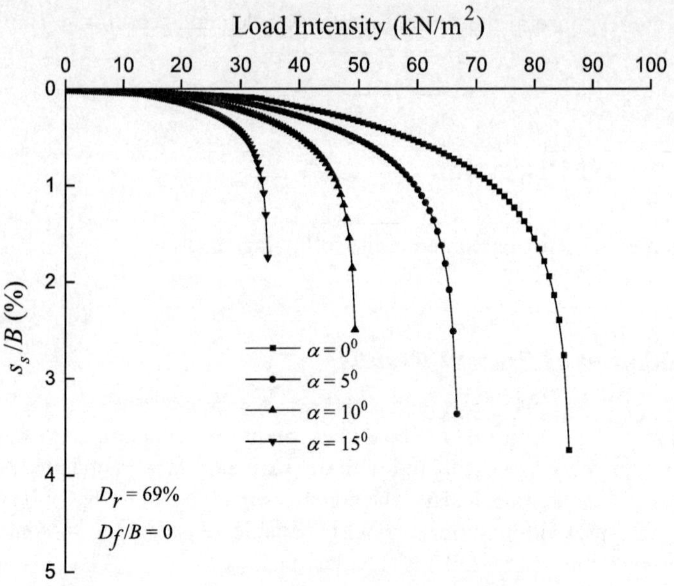

Fig. 4 Static response of the foundation under the influence of inclined load

3.2 Static Response

The load-settlement response due to only static load is illustrated in Fig. 4. It is observed that the UBC and the settlement of the footing gradually decrease with an increase in the load inclinations. The reason for this type of behavior of the foundation is already well established. From Fig. 4, the ability of the present model to produce reasonably good results is verified.

3.3 Analysis of Settlement Pattern

The settlement pattern under the influence of long-term repetitive loads is described with the help of Fig. 5. The settlement is presented in non-dimensional form, i.e., dividing the total settlement due to combined load (allowable static + cyclic) with the width of the footing. The settlement consists of two distinct parts;

(i) rapid settlement
(ii) gradual settlement.

Rapid settlement occurs up to 4×10^5 cycles. After that the settlement becomes gradual and almost ceases at the critical number of load cycles (n_{cr}). The reason for gradual settlement is that, the soil becomes sufficiently dense during the number of load cycles corresponding to the rapid settlement. From the settlement pattern,

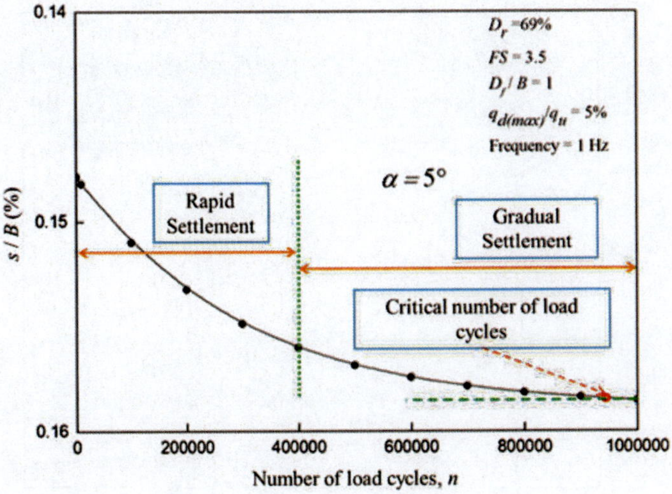

Fig. 5 Settlement pattern under the influence of cyclic excitation

it is clear that the foundation response to the first 4×10^5 cycles is crucial as the settlement during this period is a major part of the final settlement that the footing will undergo during its lifetime.

3.4 Reduction in Settlement Due to Load Inclination

The reduction in the total settlement due to load inclination is given in Figs. 6 and 7. From the plot, it is clear that;

$$[s/B(\%)]\alpha\left[\frac{q_{\mathrm{d(max)}}}{\mathrm{FS}}\right] \tag{2}$$

The foundation response is the combined response due to the static and cyclic load. With decrease in *FS*, the static load increases causing an increase in the total settlement. Also it can be seen that for lower *FS*, the reduction in settlement is more rapid, both in the case of surface foundation and the embedded foundation. The reduction in settlement for surface foundation is shown in Fig. 6. The reduction in settlement for embedded foundation is given by Fig. 7. It is also found that, for higher value of load inclination, the rate of increase in settlement due to increase in $q_{\mathrm{d(max)}}$ is not significant. From Figs. 6 and 7, it is evident that the combined settlement response is largely controlled by the amount of static load on the foundation. The static load on the foundation is a function of the *FS*. The effect of *FS* on the foundation response is given in Fig. 8. It is found that for higher values of factor of safety, the reduction rate in total settlement with increase in the angle of load inclination is less.

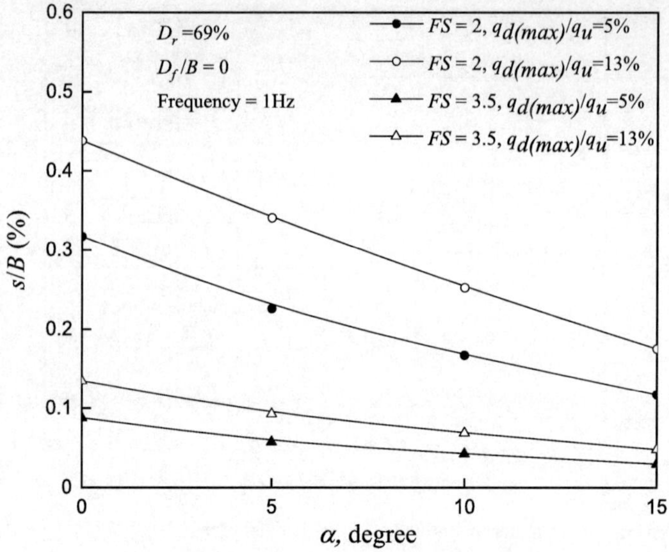

Fig. 6 Reduction in settlement due to load inclination (Surface foundation)

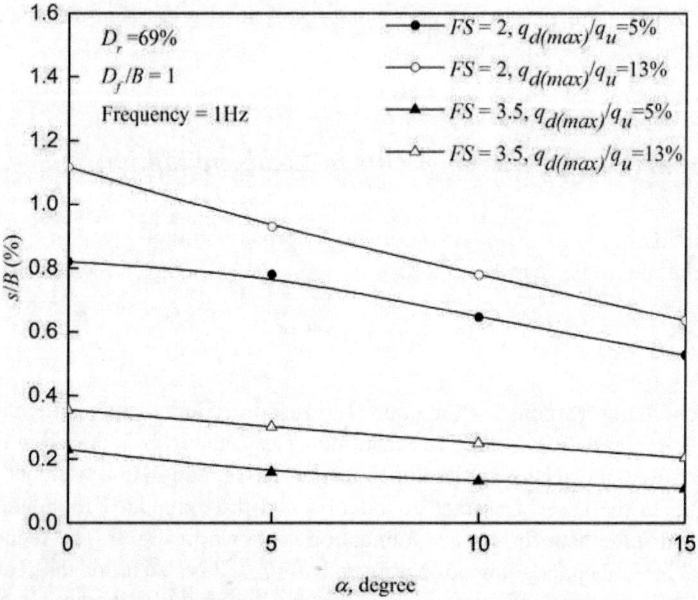

Fig. 7 Reduction in settlement due to load inclination (Embedded foundation)

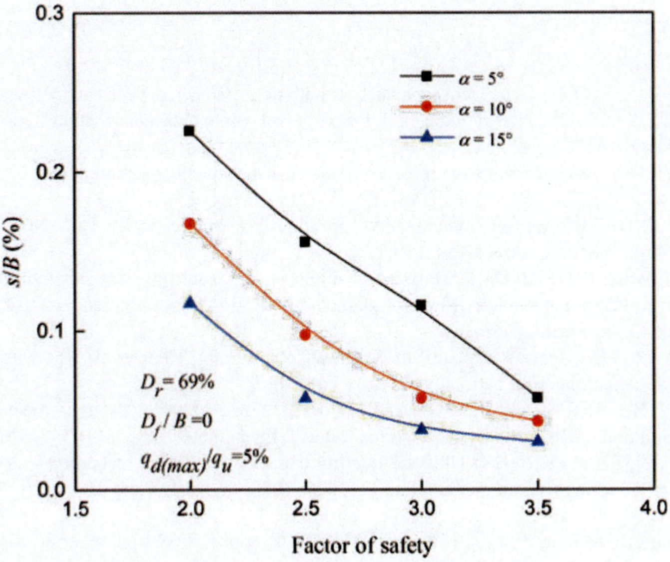

Fig. 8 Effect of factor safety on the settlement of foundation

4 Conclusions

The role of static load inclination on the settlement (due to static + cyclic loads) of a strip footing resting on and embedded in dense sand is studied, and the key foundation behaviors are explained. BNWF model is adapted to model the foundation and record its responses. The settlement pattern under long-term load is discussed. Then the angle of load inclination is gradually increased to observe the effect. It is found that the load inclination and load intensity (both static and cyclic) are closely interrelated. The following major conclusions are derived from the study;

- The major portion of the total footing settlement, for dense sand condition is caused due to the first 4×10^5 cycles. The settlement ceases at n_{cr}, beyond which the soil becomes dense enough so that further displacement is hardly visible.
- Keeping all the other parameters constant, for higher value of load inclination, the rate of increase in settlement due to increase in intensity of cyclic load is not significant.
- Keeping all the other parameters constant, the rate of reduction in total settlement with increase in the factor of safety is less, for higher range of α.

An approach to observe the role of load inclination is presented in this work. The analysis of seismic response of foundations under inclined static load can be another interesting future research.

References

1. Meyerhof GG (1953) The bearing capacity of foundations under eccentric and inclined loads. In: Proceedings of the 3rd international conference on soil mechanics and foundation engineering, vol 1, po 440–445
2. Meyerhof GG (1963) Some recent research on the bearing capacity of foundations. Can Geotech J 1(1):16–26
3. Hansen JB (1970) A revised and extended formula for bearing capacity. Bull. No. 28, Danish Geotechnical Institute, Copenhagen
4. Muhs H, Weiss K (1972) Der Einfluss von Neigung und Ausmittigkeitder Last auf die Grenztragfahigkeit flach genrundeter Einzelfundamente. DEGEBO Mitteilungen, Heft 34, Welhelm Ernst and Sohn, Berlin, Germany
5. Vesic AS (1975) Bearing capacity of Shallow foundations (Chapter 3). In: Das BM (ed) Geotechnical engineering handbook. J. Ross Publishing, Inc., USA
6. Patra CR, Behera RN, Sivakugan N, Das BM (2012) Ultimate bearing capacity of shallow strip foundation under eccentrically inclined load, Part I. Int J Geotech Eng 6(3):343–352
7. Behera RN, Patra CR (2018) Ultimate bearing capacity prediction of eccentrically inclined loaded strip footings. Geotech Geol Eng 36:3029–3080. https://doi.org/10.1007/s10706-018-0521-z
8. Raymond GP, Komos FE (1978) Repeated load testing of a model plane strain footing. Can Geotech J 15(2):190–201
9. Das BM, Yen SC, Singh G (1995) Settlement of shallow foundation on sand due to cyclic loading. In: Proceedings of the international conference on recent advances in geotechnical earthquake engineering and soil dynamics, vol 8, pp 385–388
10. Sawicki A, Swidzinski W, Zadroga B (1998) Settlement of shallow foundation due to cyclic vertical force. Soils Found 38(1):35–43
11. Tafreshi SM, Mehrjardi GT, Ahmadi M (2011) Experimental and numerical investigation on circular footing subjected to incremental cyclic loads. Int J Civil Eng 9(4):265–274
12. Allotey N, El Naggar MH (2003) Analytical moment–rotation curves for rigid foundations based on a Winkler model. Soil Dyn Earthq Eng 23(5):367–381
13. Harden C, Hutchinson T, Martin GR, Kutter BL (2005) Numerical modeling of the nonlinear cyclic response of shallow foundations. Report no. 2005/04, Pacific Earthquake Engineering Research Center (PEER), Berkeley, California
14. Gajan S, Hutchinson TC, Kutter BL, Raychowdhury P, Ugalde JA, Stewart JP (2008) Numerical models for analysis and performance-based design of shallow foundations subjected to seismic loading. Pacific Earthquake Engineering Research Center, Berkeley
15. Sasmal SK, Behera RN (2018) Prediction of combined static and cyclic load-induced settlement of shallow strip footing on granular soil using artificial neural network. Int J Geotech Eng 1–11. https://doi.org/10.1080/19386362.2018.1557384
16. Sasmal SK, Behera RN (2019) Factors influencing transient response of shallow strip footing on granular soil subjected to vertical pulse load. Adv Comput Methods Geomech 381–390
17. OpenSEES: Computer software, University of California, Berkeley
18. EPRI (1990) Manual on estimating soil properties for foundation design. Electric Power Research Institute, Palo Alto, California
19. Das BM (2016) Principles of foundation engineering Cengage learning
20. Gazetas G (1991) Formulas and charts for impedances of surface and embedded foundations. J Geotech Eng 117(9):1363–1381
21. Raychowdhury P (2008) Nonlinear Winkler-based shallow foundation model for performance assessment of seismically loaded structures, Ph. D dissertation. University of California, San Diego

A Review of Settlement Prediction Techniques for Shallow Foundations Subjected to Cyclic Loads

Suvendu Kumar Sasmal and Rabi Narayan Behera

Abstract Foundations are generally subjected to loads, viz. wind and wave load, earthquake-induced ground motion and machine vibrations. The response of a foundation depends on key factors like the intensity of loading, the soil stiffness and loading frequency. The soil properties beneath the foundation change repeatedly as a result of cyclic stress reversal. The shear modulus fluctuates with each cycle of loading, making the response of foundation under cyclic load, a complicated phenomenon. In addition, passive resistance due to the embedment of footing significantly controls the cyclic response. In engineering design, the analysis of the behavior of such foundations has significant importance. This study is a systematic approach to review available experimental as well as numerical methods for predicting settlement of the foundation. The significance of each controlling parameter is discussed. A set of constitutive relationships are presented and the merits and limitations of different methods are explained. Review of literatures, especially of last two decades, reveals a rise in analytical and numerical methods. However, the accuracy of the numerical methods is highly influenced by proper selection of soil stiffness parameters.

Keywords Foundations · Passive resistance · Shear modulus · Cyclic stress reversal · Numerical methods · Stiffness

1 Introduction

In nature, there are several structures, viz. machine foundations, railroad foundation and foundations near coastal regions which are subjected to repetitive loads. In a developing country like India, the stability analysis of these structures is very important. A critical review of available methods will not only be helpful in proper understanding of the problem but also provide scope for development of more realistic techniques. A summary of the techniques will be helpful in selecting proper methods depending on the prevailing conditions. The study summarizes effective

S. K. Sasmal · R. N. Behera (✉)
Department of Civil Engineering, National Institute of Technology Rourkela, Rourkela 769008, India

T. G. Sitharam et al. (eds.), *Earthquake Geotechnics*, Lecture Notes in Civil Engineering 187, https://doi.org/10.1007/978-981-16-5669-9_30

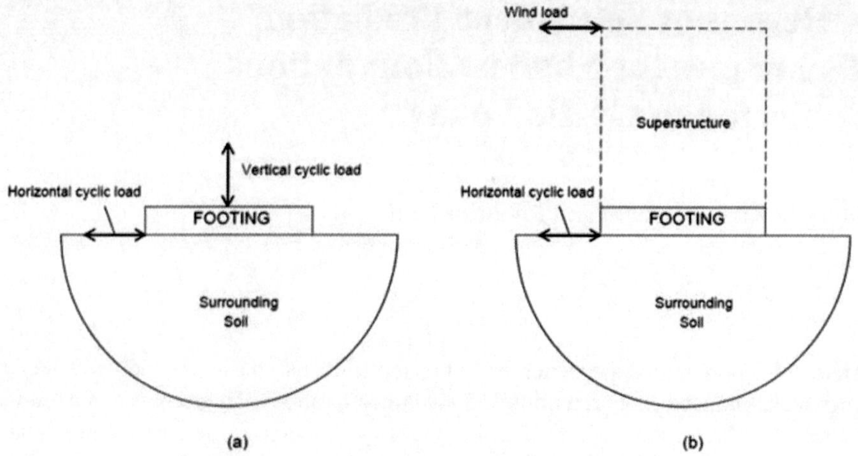

Fig. 1 Schematic representation of foundation under possible cyclic loads **a** horizontal (ground motion) + vertical, **b** horizontal (ground motion) + wind load

experimental and numerical methods along with the constitutive models those can be used to observe the settlement under different types of cyclic loads. The purpose of the review is to convey the practicing engineers as well as researchers, the merits and demerits of different methods along with the importance of loading and soil parameters. The cyclic loads, to which a foundation is generally subjected is schematically given in Fig. 1. Shallow foundations are traditionally analyzed for the static response (bearing capacity and settlement). However, under circumstances like earthquakes and machine vibrations, these are also subjected to repeated loads. The observation of the settlement of footing under repeating loads is an area of prime importance. The settlement directly influences the stability and in turn the lifetime of the superstructure. Proper understanding of the problem will be helpful in critical analysis and effective design of foundations near coastal regions, railroad foundations and foundations in industrial areas.

Soil is a complex material. The response of soil to cyclic load becomes even more complex as the soil material properties tend to fluctuate with each load cycle. Also the footing position is generally varied during the unloading and reloading. Hence, there is need to understand basic and advanced procedures and their governing principles in order to properly study the settlement response of the foundation under cyclic loads, also known as repetitive loads. This study is an approach to explore existing methods of solution for such types of foundation systems, which will be helpful in monitoring the response of above-mentioned types of foundation. Cohesionless soil is chosen because a majority of works related to the cyclic response of footing considered this type of soil beneath the foundation. Strip, square and circular footings are chosen because these are the most commonly used types of shallow foundation. The experimental methods described in the review are generalized methods (used in multiple works). The discussed numerical/FEM models are the methods that are being used to successfully predict the cyclic response in recent times. The recent trends of numerical analysis platforms are reviewed so that these can be helpful to the practicing engineers and designers in the near future. In addition to explanation of importance of soil structure interaction, a set of available constitutive relationships are also presented. The important factors to consider, while adopting numerical methods are described along with the limitations of different techniques. This paper is a collective review of available techniques on cohesionless soil which can be applied further by varying the subsoil conditions.

2 Settlement Under Cyclic Loading

Settlement of foundation under the influence of cyclic loads is a very complex phenomenon. The shape and size of footing affect the settlement patterns to a large extent, in addition to soil properties underneath the footing. It also depends on the properties of the cyclic load. Short term seismic loads may cause rapid settlements. However, long terms loads, viz. loads experienced by railroad foundations may cause rapid and steady state settlements. One of the general practices in soil dynamics is to give less importance to the transient response of the foundation. However, this cannot be neglected as these are the response of foundation to sudden loading fluctuation. Moreover, before the arrival of the cyclic loads, some settlement is caused due to only static load. Hence, the settlement of any foundation in the presence of cyclic stress reversal is the total settlement due to static load, transient response and the steady state response. In addition, the contact surface (soil and footing) controls the displacement of the footing from its original position. Given below is a compilation of methods reported to study the settlement of foundation.

2.1 Experimental Methods

Experimental methods are not only the direct way of observing the response of a foundation but also these are basis for the development of computational solutions. Experimental techniques for the analysis of foundation settlement under such loading conditions were popularized in the late 1970s. The study of rail road foundations attracted many researchers', viz. Raymond and Komos [1]. The method proposed by Raymond and Komos [1] became very popular. Experimental approach for observing cyclic load induced settlement can also be seen in the works of Das et al. [2], Sawicki et al. [3], Tafreshi et al. [4]. Later, centrifuge tests were also carried out by a number of researchers to observe the displacement of foundation under different cyclic loads. The approach of Raymond and Komos [1] are discussed below.

Surface Foundations

The procedure defined by Raymond and Komos [1] was one of the earliest techniques to find out the settlement of the foundation. The main aim was to find out the settlement/displacement of the foundation due to effect of the cyclic load (rectangular form). In the context of the cyclic load, the settlement following the first load cycle was reported. They provided graphs/charts for the foundation settlement for a wide range of cyclic load intensity which was calculated as some definite percentage of the ultimate load carrying capacity of the foundation. The important parameters influencing the settlement were the footing dimension, the intensity of static and dynamic load, the relative density of sand and duration of application of load.

Das et al. [2] carried out the experiments to study the settlement of a square foundation. They observed, after a specific number of load cycles, the settlement becomes negligible for further loading. The experiments of Sawicki et al. [3] for foundation under cyclic load suggested 10–15% (of static bearing capacity) increase in the bearing capacity of the subsoil. While studying the response of circular foundation, [4] introduced the concept of coefficient of elastic uniform compression (CEUC) of sand which attains higher value for higher density, but decreases with footing size.

Embedded Foundations

Raymond and Komos [1] did not describe the effect of embedment depth on cyclic behavior of the foundation. In the case of embedded footings, passive resistance arises due to surrounding soil. Gadre and Dobry [5] performed centrifuge tests on square footing. The soil was granular in nature (relative density = 75%). Their results indicated that the contribution of the passive side plays important role while estimating the lateral load–displacement response of the embedded foundation. Fattah et al. [6] performed experimental study to observe the effect of the rate of loading on the settlement of footing on dry sand. They observed an important phenomenon, i.e., the settlement versus rate of loading is dependent on the relative density of the soil. Fattah et al. [7] presented plots showing the variation in settlement of circular foundation. Their results indicated the occurrence of a steady rate of settlement which takes place after 100–200 cycles. Hence, it is clear that the embedment of the footing controls the settlement, as passive resistance is generated due to the embedment.

Large Scale Experiments

The experiment was aimed at studying the settlement after applying different loads in three different phases. In the first stage, single sine shaped small force-controlled cycles with low frequency (0.5 Hz) were applied to study the nonlinear soil response. In the second step, earthquake time history applied on the foundation and in the last step sine shaped displacement cycles were applied.

Negro et al. [8] reported the nonlinear behavior (settlement and rotation) of foundation (1 m × 1 m) on sand (relative density, 45% and 85%). More settlement in the case of loose sand than the dense sand was reported by them. Faccioli et al. [9] reported seismic response of shallow foundation (1 m × 1 m) on sand (relative density, 45% and 85%). More amount of settlement in the case of loose sand was also reported by Faccioli et al. [9]. Hence, from the large scale experiments, as the reported results indicate, for a square foundation the amount of settlement depends on underneath soil strength.

Shear Wall-Footing System

The deformation of shallow foundation as a response to cyclic loading and earthquake excitations was observed, with the help of centrifuge experiments by Gajan et al. [10]. Accumulation of permanent settlement with number of load cycles was reported. It was also reported that the role of embedment is to reduce the rate of accumulation. The moment-rotation-settlement of shallow foundation due to slow horizontal cyclic load was also presented. They gave the expressions for maximum shear and moment capacity which take the form as mentioned in Eqs. (1) and (2), respectively. The maximum shear capacity can be represented by;

$$H^*_{\max} = V^* k + 2p^*_0 k + p^*_p - p^*_a \tag{1}$$

And the maximum moment capacity is given by;

$$M^*_{\max} = \frac{L^* V^*}{2}\left(1 - \frac{1}{FS^*_v}\right) + p^*_p \frac{D^*}{3} + 2p^*_0 k \frac{D^*}{3} - p^*_a \frac{D^*}{3} \tag{2}$$

where V^*, L^*, $D*$ and $FS^*{}_v$ = vertical load, length of footing and depth of embedment and vertical factor of safety, respectively.

$p^*{}_0$, $p^*{}_a$, $p^*{}_p$ and k = at rest earth pressure, active earth pressure, passive earth pressure and base shear coefficients, respectively.

The response of shear wall-footing system under lateral cyclic load was reported by Gajan and Kutter [11] using centrifuge experiment tests. They compared the experimental results with results from numerical analysis based on the concept of critical contact area ratio (Eq. 3).

$$\text{Critical contact area ratio} = \frac{\text{Footing area}}{\text{Required area to support vertical load and shear load}} \tag{3}$$

Deng and Kutter [12] carried out slow cyclic tests and concluded that placement of concrete pads near edge of footing results in the considerable reduction in the settlement of the footing. Deng et al. [13] reported that, the reduction in contact area (soil and footing) leads to progressive settlement as a result of limited localized bearing failure.

2.2 Numerical Methods

Numerical methods for observing the behavior of shallow foundations are becoming popular with time. Numerical methods are based on model assumptions that determine the accuracy of the predicted outcomes. However, these techniques are time saving and suitable for rapid calculation of the foundation response. By knowing the underlying soil parameters and the load parameters the response can easily be determined. The reported techniques regarding the cyclic response of foundation, especially of last two decades are presented and discussed in this work. Chatzigogos et al. [14] , Grange et al. [15, 16] and Tistel et al. [17] were the novel researchers who used macroelement model to analyze the response/settlement of foundation. Sturm [18] presented the settlement of cyclically loaded foundation for different relative densities. The study reported, the rate of settlement is proportional to load amplitude. The study also indicated the negligible effect of loading direction on the overall average settlement. The uplift behavior of foundation was studied by Grange et al. [15], Chatzigogos et al. [14], Pecker and Chatzigogos [19].

Nayfeh and Serhan [20] studied the response of an embedded foundation. The mathematical equation for vertical vibration of embedded foundation is expressed by Nayfeh and Serhan [20], which takes the following form,

$$m^* \frac{d^2u^*}{d\tau^2} + q^* + \tilde{F}^* \text{sgn}(\dot{u}^*) - m^* g = Q^*(\tau) \tag{4}$$

where

$\text{sgn}(\dot{u}^*) = 1$, for $\dot{u}^* > 0$.
$\text{sgn}(\dot{u}^*) = -1$, for $\dot{u}^* < 0$

m^* = mass of the foundation, q^* = restoring force, $\tilde{F}^*$ = coulomb friction and $Q^*(\tau)$ = external force on the foundation.

There are different models, to incorporate the soil and footing conditions in order to simulate actual foundation like conditions. The following are mostly used numerical models in recent history to predict the cyclic response of foundation.

Macroelement Model

This model is the one of the most effective numerical model for predicting the settlement of the foundation as it captures all the possible factors that control the footing settlement. The suitability of this model has been confirmed in a number of literatures. Grange et al. [15] used the concept of the macroelement model to estimate the vertical settlement for both low density and high density soil conditions where the footing was of size 1 m × 1 m. The total displacement (S_{total}) of the footing is the combination of elastic (S_{elastic}), plastic (S_{plastic}) and the uplift (S_{uplift}) components Eq. (5). They also reported that the macroelement model uses soil plasticity and foundation upliftment to give accurate results.

$$S_{\text{total}} = S_{\text{elastic}} + S_{\text{plastic}} + S_{\text{uplift}} \tag{5}$$

Salciarini and Tamagnini [21] formulated a macroelement model for observing the response of foundation on sand. They reported the accumulated horizontal and vertical settlements with increase in number of load cycles. They also presented the cyclic load versus displacement of footing from which the prediction accuracy of the macroelement model can be ascertained. The structure and model formulation of macroelement model was briefly discussed in Chatzigogos et al. [14]. They reported the elastic and macroelement response in the case of quasistatic vertical loading. Also they provided results showing foundation response to quasistatic cyclic loading and acceleration time history. Chatzigogos et al. [22] extended their macroelement model considering three different mechanisms, i.e.

1. Sliding
2. Soil yielding
3. Uplift.

Soil structure interaction responses were successfully captured by the proposed model; however, the results were only valid in the case of negligible embedment.

Finite Element Model (FE model)

Finite element models are popular for study almost all kind of problems. The main aim is to divide specific domain into several parts and then to solve each part. Niemunis et al. [23] presented FE model for studying the cyclic response of strip footing on sand. They successfully simulated the differential settlement of the foundation using the FE technique. Anastasopoulos et al. [24] used FE model to observe the nonlinear response of the foundation and compared the FE results with the results obtained from the experimental study. They reported the suitability of FE model in predicting the settlement-rotation and moment-rotation response of shallow foundations. They also provided horizontal force versus displacement response of the footing. The settlement patterns of loose sand (LD) and dense sand (HD) as reported by Anastasopoulos et al. [24] indicate that the settlement is inversely proportional to the relative density of the soil. Pasten et al. [25] presented the long term response of foundation under the influence of repetitive loads using FEM. A flowchart was provided by Pasten et al. [25] for the numerical procedures. They reported displacements (vertical and horizontal) of footing with number of load cycles. The response of rigid footing under eccentric cyclic load was also reported in their study. Chong [26] observed the offshore foundation response under the influence of long term repetitive loads with the help of finite elements. They reported the vertical and horizontal settlement of foundation with the increase in the number of load cycles. The occurrence of pronounced settlement in early stage (i.e., for number of load cycles < 100) was reported.

Beam on Nonlinear Winkler Foundation (BNWF) Model
The BNWF method is one of the key emerging methods in the field of soil dynamics. This is a method based on the nonlinear behavior of the mechanistic springs those capture the interface behavior. The nonlinear response of shallow foundation was observed using the model by Allotey and Naggar [27], Harden et al. [28], Allotey and Naggar [29], Raychowdhury [30]. The parameter sensitivity study for shallow foundation was explained in Raychowdhury and Hutchinson [31]. The settlement of strip footing on granular soil was reported by Sasmal and Behera [32]. The transient response of foundation due to vertical pulse load was observed by Sasmal and Behera [33]. The behavior of soil supporting the foundation is highly nonlinear under dynamic loads. Generally, under cyclic loads, the settlement of a footing is significantly controlled by the behavior of the part of footing directly in contact with the underneath soil. In other words, during the event of cyclic load, the nearest part of the footing to the soil is most affected by a change in soil behavior. These points justify the significant role of the soil-footing interface in determining the overall response. In this technique, the near field (plastic), far field (elastic) responses are captured by the springs which are shown in Fig. 2. The use of drag and closure combination represent the gap component. The stiffness of the springs is generally calculated using the formulations given in Gazetas [34]. The expressions for the nonlinear springs were given by Boulanger et al. [35].

The nonlinear portion of backbone curve is represented by;

$$q = q_{\text{ult}} - (q_{\text{ult}} - q_o)\left[\frac{cz_{50}}{cz_{50} + \left|z^p - z_0^p\right|}\right]^n \tag{6}$$

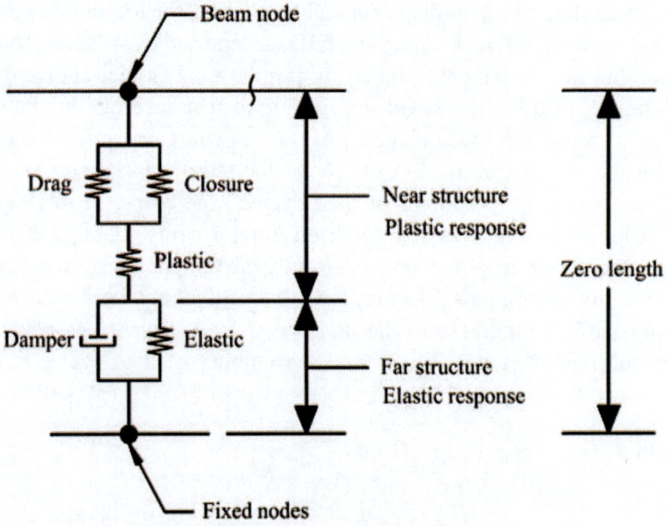

Fig. 2 Components of springs, redrawn after Raychowdhury [36]

The total displacement is given by;

$$z = \text{elastic component} + \text{plastic component} + \text{gap component}$$
$$= z^e + z^p + z^g \quad (7)$$

where q_{ult} = ultimate load, q_0 = load at yield point, z_{50} = displacement at which 50% ultimate load is mobilized, c and n = constitutive parameters, z_0 = displacement at yield point.

In general, three different types of springs are generally considered in the BNWF model, which are;

(a) QzSimple—for vertical response
(b) PySimple—for passive resistance (viz. embedded foundation)
(c) TzSimple—for sliding response.

It shall be noted that use of a minimum number of 25 springs is desirable along the length for reasonable prediction accuracy according to Gajan et al. [37].

Contact Interface Model (CIM)
The CIM model was popularized Gajan and Kutter [11]. This is the technique that gives the load settlement response of footing under several loading conditions without the use of mechanistic springs. The interface (soil—footing) behavior is studied in this technique to find out the displacement and the rotation of the foundation. However, it has been reported that, the permanent settlement of the footing is over predicted by this model for small amplitudes of rotation, which seems to be the only major limitation. The suitability of the methods in observing the response of foundation in saturated sandy soil as well as pure sliding condition needs to be explored.

3 Concluding Remarks and Scope for Future Research

This paper discussed the methods available for predicting the settlement of shallow foundations due to the cyclic loads. A compilation of experimental and numerical procedures, especially of last two decades is presented. The outputs from experimental studies are the most accurate and reliable. Amongst numerical models, macroelement model predicts the response most accurately. Despite the inherent demerit of the Winkler model, i.e., lack of coordination between the springs, the BNWF model also predicts the settlement to reasonable accuracy with proper spring spacing. The major inferences drawn from the review are as follows;

- Apart from loading and soil parameters, the passive resistance plays a vital role and is the most important factor that controls the settlement of embedded foundations. These are the factors which influence the selection of appropriate methods.

- The total settlement of the foundation is the combination of elastic settlement, plastic settlement and uplift. The results obtained from experimental methods rely on the strength of underneath soil.
- Proper selection of soil parameters is necessary while adapting numerical models.
- Numerical models based on Winkler model are influenced by the spring spacing. The selection of proper strength/stiffness parameters is the key factor that determines the prediction accuracy of the model.

The review gives an idea that the laboratory tests are generally conducted in ideal conditions. The review reveals both merits and limitations of different techniques. Experimental methods are accurate yet time consuming. Numerical methods largely depend on selection of soil parameters. In practice, neither the static load nor the cyclic load is applied on the center of gravity (CG) of any system. Furthermore, recent architectural structures supported by foundations are vertically irregular in nature. There is lack of both experimental and numerical methods to simulate these real time behavior of structure. Effect of cyclic load on the settlement behavior of shallow foundations resting over layered soil in place of homogenous soil can be another area of research. Moreover, at a particular section, the rate of loading (i.e., frequency) is not constant throughout. All these conditions when combined, will constitute a set of complex soil behavior underneath the structure. Lastly, the effect of water table on the cyclic response is an area that can be explored. Development of techniques to analyze these conditions can be scope for practicing researchers and engineers in future.

References

1. Raymond GP, Komos FE (1978) Repeated load testing of a model plane strain footing. Can Geotech J 15(2):190–201
2. Das, B.M., Yen, S.C., Singh, G.: Settlement of shallow foundation on sand due to cyclic loading. Proc. of the International Conference on Recent Advances in Geotechnical Earthquake Engineering and Soil Dynamics. 8, 385–388 (1995).
3. Sawicki A, Swidzinski W, Zadroga B (1998) Settlement of shallow foundation due to cyclic vertical force. Soils Found 38(1):35–43
4. Tafreshi SM, Mehrjardi GT, Ahmadi M (2011) Experimental and numerical investigation on circular footing subjected to incremental cyclic loads. International Journal of Civil Engineering 9(4):265–274
5. Gadre A, Dobry R (1998) Lateral cyclic loading centrifuge tests on square embedded footing. Journal of geotechnical and geoenvironmental engineering 124(11):1128–1138
6. Fattah MY, Karim HH, Al-Qazzaz HH (2017) Cyclic Behaviour of Footings on Dry Sand under Different Rates of Loading. International Journal of Construction Engineering and Management 6(6):240–253
7. Fattah, M.Y., Salim, N.M., Alwan, K.K.: Contact pressure distribution under circular shallow foundation subjected to vertical and rocking vibration modes. Journal of Building Engineering 26, 100908 (2019).
8. Negro, P., Paolucci, R., Pedretti, S., & Faccioli, E.:. Large-scale soil-structure interaction experiments on sand under cyclic loading. In Proceedings of the 12th world conference on earthquake engineering (2000).

9. Faccioli, E., Paolucci, R., Vivero, G.: Investigation of seismic soil-footing interaction by large scale tests and analytical models. In Fourth International Conference on Recent Advances in Geotechnical Earthquake Engineering and Soil Dynamics, 1–12 (2001).
10. Gajan S, Kutter BL, Phalen JD, Hutchinson TC, Martin GR (2005) Centrifuge modeling of load-deformation behaviour of rocking shallow foundations. Soil Dyn Earthq Eng 25(7–10):773–783
11. Gajan S, Kutter BL (2009) Contact interface model for shallow foundations subjected to combined cyclic loading. Journal of Geotechnical and Geoenvironmental Engineering, ASCE 135(3):407–419
12. Deng L, Kutter BL (2012) Characterization of rocking shallow foundations using centrifuge model tests. Earthquake Eng Struct Dyn 41(5):1043–1060
13. Deng L, Kutter BL, Kunnath SK (2011) Centrifuge modeling of bridge systems designed for rocking foundations. Journal of geotechnical and geoenvironmental engineering 138(3):335–344
14. Chatzigogos CT, Pecker A, Salencon J (2009) Macroelement modeling of shallow foundations. Soil Dyn Earthq Eng 29(5):765–781
15. Grange S, Kotronis P, Mazars J (2008) A macro-element for a shallow foundation to simulate soil–structure interaction considering uplift. Comptes Rendus Mécanique 336(11–12):856–862
16. Grange S, Kotronis P, Mazars J (2009) A macro-element to simulate dynamic soil-structure interaction. Eng Struct 31(12):3034–3046
17. Tistel, J., Grimstad, G., Eiksund, G.R.: A macro model for shallow foundations on granular soils describing non-linear foundation behaviour. Computers & Structures 1–18 (2017).
18. Sturm H (2009) Numerical investigation of the stabilisation behaviour of shallow foundations under alternate loading. Acta Geotech 4(4):283–292
19. Pecker A, Chatzigogos CT (2010) Non linear soil structure interaction: impact on the seismic response of structures. In Earthquake Engineering in Europe Springer, Dordrecht, pp 79–103
20. Nayfeh AH, Serhan SJ (1989) Vertical vibration of machine foundations. Journal of Geotechnical Engineering 115(1):56–74
21. Salciarini D, Tamagnini C (2009) A hypoplastic macroelement model for shallow foundations under monotonic and cyclic loads. Acta Geotech 4(3):163–176
22. Chatzigogos CT, Figini R, Pecker A, Salençon J (2011) A macroelement formulation for shallow foundations on cohesive and frictional soils. Int J Numer Anal Meth Geomech 35(8):902–931
23. Niemunis A, Wichtmann T, Petryna Y, Triantafyllidis T (2005) Stochastic modelling of settlements due to cyclic loading for soil-structure interaction. In Proc. Int. Conf. Structural Damage and Lifetime Assessment, Rome
24. Anastasopoulos I, Gelagoti F, Kourkoulis R, Gazetas G (2011) Simplified constitutive model for simulation of cyclic response of shallow foundations: validation against laboratory tests. Journal of Geotechnical and Geoenvironmental Engineering 137(12):1154–1168
25. Pasten C, Shin H, Santamarina JC (2013) Long-term foundation response to repetitive loading. Journal of Geotechnical and Geoenvironmental Engineering 140(4):04013036
26. Chong SH (2017) Numerical simulation of offshore foundations subjected to repetitive loads. Ocean Eng 142:470–477
27. Allotey N, El Naggar MH (2008) Generalized dynamic Winkler model for nonlinear soil–structure interaction analysis. Can Geotech J 45(4):560–573
28. Allotey N, El Naggar MH (2003) Analytical moment–rotation curves for rigid foundations based on a Winkler model. Soil Dyn Earthq Eng 23(5):367–381
29. Harden, C., Hutchinson, T., Martin, G.R., Kutter, B.L.: Numerical modeling of the nonlinear cyclic response of shallow foundations. Rep. No. 2005/04, Pacific Earthquake Engineering Research Center (PEER), Berkeley, California (2005).
30. Raychowdhury, P.: Seismic response of low-rise steel moment-resisting frame (SMRF) buildings incorporating nonlinear soil–structure interaction (SSI) Engineering Structures 33(3), 958–967 (2011).
31. Raychowdhury P, Hutchinson TC (2010) Sensitivity of shallow foundation response to model input parameters. Journal of Geotechnical and Geoenvironmental Engineering 136(3):538–541

32. Sasmal, S.K., Behera, R.N.: Prediction of combined static and cyclic load-induced settlement of shallow strip footing on granular soil using artificial neural network. Int. J. Geotech. Eng., 1–11 (2018). https://doi.org/10.1080/19386362.2018.1557384
33. Sasmal, S.K., Behera, R.N.: Factors Influencing Transient Response of Shallow Strip Footing on Granular Soil Subjected to Vertical Pulse Load, Advances in Computer Methods and Geomechanics, 381–390 (2019).
34. Gazetas G (1991) Formulas and charts for impedances of surface and embedded foundations. Journal of Geotechnical Engineering 117(9):1363–1381
35. Boulanger RW, Curras CJ, Kutter BL, Wilson DW, Abghari A (1999) Seismic soil-pile-structure interaction experiments and analyses. Journal of Geotechnical and Geoenvironmental Engineering 125(9):750–759
36. Raychowdhury,P.: Nonlinear Winkler-based shallow foundation model for performance assessment of seismically loaded structures, Ph. D Dissertation. University of California, San Diego (2008).
37. Gajan S, Hutchinson TC, Kutter BL, Raychowdhury P, Ugalde JA, Stewart JP (2008) Numerical models for analysis and performance-based design of shallow foundations subjected to seismic loading. Pacific Earthquake Engineering Research Center, Berkeley

Strength Comparison of Flexible Materials for Stability of Slope

Amanpreet Tangri and Saurabh Rawat

Abstract Soil nailing is an improvement technique used to overhaul the nature of ground that ensures the security of new or existing region slants. The sustaining action of the ground is extended by methods for the nail incorporation measure which in returns increases moldable force in soil nail. Specifically, cutting and bowing may contribute little to the impediment. Different bars are brought into a slope during the top-down uncovering. The unearthing support is a convincing and effective way to deal with build an upkeep divider to help top, associate projections and turnpikes. This strategy is suitable on disproportionate soils, broken stone, shale and hard surface conditions. In this exploration, a physical model is readied that contains pressure measure, acrylic sheet, flex sensor, multimeter and steel bars. The soil for the model is gathered from the grounds of Chandigarh University. In the present study, the strain was measured by using flex sensors which were placed on the steel bars. The load was applied by use of hydraulic jack and the relation of stress and strain was studied. Two materials named as HDPE hexagonal slope protection geonet and biaxial polypropylene geonet were used with three different nail arrangements, and for each nail arrangements, biaxial polypropylene geonet is able to take more stress than the HDPE hexagonal slope protection geonet. It can be concluded that the present study will lead to an increase in the soil slopes load handling and slope stability.

Keywords Flexible facing · Physical model · Soil nailing · Strain · Stress

1 Introduction

The development of hilly and far-off areas, the dam's construction and hydroelectric projects in hard terrain need stabilized ground. In the past, many stabilization methods have been developed to combat these situations. Stabilization of the earth, by mechanical and chemical land stabilization, drilling, earthworks, crushed stone and ground gravel, is useful in providing structural- and service-oriented stability. In this

A. Tangri (✉) · S. Rawat
Department of Civil Engineering, Jaypee University of Information Technology, Waknaghat, Solan, India

T. G. Sitharam et al. (eds.), *Earthquake Geotechnics*, Lecture Notes in Civil Engineering 187, https://doi.org/10.1007/978-981-16-5669-9_31

paper, a detailed description of one such technique utilized for ground stabilization known as soil nailing is provided.

Soil nailing technology is employed to offer reinforcement for natural and steep slopes on site. If unstable existing grounds require support and nearby buildings are sensitive to deformation, the utilization of soil nailing provides stability through temporary construction or permanent support structures and take appropriate measures to decrease ground movement.

Soil nailing walls is a widely used technique for keeping vertical cuts on any slope that is even more vertical in the ground with vertical slopes and normal angles. A large part of the cost of landing is related to the construction of the reinforced concrete surface. Soil facing mechanism helps to provide the stability between soil and inserted nails. In addition, the top layer facilitates surface erosion, weathering and moisture loss from the surface of the soil nail system. In this manner, to get better stylish look of the nail structure, confronting assumes a significant job.

Soil nail walls are more flexible than gravity walls and other similar systems, so it is reasonable to expect that these walls will sustain some lateral displacement during earthquakes and that these displacements can be tolerated. Experience suggests that permanent deformations of relatively flexible walls, including soil nail walls, tend to be small when $0.5 \leq kh/PGA \leq 0.67$ and as long as the strength of soils behind the wall does not decrease significantly during strong ground motions.

Observations made after recent earthquakes indicate that soil nail walls did not show significant distress or permanent deflections; although, some of the surveyed walls were subjected to ground accelerations as large as 0.7 g (Tatsuoka et al. 1997; Tufenkjian 2002). Vucetic et al. (1993) and Tufenkjian and Vucetic (2000) observed similar trends in centrifuge tests performed on reduced-scale models of soil nail walls. These findings suggest that soil nail walls have an intrinsic satisfactory seismic performance, which is attributed to their flexibility and design redundancy.

Soil nail advancement was first used in France to collect an unending holding divider cut in sensitive shake. The assignment, endeavored in 1961, was the place steel nails were used to sustain a holding divider. The principle soil nail divider to use current soil nailing strategy was worked close Versailles in 1972. The technique included presenting high-thickness, grouted soil nails into a 60-feet-high divider and facing it with reinforced framework.

Lee et al. (2004), the main of this paper, is investigated the effects of flexible facing utilized to determine the behavior of histories. In synthesis, it comprises of bridge sustaining elements along with wrapped face. The structure has been tested against different loads. From the experiment, it has been analyzed that the bridge with flexible facing have more abutments compared to the traditional bridge. Also, 0.2 m of vertical spacing the bridge can carry higher load, and the load bearing capability of bridge is small when backfilling has been done with lower strength material.

Sanvitale et al. (2013) investigated a few tests with different confronting types, contrasting in solidness and progression, which were completed so far in 1 g little-scale physical model. The exploratory outcomes demonstrate the significance of both flexional and hub solidness of looking in controlling the misshapening of the divider amid exhuming and the most extreme additional charge relevant at the back of divider.

The test results talked about above it is conceivable to see that both flexional and hub firmness impact the execution of a dirt nailing framework in unearthing and at fall. On the off chance that the confronting has no coherence, its flexional firmness can prevent the front distortion amid removal, consequently constraining the activation of shear worry along nails. In both of the cases, toward the finish of unearthing, the framework has still an abnormal state of wellbeing in connection to the worldwide security issue. In actuality, the biggest misshapenings aggregated with unearthing can decrease the security edge.

Liu (2014) has illustrated a condition of the practice and gave an account of the improvement of soil nailing strategy and an imminent audit for its applications in the interstate development and upkeep in Ontario, Canada. Soil nailing is a detached fortifying procedure where the soundness of a soil mass is strengthened by preparing chiefly the pullout commitment of consistently divided nails embedded in the dirt mass when there is a ground development. Among different development methods, the penetrated and grouted soil nail is the most mainstream soil nailing by and by. Identified with expressway development, soil nailing has been connected much of the time, including new street cut help, existing street broadening, fixing of existing holding structures and strengthening flimsy slants. An aggregate of in excess of three hundred plan codes, rules, examine articles and case narratives have been gathered for this investigation. The new advancements of hazard classes, unwavering quality-based plan and new confronting sorts will be examined in this paper.

Godayal et al. (2018) have designed Mohr–Coulomb model for soil nail wall and investigated the failure effects by observing different parameters. The modeling has been developed using PLAXIS 2D v8 package. The effect of excavation depth, length of nail and vertical nail spacing along with the space between footings to the vertical edge has been analyzed using 72 soil nailed model. It has been observed that as the length of nail increased, the load bearing capacity also increased.

Garcia et al. (2018) have examined a 24-m-deep excavation situated at Brazil. The soil taken for this project is the residual gneiss soil, and the stability of soil has been done using soil nailing concept. The nails inserted into the soil model are of different lengths, primarily 8 m nails are used and then the size of nails decreased as per the depth. It is examined that the nails near to the excavation edges are affected less with the change in the size of the nails.

2 Materials and Methodology

Initially, the soil is collected from Chandigarh University campus and then various properties were studied on the basis of code (IS2720). The properties such as moisture content, specific gravity, compaction characteristics and unconfined compression strength have been analyzed (Table 1).

Table 1 Properties of soil

Properties	Values
Water content (%)	11
Specific gravity	2.56
Optimum moisture content (OMC) (%)	7.3
Maximum dry density (MDD) (g/cc)	1.98
Coefficient of uniformity	5.6
Coefficient of curvature	0.64
Unconfined compressive strength (Kg/cm^2)	0.118

From the above results, the soil was categorized as poorly graded sand.

Physical Model

A physical model was prepared with the Perspex sheet (acrylic sheet) of thickness 8 mm. The dimensions of the model such as length, width and height were taken as 60 cm, 40 cm and 50 cm, respectively. Three types of nail arrangements were used namely as square, rectangular and staggered. The dimensions and nail length were selected according to the soil nail manual.

The model drawing is shown is Fig. 1.

The various components used in the physical model are described below.

Pressure hydraulic jack (bottle jack)

Pressure hydraulic jack of bottle jack having weight 5 ton is used and placed in between the Perspex sheet. It is used to lift heavy loads and works on the principle of screw action.

During the research, hydraulic jack is placed in between the Perspex sheets of 8 mm thick as shown in Fig. 2

Fig. 1 Physical model drawing

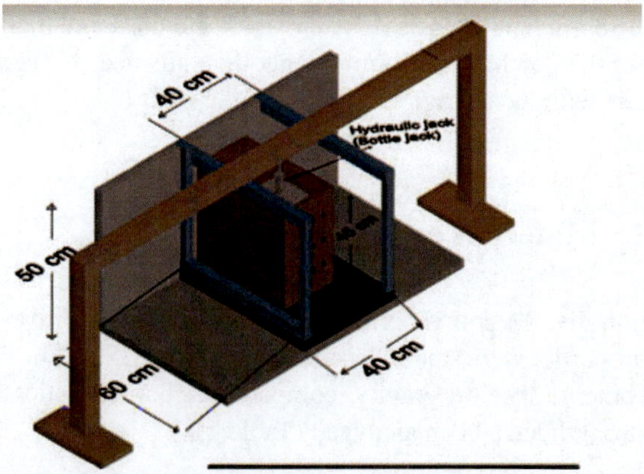

Fig. 2 Perspex sheet model and hydraulic jack

Flex sensor

The flexible sensor is just like a variable resistor. As the nails bend, the resistance of the flexible sensor increases. A multimeter is attached to the switch to measure the resistance sensed by flex sensor and then strain is calculated by using the formula written below:

$$\text{Strain} = \frac{\Delta R/R}{\sum_c}$$

Here, $\sum_c$ stands for gauge factor; its value lies between 2 and 2.5, and for 200 KΩ, the value of gauge factor is 2.1.Copper wire is attached with each nail and resistance is measured using digital multimeter.

In the present research, load was applied by the use of hydraulic jack. The copper wire is extended and connected to the switch board that comprises of four switches. Each switch is connected to an individual nail. Multimeter is connected to the individual switch, and hence measured the resistance. The strain of individual nail is determined by putting remaining switched off. The initial resistance was measured before applying load and change in resistance was measured after applying the load by the use of digital multimeter. The strain was calculated by using the above formula. Thereafter, the testing on this model was conducted by using various flexible materials by using three nail arrangements.

Various flexible materials used in this research work are as follow

- HDPE hexagonal slope protection geonet
- Biaxial polypropylene geonet.

These materials are chosen for the research work as these materials have tensile strength which is required to balance the tensile stresses that are acting on the slope.

3 Results and Discussions

As mentioned above, two materials were used for this research and three types of arrangements, i.e., square, rectangular and staggered are discussed for both the materials (Tables 2, 3 and 4; Figs. 3, 4, 5, 6, 7 and 8).

Discussions

We require facing material so as to hinder deformation, thus limiting the mobilization of shear stress along nails. Flexible facing is designed to provide the necessary restrains to areas of slope between the bearing plates as well as erosion control.

From the above Tables, it is found that biaxial polypropylene geonet have taken up maximum stresses as compared to the HDPE hexagonal slope protection geo net. In general, both the flexional and axial stiffness of facing controls the deformation.

In the present research work, we had studied three types of arrangements for soil nailing, i.e., square, rectangular and staggered arrangements, and from the Tables 5, 6 and 7, it can be concluded that maximum stresses are taken up by the biaxial polypropylene geonet as this particular material have higher stiffness as compared to other material, and further, there is increase in normal force and shear resistance along the potential slip surface in frictional soil.

Table 2 Stress and strain reading for HDPE hexagonal slope protection geonet using four nails in square pattern

Stress (N/mm^2)	Strain in N1	Strain in N2	Strain in N3	Strain in N4
0	0	0	0	0
0.49	0.021	0.028	0.020	0.034
0.98	0.072	0.071	0.057	0.071
1.47	0.097	0.109	0.161	0.173
1.96	0.308	0.219	0.430	0.351
3.43	0.427	0.358	0.665	0.499

Table 3 Stress and strain reading for HDPE hexagonal slope protection geonet using six nails in rectangular pattern

Stress (N/mm^2)	Strain in N1	Strain in N2	Strain in N3	Strain in N4	Strain in N5	Strain in N6
0	0	0	0	0	0	0
0.98	0.039	0.019	0.002	0.016	0.044	0.041
1.96	0.044	0.006	0.005	0.021	0.060	0.041
2.94	0.077	0.019	0.005	0.029	0.062	0.136
3.92	0.103	0.024	0.003	0.031	0.177	0.142
4.9	0.130	0.049	0.019	0.067	0.224	0.179
5.04	0.282	0.093	0.096	0.132	0.329	0.260

Table 4 Stress and Strain reading for HDPE hexagonal slope protection geonet using eight nails in staggered pattern

Stress (N/mm^2)	Strain in N1	Strain in N2	Strain in N3	Strain in N4	Strain in N5	Strain in N6	Strain in N7	Strain in N8
0.000	0.000	0.000	0.000	0.000	0.000	0.000	0.000	0.000
0.980	0.006	0.002	0.002	0.004	0.002	0.004	0.004	0.014
1.960	0.013	0.017	0.023	0.011	0.013	0.009	0.040	0.022
2.940	0.019	0.026	0.028	0.044	0.019	0.026	0.050	0.032
3.920	0.027	0.045	0.042	0.044	0.023	0.032	0.050	0.017
4.900	0.037	0.071	0.065	0.061	0.029	0.037	0.061	0.026
5.880	0.050	0.069	0.083	0.065	0.064	0.077	0.097	0.049
6.250	0.062	0.074	0.085	0.068	0.086	0.100	0.141	0.095

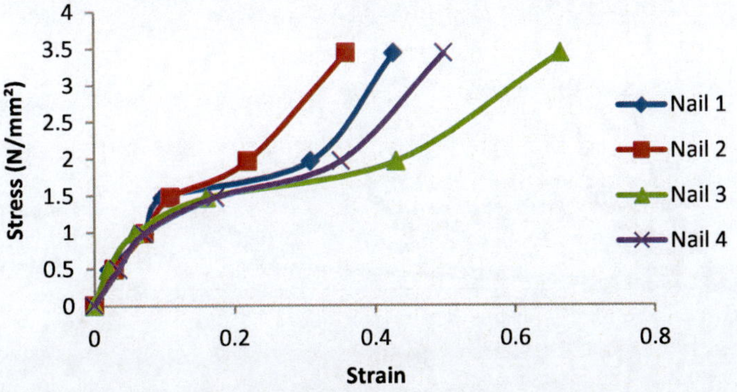

Fig. 3 Stress versus strain graph for HDPE hexagonal slope protection geonet using four nails in square pattern

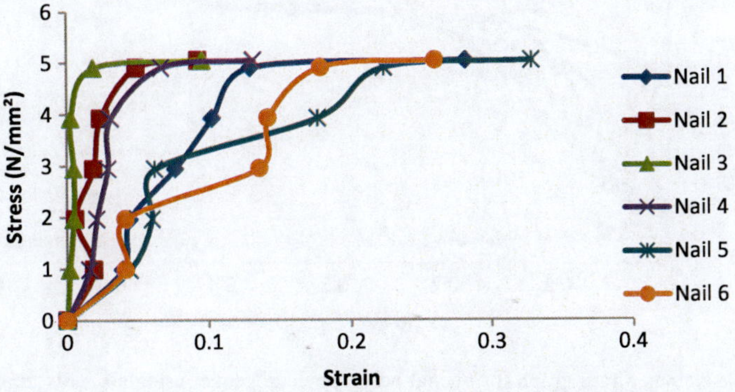

Fig. 4 Stress versus strain graph for HDPE hexagonal slope protection geonet using six nails in rectangular pattern

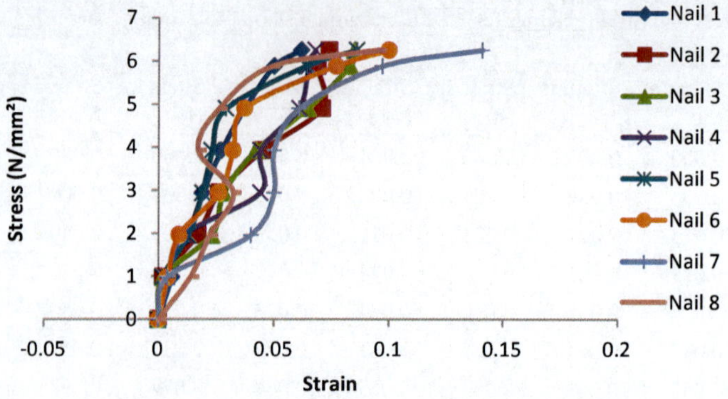

Fig. 5 Stress versus strain graph for HDPE hexagonal slope protection geonet using eight nails in staggered pattern

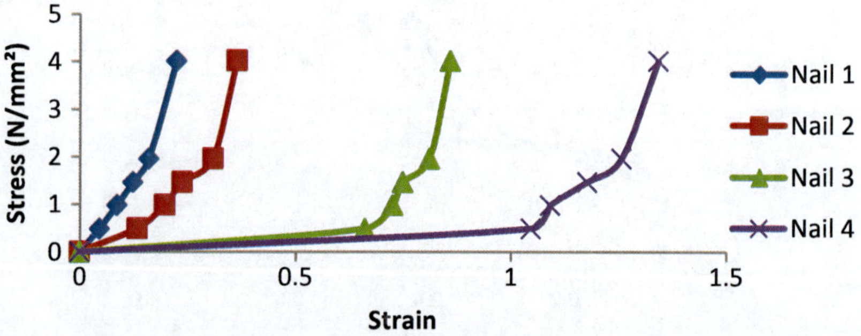

Fig. 6 Stress versus strain graph for biaxial polypropylene geonet using four nails in square pattern

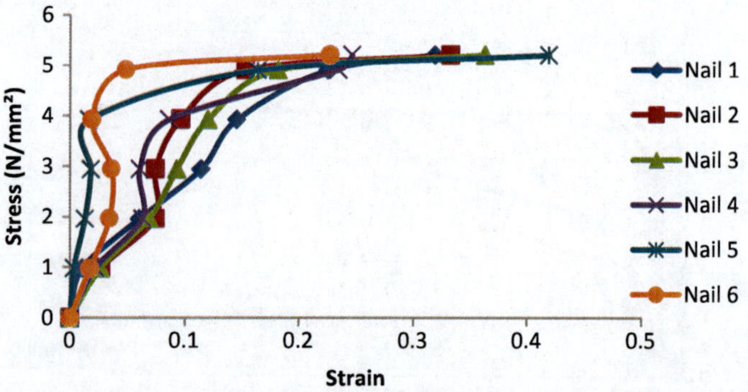

Fig. 7 Stress versus Strain graph for biaxial polypropylene geonet using six nails in rectangular pattern

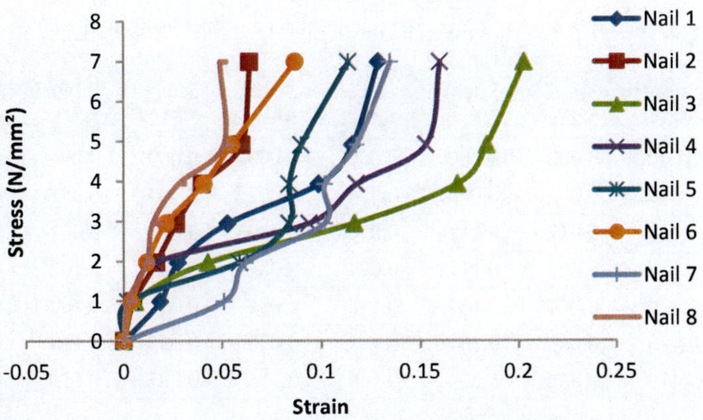

Fig. 8 Stress versus strain graph for biaxial polypropylene geonet using eight nails in staggered pattern

Table 5 Stress and strain reading for biaxial polypropylene geonet using four nails in square pattern

Stress (N/mm^2)	Strain in N1	Strain in N2	Strain in N3	Strain in N4
0	0	0	0	0
0.49	0.047	0.135	0.662	1.047
0.98	0.088	0.198	0.725	1.092
1.47	0.125	0.238	0.748	1.178
1.96	0.162	0.31	0.81	1.258
4.01	0.225	0.365	0.858	1.342

Table 6 Stress and strain reading for biaxial polypropylene geonet using six nails in rectangular pattern

Stress (N/mm^2)	Strain in N1	Strain in N2	Strain in N3	Strain in N4	Strain in N5	Strain in N6
0	0	0	0	0	0	0
0.98	0.0107	0.0269	0.0262	0.0219	0.0033	0.0173
1.96	0.0607	0.0748	0.0692	0.0632	0.0130	0.0342
2.94	0.1142	0.0748	0.0935	0.0610	0.0181	0.0360
3.92	0.1451	0.0966	0.1201	0.0865	0.0164	0.0190
4.9	0.2252	0.1526	0.1816	0.2341	0.1636	0.0486
5.20	0.3193	0.3329	0.3633	0.2459	0.4195	0.2268

Further, we can conclude that flexible facing approach also represents environmental benefits as it would allow the growth of vegetation, and use of flexible materials may also provide significant savings. In addition to this, if one nail becomes

Table 7 Stress and Strain reading for biaxial polypropylene geonet using eight nails in staggered pattern

Stress (N/mm^2)	Strain in N1	Strain in N2	Strain in N3	Strain in N4	Strain in N5	Strain in N6	Strain in N7	Strain in N8
0	0.000	0.000	0.000	0.000	0.000	0.000	0.000	0.000
0.98	0.019	0.005	0.006	0.004	0.002	0.004	0.051	0.002
1.96	0.028	0.017	0.043	0.016	0.059	0.012	0.062	0.012
2.94	0.053	0.027	0.117	0.094	0.084	0.022	0.102	0.015
3.92	0.099	0.040	0.169	0.118	0.084	0.041	0.102	0.028
4.9	0.116	0.060	0.184	0.153	0.090	0.055	0.118	0.051
6.98	0.129	0.064	0.203	0.160	0.114	0.087	0.135	0.049

overstressed for any reason, it will not cause failure in the entire system, but it would redistribute its overstress to the adjoining nails.

The performance of soil nail walls during earthquakes has been historically observed to be much better than that of gravity retaining structures. The soil nail walls are more flexible than gravity walls and other similar systems; it is reasonable to expect that these walls will sustain some lateral displacement during earthquakes, and that these displacements can be tolerated.

4 Conclusions

- From the experiment performed on design model using different materials, it can be concluded that the present study will lead to an increase in the soil slopes load handling and slope stability
- This research can be used as construction technique that even reflects cost-effective and low environmental impact.
- Maximum stresses are taken up by HDPE hexagonal slope protection geonet for all the nail arrangements due to its higher stiffness.

Conflict of Interest The author declares that I have no conflicts of interest.

References

1. Byrne RJ, Cotton D, Porterfield J, Wolschlag C, Ueblacker G (1996) Manual for design and construction monitoring of soil nail walls (No. FHWA-SA-96-069)
2. Jewell RA (1990) Strength and deformation in reinforced soil design. In: Proceedings of 4th international conference on geotextiles, geomembranes and related products. Rotterdam, The Netherlands, pp 913–946

3. Shaw-Shong L (2005) Soil nailing for slope strengthening. Geotechnical Engineering, Gue & Partners Sdn Bhd, Kuala Lumpur, Malaysia, pp 30–31
4. Milligan GWE, Chang KT, Morris JD (1997) Pull-out resistance of soil nails in sand and clay. In: Davies MCR, Scholsser F (eds) Proceeding of the 3rd international conference on ground improvement geosystems, London, UK, 3–5 June 1997. Thomas Telford, New York. pp 414–422. Google Scholar
5. Yin JH, Su LJ (2006) An innovative laboratory box for testing nail pull-out resistance in soil. Geotech Test J 29(6):451–461
6. Pei H, Yin J, Zhu H, Hong C (2012) Performance monitoring of a glass fiber-reinforced polymer bar soil nail during laboratory pullout test using FBG sensing technology. Int J Geomech 13(4):467–472
7. Byrne RJ, Cotton D, Porterfield J, Wolschlag C, Ueblacker G (1998) Manual for design and construction monitoring of soil nail walls (revised edition) (no. Fhwa-sa-96–069r)
8. Budania R, Arora RP (2016) Soil nailing for slope stabilization: an overview. Int J Eng Sci 3877
9. Liew SS, Liong CH (2006) Two case studies on soil nailed slope failures. In: Submitted international conference on slopes, Malaysia
10. Palmeira EM, Tatsuoka F, Bathurst RJ, Stevenson PE, Zornberg JG (2008) Advances in geosynthetics materials and applications for soil reinforcement and environmental protection works. Electron J Geotech Eng 13:1–38
11. Shaw-Shong L (2005) Soil nailing for slope strengthening. Geotechnical Engineering, Gue & Partners Sdn Bhd, Kuala Lumpur, Malaysia, pp 30–31
12. Taib SNL (2010) A review of soil nailing design approaches. J Civil Eng Sci Technol 1(2):1–6
13. Jewell RA, Pedley MJ (1992) Analysis for soil reinforcement with bending stiffness. J Geotech Eng 118(10):1505–1528
14. Raju GVR (1996) Behaviour of nailed soil retaining structures (Doctoral dissertation, Nanyang Technological University, School of Civil and Structural Engineering)

Effect of Number of Reinforcement Layers and Loading Magnitude on the Performance of Square Footing Embedded in Geogrid Reinforced Flyash Beds Under Cyclic Loading

K. Bindiya and H. C. Muddaraju

Abstract Flyash is a very fine powder obtained from the gases of burning coal during the production of electricity. Handling and disposal of the flyash is a serious problem and a large storage area is required for the safe disposal of flyash. Hence flyash can be utilized in a geotechnical applications like as a backfill material in the construction of embankments, as a sub-base material etc. rather than dump it. The subgrade properties of soil can be improved by using geosynthetics in the subgrade. In order to overcome the problem of disposal of flyash, and to increase the strength of flyash as a subgrade, an attempt was made to improve the flyash subgrade properties by using geogrid. Square footing resting on flyash bed with geogrid performs better than its counterpart resting on flyash bed without geogrid and optimum number of geogrid layers is three. Square footing performs better when they are subjected to lower loading magnitude compare to the higher loading magnitude both in 1B and 2B embedment depth of square footing.

Keywords Flyash · Geogrid · Loading magnitude

1 Introduction

Flyash is a very fine powder obtained from the gases of burning coal during the production of electricity. Handling and disposal of the flyash is a serious problem and a large storage area is required for the safe disposal of flyash. Hence flyash can be utilized in a geotechnical applications like as a backfill material in the construction of embankments, as a sub-base material etc. rather than dump it.

The subgrade properties of soil can be improved by using geosynthetics in the subgrade. The geosynthetics is the cost effective method of ground improvement technique, and the properties of soil reinforcement depend on the tensile strength, frictional and the adhesion properties of the reinforcement. The different types of

K. Bindiya (✉)
JNNCE, Shivamogga, Karnataka, India

H. C. Muddaraju
UVCE, Bangalore, Karnataka, India

T. G. Sitharam et al. (eds.), *Earthquake Geotechnics*, Lecture Notes in Civil Engineering 187, https://doi.org/10.1007/978-981-16-5669-9_32

Table 1 Chemical properties of flyash

Chemical configuration	Values
Silica	55–65%
Alumina	22–25%
Calcium	5–6%
Unburnt carbon	1–5%
Magnesium	1%
Manganese	1%
Phosphorous	1%
Sulphate	0.1%
Potassium	0.9%
Titanium	Traceable

geosynthetics include geotextiles, geogrids, geomembranes, geonets, geosynthetic clay liners etc.

In order to overcome the problem of disposal of flyash, and to increase the strength of flyash as a subgrade, an attempt was made to improve the flyash subgrade properties by using geogrid.

2 Materials and Methods

2.1 Backfill Material: Flyash

The flyash is collected from open dry dumps of Raichur thermal power plant, Karnataka. According to ASTM classification, it belongs to class "C" (non-pozzolanic). The chemical and physical properties are presented in Tables 1 and 2.

2.2 Reinforcement

The biaxial geogrid with square aperture is used as a reinforcement material. The physical properties are presented in Table 3.

2.3 Footing

The square footing is made up of mild steel. The size of the footing is 100 × 100mm and of 4 mm thickness.

Table 2 Physical properties of flyash

Physical properties	Values
Colour	Grey
Specific gravity	2.07
Grain size distribution	
Sand fraction	54%
Silt and clay fraction	46%
Co-efficient of curvature	0.694
Co-efficient of uniformity	7.46
Atterberg's limits	
Liquid limit	31.80%
Plastic limit	–
Compaction characteristics	
Maximum dry density	12.70 kN/m^3
Optimum moisture content	23%
Unconfined compressive strength	51.40 kN/m^2

Table 3 Physical properties of geogrid

Physical properties		Test results
Aperture size	Machine direction	34 mm
	Cross direction	32 mm
Roll width	–	3.9 m
Roll length	–	50 m
Ultimate tensile strength	Machine direction	33.2 kN/m
	Cross direction	31.1 kN/m
Strain at ultimate	Machine direction	14.4%
	Cross direction	6.9%

2.4 *Testing Tank*

The mild steel tank is 500 mm in diameter and 390 mm in height.

2.5 *Methodology*

Repeated load is applied to the footing by using Automated Dynamic Testing Apparatus (ADTA). The machine is a sophisticated computer controlled device, runs on software, Movicon. The cyclic pressure of different intensity and frequency of 0.2 Hz are fed into the computer. According to the given input, the load is applied to the footing through load cell and the corresponding settlements after each cycle of loading are recorded in the data acquisition system through three LVDT's. Analog to digital conversion takes place through control unit.

The Automated Dynamic Testing Apparatus is used for applying the repeated load to the footing resting on the flyash bed with and without geogrid. The cyclic pressure of different magnitude of 250, 350 and 450 kPa are selected and the frequency of 0.2 Hz is fed into the computer. Three LVDT's are used to measure the corresponding settlements. The LVDT's and the load cell are connected to the control unit. The input and output data are recorded in the data acquisition system (Figs. 1 and 2).

Flyash bed is prepared in the test tank at its maximum dry density and optimum moisture content with and without geogrid by manual compaction. For unreinforced flyash bed, flyash is compacted in three layers of 120 mm thick for a height of 360 mm.

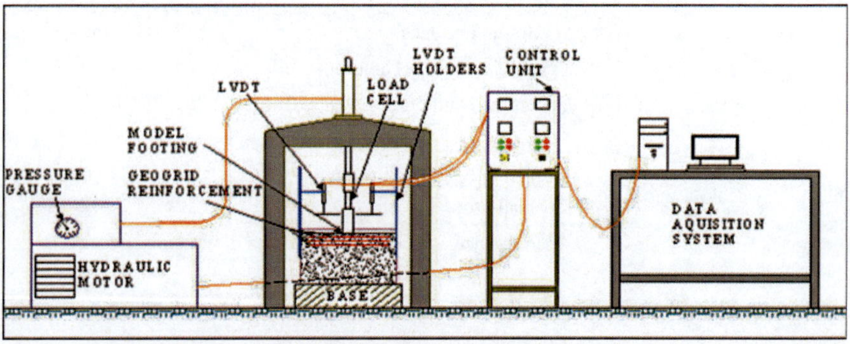

Fig. 1 Schematic diagram of experimental setup

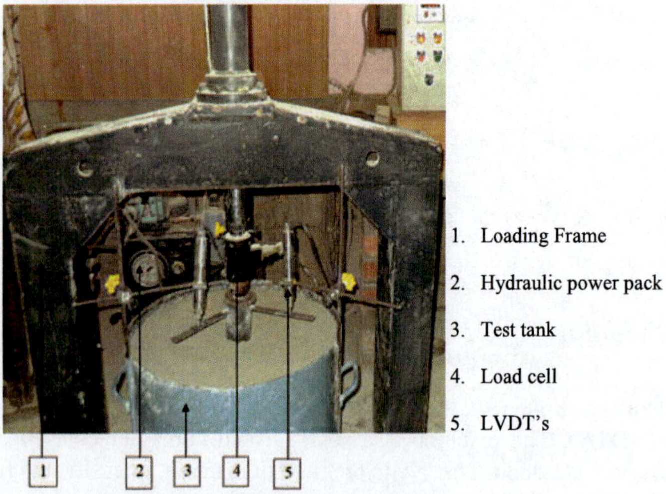

Fig. 2 Photographic view of experimental setup

For reinforced flyash bed, the geogrid was provided in circular shape. A 5 mm clearance is provided between the inner surface of tank and the geogrid edges to avoid the friction generation between geogrid and the tank wall. The reinforcements are placed at a spacing of 0.3 times of the width of footing (30 mm) in all the experiments [1]. Also the first layer of reinforcement is placed at a minimum depth of 0.3 times the width of footing [2], as this minimum thickness becomes essential for confinement effect. The critical depth of reinforcement is about 1.33times the width of footing, and the critical width of reinforcement layers is about 4 times the width of footing [3]. Geogrid is placed at pre-determined spacing in the flyash layers and remaining height is compacted by manual compaction.

3 Results and Discussions

3.1 *Effect of Number of Geogrid Layers on 1B Embedment Depth*

It is seen from Figs. 3, 4 and 5 that, the footing in flyash bed without geogrid taken less number of load cycle as compare to flyash bed with geogrid at any settlement. Two, three and four layers of geogrid in flyash bed show 4000, 40,000 and 20,000 times improvement in the performance of square footing, respectively.

This indicates the performance of flyash bed with geogrid is better than flyash bed without geogrid and the optimum number of geogrid layers is three.

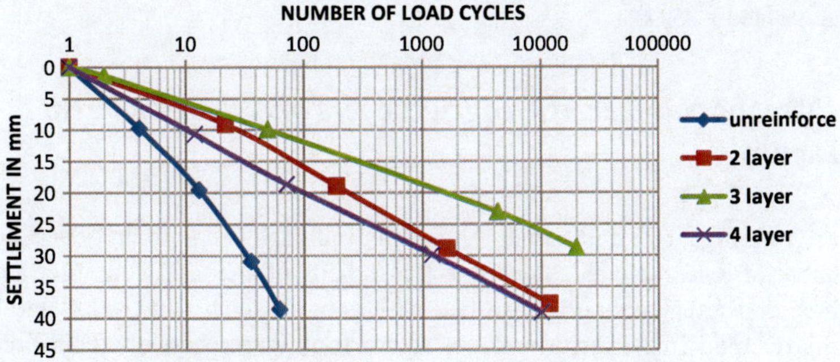

Fig. 3 Effect of number of geogrid layers on the performance of square footing subjected to a loading magnitude of 250 kPa

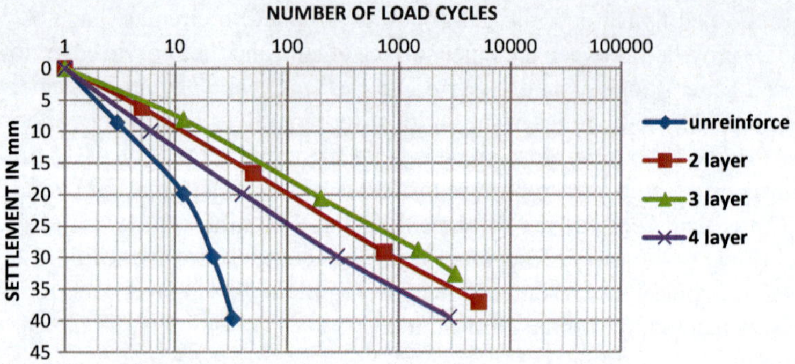

Fig. 4 Effect of number of geogrid layers on the performance of square footing subjected to a loading magnitude of 350 kPa

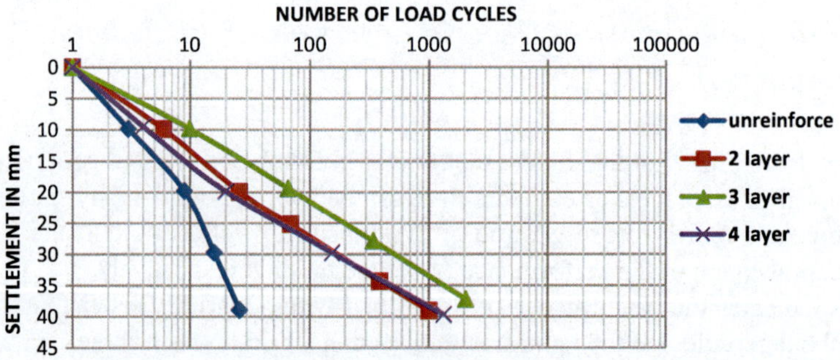

Fig. 5 Effect of number of geogrid layers on the performance of square footing subjected to a loading magnitude of 450 kPa

3.2 *Effect of Number of Geogrid Layers on 2B Embedment Depth*

It is seen from Figs. 6, 7 and 8 that, the footing in flyash bed without geogrid taken less number of load cycle as compare to flyash bed with geogrid at any settlement. Two, three and four layers of flyash bed with geogrid show 1400, 6000 and 2000 times improvement in the performance of square footing, respectively. This indicates the performance of flyash bed with geogrid is better than flyash bed without geogrid, and the optimum number of geogrid layers is three.

It is seen from Figs. 3, 4, 5, 6, 7 and 8 that at any settlement, the flyash bed with geogrid experiences a more number of load cycles compared to the flyash bed without geogrid. In flyash bed, flyash bed with three layer geogrid exhibited the best performance for square footing embedded at 1B and 2B depth.

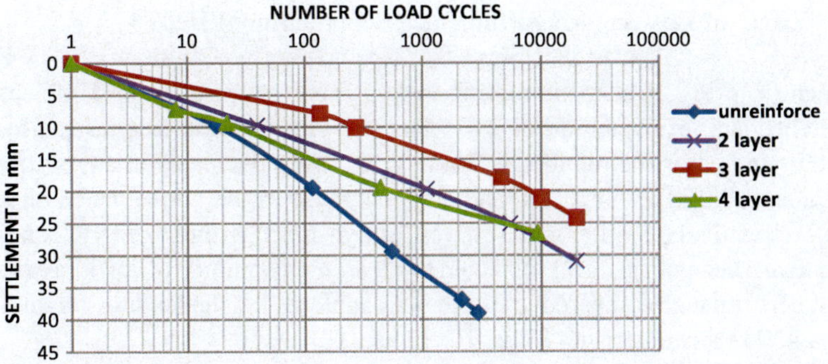

Fig. 6 Effect of number of geogrid layers on the performance of square footing subjected to a loading magnitude of 250 kPa

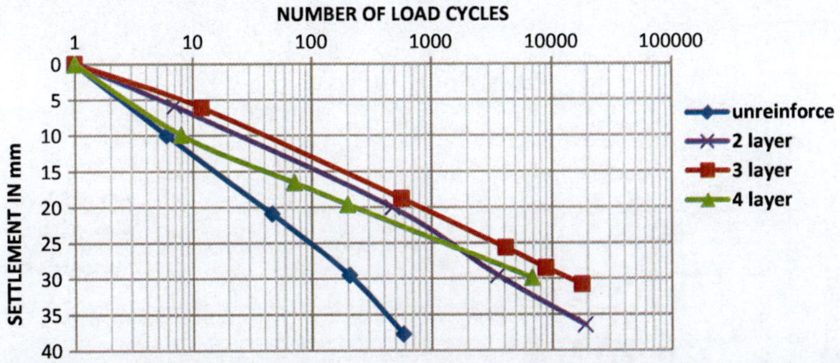

Fig. 7 Effect of number of geogrid layers on the performance of square footing subjected to a loading magnitude of 350 kPa

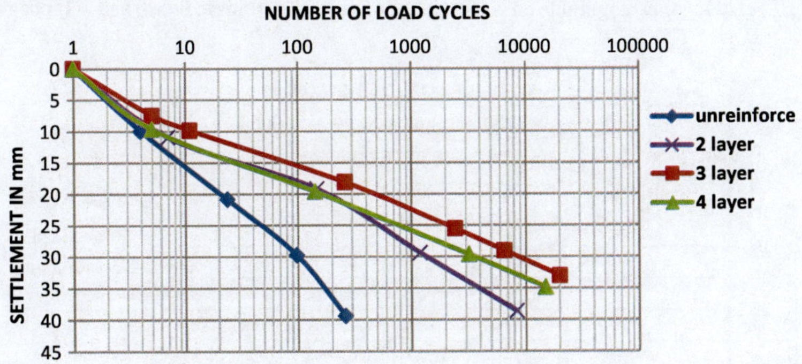

Fig. 8 Effect of number of geogrid layers on the performance of square footing subjected to a loading magnitude of 450 kPa

3.2.1 Effect of Loading Magnitude on 1B Embedment Depth

It is seen from Fig. 9 that, the footing in flyash bed without geogrid subjected to a different loading magnitude of 250 kPa, 350 kPa and 450 kPa, as the loading magnitude increases the number of load cycles taken decreases at any settlement. 350 kPa and 450 kPa show 30% and 50% decrease in the performance of square footing, respectively. Similar to the curves for flyash bed without geogrid conditions even in flyash bed with geogrid subjected to a loading magnitude of 250 kPa exhibited the best performance compared to flyash beds subjected to the loading magnitude of 350 and 450 kPa counterparts (Figs. 10, 11 and 12).

It is seen from Figs. 9, 10, 11 and 12 that at any settlement, flyash bed with and without geogrid subjected to lower magnitude of loading experiences more number of load cycles compared to the conterparts subjected to the higher magnitude of loading.

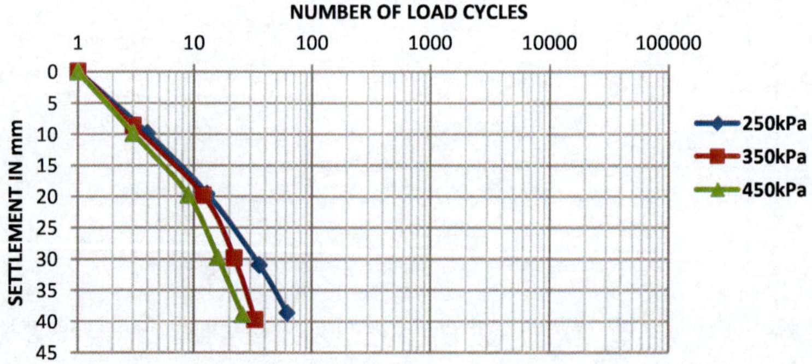

Fig. 9 Effect of loading magnitude on performance of square footings in flyash bed without geogrid

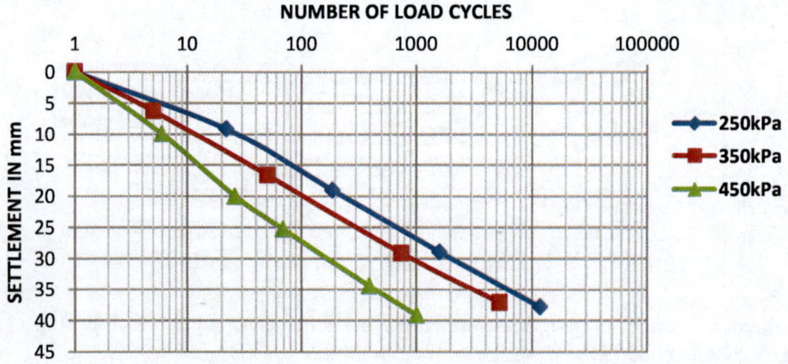

Fig. 10 Effect of loading magnitude on performance of square footings in flyash bed with two layers geogrid

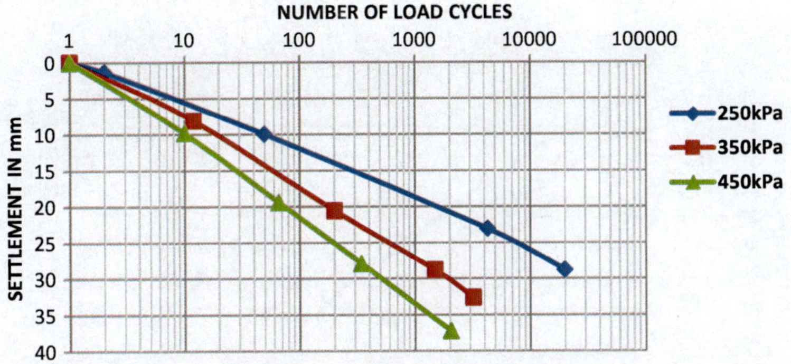

Fig. 11 Effect of loading magnitude on performance of square footings flyash bed with three layers geogrid

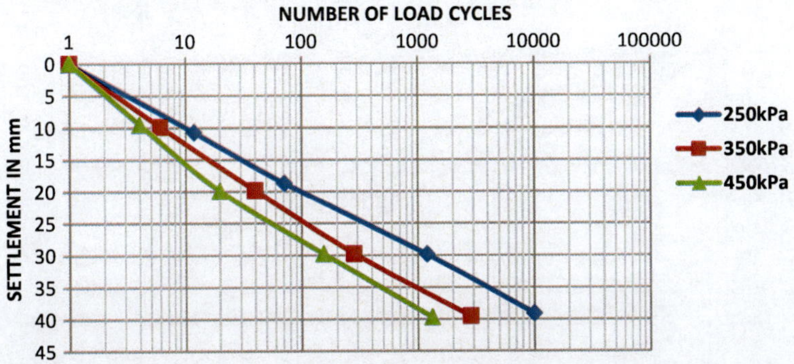

Fig. 12 Effect of loading magnitude on performance of square footings in flyash bed with four layers geogrid

3.3 *Effect of Loading Magnitude on 2B Embedment Depth*

It is seen from Fig. 13 that, the footing in flyash bed without geogrid, as the loading magnitude increases the number of load cycles taken decreases at any settlement. 350 kPa and 450 kPa show 60% and 85% decrease in the performance of square footing, respectively.

It is seen from Figs. 13, 14, 15 and 16 that at any settlement, flyash bed without geogrid and with geogrid subjected to lower magnitude of loading experiences more number of load cycles compared to the counterpart subjected to the higher magnitude of loading.

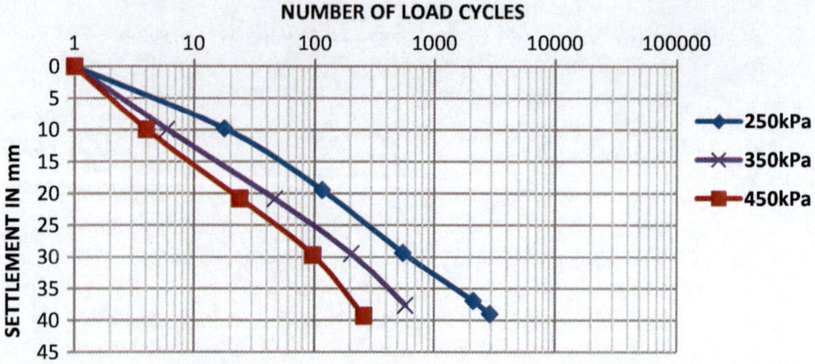

Fig. 13 Effect of loading magnitude on performance of square footings in flyash bed without geogrid

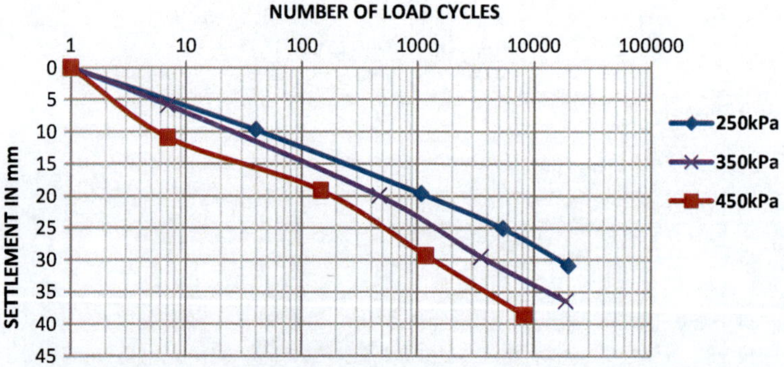

Fig. 14 Effect of loading magnitude on performance of square footings in flyash bed with two layers geogrid

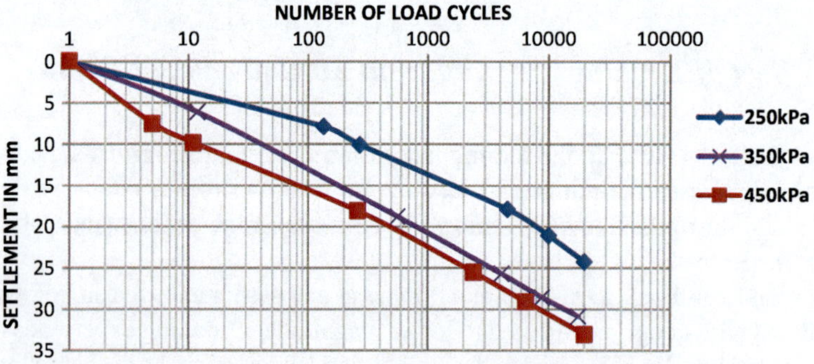

Fig. 15 Effect of loading magnitude on performance of square footings in flyash bed with three layers geogrid

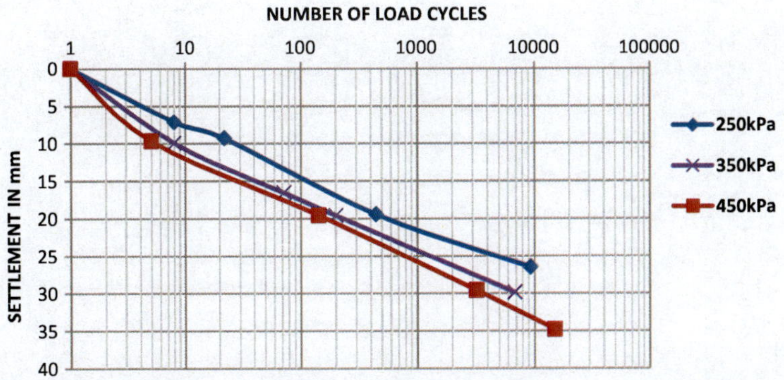

Fig. 16 Effect of loading magnitude on performance of square footings in flyash bed with four layers geogrid

4 Conclusions

The following conclusions are drawn based on the results of experiments conducted,

1 Square footing resting on flyash bed with geogrid performs better than its counterpart resting on flyash bed without geogrid.
2 Square footing resting in flyash bed with three layers geogrid perform much better by taking more number of load cycles and undergoing less settlement when compared to the square footing resting in flyash bed with two and four layers geogrid, both at 1B and 2B embedment depth.
3 Footing performs better when they are subjected to lower loading magnitude by taking more number of loading cycles and undergoing less settlement when compare to the higher loading magnitude both in 1B and 2B embedment depth of square footing.

References

1. El-Mahallawy NA, Rashed AS (2012) Experimental study for the behaviour of footings on reinforced sand beds overlying soft clay zone. J Am Sci 8(1)
2. Gangadara S, Muddaraju HC (2013) Effect of reinforcement spacing on the performance of embedded circular footing in reinforced fly ash. Int J Sci Eng Res 4(5) 2013. ISSN 2229-5518
3. Shin EC, Das BM (1999) Dynamic behavior of geogrid-reinforced sand. Geotech Eng 3(4):379–386

A Study on Shake Table Test and Centrifuge Test with a Scaled Model in Geotechnics

M. K. Pradhan, P. Kumar, V. S. Phanikanth, and K. Srinivas

Abstract To study the dynamic responses of a civil engineering structure, shake table tests and centrifuge tests are more commonly used. Due to various limitations, sometimes it is not possible to carry out the required tests on original structures, but it becomes inevitable to perform experiments by changing the scale (scaled up or reduced scale) in a model from the prototype. The geometrical shape and materials of the scaled model may be different from the prototype. In the present study, a detailed presentation is made about the design philosophy, limitations, advantages of shake table tests, as well as centrifuge tests. It is also detailed that the principles and approaches used in the development of the model in the scaled format of a prototype. Stepwise, it is illustrated the development of a scaled model, extrapolating the results in a prototype. The present paper will be useful to researchers performing dynamic experiments using shake table tests or centrifuge tests with a scaled model.

Keywords Shake table test · Centrifuge tests · Similitude rule · Buckingham pi theorem · Prototype

1 Introduction

The first step in the design of any structures is to conduct tests (field or laboratory) to study the suitability as well as the accuracy of the design considering intend loads. Due to the access to advanced software, the development of the numerical model of a complex structure has become easy. However, the validations of their results through extensive sets of tests before finalizing are required. Various types of tests are performed in structural engineering or geotechnical engineering to validate and investigate the behaviours and various responses under different sites and loading conditions. The test may be performed at site (in-situ) or in laboratory using the model

M. K. Pradhan (✉) · P. Kumar
Homi Bhabha National Institute & BARC, Trombay, India
e-mail: mpradhan@barc.gov.in

M. K. Pradhan · P. Kumar · V. S. Phanikanth · K. Srinivas
Bhabha Atomic Research Centre, Trombay, India

T. G. Sitharam et al. (eds.), *Earthquake Geotechnics*, Lecture Notes in Civil Engineering 187, https://doi.org/10.1007/978-981-16-5669-9_33

of same size of prototype. Also, the test can be performed in laboratory using scaled model i.e. small-scale or scaled-up model. A full-scale test of real-size structures is being expensive, time-consuming, and sometimes not feasible (slope stability of a hill, earthquake condition, etc.). The testing with real-size structures with creating actual working conditions for offshore structures, dams, tall buildings and bridges is not possible. The prediction of the occurrence of an earthquake is not possible, and all structures cannot be instrumented to observe the responses. Also due to the limitations of cost and time-consuming model, tests are sometimes preferable to the full-size tests. The most common and important geotechnical test to study the soil–structure interactions is performed in the laboratory under dynamic conditions using a shake table (at 1-*g*) or centrifuge (multi-*g*).

Physical modelling reduces the size of a large and complex structure to a simple structure that maintains the same physical characteristic and the main interaction in the system to reproduce the true behaviours of the prototype in the relevant set of loading conditions. However, thorough analyses of similarity laws for material deformation behaviour and failure of soil and underground structures subjected to dynamic loads have found very limited attention. More comprehensive studies are required.

In the present paper, the advantages and limitations of shaking table tests and centrifuge tests are presented. The principles of similitude with different approaches followed by various researchers are also explained. The approaches in scaling laws for geometrical variation, material variation between prototype and model piles are detailed. Various instruments used during the model tests and extracting various results (accelerations, stains, pore pressure and displacements) and deriving parameters (bending moment, pressure and force) of soil and structures (viz. pile) also highlighted.

2 Similitude Rules or Scaling Laws in the Design and Developments of a Scaled Model

The model tests can be divided into two categories, namely those performed under the gravitational field of the earth (shaking table tests) and those performed under a higher gravitational field (centrifuge tests). Both the shaking table and centrifuge model tests have certain advantages and limitations. For performing the scaled model test using a shake table or centrifuge, proper model design and development are required.

The full-scale or real-size system is called a prototype, and the scaled-up/down version of the prototype is called a model. The link between the prototype and model is governed by an analytical expression between their relevant geometrical and physical parameters are known as similitude laws. It is also known as laws of similarity which describes how to design a model and carry out a model test that resembles the real-size structures. Similitude theory provides the conditions to

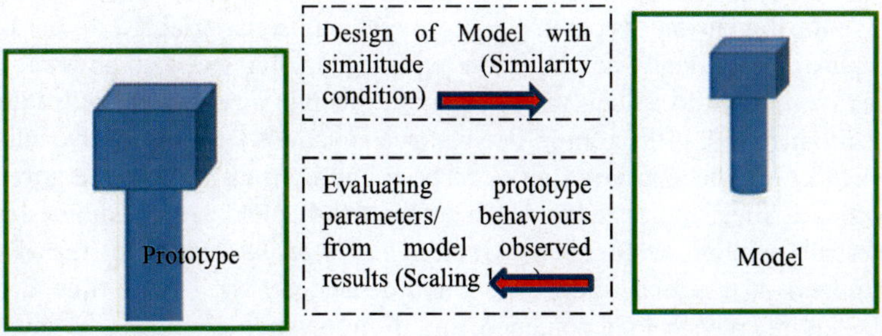

Fig. 1 Flow diagram of design a model and results interpretation for the prototype from model using similitude rule

design the scaled-up or scaled-down model of a prototype to predict the prototype's structural responses from the results of the scaled tests. The design of a model resembles the prototype and the steps for transferring the model test results to generate an expected output of the prototype is presented in Fig. 1. A perfect replica model can never be realized in real practice because of several limitations such as design, availability of material and available test facility. However, in civil engineering, a full-scale test is very expensive, time-consuming and sometimes not feasible. Hence, it is preferably performed through a scaled-down model test. Rocha had developed a similitude model for the soil for static conditions under 1-*g* gravitational field [1]. The application of similitude rules to a reduced-scale soil model has also been described by Roscoe [2]. A similitude of soil structure under dynamic loading was studied by Kagwa by using ratios of forces [3].

Nonlinear dynamic response of ground was examined using Buckingham's theorem by Kokusho et al. [4]. Iai had derived the similarity law for dynamic shaking table tests in a 1-*g* gravitational field [5]. Iai et al. had proposed a two-stage scaling relationship called generalized scaling relationship for centrifuge test [6]. The extensive needs of experimental tests due to the disadvantages and limitations of full-scale tests have resulted in rapid growth in the application of similitude methods [7]. Researchers developed similitude based on the governing equations for soil–structure interaction, for shaking table test and centrifuge tests and also for liquefied soils [5, 6, 8]. The similitude and shaking table tests may be used for a simple problem like static conditions or also maybe for a complicated (structural or geotechnical) problem. Dimensional analysis is the key aspect in developing the similitude or scaling laws in model tests. Derivation of the proper scaling laws is the first step in experimental tests.

2.1 Dimensional Analysis Using Buckingham Pi (Π) Theorem

Dimensional analysis is done by using the most common methods like Rayleigh's method and Buckingham *pi* (π) technics. Buckingham *pi* technics is a method to determine the dimensionless parameters formed by a list of variables. As per Bunkingham, any dimensionally homogeneous equation involving certain physical quantities can be reduced to an equivalent equation involving a complete set of dimensionless products [9]. If the appropriate balance equations and constitutive relations have been set up, the similarity laws can be found by transforming the governing equations in a dimensionless form which automatically yields a set of dimensionless characteristic numbers or functions. Buckingham's *pi theorem* states that if there are n numbers of variables (independent and dependent, viz. length, time, density, gravity, etc.) in a physical phenomenon, and if these variables contain 'm' numbers of fundamental dimensions (viz. force, length, time), then the variables are arranged into (n-m) dimensionless (π) groups. If physical phenomena are expressed as $f(x_1, x_2, x_3, \ldots, x_m)$ where x_1, x_2, x_3 and x_m are dimensional variables, then the physical phenomena can be expressed as $f(\pi_1, \pi_2, \pi_3, \ldots, \pi_n)$ where π are the dimensionless constants and $n = m - k$, where k is minimum numbers of fundamentals dimensions involved.

The objective of dimensional analysis is to identify the non-dimensional groups and represent the expression using the non-dimensional group of models and prototypes are as equal. For better understanding, it can be referred to as Bhattacharya et al. [10] identified several non-dimensional groups and expressed them as Eqs. 1–3. Considering stain field in soil around the laterally loaded pile and the cyclic stress ratio as;

$$\left(\frac{P}{\mathrm{GD}^2}\right)_{\text{model}} = \left(\frac{P}{\mathrm{GD}^2}\right)_{\text{prototype}} \tag{1}$$

where P is an external load on the pile, G is the shear modulus of soil and D is the diameter of the pile. Similarly, considering the rate of loading, the expression of model and prototype as;

$$\left(\frac{k_h}{f_1 D}\right)_{\text{model}} = \left(\frac{k_h}{f_1 D}\right)_{\text{prototype}} \tag{2}$$

where k_h is soil permeability, f_1 is the frequency of the force and D is the diameter of the pile. Further, considering the bending strain in the monopole, the expression developed as;

$$\left(\frac{P.y}{ED^2 t_w}\right)_{\text{model}} = \left(\frac{P.y}{ED^2 t_w}\right)_{\text{prototype}} \tag{3}$$

where P is the horizontal force acting at a distance y above the foundation level, E is the Young's modulus of the pile, D is the diameter of pile and t_w is the pile wall thickness.

2.2 Geometric Similitude

In the geometric similitude, the geometric characteristics are equally proportional as;

$$\frac{L_p}{L_m} = \frac{b_p}{b_m} = \frac{d_p}{d_m} = \lambda \tag{4}$$

Similarly, if A is the area of a surface in the prototype and a is the area of the same surface in the model, then following the geometrical scaling, $A = \lambda^2 a$. If the V is the volume in the prototype and v is the volume of model, then $V = \lambda^3 v$.

2.3 Kinematic Similitude

In kinematic problems, the fundamental quantities are length and time. Hence, in kinematic scaling along with the geometric similarity, there should be also the time scale similarity, viz. $T = \tau t$. Velocity and acceleration at a point of the prototype can be evaluated if the same in the models is known using kinematic scaling as;

$$\frac{V_{p1}}{V_{m1}} = \frac{V_{p2}}{V_{m2}} = \frac{V_{p3}}{V_{m3}} = V_r \tag{5}$$

$$\frac{a_{p1}}{a_{m1}} = \frac{a_{p1}}{a_{m1}} = \frac{a_{p1}}{a_{m1}} = a_r \tag{6}$$

2.4 Dynamic Similarity and Constitutive Similitude

Similarly, in dynamic similitude, homogeneous parts are subject to the homogeneous net force. Further, a constitutive similitude is achieved if there is a similitude between the material stress–strain curve of prototype and model or between constitutive properties between them [11].

2.5 Additional Mass

A prototype is made of concrete with a density of 25,000 N/m^3 as shown in Fig. 2. The top platform diameter and thickness of the prototype are 10 m and 1.00 m, respectively. The diameter and length of the cylindrical support in the prototype are 1.0 m and 10.00 m, respectively. With adopting the geometrical scale λ as 10, the

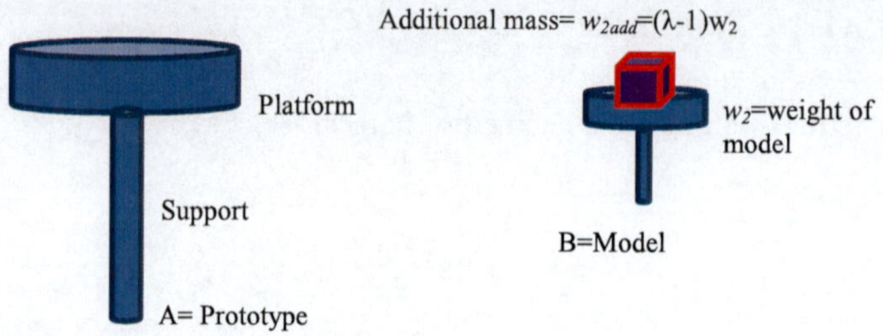

Fig. 2 Prototype and model with additional mass

diameter and thickness of the platform in model are 1.00 m and 0.1 m, respectively. Similarly, the diameter and the length of the support of the model are 0.1 m and 1.00 m, respectively. The stress in the cylindrical support of the prototype is 2500 kN/m^2, whereas the stress in the support part of model is 250 kN/m^2 and is not equal. The stress in both prototypes and models can be matched by either increasing the density in model material or increasing the gravity, or adding mass in the model as shown in Fig. 2. In the present case considering the geometrical scaling as 10, the additional weight will be $w_{2\text{add}} = (\lambda - 1)w_2$ which is equal to nine times the weight of the model (w_2).

Hence, to obtain the proper results, some additional mass is restored in the model without changing the gravitational acceleration, dimension and stiffness of the model. It is to be noted that if the multi-storied building or structure is to be modelled, then the mass is to be distributed on each floor at a rate of $\lambda - 1$ times the floor weight.

2.6 *Scaling Laws with Different Material of Model and Prototype*

Sometimes it becomes essential to develop the model with different materials than the prototype to compliance the stress–strain relation in both prototype and model, viz. aluminium model to represent the RCC pile. In such cases, the similitude can be achieved as Eq. (7).

$$\frac{(\text{EI})_p}{(\text{EI})_m} = \lambda^5 \tag{7}$$

where '*E*' is Young's modulus, '*I*' is the moment of inertia and 'λ' is geometrical scaling (ratio of the prototype to model). The various possible model configurations for a typical prototype RCC pile of 1000 mm diameter considering the geometrical scaling factor of 20 is presented in Sect. 7.1. Unsever et al., developed an aluminium model using a scale ratio of 30 for a prototype concrete pile having 0.6 m diameter and 7.65 m length. The length, outer diameter and the wall thickness of the hollow circular aluminium model pile were 255 mm, 20 mm and 1.1 mm, respectively. Haeri et al. used an aluminium hollow pile of 1.25 m height, 5.00 cm outer diameter and 4.74 m inner diameter as a model pile to investigate the responses of pile group under lateral spreading using a scale ratio of 8 [12]. The approach proposed by Alves is also a very useful guide when models and prototypes are made of different materials [13].

It is to be accepted that the complete replication of prototype in model tests never be realized due to the limitations of availability of materials, availability of tests facility and complexity in problem and design. Hence, it is essential to accept some compromises between the prototype and model. However, care should be taken in the design and developing the model that these compromises will not affect the outcome of the physical models.

3 Shake Table Test

The shaking tables are used by researchers for the assessment of the dynamic and seismic behaviour of civil engineering structures, to study the soil–structure interaction, soil settlements, soil liquefaction, and lateral load problems on underground structures. Shaking table tests have various advantages, viz. well-controlled large-amplitude, multi-axis input motions and easier experimental measurements. The use of shake table tests is justified if the purpose of the test is to validate the numerical model or to understand the basic failure mechanisms. A schematic presentation of a prototype structure and shake table test in a model form which is more commonly used to study the responses of pile foundation with superstructure is presented in Fig. 3.

In shake table tests of reduced-scale models, the similarity rule in terms of stress and strain against the prototype cannot be satisfied because of the stress dependency of the stress–strain soil behaviour. Though higher gravitational stresses cannot be produced in a shaking table test, the contractive behaviour associated with high normal stresses at significant depths can be simulated by placing soil very loosely during model preparations. Using undrained triaxial test results on Toyoura sand, Verdugo and Ishihara [14] showed that contractive behaviour at low confining pressures can be achieved by placing soil loosely. Several advanced shake tables are available worldwide for experiments. The size and capacity of some important shake table test facilities are listed in Table 1.

The design of the model, the parameter considered to be observed and instrumentations during experiments depend upon the complexity of the problem for the study. The important aspects of conducting the shake table tests are shake table of requires size and capacity, design of box contained (if embedded parts in soil), design of scaled model using similitude rule/scaling laws. The loading platform and box contained should be designed suitably considering natural frequency, so that the resonance conditions can be avoided during model testing under earthquake loading or dynamic loadings. Adopting low-scaling factors resulted in a high mass of the model,

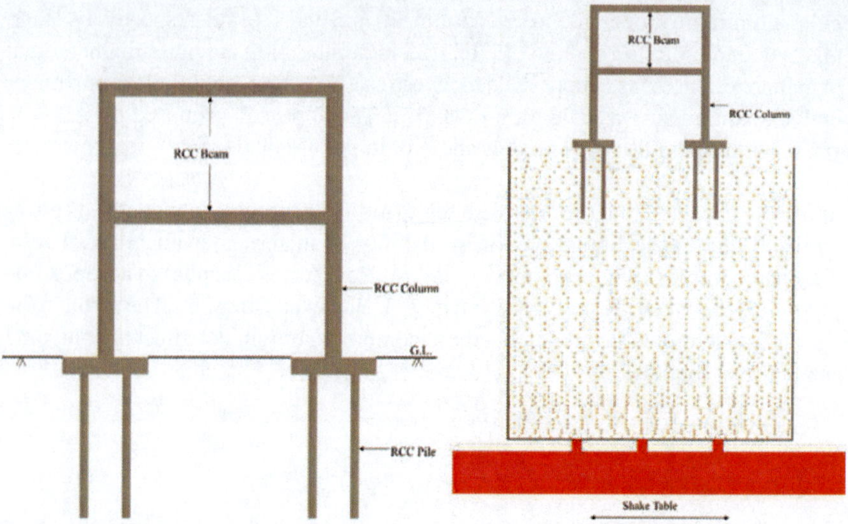

Fig. 3 Schematic representation of prototype structures and scaled-down model for pile foundation with superstructure using a shake table

Table 1 Important shake table testing facilities with details

Location	Size	Payload Capacity
E-defence Miki city in japan	20 feet × 150 feet	1200 MT
University of California UCSD	20 feet × 40 feet	6800 kN
Sharif University of Technology	4.0 m × 4.0 m	500 kN
Buffalo university	7.0 m × 7.0 m	50 MT
IISC Bangalore	1.2 m × 1.2 m	30 kN
IGCAR	6.0 m X 6.0 m	100 MT

whereas small-scaling factors scaling the time results in a high frequency of input motion in the model test. In designing the scaled model, it is very essential to keep in mind the desired outcomes, expected interaction and expected failure mechanism. The various important components of the shake table experiment set up are briefed in the subsequent section.

3.1 Actuator

Shaking tables are usually driven by servo-hydraulic actuators operated with hydraulic pumps. The hydraulic power pack is also to delivering of pressurized oil flow at a rated pressure for the shaking table system. The shake table is operated for required dynamic motion, viz. pure sinusoidal, variable amplitude harmonic or random real earthquake motion.

3.2 Box Container

The geotechnical model such as a pile foundation cannot be directly mounted on a shaking table and placed inside box containers because of the requirements of confinement. To simulate the free-field response of soil, various soil tanks have been proposed by researchers to minimize the boundary effect of the finite tank. The tanks are of

1. Rigid tank with sufficiently large size [15]'
2. Rigid tank packed with foam at sides of tank [16, 17]'
3. Laminated soil tank [18, 19],
4. Soil tank with walls having a hinged-base Iwasaki [20] and
5. Flexible container [21].

In the rigid tank, the wall is fixed in nature; however, in the flexible tank, the flexibility of the wall increased. The flexibility of the wall is increased by setting each layer on the other within the frame which can easily move on each other or applying a rubber wall reinforced with rigid frame/rings. In a flexible tank, the stiffness of the wall is proportional to the stiffness of the tank. In a typical flexible tank, laminar box is a large-sized shear box consisting of several horizontal layers, built in such a fashion that the friction between the layers is minimum. Hence, the layers move relative to one another in accordance with the deformation of the soil. In a rigid box container during excitations, the region near the boundary of the tank is more affected than the region far from the boundary. There is no conclusion drawn that which particular tank system is better than others. As proposed by Lu et al., the model should be placed in the container in the middle region that has little influence from the boundary [22]. Flexible materials are provided by various researchers during their experiments to minimize the wave reflection from a rigid boundary [23, 24]. Various typical soft materials, viz. sponge and expanded polystyrene (EPS), used along the end walls of the container used by researchers to modified the boundary condition of rigid tanks [23]. To minimize the reflection of waves into the soil from a rigid boundary, a flexible boundary can be provided by using foam board which is glued on the vertical side in the shaking direction of the tank as a typical presentation is made in Fig. 4.

Also, for better propagation of shear wave velocity through the medium of soil, it is recommended to glue the coarse sand on the foam face [23, 24].

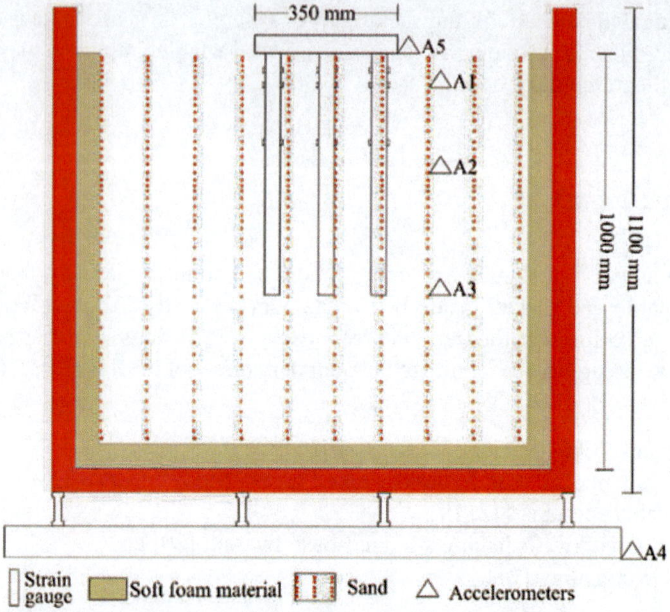

Fig. 4 Provision of foam material in rigid box to minimize wave reflections from the boundary

3.3 Soil Sample Preparation

Soil may show dilatative behaviours at low stress and shows contractive behaviours under high normal stress. Hence, it is very essential to model geotechnical problems that must be accounted for the stress-dependent behaviours. High gravitational stress cannot be developed in a reduced-scale model using 1-*g* shake table test. Hence, stress dependency problems can be addressed by using changing the density of model soil in 1-*g* tests. As suggested by Kelly et al. and Leblane et al., the issues can be addressed by preparing the soil model of low density by using suitable pouring methods [25, 26]. For physical model tests conducted for the study of liquefaction and associated large ground deformations such as lateral spreading, to employ extremely loose sand which is much looser than the in-situ sand, so that the effect of reduced stress in the scaled model would be compensated for reduced density [12]. Soil samples are prepared using various techniques such as wet or dry pluviation to achieve uniform samples of desired density. A typical presentation of Leighton buzzard sand comparing the relative density and peak frictional angle of soil to achieve required effective stress is presented by Bhattacharya et al. [27]. Based on this approach, if 120 kPa of mean effective prototype stress at 50% relative density is to be modelled in a small-scale laboratory model at 25 kPa stress, the sand is to be poured at about 39% relative density ensuring that the peak friction angle is the same. There is a limitation of approach in this case that there is a minimum density beyond which sand cannot be poured and other secondary effects become important.

4 Centrifuge Tests

The centrifuge model test is an advanced physical modeling technique that provides physical data for investigating mechanisms of deformation and failure of geotechnical problems for validating analytical and numerical methods. The centrifuge (*N*-*g*) tests are considered as special tests in which the gravity effect is artificially increased using the centrifuge principle (recreate stress condition similar to prototype) as they are spun around in the centrifuge as shown in Fig. 5 to develop true similarity between stress–strain between model and prototype. In a high gravitational acceleration field produced by a centrifuge, realistic self-weight-induced stresses in a small-scale model can be achieved. Hence, the stress and strain distributions in the model is similar to that of a field situation. The centrifuge modelling provides advantages for geotechnical engineering in which a full-scale structure can be examined by a scaled-down model, while the required stress states describe appropriate soil behaviours or field situation properties of interest, viz. slope stability, retaining wall structures, tunnel stability and lateral earth pressure stability.

Stress replication in the $1/\lambda$ scaled model is approximately achieved by subjecting model components to an elevated "*gravitational*" acceleration, which is provided by centripetal acceleration ($R\omega^2 = ng$), where 'R' is the radius and 'ω' is the angular velocity of the centrifuge. For example, the stress beneath a 0.2-m-deep (0.6 ft) layer of model soil spun at a centrifugal acceleration of 50 g produces stresses equivalent to those beneath a 10-m-deep (32 ft) prototype layer of soil in earth's gravity. The shaking like earthquake scenarios can be developed in the centrifuge models using earthquake simulators. Geotechnical materials such as soil and rock have nonlinear mechanical properties that depend on the effective confining stress and stress history. The one to one scaling of stress enhances the similarity of geotechnical models and makes it possible to obtain accurate data to help solve complex problems such as earthquake-induced liquefaction, soil–structure interaction. The working principle is the ratio of effective stress in the soil in a model to prototype is unity.

Fig. 5 Schematic presentation of the centrifuge test

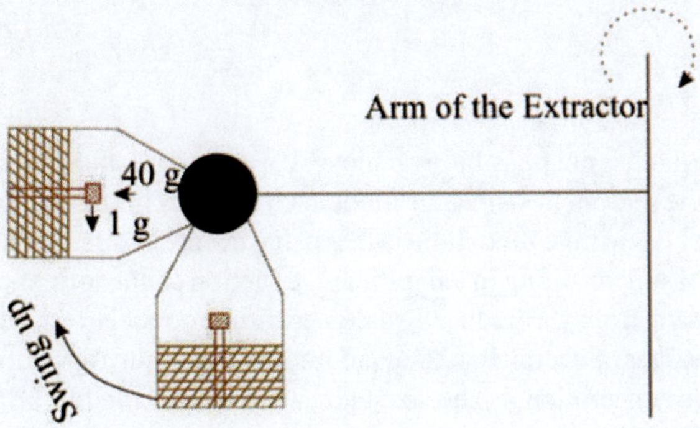

$$\frac{\sigma'_m}{\sigma'_p} = 1 \tag{8}$$

$$\sigma' = \sigma^t - U \tag{9}$$

σ^t is the total stress ($\rho g H$), and U is the pore water pressure ratio. In dry soil, U is zero.

$$\frac{(\rho g H)_m}{(\rho g H)_p} = 1 \tag{10}$$

If the same soil material is used in both prototype and model, then the ratio of density is unity. Hence,

$$\frac{(g)_m}{(g)_p} = \frac{(H)_m}{(H)_p} \tag{11}$$

Hence, if the geometrical scaling in model tests adopted as λ, then the gravitational force in the model is to be increased by the same scale to achieve the required condition of effective stress of both prototype and model. In the case of dynamic problems, all accelerations are also to be scale as gravity. A typical application of centrifuge is the case of soil stability analysis where the stability number N_s as expressed in Eq. (12) of model and prototype is essential to be the same.

$$N_s = \frac{C_u}{\gamma H} = \frac{C_u}{\gamma g H} \tag{12}$$

where C_u is the shear strength of soil and H is the height of fill. To achieve the same, in the 1-g tests reduced-scale model with a scaling factor of λ, the material used in the model should be such that

$$\frac{C_u}{\gamma} = 1/\lambda \tag{13}$$

which is not possible to achieve. In such a case, the gravity of the model is increased in the scale of λ, so that the model will behave similarly to the prototype. Another benefit of centrifuge modelling is that using a small-scale model that shortens drainage paths of soil, resulting in a significant reduction of consolidation time by $1/\lambda^2$. The observed data from the reduced-scale centrifuge models test can be scaled up to prototype values. Several big sizes and advanced centrifuge tables are available worldwide for experiments. The size and capacity of some important facilities are presented in Table 2.

Table 2 Available centrifuge testing facilities and details

Location	Details
University of California (UC), Davis	240 g tonne, 9 m radius, 4500 kg payload mass, up to 75 g
Hong Kong University of Science and Technology (HKUST)	400 g t, Bi-axial, 4.47 m radius, 150 g, 4 MT
Colorado Centrifuge Laboratory	5.6 m radius, 200 g, 2 MT
C-CORE, Memorial University of Newfoundland	5.5 m radius, 200 g, 2.2 MT
University of Delft	1.2 m radius, 300 g, 0.04 MT
Schofield Centre Cambridge	4.125 m radius, 150 g, 1 MT
ETH Zurich Geotechnical Drum Centrifuge	1.1 m radius, 440 g, 2 MT
RPI Geotechnical Centrifuge Research Centre	3.0 m radius, 200 g at 0.5 MT
Geotechnical Centrifuge Laboratory, IWHR (Beijing, China)	5.03 m radius, 300 g, 1.5 MT
IIT Bombay	4.5 m radius, 200 g, 2.5 MT, 250 g-Tone

5 Instrumentations During Tests

The objectives of the experiments are to measure various parameters such as strain, acceleration, velocity, pressure and pore pressure. Various instruments used in the model tests are given in Table 3.

To get the desired acceleration of the shake table, the input amplitude and frequencies in case of sinusoidal excitations are set as presented below.

$$X = A \times \mathrm{Sin}(\omega t) \tag{14}$$

Table 3 Instruments used to measure various parameters during experiments

Parameters	Instruments
Stain	stain gauges
Acceleration	Accelerometers
Velocity	Linearly varying differential transformer (LVDT)
Displacements	LVDT
Pressure	Pressure transducers
Pore pressure	Pore pressure transducers

$$\dot{X} = \omega \times A \times \mathrm{Cos}(\omega t) \tag{15}$$

$$\ddot{X} = -\omega^2 \times A \times \mathrm{Sin}(\omega t) = -(2\pi f)^2 \times A \times \mathrm{Sin}(2\pi f) \times t \tag{16}$$

where $\ddot{X}$ is the desired acceleration, f is base excitation frequency and A is excitation amplitude.

6 Extraction of Experimental Results

Accelerations as well as displacements of soil and pile, pore pressure in the soil at various points during tests can be observed directly from the readings of the instrument. However, the pile bending moment can be evaluated from the stain gauge readings as explained below.

$$\text{Pile bending Moment} = M = \frac{\mathrm{EI}}{R} = \mathrm{EIK} = \frac{\varepsilon_t - \varepsilon_c}{2y}\mathrm{EI} \tag{17}$$

The lateral pressure $P(z, t)$ on the pile can be extracted from the above-evaluated bending moment using Eq. (18).

$$P(z,t) = \frac{\mathrm{d}^2(M(z,t))}{\mathrm{d}z^2} \tag{18}$$

The total lateral force exerted on the pile $F(t)$ can be evaluated by the integration of the soil pressure along the pile as;

$$F(t) = \int_{Z=0}^{Z=L} P(Z,t)\,\mathrm{d}z \tag{19}$$

Displacement of the pile along the depth at depth z can be evaluated as;

$$y_p(Z,t) = \iint \frac{(M(Z,t))}{\mathrm{EI}}\mathrm{d}^2 z \tag{20}$$

7 Practical Application

It is required to extrapolate the prototype results from the model test results for practical application. It is very tricky to scale up the test results and evaluate the same for the prototype from the basics of scaling laws. In the present paper, a case study is also carried out in design the scaled model. Further, it is also presented extrapolating the prototype results from the models.

7.1 Design of the Scaled Model

Model tests are derived such that they are useful in practical applications. Hence, model test should replicate the behaviour and mechanism of the prototype. However, it is to be noted that the model is of a completely different structure may be made of different materials other than a prototype, different geometry and also different dynamic responses. Aluminium, steel, plastic and micro-concrete are used by researchers in model pile for RCC prototype piles during the model test. In the present study, various option of model pile for a prototype RCC pile of 1000 mm diameter has investigated. The graph showing the circular section model details of various materials (micro-concrete ($E = 25$ GPa), steel ($E = 200$ GPa), aluminium ($E = 69$ GPa), plastic ($E = 2.5$ GPa)) for a prototype RCC pile ($E = 27.5$ GPa) of 1000 mm are presented in Fig. 6.

The details of one accepted model pile and soil for a typical case to perform model test with scaling factor of 20 are given in Table 4.

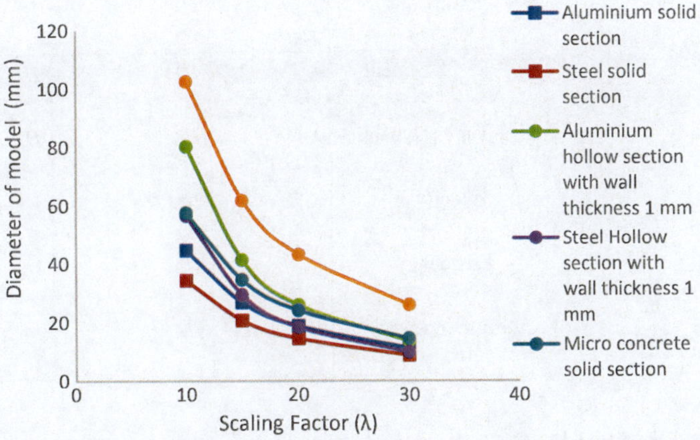

Fig. 6 Variation of the diameter of model piles of different materials for an RCC pile of 1000 mm diameter prototype pile with geometrical scaling

Table 4 Details of a typical case representing prototype and model

	Parameters	Prototype	Model
Soil	Depth	20 m	1.00 m
	Shear wave velocity	200 m/s	94.57 m/s
Pile	Material	Concrete	Aluminium
	Section	Solid	Hollow circular
	Ep	27.5 GPa	69 GPa
	Diameter	1000 mm	27 mm (OD), 1.2 mm wall thickness
	Moment of Inertia (*I*)	4.9×10^{10} mm^4	8.1×10^4 mm^4
	Length	15 m	0.75 m

Table 5 Scaling factor for the shake table test [28]

Quantity	General	1-g (Laboratory)	N-g (centrifuge)
Length	λ_l	$1/\lambda$	$1/\lambda$
Density	λ_ρ	1	1
Acceleration	λ_g	1	1
Stiffness	λ_G	$1/\lambda^\alpha$	1
Stress	$\lambda_\rho \lambda_g \lambda_l$	$1/\lambda$	1
Force	$\lambda_\rho \lambda_g \lambda_l^3$	$1/\lambda^3$	$1/\lambda^2$
Strain	$\lambda_\rho \lambda_g \lambda_l / \lambda_G$	$1/\lambda^{1-\alpha}$	1
Displacement	$\lambda_\rho \lambda_g \lambda_l^2 / \lambda_G$	$1/\lambda^{2-\alpha}$	$1/\lambda$
Time (Dynamic)	$\lambda_l (\lambda_\rho / \lambda_G)^{1/2}$	$1/\lambda^{1-\alpha/2}$	$1/\lambda$
Velocity	$\lambda_g \lambda_l (\lambda_\rho / \lambda_G)^2$	$1/\lambda^{1-\alpha/2}$	1
Frequency	$(\lambda_G / \lambda_\rho)^{1/2} / \lambda_l$	$1/\lambda^{1-\alpha/2}$	$1/\lambda$
Shear wave velocity	$(\lambda_G / \lambda_\rho)^{1/2}$	$1/\lambda^{\alpha/2}$	1

7.2 Results and Discussion

As per Muir Wood [28], the scaling factors and magnitude using scaling laws for the geotechnical model are given in Table 5.

The results of various parameters can be evaluated from model test results using the scaling factor presented in Table 5. The displacement observed in the model is to be multiplied by a geometrical scaling factor (λ) to obtain the same in the prototype. Similarly, the natural frequency of the prototype is evaluated as λ times the frequency of the model.

8 Conclusions

It is observed that there are certain advantages such as simplification, cost-saving and time-saving in the model tests. It can be adopted in a more complex problem where analytical model development is not possible. It is a universally accepted method to validate the theoretical models. However, some limitations are inbuilt in model tests. Full compliance of similarity in the model is not feasible. Some important but unknown parameters and interactions may not have captured during tests. The availability of a test facility may be one of the limitations in the design of the model. The present study will useful to the researchers, practitioners and designers in developing and planning model tests.

Acknowledgements The authors are grateful to Homi Bhabha National Institute (HBNI) for providing academic support and BARC for providing technical support for the present study.

References

1. Rocha M (1957) The possibility of solving soil mechanics problems by use of models. In: Proceeding of 4th.I.C.S.M.F.E., vol 1, pp 183–188
2. Roscoe (1968) Soil and model test. J Strain Anal 3(1):57–64
3. Kagwa T (1978) On the similitude in model vibration tests of earth structures. Proc Japan Soc Civil Eng 275:69–77
4. Kokusho T, Iwatate T (1979) Scaled model tests and numerical analyses on nonlinear dynamic response of soft grounds. Proc Japan Soc Civil Eng 285:57–67
5. Iai S (1989) Similitude for shaking table tests on soil-structure-fluid model in 1g gravitational field. Soil Found Jpn Soc Soil Mech Found Eng 29(1):105–118
6. Iai S, Tobita T, Nakhara T (2005) Generalized scaling relations for dynamic centrifuge test. Geotechnique 55(5):355–362
7. Casaburo A, Petrone G, Fransco F, Rosa S (2019) Areview of similitude methods for structural engineering. Appl Mech Rev 71(3):1–32
8. Iai S, Tobita T (2004) Centrifuge model tests on group piles in liquefiable and nonliquefiable ground. In: 13 th world conference on earthquake engineering, Vancauver, Canada, Paer no. 652
9. Buckingham E (1914) On physically similar systems: illustrations of the use of dimensional equations. Phys 4:345–376
10. Bhattacharya S, Wood DM, Lombardi D (2011) Similitude relationships for physical modelling of Monopile-supported offshore turbine. Int J Phys Model Geotech 11(2):58–68
11. Baker WE, Westine PS, Dodge FT (1991) Similarity methods in engineering dynamics: theory and practice of scale modeling, Rev ed. Elsevier, Amsterdam, The Netherlands
12. Haeri S, Kavand A, Rahami I (2012) Response of a group of piles to liquefaction-induced lateral spreading by large scale shake table testing. Soil Dyn Earthq Eng 38:25–45
13. Alves M, Oshiro R (2006) Scaling impacted structures when the prototype and the model are made of different material. Solid Struct 43:2744–2760
14. Verdugo R, Ishihara K (1996) The steady state of sandy soils. Soils Found 36(2):81–91
15. Sato H, Kouda M, Yamashita T (1996) Study on nonlinear dynamic analysis method of pile subjected to ground motion, Part-2: comparison between theory and experiment. In: Proceeding of the 11th world conference on earthquake engineering, Paper no. 1289

16. Mizuno H, Sugimoto M, Mori T, Iiba M, Hirade T (2000) Dynamic behaviour of pile foundation in liquefaction process-shaking table test utilising big shear box. In: Proceedings of the 12th world congress in earthquake engineering, Paper no. 1883
17. Finn WD, Gohl WB (1992) Response of model pile groups to strong shaking, pile in dynamic loads. Geotech Special Publ ASCE 34:27–55
18. Tamura S, Suzuki Y, Tsuchiya T, Fujii S, Kagawa T (2000) Dynamic responses and failure mechanism of a pile foundation during soil liquefaction by shaking table test with a large-scale laminar shear box. In: Proceeding of the 12th world congress on earthquake
19. Yasuda S, Ishihara K, Morimoto I, Orense R, Ikeda M, Tamura S (2000) Large scale shaking table tests on pile foundation in liquefied ground. In: Proceedings of the 12th world congress on earthquake engineering, Paper no. 1474
20. Iwasaki T (1986) Soil liquefaction studies in Japan, state of the art. Soil Dyn Earthquake Eng 5:28–33
21. Meymand PJ (1998) Shaking table scale model test on nonlinear soil-pile-superstructure interaction in soft clay. PhD dissertation, University of California, Berkeley
22. Lu XL, Chen Y, Chen B, Huang W, Zhao L (2000) Shaking table testing of dynamic soil-structure interaction system. Earthquake Eng Eng Vibr 4:20–29
23. Bhattacharya S, Lombardi D, Dihoru L, Dietz MS, Crew AJ (2012) Model container design for soil-structure interaction studies. In: Geotechnical, geological and earthquake engineering series workshop. Springer, pp 135–158
24. Lombardi D, Bhattacharya S, Scarpa F, Bianchi M (2014) Dynamic response of a geotechnical rigid model container with absorbing boundaries. Soil Dyn Earthquake Eng 46–56. https://doi.org/10.1016/j.soildyn.2014.09.008
25. Kelly R, Houlsby G, Byrne B (2006) A comparison of field and lab tests of caisson foundation in sand and clay. Géotechnique 56(9):617–626
26. Leblanc C, Byrne B, Houlsby G (2010) Response of stiff piles to random two-way lateral loading. Géotechnique 60(9):715–721
27. Bhattacharya S, Lombardi D, Dihoru L, Dietz M, Crewe A, Taylor C (2012) Model container design for soil-structure interaction studies. In: Role of seismic testing facilities 135, geological and earthquake engineering, vol 22. https://doi.org/10.1007/978-94-007-1977-4_8
28. Wood DM (2004) Geotechnical modelling. Oxfordshire

Liquefaction Potential Index for Lateritic Soils in Goa

Leonardo Souza and Purnanand Savoikar

Abstract Earthquakes carry with them the hazard of earthflows due to liquefaction. The evaluation of liquefaction phenomenon and related ground failures is an important aspect of geotechnical earthquake engineering. The tendency of soils with high sand content to suddenly lose all its strength due to violent shaking caused by earthquakes is measured using a liquefaction potential index (LPI). The factor of safety against liquefaction has been widely researched by Seed, Idris, Youd, Iwasaki, etc. This paper will attempt to estimate the LPI for different cities in Goa. Liquefaction resistance can be measured using standard penetration test (SPT), cone penetration test (CPT), and shear wave velocity (Vs). This paper will use SPT-based calculations for Goa region.

Keywords Liquefaction · Liquefaction potential index · LPI · Lateritic soils · Seismic hazard Goa · SPT

1 Soil Liquefaction

1.1 Liquefaction

Soil liquefaction is loss of strength and/or stiffness due to undrained loading which causes the soil to flow like liquid. It is a major factor in earthquake damages. Liquefaction occurs in saturated soils. The water in the pores exerts the pressure on soil particles called pore pressure that influences how tightly the soil particles are pressed together. Prior to an earthquake, the pore pressure is low. Earthquake shaking causes pore pressure to increase due to the undrained condition. Pore pressure equals the ratio of overburden press to confining stress, and the effective stress is zero, i.e., there is no shear strength! Soil particles start to readily move with respect to each other due to zero shear strength and liquefaction occurs!!

L. Souza · P. Savoikar (✉)
Department of Civil Engineering, Goa Engineering College, Farmagudi 403 401, Goa, India

T. G. Sitharam et al. (eds.), *Earthquake Geotechnics*, Lecture Notes in Civil Engineering 187, https://doi.org/10.1007/978-981-16-5669-9_34

Fig.1 Liquefaction of soil

In an extreme case, the pore-water pressure may become so high that many of the soil particles lose contact with each other. In such cases, the soil has very little strength and behaves like a liquid; this process is called "Liquefaction" (Fig. 1).

1.2 Manifestation of Liquefaction

It manifests as

- Sand boils
- Ground cracks
- Slumping of embankments
- Lateral spreading
- Ground oscillations
- Flow slides.

1.3 Types of Liquefaction

Liquefaction is always statically or dynamically triggered. It occurs in waterlogged soils. There are two types of liquefactions such as

1. Cyclic (seismic) liquefaction—caused by zero effective stress (during cyclic loading)
2. Flow (static) liquefaction—caused by strain softening response.

Following the 1964 earthquakes in Alaska and Niigata, Japan, the "simplified procedure" was developed by Seed and Idriss [1] for evaluating seismic demand and liquefaction resistance of sands.

- CSR = 0.65 $(a_{\max}/g)$ (s_v/s'_v) r_d
- CRR = as a function of penetration resistance (SPT & CPT).

A NCEER Workshop to develop consensus update paper [2] was conducted which is the basis of most liquefaction studies in the world today. The same method is recommended by IS:1893 (Part-1)-2016 and is used in this paper.

2 Liquefaction Potential for Soils in Goa

2.1 Liquefaction Potential in Goa

The soils of Goa are lateritic in nature. They widely vary from place to place and even within 10 m of each bore. Sample calculation has been done for representative soil samples as study of all soils is beyond the scope of this paper. Calculations for this paper have been done on the basis of [2] as recommended by IS:1893 (Part-1)-2016.

Horizontal acceleration due to earthquake is

$$a_h = \frac{ZIS_a}{2Rg} \tag{1}$$

Goa lies in zone III, so z is 0.24

For TGSB revetment I, value can be taken as 1.5 (Clause 6.4.2 in IS:1893 (Part-1)-2016 as it is an important structure); for other uses, it is taken as 1.

Response reduction factor R is usually for buildings. The value I/R is taken as 1 (Clause 6.4.2 IS:1893 (Part-1)-2016).

Average acceleration response coefficient S_a depends on the period of the earthquake, and type of soil is given by 1.36 T. For Goa, consider medium soils, period can be taken as 0.71, and S_a/g is 0.97.

The maximum possible horizontal acceleration for Goa region is $a_h = 0.11$ g, so the horizontal acceleration factor $k_h = 0.11$.

The stress in the soil caused by the earthquake is found out first.

$$\text{CSR} = \frac{\tau_{av}}{\sigma'_{vo}} = 0.65\frac{a_{\max}}{g} r_d \frac{\sigma_{vo}}{\sigma'_{vo}} \tag{2}$$

$$r_d = 1.174 - 0.0267z\text{for } 9.15 < \text{z} < 23 \text{ m} \tag{3}$$

Cyclic resistance is found next from the standard penetration test (SPT-N), cone penetration test (CPT-q_u), or shear wave velocity (V_s) plots.

Factor of safety against liquefaction is given by the ratio of resistance to strength.

$$Fs = \frac{\text{CRR}}{\text{CSR}}\text{MSF} \tag{4}$$

MSF for lateritic soils can be found out from Table 1 considering an earthquake of magnitude 5.5.

Table 1 Magnitude correction factor

Earthquake magnitude	8.5	8.0	7.5	7.0	6.5	6.0	5.5
MSF	0.72	0.84	1.00	1.19	1.44	1.76	2.20

Table 2 LPI classification

LPI	Occurrence potential classification
0–2	Low
2–5	Moderate
5–15	High
> 15	Very high

2.2 *Liquefaction Potential Index (LPI)*

The potential for liquefaction can be defined using the liquefaction potential index (LPI) suggested by Iwasaki et al. [3]. It depends on the factor of safety (F_s) taken for every (max) 10 m and effective depth of layer (w_z) normalized for 10 m (Table 2).

Liquefaction potential index is given by

$$\text{LPI} = \int_0^z F_L w_z dz \tag{5}$$

$$w_z = 10 - \frac{z}{2} \geq 1 \tag{6}$$

$$\begin{aligned} F_L &= 1 - F_s \text{for} F_s \leq 1 \\ &= 0 \text{for} F_s > 1 \end{aligned} \tag{7}$$

2.3 *Study Area*

Based on the above, the liquefaction potential index was found out for seven different sites in Goa (Fig. 2). CPT and V_s are usually found by co-relationships when test results are not available [4–7]. They are dependent on soil properties and energy of SPT. Borehole data were collected about the SPT-N values at 0.45, 5, 10, 15, and 20 m, while CPT shear wave velocity was found by suitable correlations (Table 3). The sites chosen in Vasco, Ponda, and Canacona Goa are for studying the liquefaction potential index of plateau areas, while the sites in Mapusa, Panaji, and Margao Goa are for studying the liquefaction potential index of lower lying areas. Bicholim site is chosen for studying the liquefaction potential index of higher lying areas. Based on this study, a rough idea of liquefaction potential for the State of Goa can be estimated.

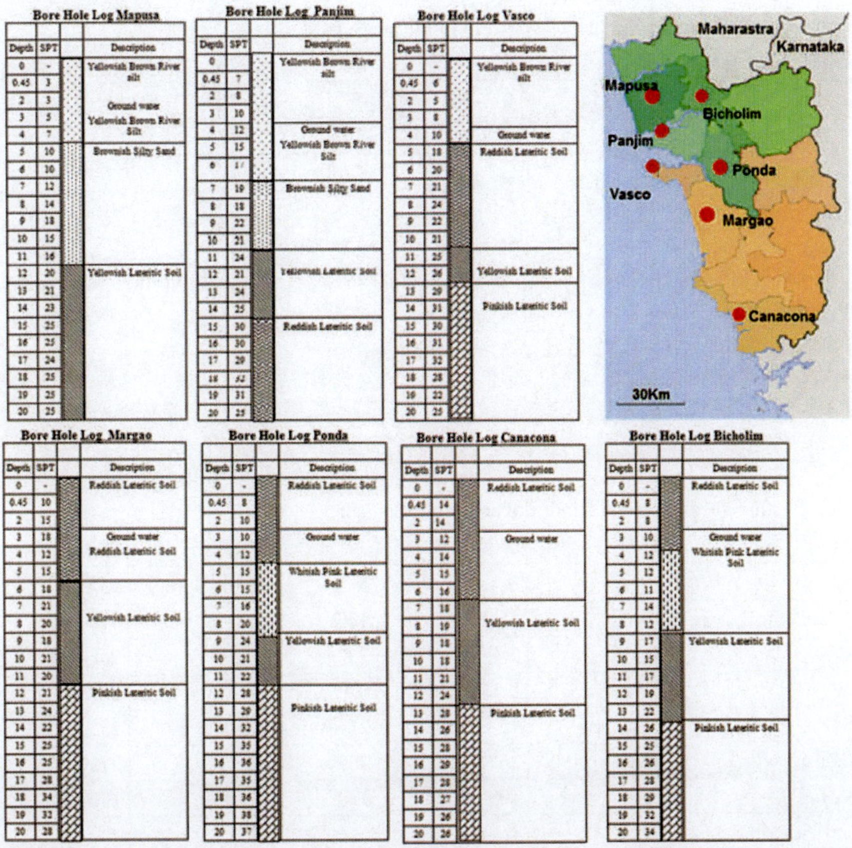

Fig. 2 Map showing locations of boreholes and typical borehole logs at Mapusa, Panaji, Vasco, Margao, Ponda, and Canacona and Bicholim

Table 3 Co-relationships between SPT, CPT, and versus assumed for different cities in Goa

Site	q_u (kN/m^2)	V_s (m/s)
Mapusa	2.40 $N^{0.011}$	60 + 3.00 N
Panaji	2.60 $N^{0.011}$	60 + 2.85 N
Vasco	2.60 $N^{0.011}$	60 + 2.85 N
Margao	2.60 $N^{0.011}$	60 + 2.39 N
Ponda	2.89 $N^{0.011}$	60 + 3.00 N
Canacona	2.46 $N^{0.011}$	60 + 1.80 N
Bicholim	2.51 $N^{0.011}$	60 + 2.54 N

Table 4 Liquefaction potential index of soil profile at Mapusa

z (m)	0.45	5	10	15	20
z_w (m)	0	0	5	10	15
γ (kN/m^3)	15.2	15.6	16.8	17.5	17.3
G_s	2.25	2.3	2.48	2.5	2.49
e_0	0.54	0.53	0.6	0.56	0.55
STP-N_{60}	3	10	15	25	25
CPT-q_u	2430	2450	2470	2490	2485
V_s (m/s)	70	90	105	135	135
LPI-N_{60}	−32.212	$< 1 = 0$			
LPI-q_u	−39.685	$< 1 = 0$			
LPI-V_s	−34.319	$< 1 = 0$			

Table 5 Liquefaction potential index of soil profile at Panaji

z (m)	0.45	5	10	15	20
z_w (m)	0	0	5	10	15
γ (kN/m^3)	16.5	16.7	17.2	17.6	17.4
G_s	2.4	2.4	2.5	2.4	2.43
e_0	0.52	0.53	0.55	0.58	0.55
STP-N_{60}	7	15	21	30	25
CPT-q_u	2660	2650	2690	2700	2750
V_s (m/s)	80	105	120	145	130
LPI-N_{60}	−51.307	$< 1 = 0$			
LPI-q_u	−49.422	$< 1 = 0$			
LPI-V_s	−52.407	$< 1 = 0$			

3 Liquefaction Potential Index of Soil Profiles in Goa

Tables 4, 5, 6, 7, 8, 9, and 10 give the calculation for liquefaction potential index for soils of Goa from three criteria: SPT - N_{60}, CPT - q_u, and V_s.

4 Conclusions

It is found that the liquefaction potential for all sites studied in Goa considering an earthquake of 5.5 magnitude was less than zero. As the State of Goa has not experienced earthquake greater than 1 magnitude so far, this is much above the factor of safety limits. This indicates that lateritic soils of Goa are not liquefiable, for the sites chosen in the present study.

Table 6 Liquefaction potential index of soil profile at Vasco

z (m)	0.45	5	10	15	20
z_w (m)	0	0	5	10	15
γ (kN/m^3)	16.5	17	17.7	18.4	18.6
G_s	2.33	2.35	2.42	2.4	2.45
e_0	0.48	0.5	0.55	0.5	0.55
STP-N_{60}	6	18	21	25	27
CPT-q_u	2650	2680	2680	2690	2700
V_s (m/s)	75	110	120	130	135
LPI-N_{60}	−49.302	$< 1 = 0$			
LPI-q_u	−49.147	$< 1 = 0$			
LPI-V_s	−49.767	$< 1 = 0$			

Table 7 Liquefaction potential index of soil profile at Margao

z (m)	0.45	5	10	15	20
z_w (m)	0	0	5	10	15
γ (kN/m^3)	17.5	16.9	17	17.6	18
G_s	2.36	2.41	2.42	2.45	2.43
e_0	0.52	0.53	0.56	0.55	0.57
STP-N_{60}	10	15	21	25	28
CPT-q_u	2670	2680	2690	2690	2700
V_s (m/s)	85	95	110	120	125
LPI-N_{60}	−46.898	$< 1 = 0$			
LPI-q_u	−46.694	$< 1 = 0$			
LPI-V_s	−47.060	$< 1 = 0$			

Table 8 Liquefaction potential index of soil profile at Ponda

z (m)	0.45	5	10	15	20
z_w (m)	0	0	5	10	15
γ (kN/m^3)	16.5	16.8	17.2	18.8	19.2
G_s	2.4	2.4	2.45	2.51	2.55
e_0	0.52	0.5	0.48	0.49	0.53
STP-N_{60}	8	15	21	35	37
CPT-q_u	2950	2970	2990	3005	3010
V_s (m/s)	84	105	123	165	171
LPI-N_{60}	−72.680	$< 1 = 0$			
LPI-q_u	−72.487	$< 1 = 0$			
LPI-V_s	−72.408	$< 1 = 0$			

Table 9 Liquefaction potential index of soil profile at Canacona

z (m)	0.45	5	10	15	20
z_w (m)	0	0	5	10	15
γ (kN/m^3)	15.2	15.8	17.5	18.3	18.6
G_s	2.22	2.36	2.4	2.53	2.54
e_0	0.5	0.53	0.55	0.54	0.58
STP-N_{60}	14	15	18	28	26
CPT-q_u	2530	2530	2540	2550	2550
V_s (m/s)	85	90	95	105	105
LPI-N_{60}	−45.364	$< 1 = 0$			
LPI-q_u	−45.806	$< 1 = 0$			
LPI-V_s	−45.628	$< 1 = 0$			

Table 10 Liquefaction potential index of soil profile at Bicholim

z (m)	0.45	5	10	15	20
z_w (m)	0	0	5	10	15
γ (kN/m^3)	16.6	16.5	16.8	17.6	17.9
G_s	2.32	2.38	2.4	2.45	2.5
e_0	0.52	0.49	0.51	0.52	0.55
STP-N_{60}	8	12	15	25	34
CPT-q_u	2570	2580	2585	2600	2610
V_s (m/s)	80	90	100	120	145
LPI-N_{60}	−43.541	$< 1 = 0$			
LPI-q_u	−43.428	$< 1 = 0$			
LPI-V_s	−44.133	$< 1 = 0$			

References

1. Seed HB, Idriss IM (1971) Simplified procedure for evaluating soil liquefaction potential. J SoilMech Found Div ASCE 97(SM8):1249–1274
2. Youd TL et al (2001) Liquefaction resistance of soils: Summary report from the 1996 NCEER and 1998 NCEER/NSF workshops on evaluation of liquefaction resistance of soils. J Geotech Geoenviron Eng 127(10):817–833. https://doi.org/10.1061/(ASCE)1090-0241(2001)127:10(817)
3. Iwasaki T, Tokida K, Tatsuko F, Yasuda, S (1978) A practical method for assessing soil liquefaction potential based on case studies at various sites in Japan. In: Proceedings of 2nd international conference on microzonation, San Francisco, 885–896, 1978
4. Robertson PK, Campanella RG, Wightman A (1983) SPT-CPT Correlations. J Geotechn Eng 109(11)
5. Tarawneh B (2014) Correlation of standard and cone penetration tests for sandy and silty sand to sandy silt soil. EJGE Bund W 19:6717–6727
6. Guen O (2014) Correlation between SPT and CPT, Geotechnics and Geohazards, master's thesis NTNU, Norwagien University of Science and Technology

7. Maibam TS, Rastogi BK, Dogra NN, Pancholi V(2017) Empirical correlation of SPT blow counts versus shear wave velocity for different types of soils in Dholera, Western India, Natural Hazards 86(3) January 2017 https://doi.org/10.1007/s11069-017-2744-3
8. Seed HB (1979) Soil liquefaction and cyclic mobility evaluation for level ground during earthquakes. J Geotech Eng Div ASCE 105(GT2):201–255
9. Seed HB, Idriss IM (1982) Ground motions and soil liquefaction during earthquakes. Berkeley, CA: Earthquake Engineering Research Institute, p 134
10. Seed HB, Idriss, IM, Arango I (1983) Evaluation of liquefaction potential using field performance. J Geotech Eng, ASCE 109(3):458–482
11. Seed HB, Tokimatsu K, Harder LF, Chung, RM (1984) The influence of SPT procedures in soil liquefaction resistance evaluations. Earthquake Engineering Research Center Rep. No. UCB/EERC-84/15, Univ. of California, Berkeley, Calif
12. Idriss IM, Boulanger RW (2006) Semi-empirical procedures for evaluating liquefaction potential during earthquakes. Soil Dyn Earthquake Eng 26:115–130
13. Idriss M (1999) An update to the Seed-Idriss simplified procedure for evaluating liquefaction potential. Proc., TRB Workshop on New Approaches to Liquefaction, January, Publication No. FHWA-RD-99-165, Federal Highway Administration
14. IS 1893 (Part I) Indian Seismic Code Bureau Of Indian Standards, New Delhi

Need for Artificial Intelligence in Geotechnical Earthquake Engineering

Leonardo Souza and Purnanand Savoikar

Abstract This paper reviews the use and necessity of artificial intelligence (AI) in today's environment of geotechnical earthquake engineering (GEE). As population boom in earthquake sensitive zones the loss of life and property is on the increase. There are an increasing number of populated problematic sites and a limited number of human experts in geotechnical earthquake engineering (GEE). Automated intelligence monitored interpretation, warning and mitigation systems are required. This paper attempts to review knowledge-based system to enable us to put in place an automatic process of interpretation and determining of the risk potential then alerting the stakeholders and initiating mitigation operations. Remote monitoring of slopes movements and deformations, and automated linkage of seismographs together with coordination with state and national disaster relief teams can help in timely warning and damage reduction. Scenarios of modeling, simulation and nonlinear seismic analysis can be applied to damage models for typical of the existing urban infrastructure of India. Data mining techniques, decision trees, Geographical Information Systems and existing open source Software modules can be explored for developing a reliable model for India, incorporating machine learning in GEE. This paper will review some work already done in this area to underline the need for AI in GEE.

Keywords Artificial Intelligence · Geotechnical earthquake engineering · Disaster mitigation · Earthquake hazard mitigation · Artificial neural networks

1 Artificial Intelligence or AI

1.1 Introduction to AI

With the invention of computers their capability and power of computer systems in diverse working domains grew exponentially with increasing speed, and reducing size and time. With it grew the desire to make computers as human as possible.

L. Souza · P. Savoikar (✉)
Department of Civil Engineering, Goa Engineering College, Farmagudi 403 401, Goa, India

T. G. Sitharam et al. (eds.), *Earthquake Geotechnics*, Lecture Notes in Civil Engineering 187, https://doi.org/10.1007/978-981-16-5669-9_35

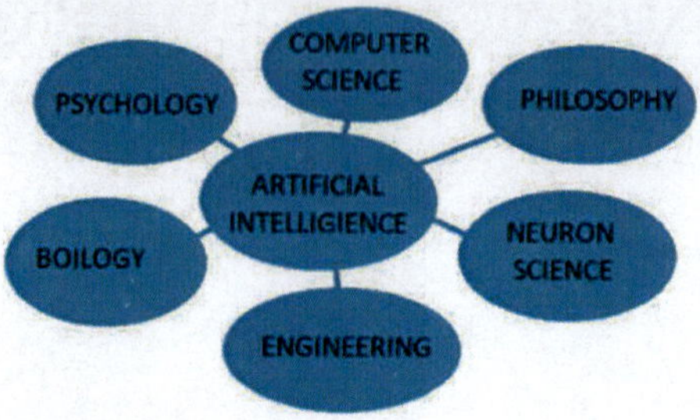

Fig. 1 Different technologies contributing to AI

Artificial intelligence (AI) is a branch of Computer Science aimed at creating the computers as intelligent as human beings. John McCarthy is credited with the concept of AI as early as 1940's, yet the full understanding of the AI is yet an unexplored domain which engineers are now beginning to apply thanks to the power unleashed by the internet. The aim of AI is to make computer/machines think like humans do not to supplement but to complement humans.

AI is a hotchpotch of various technologies and sciences (see Fig. 1), from computer sciences to biology to psychology to philosophy.

The applicability of AI is vast as human imagination and capacity from Robotics to finding our way using Google Maps. Today we see the applications already in day to day life. This paper will attempt to review some select papers to identify the need for AI in geotechnical earthquake engineering (GEE).

1.2 Beginners Introduction to AI

Many websites and online courses are available for a beginners understanding of AI.

Tutorial point (India) offers one such resource that deals with concept, goals and philosophy and techniques of AI. It explains intelligent systems, research areas of AI, environments, algorithms, fuzzy logics systems, natural language processing, expert systems, and neural networks [tutorial point] [1].

1.3 Understanding AI and Neural Networks

Kriesel [2] gives a detailed explanation of neural networks and its contribution to artificial intelligence. A simple 100 step rule can be used to program a network to behave like a learning human brain is explained with examples. The similarity in working to biological networks is elucidated. The various parts of neural networks connections, propagation, activation, threshold value, activation functions, output and learning strategy are dealt with in simplicity. We learn of synchronous and asynchronous activation. Supervised and unsupervised learning paradigms are explained with all their features. The additional features needed for neural networks like excursi, appendices and registers are also clearly clarified. All the necessary equations needed for neural networks can be obtained from this detailed work.

The most common type of networks are the feed forward network (see Fig. 2) where the neurons are grouped and completely linked in input layer, hidden (processing) layer and the output layer. Then, there are recurrence networks where a layer can influence itself directly or indirectly. The linkages can be lateral. The neurons are biased or repressing type. The order of receiving and processing the information hence becomes important.

Krose and Smagt [3] have also dealt extensively with the notations the treminology and theory of neural networks. They deal with perceptrons, adaptive linear element (adaline), back-propagation, recurrent networks, multilayered feed forward, self-organizing networks and reinforcement learning. They also briefly give the application to robotics and the hardware needed for the networks to work.

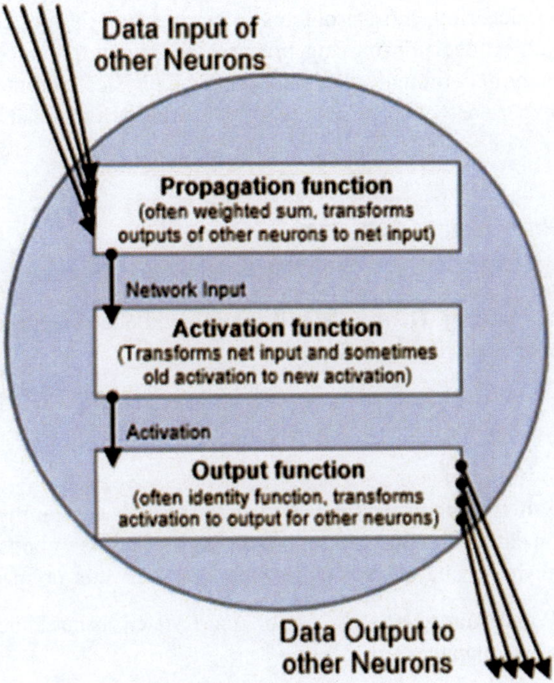

Fig. 2 Data processing of neuron where the activation function means threshold value [2]

2 AI in Civil Engineering

2.1 AI in Civil Engineering

Artificial Intelligence methodologies have extensively usage in the field of civil engineering. It can be used to optimize construction management, building materials, hydraulics, irrigation, geotechnical and transportation engineering and surveying.

It is mainly used in evaluation, diagnosis, prediction, decision-making, design and optimization. AI-based computational techniques use adaptive neuro-fuzzy inference systems which are particularly suitable for modeling complex systems with known or unknown input–output data sets especially to study the behavior of natural materials undergoing single, dual, or multiple damage factors. AI also helps in development of robotic automated systems and smart materials for use in Civil engineering [4]. The neural network is trained using cases collected from several engineering and construction firms and owner firms of industrial process plants. In the field of civil engineering, most problems in engineering design, construction management, and project decision-making, are affected by multiple uncertainties. These can be solved only in combination of mathematics, physics, and mechanics calculations with the experience of practitioners [5]. This difficulty is overcome by AI.

2.2 Smart Materials

Structural Health Monitoring—Embedded sensors within structures help to monitor stress and damage.

Structural Engineering—Fiber-optic sensors embedded in the structures identify the trouble areas. They can also be used to assess durability and environmental degradation.

Self-Repair materials—embedded tubules containing materials which are released on damage to fills and set cracks. Self-repair is important in inaccessible and hostile environments such as underground, underwater or space.

Waste management—use of smart materials could help to automate the process of waste management.

2.3 Artificial Neural Networks (ANNs)

Modular Construction Decision-Making: to help decide whether to use a conventional method or to use some degree of modularization in a particular project. Based on several decision attributes: organization, location, environment, labor, capital, characteristics, and risks.

Concrete mix design: Concrete mix design is sensitive to quality of the constituents, workability, desired strength and durability. AI makes it easier to design concrete mixes in a batching plant.

Estimation: ANNs are most suited for developing decision aids-based problem-solving in estimation and quantity surveying.

Initial RCC Design Process: The preliminary RCC design is of vitally important in construction planning. The ANN can predicts a good initial design (i.e., tensile reinforcement, beam/footing/slab depth, beam/column width, cost per meter and the moment capacity) for a given input parameter set (span, loads, concrete and steel grade) making the project efficient.

Planning Construction Projects: AI can generate plans in all the phases of the project planning.

Transportation Planning: In planning of a new rail river or road network, possible locations of bridges, tunnels, docks warehouses stations, etc., for maximum optimization, best alignment, effective location can be achieved by ANNs.

Rail/Pavement Maintenance: ANNs assist decision makers in selecting appropriate maintenance and repair strategy for railways and pavement by integrated collection of data, analysis and selecting appropriate maintenance and repair procedures.

Tidal Forecasting: Tidal level record is important in construction activity in maritime areas. It could be easily determined from only a small amount of historical tidal records and can be used for tide level forecasting using ANNs.

Earthquake prediction: During earthquakes civil structures (buildings, highway embankments and retaining structures) are damaged or completely destroyed. Therefore, estimation of the earthquake- is essential for the civil engineers.

2.4 Different Techniques Used in Civil Engineering

Due to a lot of uncertain and complicated influence factors in civil engineering, each project has its unique generality and character. This necessitates the function of special expert system in cases for a notable effect.

Genetic algorithms (GAs) are evolutionary algorithms which simulate the Darwinian principle of evolution are used for multi-objective optimization model for the scheduling of linear construction projects. Artificial immune system (AIS) stimulates the adaptive immune system of a living creature for reservoir management. Fuzzy polynomial neural networks (FPNN) can be to predict the compressive strength of concrete. Particle swarm optimization (PSO) is a population-based global optimization technique that enables a number of individual solutions, called particles, to move through a hyper dimensional search space to search for the optimum. Ant colony optimization (ACO) algorithm mimics the behavior of real ants living in colonies. Chaos Theory, firefly algorithm and Cuckoo Search are other innovative methods used for civil engineering projects.

3 AI in Geotechnical Engineering

3.1 AI in Geotechnical Engineering

Geotechnical engineering deals with natural materials. Their engineering properties show complex and uncertain behavior due to their varied formation processes. Some sources of uncertainty are errors in soil boring, incoherent soil composition, sampling procedures, in-situ testing, laboratory testing, loading effects, time effects, construction effects and human error. Such complexity and related uncertainties of geotechnical engineering can best be handled by using AI.

Neural networks have many important functions that help in modeling complex geotechnical behavior: universal function approximation capability, good generalization capability, accommodation of multiple nonlinear variables for unknown interactions and resistance to noisy or missing data sets [6]. In the field of geotechnical engineering, some problems are very complex and poorly understood. Classical modeling is unable to properly simulate the behavior of geomaterials [7].

3.2 Geotechnical Applications of AI and ANNs

Site characterization: The main objective of site characterization is the prediction of in-situ soil properties and prediction of spatial patterns in the depth to bedrock at any half-space point at a site based on limited tests.

Soil properties and behavior: There is an increasing in employment of ANNs in modeling soil properties and its behavior, correlation between the relative density and the cone resistance from cone penetration test (CPT), stress–strain relationship of sands with varying grain size, develop prediction models, etc.

Pile capacity Evaluation pile bearing capacity using empirical data of actual field case records to training the neural network.

Bearing capacity and settlement of foundations: The criteria governing foundation design are bearing capacity and settlement; neuro-fuzzy models can be used for prediction of shallow foundations.

Load-settlement response modeling: Actually, soil resistance and settlement are influenced by each other, and the design of foundations should consider the bearing capacity and settlement inseparably. The recurrent neural network (RNN) model can capture the highly nonlinear load-settlement response reasonably well.

Slope stability: Landslides or slope failures are important because they can lead to the loss of life and property. Limited data and unclearly defined problems often complicate the study of landslides. The results of the neural network showed better predicting when compared with the analytical model results.

Earth retaining structures: The bracing of excavation is one of the important matter in geotechnical engineering, could be temporary during the construction stage, or permanent as in underground structures or that adjacent to soil ramp.

Tunnels and underground openings: ANN can predict settlements and deflections in tunnels and underground opening accurately.

4 AI in Earthquake Engineering

4.1 AI in Earthquake Prediction

Earthquakes are generally the result of propagating seismic waves underneath the surface of the earth, recorded by the seismometers installed in different geographical positions. This is a most complex scenario as it involves the study of geophysical layers, its actual deformations and alteration in structural parameters with respect to different time points. This is best accomplished by AI. Bojórquez et al. [8] presented an ANN model is to estimate inelastic response spectra for earthquake acceleration records. The back-propagation algorithm was used to train the Network.

4.2 Earthquake Patterns

Earthquake patterns have been classified with respect to identified features using ANN-based networks. Both supervised and unsupervised techniques are utilized for prediction purpose. Using eight different identified seismic indicators, a probability model is built for future earthquakes by using a probabilistic neural network approach. The higher magnitude is predicted by slowly increasing the threshold value by 0.5. Each time, on encountering earthquake, values and matrixes related to ANN is updated in for each defined area. Sigmoid activation function is selected for experimentation under feed forward network topology. Back propagation-based learning algorithm is used for training the data [9]. Artificial Bee Colony algorithm has proved useful for solving different complex problems. Advanced inversion techniques using Particle Swarm optimization algorithm are used to find the local earthquake locations.

4.3 Liquefaction

During earthquakes liquefaction can cause ground failure and very dangerous damage to civil engineering structures. The neural networks are trained using actual field records from earthquakes that occurred during the preceding 100 year period. It gives 85 to 90% similar predictions to manual Codal methods.

4.4 Risk Mitigation

By considering geospatial data, it is possible to develop alternative scenarios to visualize the structural vulnerability of the urban built infrastructure. Risk assessment due to natural hazards such as earthquakes, hurricanes and floods is location dependent. Atanasiu et al. [10] present a GIS-based method to monitor seismic performance, considering the impact calculated during scenarios, to identify the seismic serviceability of the structure. A method for the simulation of structural seismic responses based on neural networks is given by Li et al. [11] modeling a linear system. Ornthammarath et al. [12] discuss the seismic response of tunnels to predict the earthquake-induced strain field at depth in near-fault conditions.

5 Need for AI in Geotechnical Earthquake Engineering

5.1 AI in Geotechnical Earthquake Engineering

Geotechnical earthquake engineering (GEE) mainly deals with the effect of earthquakes on geotechnical structures, slope stability and liquefaction. There are a number of earthen dams used all over the world and in India. There are numerous earthen embankments built for transportation infrastructure. There are many historic geotechnical structures existing in India like Traditional Goan Saraswat Bunds. India today has a vast network of tunnels and underground sewerage systems. All of these need to be evaluated for seismic stability before and after earthquakes. The regular monitoring has to be combined with continuous analysis and timely interventions. A number of possible scenarios have to be evaluated and the best one needs to be picked up. This is a laborious and time consuming work to do humanly. But it can easily be done using machine learning.

5.2 *Need for AI in Geotechnical Earthquake Engineering*

There are an increasing number of populated problematic sites and a limited number of human experts in GEE. Automated intelligence monitored interpretation, warning and mitigation systems are required. We need to put in place an automatic process of interpretation and determining of the risk potential then alerting the stakeholders and initiating mitigation operations. Remote monitoring of slopes movements and deformations, and automated linkage of seismographs together with coordination with state and national disaster relief teams can help in timely warning and damage reduction. Scenarios of modeling, simulation and nonlinear seismic analysis can be applied to damage models for typical of the existing urban infrastructure of India. Data mining techniques, decision trees, Geographical Information Systems and existing open source Software modules can be explored for developing a reliable model for India, incorporating machine learning in GEE.

6 Conclusions

AI has amply demonstrated its influence to outperform the conventional methods. AI has a number of significant benefits that makes it a practical powerful tool for problems solving in civil engineering to reduce the efforts, cost and management. The success of artificial intelligence applications has been demonstrated in geotechnical engineering. Based on the results of the research above, it can be said that artificial intelligence techniques have performed better than or at least approximated conventional methods. Despite the success of ANNs in geotechnical engineering and other disciplines, they suffer from some shortcomings that need further attention in the future including model transparency and knowledge extraction, extrapolation and uncertainty. Until recently, artificial intelligence-based techniques were widely used for earthquake time series prediction. They are also used to estimate the possibility of potential damage likely to be caused by a certain earthquake.

References

1. Tutorials Point (I) Pvt. Ltd (2015) Artificial Intelligence. www.tutorialspoint.com
2. Kriesel D (2005) A brief introduction to neural networks. http://www.dkriesel.com/en/science/neural_networks
3. Krose B, van der Smagt P (1996) An introduction to neural networks, The University of Amsterdam
4. Patil A, Patted L, Tenagi M, Jahagirdar V, Patil M, Gautam R (2017) Artificial intelligence as a tool in civil engineering—a review. IOSR J Comput. Eng. (IOSR-JCE) 36–39. e-ISSN: 2278–0661,p-ISSN: 2278–8727. www.iosrjournals.org. National Conference On Advances In Computational Biology, Communication, And Data Analytics

5. Lu P, Chen S, Zheng Y (2012) Artificial intelligence in civil engineering, mathematical problems in engineering. In: F. Kang, Hindawi Publishing Corporation, vol 2012, Article ID 145974, 22 pages. https://doi.org/10.1155/2012/145974
6. Juwaied N S (2018) Applications of artificial intelligence in geotechnical engineering. ARPN J Eng Appl Sci 13(8). ISSN 1819-6608
7. Shahin MA, Jaksa MB, Maier HR (2008) State of the art of artificial neural networks in geotechnical engineering EJGE Boquet 08
8. Bojórquez J, Ruiz S E, Bojórquez E (2012) Estimation of inelastic response spectra using artificial neural networks. In: World conference on earthquake engineering, Lisbon 2012
9. Azam F, Sharif M, Yasmin M, Mohsin S (2014) Artificial intelligence based techniques for earthquake prediction: a review. Sci. Int. (Lahore) 26(4):1495–1502. ISSN 1013–5316; CODEN: SINTE 8
10. Atanasiu G M, Leon F, Popa B F, (2008) Seismic risk mitigation in urban areas based on artificial intelligence methods. In: The 14th world conference on earthquake engineering Octber 12–17, 2008, Beijing, China
11. Li Y, Zhang F, Wu B, Yang Q (2004) Simulation of structural seismic responses based on neural networks. In: 13th World conference on earthquake engineering Vancouver, B.C., Canada August 1–6, 2004 Paper No. 1534
12. Ornthammarath T, Corigliano M, Lai CG (2008) Artificial neural networks applied to the seismic design of deep tunnels. In: The 14th World conference on earthquake engineering October 12–17, 2008, Beijing, China

Comparison of Safe Bearing Capacity Based on Laboratory and Field Tests

Manish S. Dixit and Madhuri N. Mangulkar

Abstract Soil is considered by the engineer as a complex material produced by weathering of the solid rock. Soil is the most important material, which is used for construction of civil engineering structures. Among all parameters, the bearing capacity of soil to support the load coming over its unit area is very important. There are various methods for calculation of bearing capacity of soil put forth by scientists like Prandtl, Terzaghi, Meyerhoff, Hansen, Vesic, and others. Principal factors that influence ultimate bearing capacities are type of soil, width of foundation, soil weight in shear zone, and surcharge. Structural rigidity and the contact stress distribution do not greatly influence bearing capacity. Bearing capacity analysis assumes a uniform contact pressure between the foundation and underlying soil. With other factors unchanged, the type of failure of soil, depth of foundation, and effect of water table also govern the bearing capacity of soil. Field tests are very important in knowing the actual performance of soil after loading. Commonly used field tests are plate load test and standard penetration test. In the present study, standard penetration tests are conducted at site. Pit was excavated for different depths, i.e., 1.5, 2.0, 2.5, and 3.0 m, and the soil samples from these depths were collected for determination of its geotechnical properties. Based on these properties, safe bearing capacities were determined by IS code method. At each depth, standard penetration tests were conducted to know the number of blow count "N." Based on the blow count "N," safe bearing capacities were determined by empirical equation. The values of safe bearing capacities obtained at different depths by both methods are compared. They show the fairly good agreement at lesser depth of foundation, but as depth of foundation increases, the safe bearing capacity based on blow count "N" is found to be higher.

Keywords Standard penetration test · Depth · Moisture content · Density · Cohesion

M. S. Dixit (✉) · M. N. Mangulkar
Civil Engineering Department, GSM's, Marathwada Institute of Technology, Aurangabad 431005, Maharashtra, India

T. G. Sitharam et al. (eds.), *Earthquake Geotechnics*, Lecture Notes in Civil Engineering 187, https://doi.org/10.1007/978-981-16-5669-9_36

1 Introduction

Practicing civil engineers determine the value of bearing capacity by different methods such as (i) by laboratory testing—standard proctor test and direct shear tests are conducted to determine maximum dry density, optimum moisture content, cohesion, and angle of internal friction. Then, by using Terzaghi's equation, or by IS code method bearing capacity is found out. (ii) Empirical knowledge—as a thumb rule, depending upon type of soil, civil engineers assume the value of bearing capacity of soil. (iii) Field testing—by conducting plate load test and standard penetration test. These tests are labor-intensive, time consuming, and expensive.

Plate load test, standard penetration test, and pressure meter are commonly used field tests for determination of ultimate bearing capacity. Based on the observations of the field tests, safe bearing capacity or allowable soil pressure is determined. The standard penetration test (SPT) is firmly established in the geotechnical engineering practice as a routine procedure for in situ investigation of soil deposits. Due to long time and worldwide use of this test, comprehensive data has been accumulated in the past. Some investigators have tried to correlate this data for soil characterization. Generally, these correlations may be classified in to two groups: one derived from field and laboratory tests on undisturbed or disturbed soils and other based on field performances of soil deposits with known SPT values. It has been known that the penetration resistance depends on the grain size of soils and that fines containing sands have smaller SPT values than clean sands. The density of cohesive and c-ϕ soils and relative density of sandy soils are the most important parameter, and therefore, to establish a correspondence between the field and laboratory parameters, the first step is to provide the link between the SPT N-values and density/relative density.

In the present study, at different depths, SPT tests were conducted. The site chosen was near petrol pump at Sundarwadi on Aurangabad, Jalna road. By using the following empirical equation, the values of safe bearing capacities are calculated for local shear failure case.

$$q_u = c'N_c' + \gamma D_f\left(Nq' - 1\right) + 0.5\gamma B N_\gamma' \quad (1)$$

Similarly, for general shear failure case, the safe bearing capacity is calculated by the following equation.

$$q_u = cN_c + \gamma D_f(Nq - 1) + 0.5\gamma B N_\gamma \quad (2)$$

2 Literature Review

Terzaghi [1] formed a semi-empirical equation for computing the ultimate bearing capacity of a foundation.

He was the first to propose a bearing capacity expression on the consideration of general shear failure in the soil below a rough, rigid strip footing. Using the principle of superposition, he demonstrated the effects of soil cohesion, its angle of internal friction, surcharge (soil lying above the level of footing base), soil unit weight, and footing width on the ultimate bearing pressure. Later, Meyerhof [2] proposed a general bearing capacity equation similar to that of Terzaghi which included different shape and depth factors. He took into account the shear strength of the soil above the base level of the footing. Thereafter, later on, Hansen [3] introduced the factors that accounted for footing shape and load inclination in the bearing capacity equation. Vesic [4] used an equation very similar to that suggested by Hansen [3]. However, there are some restrictions and assumption in all of these classical formulations; therefore, they do not always give reasonable results compared to available experimental data. Vidal [5] introduced the modern forms of earth reinforcement. His concept was based on composite material formed from flat strips, laid horizontally in frictional soil. Since then, the concept of "Reinforced Earth" has caught the attention of practicing engineers and researchers around the world. More et al. [6] carried out the studies of bearing capacity for two cases by using plate load test. Erickson and Drescher [7] carried out the study for theoretical prediction of the ultimate soil bearing capacity beneath rigid circular footings subject to axial static load. Sakleshpur and Satyanarayana reddy carried out the study on the bearing capacity of strip, square, circular footings and raft foundations in sand, estimated from the standard penetration resistance, N and angle of shearing resistance, ϕ. The net safe bearing capacity estimated directly from N using Teng's equation is compared with that obtained from ϕ as per IS:6403. A parametric study quantifies the effects of the standard penetration resistance and the size and depth of foundation on the net safe bearing capacity and the net safe settlement pressure of strip, square, circular footings and raft foundations in sand. Interactive charts were prepared by Misir et al. [9] A logarithmic model has been developed by using regression analysis to estimate the bearing capacity of a circular footing resting on granular fill at any settlement ratio, using all possible regression techniques based on 342 field test data, to select the significant subset of the predictors. The results indicate that the logarithmic model serves a simple and reliable tool to predict the bearing capacity of circular footings placed on a granular fill with different thicknesses above a soft clay soil. Dussa et al. [10] A new way of finding bearing capacity of soil by combination of direct shear test (lab test) and analytical bearing capacity equations like Terzaghi's and IS code equations bearing capacity of soil can be determined. In this work, a new approach is shown which may be the addition in the literature of geotechnical engineering field. For this work, a site from Naldurg (district Osmanabad) is selected and the number of tests was conducted over c–φ soil. Based on the interpretation of results, the value of bearing capacity is obtained which can be useful for further subsoil investigation and construction activity.

Because of the uncertain nature of soils and the difficulties of experimental tests in laboratory and in situ, there is an increasing tendency to seek alternative bearing capacity prediction methods, other than the traditional computing techniques, to obtain more accurate results.

3 Laboratory Tests

The soil used in the study is collected from Sundarwadi on Aurangabad, Jalna road, 10 km from Aurangabad city. The soil samples were collected from depths of 1.5, 2.0, 2.5, and 3.0 m. The aim of this work was to study the effect of different parameters on bearing capacity of the soil. Experimental work was planned to study the properties of soils collected for determination of ultimate bearing capacity of the soil. For the soils, the basic properties such as specific gravity were determined, sieve analysis was carried out, and consistency limit, i.e., liquid limit and plastic limits, was determined for classification of the soil.

Standard proctor test and direct shear tests were conducted to determine maximum dry density, optimum moisture content, cohesion, and angle of internal friction (ϕ). The test results of soils tested for these properties are as shown in Table 1.

Based on the laboratory test results and as per the Bureau of Indian Standards, the soil is classified as CH, i.e., clayey soil with high compressibility.

3.1 *Experimental Analysis of Soil Collected from Sundarwadi*

Based on the laboratory experimentation carried out, the values of γ c and ϕ were obtained. These parameters are important to determine ultimate and safe bearing capacity of soil. Based on this data, the effect of depth of footing is studied and is discussed in the following paragraphs.

Table 1 Geotechnical properties of soil collected from Sundarwadi

S. No.	Properties of soil samples	Depth in meter			
		1.5	2.0	2.5	3.0
1	Grain size distribution Gravel size (4.75 mm to 80.00 mm) in % Sand size (0.075 mm to 4.75 mm) in % Silt and clay size (below 0.075 mm) in %	– – 12.41 87.59	– – 13.89 86.11	1.30 12.60 86.10	2.35 15.20 82.45
2	Consistency limits Liquid limit % Plastic limit % Plasticity index %	68.00 34.00 34.00	64.50 31.25 33.25	59.80 29.50 30.30	52.50 28.50 24.00
3	Specific gravity of soil solids (G_S)	2.64	2.68	2.71	2.77
4	Compaction characteristics Maximum dry density in gm/cm^3 Optimum moisture content %	1.62 26.20	1.64 24.90	1.66 22.60	1.68 21.40
5	Shear parameters (DS) Cohesion in kN/m^2 Angle of internal friction (ϕ)	29.43 10.92	26.97 11.30	24.52 12.02	19.62 12.73
6	Classification of soil	CH	CH	CH	CH

3.1.1 Effect of Depth of Footing on Bearing Capacity

The depth of footing is important parameters, which governs the ultimate bearing capacity of the soil. Ultimate bearing capacity is determined by using equation given by IS 6403–1981. The shape of footing taken is rectangular footing with length = 1.5 m and width = 1 m.

The values of ultimate bearing capacities determined by IS code method are as shown in Table 2.

Figure 1 shows effect of depth of footing on ultimate bearing capacity for soil for local shear failure case. In Table 2, the values in parenthesis indicate the percentage increase in ultimate bearing capacity in comparison with 1.5 m depth of foundation. Thus, for soil, which is clayey soil with high compressibility having good value of cohesion and lesser angle of internal friction, the percentage increase in ultimate bearing capacity in comparison with 1.5 m depth, for depths of 2 m, 2.5 m, and 3.0 m are found to be 0.35%, 2.64%, and 4.77%, respectively.

Figure 2 shows effect of depth of footing on safe bearing capacity for soil for local shear failure case. In Table 3, the values in parenthesis indicate the percentage increase in safe bearing capacity in comparison with 1.5 m depth of foundation. Thus, for soil, which is clayey soil with high compressibility having good value of cohesion and lesser angle of internal friction, the percentage increase in safe bearing capacity in comparison with 1.5 m depth, for depths of 2 m, 2.5 m, and 3.0 m, is found to be 20.69%, 28.26%, and 35.80%, respectively.

Table2 Effect of depth of rectangular footing on ultimate bearing capacity

Type of soil	Depth of foundation in meter				Type of failure
	1.5	2.0	2.5	3.0	
Blackish Clayey soil	198.61	199.32 (0.35)	203.87 (2.64)	208.09 (4.77)	Local shear failure

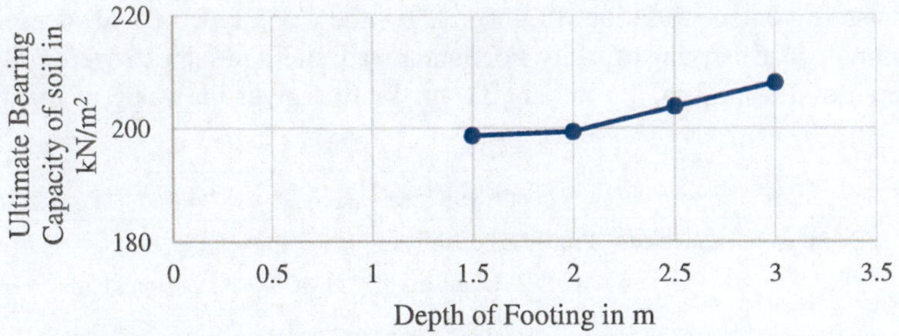

Fig. 1 Effect of depth of footing on ultimate bearing capacity of soil

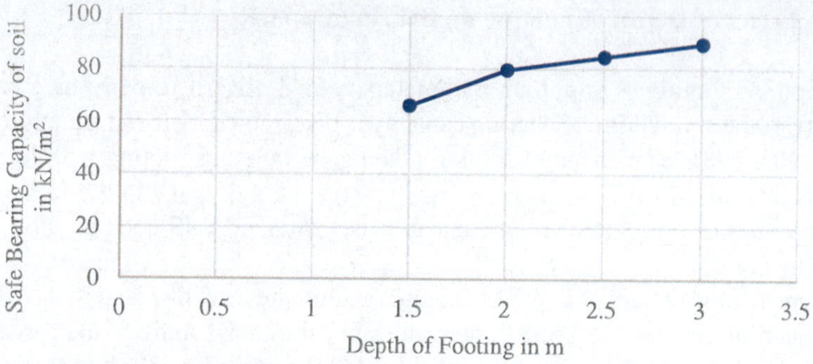

Fig. 2 Effect of depth of footing on safe bearing capacity of soil

Table 3 Effect of depth of rectangular footing on safe bearing capacity

Type of soil	Depth of foundation in meter				Type of failure
	1.5	2.0	2.5	3.0	
Blackish Clayey soil	65.92	79.56 (20.69)	84.55 (28.26)	89.52 (35.80)	Local shear failure

3.1.2 Laboratory and Field Testing of Soil at Sundarwadi

The soils samples are collected at a depth of 1.5 m, 2 m, 2.5 m, and 3.0 m. Depths to carry out the laboratory testing for the determination of safe bearing capacity values should be cancelled or omitted. Similarly, standard penetration test was conducted at these depths of 1.5 m, 2 m, 2.5 m, and 3.0 m. From standard penetration test, the number of blow count "N" is obtained. These values of "N" are used to determine safe bearing capacity values by using empirical equation. Table 4 shows properties of soils at different depths at petrol pump site Sundarwadi.

The load carrying capacity of soil is governed by the type of soil, its arrangement, and moisture content variation. In general, for the same type of soil, denser soil possess more load carrying capacity. For Sundarwadi site, soil samples were collected at a depth of 1.5 m, 2 m, 2.5 m, and 3.0 m. Table 4 gives the value of maximum

Table 4 Properties of soils at different depths at Sundarwadi

S. No.	Depth of soil collected (m)	Properties of soil				
		MDD g/cm^3	Unit weight kN/m^3	OMC%	Cohesion kN/m^2	ϕ (°)
1	1.5	1.62	16.2	26.20	29.43	10.92
2	2.0	1.64	16.4	24.90	26.97	11.30
3	2.5	1.66	16.6	22.60	24.52	12.02
4	3.0	1.68	16.8	21.40	19.62	12.73

dry density, optimum moisture content, unit weight of soil, cohesion, and angle of internal friction for soils collected from different depths. From Table 4, it is found that as depth of foundation increases maximum dry density goes on increasing. The maximum dry density for soil samples collected for different depths 1.5 m, 2 m, 2.5 m, and 3.0 m is found to be 1.62 gm/cm^3, 1.64 gm/cm^3, 1.66 gm/cm^3, 1.68 gm/cm^3, respectively. Similarly, the optimum moisture content for this depth, i.e., 1.5 m, 2 m, 2.5 m, and 3.0 m, is found to be 26.20%, 24.90%, 22.60%, and 21.40%, respectively. Thus, from the table, it is found that as depth of foundation increases from 1.5 m to 3 m, the maximum dry density gradually increases from 1.62 gm/cm^3 to 1.68 gm/cm^3. The percentage increase in maximum dry density in comparison with 1.5 m depth, i.e., 1.62 gm/cm^3 for 2 m, 2.5 m, and 3.0 m depth is found to be 1.21%, 2.36%, and 3.57%, respectively. Similarly, the percentage reduction in optimum moisture content in comparison with 1.5 m depth, i.e., 26.20% for 2 m, 2.5 m, and 3.0 m depth is found to be 4.96%, 13.74%, and 18.32%, respectively.

Thus, in general, as the depth of foundation increases, the maximum dry density increases and optimum moisture content decrease. Similarly, cohesion for the depth 1.5 m, 2 m, 2.5 m, and 3.0 m is found to be 29.43 kN/m^2, 26.97 kN/m^2, 24.52 kN/m^2, and 19.62 kN/m^2, respectively.

Thus, from Table 4, it is found that at depths, i.e., 1.5 m, 2 m, 2.5 m, and 3.0 m, the percentage reduction in cohesion in comparison with 1.5 m depth, i.e., 29.43 kN/m^2 for 2 m, 2.5 m and 3.0 m, is found to be 3.35%, 16.68%, and 33.33%, respectively. Similarly, the angle of internal friction for the depths of 1.5 m, 2 m, 2.5 m, and 3.0 m is found to be 10.92^0, 11.30^0, 12.02^0, and 12.73^0, respectively. Thus, from Table 4, it is found that as depth of foundation increases from 1.5 m to 3.0 m the angle of internal friction increases gradually from 10.92 to 12.73°. The percentage increase in angle of internal friction in comparison with 1.5 m depth, i.e., 10.92° for 2 m, 2.5 m, and 3.0 m depth, is found to be 3.36%, 9.15% and 14.21%, respectively. Thus, in general as depth of foundation increases, cohesion decreases and angle of internal friction increases.

Based on the laboratory test results obtained for soil samples collected from different depths, the safe bearing capacities are calculated by IS code method. The factor of safety equal to 3 is taken for determination of safe bearing capacity of soil. Similarly, standard penetration tests were conducted in a pit at different depths, i.e., 1.5 m, 2 m, 2.5 m, and 3.0 m depths, and the numbers of blow counts "N" are found to be 9, 11, 15, and 17 for these depths, respectively. Thus, it is clearly found that at depth 1.5 m, the number of blow counts is found to be 9 whereas that increased up to 17 for 3.0 m depths.

From Table 5, it is found that by using laboratory test results, the value of safe bearing capacity calculated by IS code method for different depths, i.e., 1.5 m, 2 m, 2.5 m, and 3.0 m, is found to be 68.88 kN/m^2, 72.53 kN/m^2, 77.54 kN/m^2, and 78.55 kN/m^2, respectively.

In comparison with 1.5 m depths safe bearing capacity value, i.e., 68.88 kN/m^2, the percentage increase in safe bearing capacity values for different depths, i.e., 2 m, 2.5 m, and 3.0 m, is found to be 5.03%, 11.16%, and 12.31%, respectively.

Table 5 Comparison of SBC. Based on Laboratory and Field Test at Sundarwadi site

S. No.	Depth of soil collected (m)	N	Safe bearing capacity in kN/m^2	
			IS code method (Laboratory test)	Empirical equation (Field test)
1	1.5	9	68.88	61.73
2	2.0	11	72.53	81.82
3	2.5	15	77.54	108.13
4	3.0	17	78.55	132.36

From Table 5, it is found that by using the field test results the value of safe bearing capacity calculated by empirical equation for different depths i.e. 1.5 m, 2.0 m, 2.5 m, 3.0 m, are found to 61.7 3kN/m^2, 81.82 kN/m^2, 108.13 kN/m^2 and 132.36 kN/m^2 respectively. In comparison with 1.5 m depths safe bearing capacity value, i.e., 61.73 kN/m^2, the percentage increase in safe bearing capacity for different depths, i.e., 2.0 m, 2.5 m, and 3.0 m is found to be 24.55%, 42.91%, and 53.36%, respectively. For depth 1.5 m as per IS code method, the safe bearing capacity is found to be 68.88 kN/m^2 for the same depth based on standard penetration test results that is the number of blow count "N," and by using the empirical equation, the safe bearing capacity is found to be 61.73 kN/m^2. Thus, for 2 m depth in comparison with IS code method, the safe bearing capacity found by empirical equation on "N" is found to be reduced by 10.38%. For 2 m, 2.5, and 3.0 m depth, the percentage increase by empirical equation is 11.35%, 28.29, and 40.65%, respectively (Figs. 3 and 4).

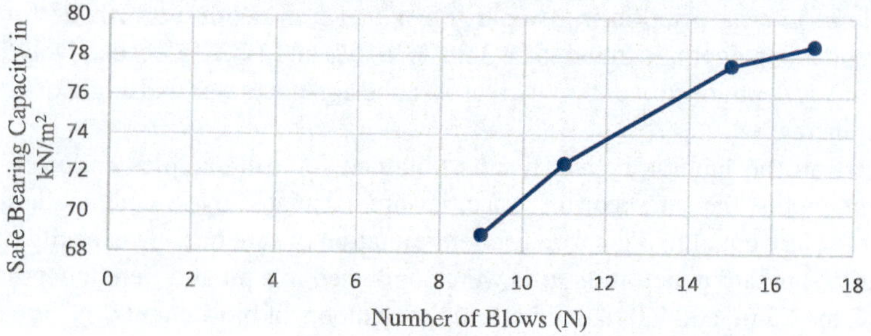

Fig. 3 Number of blow count and safe bearing capacity of soil by I S code method

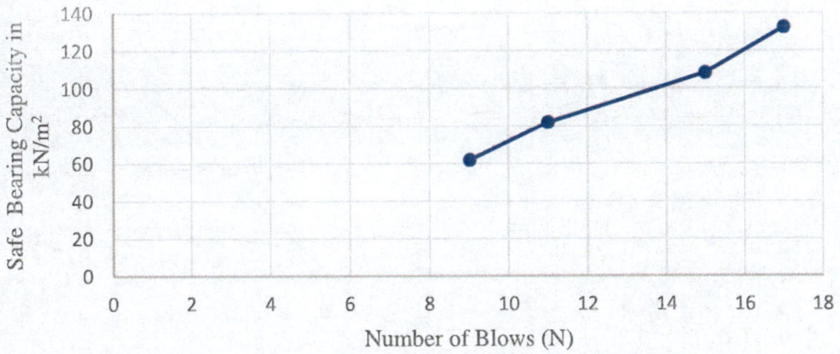

Fig. 4 Number of blow count and safe bearing capacity by empirical equation

4 Statistical Analysis

4.1 Comparison of Experimental and Statistical Results

Safe bearing capacity of soil is an important parameter, which governs design of foundation. The safe bearing capacity of soil depends on the type of soil and its properties, depth of foundation, type of footing decided. Depending upon the type of soil and other factors, geotechnical engineer has to use a suitable value of factor of safety. The factor of safety may range from 1.5 to 3.0. In case of uncertainty, one has to use more value of factor of safety that is up to 3, and if assurance of properties of soil and its behavior is possible, then lower value of factor of safety is used.

Table 6 shows comparison of observed and predicted values of ultimate bearing capacity for different depths of footing of soil.

From Table 6, it is found that there is a close agreement between observed and predicted values of ultimate bearing capacity obtained by derived expression. The linear expression obtained from the graph is $y = 6.598x + 187.6$. The value of R^2 is found to be 0.932 (Fig. 5).

Table 7 shows comparison of observed and predicted values of safe bearing capacity of soil.

Table 6 Observed and predicted values of ultimate bearing capacity of soil

Depth of footing in meter	Observed ultimate bearing capacity in kN/m^2	Expression	Predicted ultimate bearing capacity in kN/m^2
1.5	198.61	$y = 6.598x + 187.6$ $R^2 = 0.932$	197.52
2.0	199.32		200.82
2.5	203.87		204.2
3.0	208.09		207.42

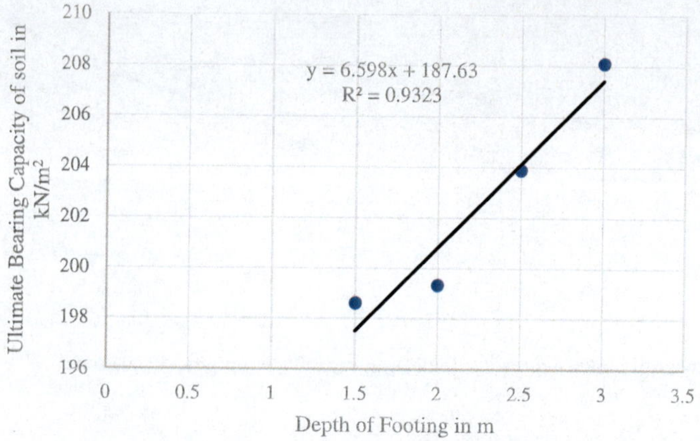

Fig. 5 Predicated values of ultimate bearing capacity of soil

Table 7 Observed and predicted values of safe bearing capacity of soil

Depth of footing in meter	Observed safe bearing capacity in kN/m^2	Expression	Predicted safe bearing capacity in kN/m^2
1.5	65.92	$y = 15.15x + 45.78$ $R^2 = 0.927$	68.51
2.0	79.56		76.09
2.5	84.55		83.67
3.0	89.52		91.25

From Table 7, it is found that there is a close agreement between observed and predicted safe bearing capacity values obtained by derived expression for all depths. The linear expression obtained from the graph is $y = 15.15x + 45.78$. The value of R^2 is found to $R^2 = 0.9273$ (Fig. 6).

Table 8 shows comparison of observed and predicted values of safe bearing capacity for different depths by IS code method at Sundarwadi site.

From Table 8, it is found that there is a close agreement between observed and predicted safe bearing capacity values obtained by derived expression for 1.5 m, 2 m,

Table 8 Observed and predicted values of safe bearing capacity by IS code method at Sundarwadi site

Depth of footing in meter	Safe bearing capacity in kN/m^2	Expression	Predicted safe bearing capacity in kN/m^2
1.5	68.88	$y = 6.804x + 59.06$ $R^2 = 0.947$	69.50
2.0	72.53		71.93
2.5	77.54		76.80
3.0	78.55		79.24

2.5 m, and 3.0 m depths. The linear expression obtained from the graph is y = 6.804x + 59.06. The value of R^2 is found to be 0.947 (Fig. 7).

Table 9 shows comparison of observed and predicted values of safe bearing capacity for different depths by empirical equation at Sundarwadi site.

From Table 9, it is found that there is a close agreement between observed and predicted safe bearing capacity values obtained by derived expression for 1.5 m, 2 m, 2.m, and 3.0m depths. The linear expression obtained from the graph is y = 47.64x − 11.18. The value of R^2 is found to be 0.997 (Fig. 8).

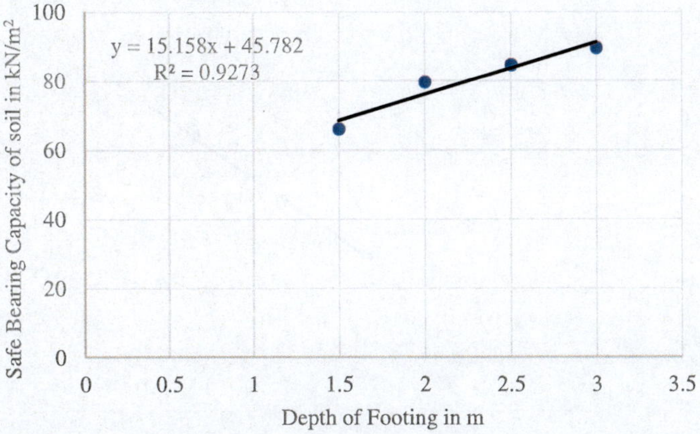

Fig. 6 Predicated values of safe bearing capacity of soil

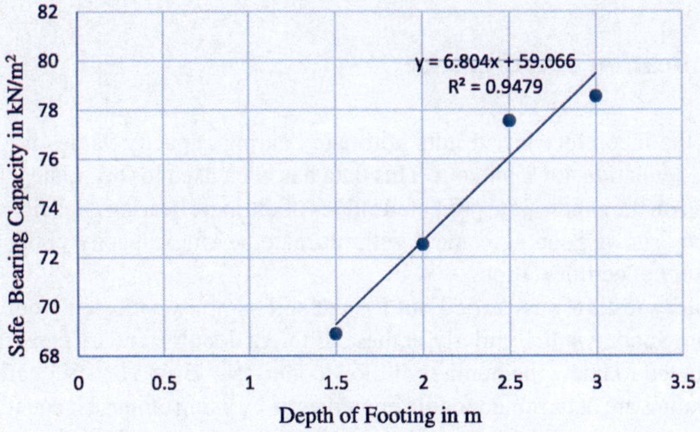

Fig. 7 Depth of footing and safe bearing capacity by IS code method

Table 9 Observed and predicted values of safe bearing capacity by empirical equation at Sundarwadi site

Depth of footing in meter	Safe bearing capacity in kN/m^2	Expression	Predicted ultimate bearing capacity in kN/m^2
1.5	61.73	$y = 47.64x - 11.18$ $R^2 = 0.997$	62.49
2.0	81.82		79.25
2.5	108.13		112.76
3.0	132.36		129.52

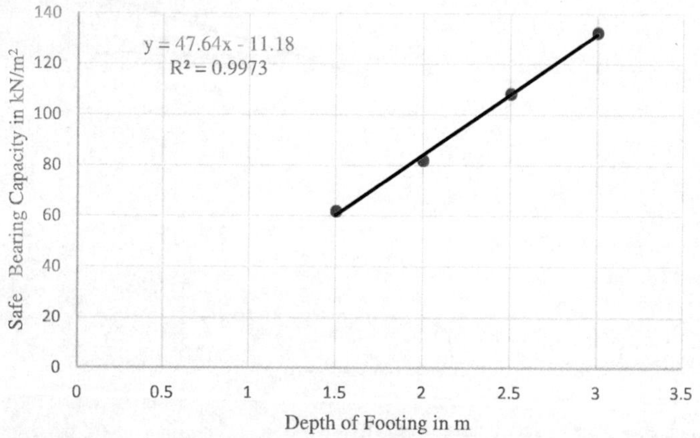

Fig. 8 Depth of footing and safe bearing capacity by empirical equation

5 Justification of Difference

Based on the laboratory test results, ultimate bearing capacity values for different depths of foundation are found out. This data has been used to formulate the expression. Based on the expression, predicted values of ultimate bearing capacity are found out and they shows good agreement with ultimate bearing capacity obtained based on laboratory experimentation.

Laboratory testing was carried out for the soil samples collected from different depths from Sundarwadi. Similarly, at these different depths standard penetration test was conducted to know the number of blow count "N." Based on "N," safe bearing capacity values are determined for different depths by using empirical equation. With this obtained data, expressions are obtained by "regression analysis," and values of safe bearing capacity are predicted. The observed values and predicted values of safe bearing capacity show good agreement.

6 Conclusions

1. The important parameters, which govern the bearing capacities of soil, are: cohesion, unit weight of soil, depth of proposed foundation, width of foundation and angle of internal friction.
2. As the depth of foundation increases, ultimate bearing capacity of soil increases. The effect of increase in depth on safe bearing capacity is predominant due to increase in surcharge weight.
3. The standard penetration test results are useful for determining the safe bearing capacity of soil by using empirical equations.
4. Based on laboratory test results by IS code method for petrol pump site, safe bearing capacity increases with increase in depth of foundation. The standard penetration test results are useful for determining the safe bearing capacity of soil by using empirical equations. In comparison with 1.5 m depths safe bearing capacity value, i.e., 68.88 kN/m^2, the percentage increase in safe bearing capacity values for different depths, i.e., 2 m, 2.5 m, and 3.0 m, is found to be 5.03%, 11.16%, and 12.31%, respectively.
5. Thus, for 2 m depth in comparison with IS code method, the safe bearing capacity found by empirical equation on "N" is found to be reduced by 10.38%. For 2 m, 2.5, and 3.0 m depth, the percentage increase by empirical equation is 11.35%, 28.29, and 40.65%, respectively.

References

1. Terzaghi K (1943) Theoretical soil mechanics. Wiley, New York, USA
2. Meyerhof GG (1965) Shallow Foundation. J Soil Mech Found Div ASCE 21–31
3. Hansen B (1961) A general formula for bearing capacity. Danish Geotechnical Institute, Bulletin No. 11, Copenhagen, Denmark, pp. 38–46
4. Vesic AS (1973) Analysis of ultimate loads of shallow foundations. J. Soil Mech. Foundation Div, ASCE 99(1):45–76
5. Vidal H (1966) Reinforced earth annals. De Institute Technique Du Batiment Et Des Travanx Public, vol 19, pp 233–244
6. More DM, Pathade NK, Kulkarni AA (2006) Bearing capacity by plate load test. In: National conference on corrective engineering practices in trouble some soils, Kakinada, pp 95–96
7. Erickson HL, Drescher A (2002) Bearing capacity of circular footings. J Geotechn Geo Environ Eng 128(1):38–43
8. Sakleshpur VA, Satyanarayana Reddy CNV (2017) A comparative study on bearing capacity of shallow foundations in sand from N and ϕ". J Inst Eng India Ser A 98(4):355–365
9. Misir G, Laman M (2016) A modern approach to estimate the bearing capacity of layered soil. Periodica Polytechnica Civil Eng 61(3):434–446
10. Dussa PS, Mujawar KC (2016) New approach for determination of bearing capacity of soil using direct shear test (Lab test). Int J Innov Eng Technol (IJIET) 7:426–430
11. I.S. 6403–1981 (1957) Code of practice for determination of bearing capacity of shallow foundations. Bureau of Indian Standard, New Delhi
12. Nayak NV (2001) Foundation design manual. Dhanpat Rai Publications Private Limited, New Delhi

13. Punmia BC (1996) Soil Mechanics and foundation engineering. Laxmi Publications, New Delhi, pp 661–724
14. Gulhati SK, Datta M (2007) Geotechnical engineering. Tata McGraw-Hill, New Delhi
15. Ranjan G, Rao ASR (2000) Basic and applied soil mechanics. New Age International, New Delhi
16. Purushothama Raj P (2008) Soil mechanics & foundation engineering. Pearson Education, New Delhi
17. Venkatramaiah C (2006) Geotechnical engineering. New Age Publishers, New Delhi

Integrating Landslide and Seismic Hazards: A Case Study for Lower Indian Himalaya

Ritu Raj Nath

Abstract The Himalaya is considered as a global hotspot for landslides and earthquakes. The impacts of seismically induced landslides are astounding in terms of loss of life and property. An endeavour has been made in this present study to prepare a seismically induced landslide hazard zonation map for a part of lower Indian Himalaya. Eight landslide causative factors are identified for the study area, out of which seven are static in nature and preparatory in contribution. Earthquake is considered as the eighth causative parameter, which is recognized as the triggering factor of landslides in the study area. Peak ground acceleration at bedrock level, generated through probabilistic seismic hazard assessment for 10% exceedance probability in 50 years, is considered as a scenario earthquake in the study area. Analytic hierarchy process is used to allocate the weights to different landslides causative factors judiciously. The results of the study show that the probability of landslides increases by 35%, when earthquake scenario is integrated into the landslide hazard zonation map prepared under the preparatory causative factors. The study demonstrates that for seismically active mountain zones, incorporation of seismic factor becomes the most crucial parameter in assessing the prevailing landslide hazard.

Keywords Himalaya · Landslides · Earthquakes · Analytic hierarchy process

1 Introduction

The Himalaya is the long-term consequence of the continent–continent collision of the Indian and the Eurasian plates, which had started around ~50 Myr ago [1–3]. This active collision orogeny of prompt crustal shortening and thickening has resulted in the formation of the most active seismogenic sources in the Himalaya that had often produced strong earthquakes in this region. One of the major secondary hazards associated with strong ground motion (SGM) is the widespread and devastating

R. R. Nath (✉)
Department of Civil Engineering, Assam Down Town University Guwahati, Guwahati 781026, Assam, India

T. G. Sitharam et al. (eds.), *Earthquake Geotechnics*, Lecture Notes in Civil Engineering 187, https://doi.org/10.1007/978-981-16-5669-9_37

landslides. It has been observed that losses due to seismically induced landslides are generally more than those caused directly due to shaking [4]. Marano et al. (2010) reported that out of all earthquake-related casualties, which are not caused directly by ground shaking, approximately 70% may be attributed to landslides [5]. Impacts of seismically induced landslides are of long-term nature, which manifests their damage intensity in terms of indirect socio-economic losses [6–8]. In this context, landslide hazard zonation (LHZ) mapping for the Himalayan region under seismic condition becomes very important considering its active seismicity and the number of landslide incidences. Many previous studies [9–11] have also emphasized the significance of establishing a relationship between Himalayan seismicity and its persistent landslide hazard.

The most critical aspect of carrying out seismically induced LHZ mapping is coalescing the two hazards at the same scale, i.e. integrating seismic and landslide hazards. The main challenge is the selection and applicability of a well-recognized method. Although various statistical methods [12, 13] are available for assessing the landslide susceptibility of a region, it is difficult to select any globally recognized method for earthquake-induced LHZ mapping. This is because different statistical methods, which are widely used for landslide susceptibility assessment, generally lack in incorporating seismic indicators [14]. As earthquakes are extreme and rare events, compilation of earthquake-induced landslide inventory is a difficult task. This greatly hinders the use of neural network methods for seismically induced LHZ mapping as well, as these methods require a substantial database. The choice of an adopted method is therefore governed by the aim and scope of the study. If the study is carried out post a major seismic event, distance-magnitude frequency correlation becomes an automatic choice. However, such studies are limited to the investigated region and specific strong ground motion only. For micro-scale, site-specific LHZ studies, physical-based models might represent a better estimate of co-seismic displacements, but the main disadvantage of such model is the complex and robust analytical processes associated with it [13]. Another major challenge is the selection of suitable SGM parameters for multi-hazard integration. Different earthquake characteristics, viz. amplitude, frequency content and duration, have different implications and need careful examination before being used for carrying out seismically induced LHZ mapping. These points rather indicate the challenges in seismically induced LHZ studies precisely.

In this context, an endeavour has been made in this study to generate a LHZ map for a part of lower Indian Himalaya under seismic condition. The seismic parameter considered is the peak ground acceleration (PGA) at bedrock level, which is derived probabilistically for an earthquake probability of 10% in 50 years (corresponding to a return period of 475 years) using probabilistic seismic hazard assessment (PSHA). Map combination method [15] has been used for integrating both seismic and landslide hazards at the same scale. Analytic hierarchy process (AHP) is used to reduce the subjectivity in the weight assignment process [16, 17]. The study quantifies the impact of seismicity in LHZ mapping for a part of lower Indian Himalaya.

2 Study Area

The study area covers several important and thickly populated cities of Uttarakhand and Himachal Pradesh states of India and encompasses approximately 12,000 km^2 (Fig. 1), with estimated population of more than 15 lakhs as per 2011 general census. Four National Highways, several State Highways and various auxiliary roads traverse through the study area. Parts of Jim Corbett National Park and Rajaji National Park, two bio-diverse green forests of national importance are also situated within the study area. Geologically, the study area exhibits a complex and heterogeneous amalgamation of multiple formations from different ages' (Fig. 1), viz. Lower Siwalik Group, Middle Siwalik Group, Krol, Infra-Krol and Blaini Formation, Tal Formation, Jonsar Formation, Subathu Formation, etc. The dominant rock types are Sandstones, Limestones, Carbonaceous Shales, Slate, Dolomite and Conglomerate, with locally available Quartzite, Gneiss, Amphibolite and Mudstones. Three major thrusting systems of the Himalayan arc, viz. the Main Frontal Thrust (MFT), the Main Boundary Thrust (MBT) and a portion of Main Central Thrust (MCT), passes through the study area making it tectonically very active (Fig. 1).

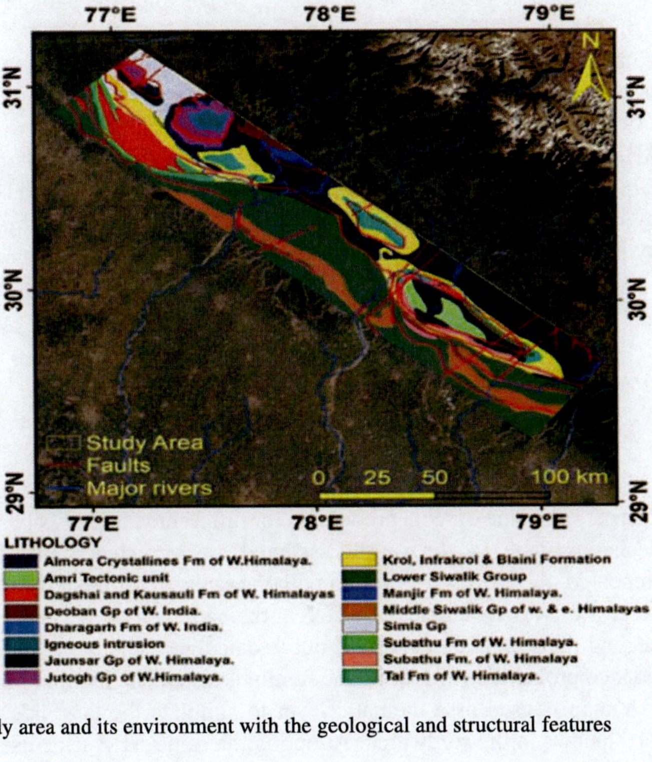

Fig. 1 Study area and its environment with the geological and structural features

3 Methodology

The present study is carried out in two phases. In the first phase, LHZ map of the study area is prepared under static conditions using map combination method [15]. The methodology incorporates collection of input data on various landslide causative factors, preparation of thematic data layers, assignment of weights to different thematic data layers, numerical integration of thematic data and generation of LHZ map. The numerical integration of weights of different thematic layers in a Geographic Information System (GIS) environment will produce a numerical value known as Landslide Potential Index (LPI) for each pixel. LPI [18] indicates the likelihood of landslide occurrence based on the prevailing causative factors in an area. Higher value of LPI indicates greater susceptibility to landslide hazard and vice-versa. Mathematically, it may be defined as,

$$\text{LPI} = \sum_{i=1}^{n}\sum_{j=1}^{m} R_i \times W_{i.j} \tag{1}$$

where R_i and W_{ij} denote the rank of the ith layer of landslide thematic layer and the weightage of jth class of ith layer, respectively.

The major disadvantage of map combination method is the subjectivity in weight assignment process. To reduce subjectivity and inconsistency in assigning weights and ranks of both landslide causative factors and their classes, analytic hierarchy process (AHP) has been used [16, 17]. AHP is a semi-quantitative method, in which decisions are taken using weights through pair-wise relative comparisons [19]. The consistency of the decision matrix is examined using consistency ratio (CR) and consistency index (CI), which are defined as,

$$\text{CI} = \frac{\lambda_{\max}}{N - 1} \tag{2}$$

$$\text{CR} = \text{CI}/\text{RI} \tag{3}$$

where $\lambda_{\max}$ is the largest normalized principal eigenvector and N is the order of the decision matrix. The value of RI is predefined for different values of N [19]. If CR ≤ 10%, the decisions made (*i.e.* the weights assigned) are consistent. If not, the process should be repeated. AHP is used to assign both the ranks of different thematic layers (R_i) and weights of various classes of different thematic layers (W_{ij}) in this study.

In the second phase, PSHA is carried out to calculate PGA at bedrock level for 10% exceedance probability in 50 years assuming Poissonian distribution [20]. The derived PGA map is used as a thematic layer to simulate the landslide triggering effect of earthquakes in the study area. Adopting the same AHP scale used for map combination method, the seismic and landslide hazards are integrated into produce the final LHZ map of the study area under seismic condition. It is to be noted that application of PSHA for generation PGA maps for different earthquake exceedance probabilities is well-documented for the Himalayan region. Considering the paucity of both scope and space, a detailed discussion on the PSHA methodology is not included in this paper.

4 Preparation of Landslide Distribution Map for the Study Area

The first and the most important point of any LHZ studies is the preparation of an accurate landslide inventory. Since the study area encompasses almost 12,000 km^2 of rugged Himalayan terrain, remote sensing methods are used for supervised identification and delineation of the observed landslides in the area. Geo-referenced LISS IV image and Google Earth platform are extensively used for that purpose, and ground-truthing is carried out at selected few locations. Based on this three-way system, 1062 landslides are identified in the study area at the scale of 1:12,000. The observed landslides are then mapped (Fig. 3) in a GIS environment for spatial analyses. Various terrain attributes are also mapped in the same GIS environment for data collation and analysis. Figure 2 shows an example of the adopted three-way system of landslide inventory preparation, where LISS IV imagery, Google Earth platform and ground reconnaissance are performed for both identification and verification of a particular observed landslide in the study area.

LISS-4 is a three-band sensor component of LISS-3 heritage (same spectral VNIR bands as LISS-3) with a spatial resolution of 5.8 m and a swath of 70 km. It can be operated in either multispectral or mono-mode. In this study, multispectral images with three bands having 10-bit quantization are used.

The prepared landslide inventory accounts for a first-hand information on the existing landslide causative factors in the study area and is used for proximity analyses. Based on extensive field study, available literature and previous knowledge of the investigated terrain, seven landslide causative factors are considered in the study: slope morphometry (angle and aspect), lithology, distance from major geological discontinuity, elevation profile, distance from drainage and distance from road. All these factors are static in nature (i.e. negligible temporal variations are observed) and are categorized as the preparatory factors of landslides [12, 13]. Seismicity is considered as a landslide triggering factor in the study area. It is to be noted that every landslide hazard study in the Himalaya has recognized these eight landslide causative factors.

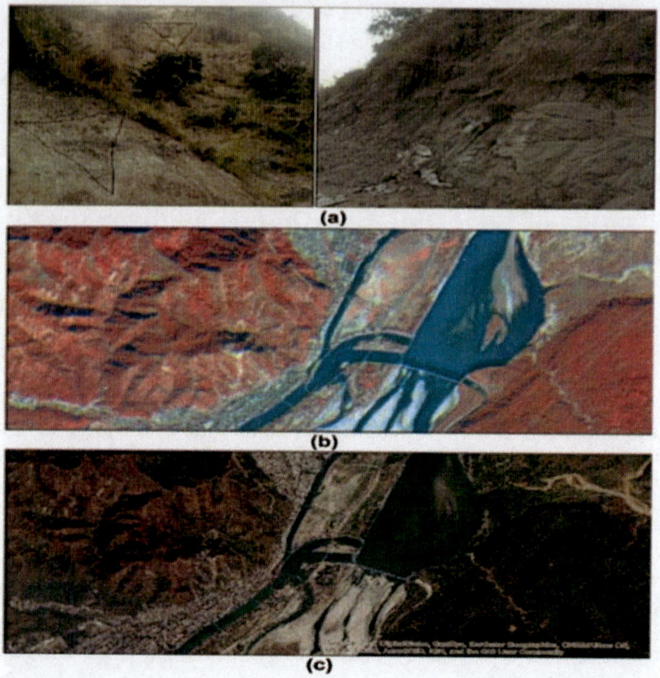

Fig. 2 Mansadevi Landslide (29.9577° N, 78.1655° E): **a** field photograph; **b** LISS IV image; **c** Google Earth image

5 Preparation of Database

5.1 Digital Elevation Model (DEM) and Its Derivatives

The surface topography of an area can be derived in the form of digital elevation model (DEM). The study area has the minimum and the maximum elevations of 201 m and 3638 m above mean sea level, respectively, which is shown in Fig. 4a. The DEM is categorized into five (5) classes at an interval of 500 m, for the ease of classification. It is observed that maximum landslide density corresponds to an elevation class of [500–1000] m, followed by the class of [1000–1500] m. The least landslide density is observed at elevations less than 500 m.

The DEM is used to derive various terrain attributes like slope angle (Fig. 4b), slope aspect (Fig. 4c) and the drainage density of the study area, which are used as input data layers for LHZ mapping. The drainage density map is further validated using Survey of India (SoI) toposheets at a scale of 1:50,000. From the drainage density map, drainage Euclidean distance (ED) map (Fig. 4d) is prepared in GIS environment.

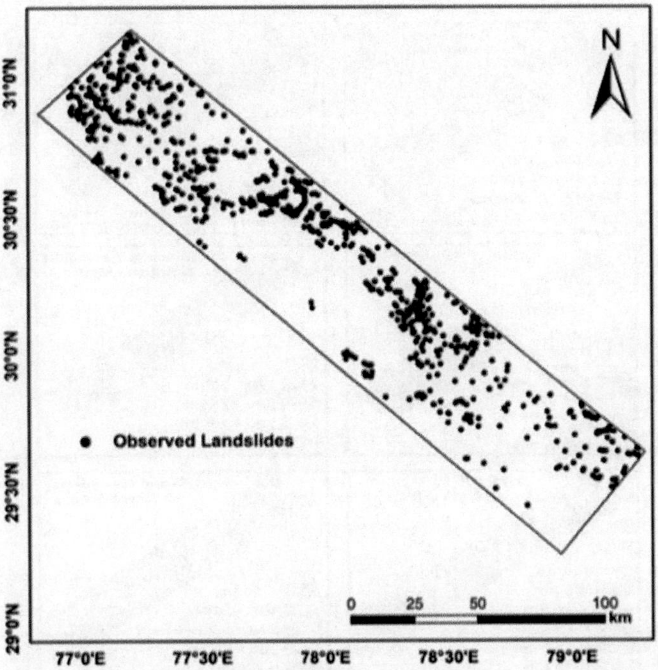

Fig. 3 Landslide distribution map of the study area

5.2 Fault Euclidean Distance (ED) Map

The active faults and lineaments presented in the study area are mapped (Fig. 1) using the Seismotectonic Atlas of India [21]. The digitized fault contours are exported in GIS to calculate the Euclidean distance of the existing landslides from the nearest source (fault/lineaments). The fault Euclidean distance map of the study area is shown in Fig. 4e. The minimum and the maximum distances are 0 km (i.e. landslides observed at the fault wall itself) and 14.9 km, respectively. The map is categorized into five classes at an interval of 3 km. The highest landslide density is observed at less than 3 km from a major discontinuity.

5.3 Road Euclidean Distance Map

There exists a sporadic distribution of road networks throughout the study area. While the urban centres of the area have a high concentration of road networks, the overall density of road is moderate due to large forest areas. The Euclidean distance from existing landslides from roads is calculated, which ranges from a minimum of 15 m

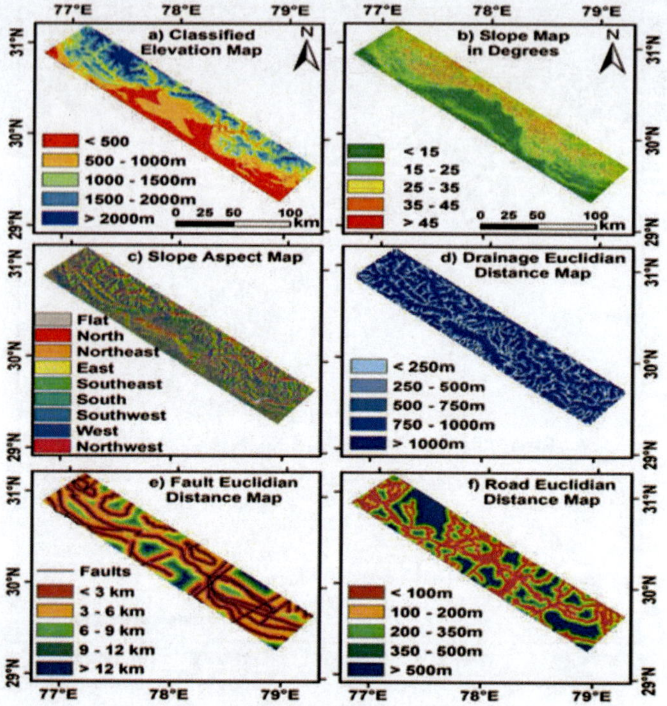

Fig. 4 Database of thematic layers prepared for the study area: **a** elevation map, **b** slope angle map, **c** slope aspect map, **d** drainage Euclidean distance map, **e** fault Euclidean distance map and **f** road Euclidean distance map

to a maximum of 750 m. The road Euclidean distance map is shown in Fig. 4f. It is further categorized into five classes.

The geological map of the study area is shown in Fig. 1, which is also used as a thematic layer in this study. The geology of the area is already discussed in Sect. 2. It is to be noted that land-use–land-cover (LULC) pattern is recognized as landslide causative parameter by various researchers. However, as the study area represents a somewhat homogenous coverage of forests and green land, with sporadically located few urban centres in between, LULC is not considered in this study as a thematic layer. Similarly, distance from folds and slope curvatures are also excluded from the landslide causative factors, considering the aim and scope of the present study.

5.4 *PSHA and Peak Ground Acceleration (PGA) Map*

The study area shares high seismicity of the north-western Himalayan region. It is falling in between the seismic gap of 1934 Bihar–Nepal earthquake and 1905

Kangra earthquake in the central Himalaya active region [20]. PSHA is carried out for a bigger area (~R = 300 km) as a standard practice (Fig. 5). Based on geological conditions, tectonic features and observed recorded earthquake events, the seismotectonic modelling has been carried out and seven seismogenic source zones (SSZ) are identified. Table 1 presents some salient tectonic features and the seismicity parameters (α and β) of each seismogenic source zone.

The result of the PSHA estimated in terms of the PGA map for earthquake exceedance probability of 10% in 50 years is shown in Fig. 6. Corresponding to that, the maximum and minimum PGAs observed at bedrock level are 0.190 g and

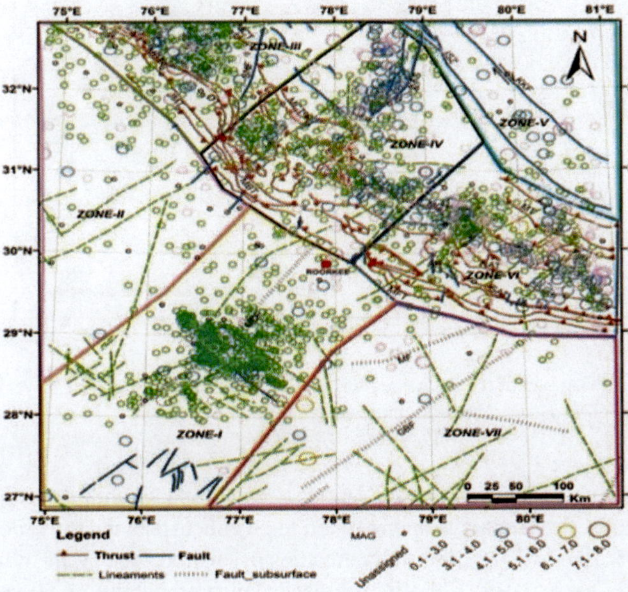

Fig. 5 Seismotectonic modelling for performing PSHA

Table 1 Important tectonic features and seismicity parameters of seismogenic source zones

SSZ	Prominent tectonic features	M_{max}	M_{min}	α	β
SSZ I	Mahendragrah Dehradun Fault (MDF)	6.5	4.5	2.23	4.26
SSZ II	No major fault system except some lineaments	5.5	4.0	1.93	3.87
SSZ III	MCT, MBT, MFT, Sunder Nagar Fault (SNF)	8.0	3.1	1.70	2.66
SSZ IV	MCT, MBT, MFT, North Almora Thrust	6.6	4.7	2.20	4.08
SSZ V	Indus Suture Zone (ISZ), Karakoram Fault (KKF)	6.0	2.7	2.00	2.67
SSZ VI	MCT, MBT, MFT, Martoli Thrust (MT), NAT, South Almora Thrust (SAT), Ramgarh Thrust (RT)	6.8	3.5	2.14	3.40
SSZ VII	Moradabad Fault (MF), Great Boundary Fault (GBF)	6.7	2.6	1.49	1.83

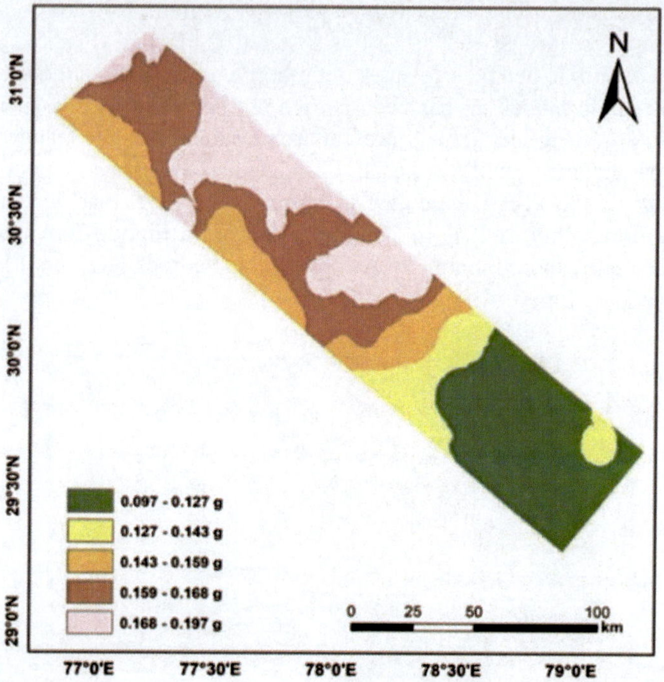

Fig. 6 Classified PGA map of the study area for 10% exceedance probability in 50 years showing the variations of PGA contours throughout the study area

0.097 g, respectively. The result of PSHA has been compared with past studies [22, 23] carried out for the Himalaya and is found to be concurrent and thus satisfactory. As per IS 1893: Part I (2016), the study area lies in seismic zone IV of India, with a maximum probable PGA of 0.24 g for 2% exceedance probability in 50 years.

6 Results and Discussion

The ranks and weights of the various landslide causative factors and their subclasses are decided based on the proximity analysis of the landslide inventory prepared for the study area. The effects of slope angle and lithology on landslide activities in the lower Himalayan belt have been observed to be the most prominent. Equally, the distance from the faults has a distinct structural impact on landslide occurrence due to the presence of MCT, MBT and MFT, along with numerous transverse lineaments in the study area. The influence of drainage density and elevation on landslide distribution has been observed to be moderate. Although road construction seems to be a major landslide causative factor, owing to a sporadically distributed road network, its effect on landslide indices is observed to be low in the study area. Similarly, slope aspect

Table 2 Assignment of ranks (Ri) and formation of the decision matrix

Parameter	Assigned rank	1	2	3	4	5	6	7	Eigenvector
Slope angle	9	1							0.355
Geology	8	0.50	1						0.242
Fault ED	7	0.33	0.50	1					0.161
Elevation	6	0.25	0.33	0.50	1				0.107
Drainage ED	5	0.20	0.25	0.33	0.50	1			0.072
Road ED	3	0.14	0.20	0.25	0.33	0.50	1		0.036
Slope Aspect	2	0.12	0.14	0.20	0.25	0.33	0.50	1	0.026

has a relatively lower impact. Based on these observations, the ranks (R_i) of different landslide causative parameters are assigned. In this context, a hazard scale is derived, where distinct linguistic variables, viz. Very High (VH), High (H), Moderate (M), Low (L) and Very Low (VL), are used to define the perceived effect of a particular causative factor. Using the AHP scale [19], five discrete values of 9, 7, 5, 3, 1 are used to describe the linguistic variables according to their degree of landslide control. Also, intermediate values (8, 6, 2) are used to denote H-VH and L-VL effects. Using this scale, the decision matrix is formed which is shown in Table 2.

For this decision matrix, 21 pairwise comparisons are made. Solution of the decision matrix yields a CR of 3.1%. Since CR <10%, the decision made is consistent; *i.e.* the assigned ranks are satisfactory. Using the same AHP scale, the weights ($W_{i,j}$) of different classes of various thematic layers are assigned. For every layer, the CR is checked and found to be satisfactory. Table 3 shows the weight assignment process for a particular thematic class as an illustrative example.

For this decision matrix, 10 pairwise comparisons are made. Solution of the decision matrix yields a CR of 5.3%. Since CR < 10%, the decision made is consistent; i.e. the assigned weights are satisfactory. Similar procedure has been adopted for assignments of weights to all the thematic classes. Once the R_i and $W_{i,j}$ are finalized, they are numerically integrated in a GIS environment to produce the static LHZ map of the study area in a GIS environment, which is shown in Fig. 7. The prepared LHZ

Table 3 Assignment of weights ($W_{i,j}$) to various classes of the slope angle thematic layer and formation of the decision matrix

Thematic class	Perceived effect on LHZ	Weight ($W_{i,j}$)	1	2	3	4	5	Eigenvector
<15°	VL	1	1					0.033
15–25°	L	3	3	1				0.063
25–35°	M	5	5	3	1			0.129
35–45°	H	7	7	5	3	1		0.262
> 45°	VH	9	9	7	5	3	1	0.513

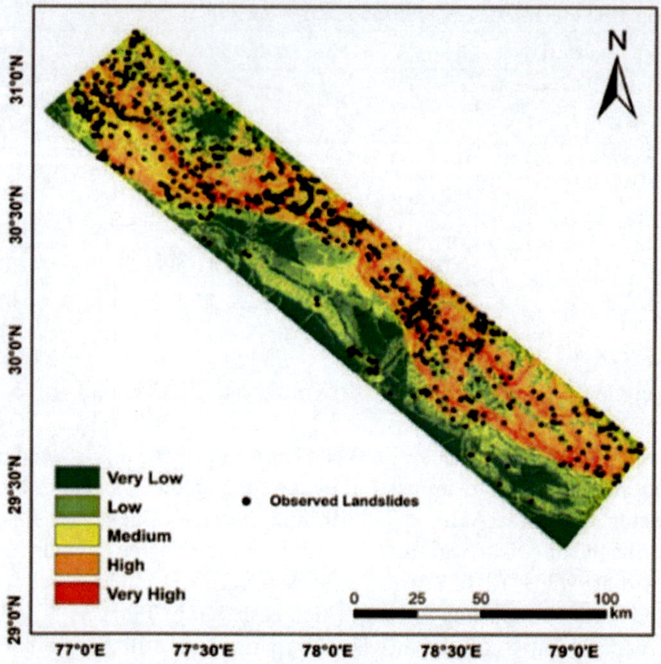

Fig. 7 Landslide hazard zonation map of the study area under static conditions (LHZ_Static)

map (under static conditions) is analysed statistically and found to yield a model success rate of 85%, which is satisfactory in the practice of LHZ mapping [15–18].

To integrate seismic and landslide hazards, the same AHP scale is used. Considering the high frequency of large magnitude earthquakes in the Himalaya, R_i value of 8 (corresponding to H-VH perceived impact) is assigned to the PGA thematic layer. The PGA thematic layer is categorized into five distinct classes (Fig. 6), for which W_{ij} are assigned following the same procedure as shown in Table 3. Integrating the seismic factor (PGA layer and classes) and the static factors, a new decision matrix is formed. After the assigned ranks and weights are found consistent (CR <10%), they are numerically integrated in a GIS environment to produce the LHZ map under seismic condition (Fig. 8). The seismic LSZ map is classified into five distinct zones of landslide hazard. The final LHZ map (Fig. 8) that incorporates both static and dynamic landslide causative factors depicts a realistic situation of the prevailing state of landslide hazard when earthquake scenario is considered. Both maps are statistically compared for a better understanding of the implication of seismicity on landslide hazard in the Himalaya. The results are shown in Table 4.

Incorporation of seismic factor has substantially increased the degree of landslide potential in the study area. It is observed that the maximum and the minimum increments in LPI values are 34.6% and 18.3%, respectively. This implies that while static factors remain the primary preparatory causes of landslide incidence,

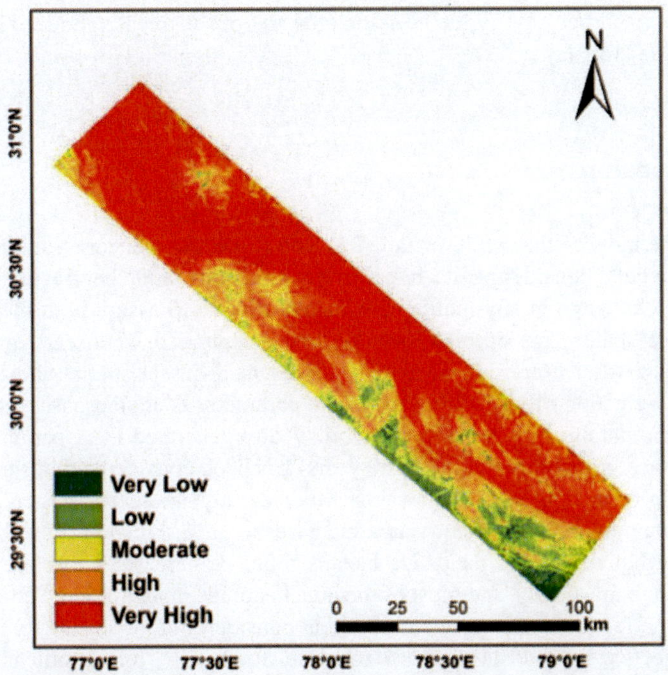

Fig. 8 Landslide hazard zonation map of the study area under seismic conditions (LHZ_Seismic)

Table 4 Comparison of LHZ maps under static and seismic conditions

LHZ map	% Change in LPI_{min}	% Change in LPI_{max}	Ratio of area of VL zone to total area	% Change in Area of VL zone	Ratio of area of VH zone to total area	% Change in Area of VH zone
LHZ_Static	18.3	34.6	0.17	−32	0.9	+40
LHZ_Seismic			0.01		51	

earthquakes induce landslides at a higher proportion. Especially, an increase of approximately 35% in maximum LPI value indicates the alarming risk of seismically induced landslides in the study area. The same conclusion can be arrived from analyses of landslide hazard zones. Introduction of seismic factor renders 51% of the total study area susceptible to a very high degree of landslide hazard, whereas the corresponding figure under static conditions was only 9%. That results in a spatial gain of 40% in Very High hazard zones, which *in-turn* is compensated by shrinkage of Very Low, Low and Moderate hazard zones. These observations validate the point that a major earthquake renders paradigm shifts in the zones of landslide hazard from lower to higher, signifying the associated risk of seismically induced landslide hazard in the Himalaya.

7 Conclusion

Seismically induced landslide hazard has been a major concern for the whole Himalayan belt, considering its hyperactive seismicity and the fragile geology. The major challenge in any multi-hazard integration is to assign both the hazards to the same scale. This problem gets further accentuated when earthquake and landslide hazards are considered as both represent different spatial and temporal variations. Selection of a suitable earthquake parameter is another issue that needs careful examination. In this study, probabilistically generated PGA parameter (for an exceedance probability of 10% in 50 years) has been considered that shows very encouraging results. The study shows that when earthquake scenario is considered, the landslide potential of the study area increases to 35%. This indicates that when a major earthquake occurs, landslide hazard zones demarcated under preparatory causes change drastically, and most of the area becomes more vulnerable to failure. Based on the results, the study concludes that consideration of seismicity is a very crucial objective for landslide hazard zonation. Especially, for tectonically active mountain chains like the Himalaya, integration of seismic and landslide hazard has practical implications in realizing the actual or potential risk associated with landslides.

References

1. Molnar P, Tapponnier P (1975) Cenozoic tectonics of Asia: effects of a continental collision. Science 189:419–426
2. Bilham R, Gaur VK, Molnar P (2001) Himalayan seismic hazard. Science 293:1441–1444
3. Harrison TM, Copeland P, Kidd WSF, Yin A (1992) Raising Tibet. Science 255:1663–1670
4. Bird JF, Bommer JJ (2004) Earthquake losses due to ground failure. Eng Geol 75(2):147–179
5. Marano KD, Wald DJ, Allen TI (2010) Global earthquake casualties due to secondary effects: a quantitative analysis for improving rapid loss analyses. Nat Hazards 52(2):319–328
6. Marui H, Nadim F (2009) Landslides and multi-hazards. In: Landslides–disaster risk reduction, pp. 435–450. Springer, Berlin, Heidelberg
7. Tang C, van Westen CJ, Tanyas H, Jetten VG (2016) Analysing post-earthquake landslide activity using multi-temporal landslide inventories near the epicentral area of the 2008 Wenchuan earthquake. Nat Hazards Earth Syst Sci 16(12):2641–2655
8. Korup O (2006) Effects of deep-seated bedrock landslides on hillslope morphology, Southern Alps, New Zealand. J Geophys Res 111(F01018):1–18
9. Champati-ray PK, Dimri S, Lakhera RC, Sati S (2007) Fuzzy-based method for landslide hazard assessment in active seismic zone of Himalaya. Landslides 4(2)101–109
10. Pareek N, Sharma ML, Arora MK (2010) Impact of seismic factors on landslide susceptibility zonation: a case study in part of Indian Himalayas. Landslides 7(2):191–201

11. Chakraborty I, Ghosh S, Bhattacharya D, Bora A (2011) Earthquake induced landslides in the Sikkim-Darjeeling Himalayas—an aftermath of the 18th September 2011 Sikkim earthquake. Report Geol Surv India, Kolkata, 1–8 (2011)
12. Kanungo DP, Arora MK, Sarkar S, Gupta RP (2009) Landslide susceptibility zonation (LSZ) mapping–a review. J South Asia Disaster Stud 2(1):81–105
13. Nath RR, Sharma ML, Tyagi A (2020) A review of the current practice on inclusion of seismicity in landslide susceptibility zonation: a case study for Indian Himalayas. Himalayan Geol 42:222–233
14. Tanyaş H et al (2017) Presentation and analysis of a worldwide database of earthquake-induced landslide inventories. J Geophys Res Earth Surf 122(10):1991–2015
15. Van Westen CJ (1997) Statistical landslide hazard analysis. ILWIS 2.1 for Windows application guide. ITC Publication, Enschede, 73–84
16. Kayastha P, Dhital MR, De Smedt F (2013) Application of the analytical hierarchy process (AHP) for landslide susceptibility mapping: a case study from the Tinau watershed, West Nepal. Comput Geosci 52:398–408
17. Pourghasemi HR, Pradhan B, Gokceoglu C (2012) Application of fuzzy logic and analytical hierarchy process (AHP) to landslide susceptibility mapping at Haraz watershed. Iran Natural Hazards 63(2):965–996
18. Sarkar S, Kanungo DP (2004) An integrated approach for landslide susceptibility mapping using remote sensing and GIS. Photogramm Eng Remote Sens 70(5):617–625
19. Saaty TL (1980) The analytic hierarchy process: planning, priority setting, resource allocation. McGraw-Hill Book Co, USA
20. Chaudhary C, Sharma ML (2017) Probabilistic models for earthquakes with large return periods in Himalaya region. Pure Appl Geophys 174(12):4313–4327
21. Dasgupta S, Narula PL, Acharyya SK, Banerjee J (2000) Seismotectonic atlas of India and its environs. New Delhi, A publication of Geological Survey of India
22. Nath SK, Thingbaijam KKS (2012) Probabilistic seismic hazard assessment of India. Seismol Res Lett 83(1):135–149
23. Mahajan AK, Thakur VC, Sharma ML, Chauhan M (2010) Probabilistic seismic hazard map of NW Himalaya and its adjoining area. India. Natural Hazards 53(3):443–457

Ground Improvement by Using Floating Granular Piles: Experimental Studies and Numerical Investigations

Murtaza Hasan and N. K. Samadhiya

Abstract Increased infrastructure expansion in urban and metropolitan areas has contributed in recent years to a significant rise in real estate prices and a lack of viable construction sites. In order to artificially enhance the soil properties at these sites, a relatively recent innovation has been the ground improvement of soft cohesive soils using granular piles, becoming common in the last forty years. This paper describes the findings on floating granular piles installed in soft clay from laboratory model studies and numerical analyses. By simulating the idea of unit cells, short-term and displacement-controlled laboratory model experiments were performed. The impacts on the performance of granular piles of multiple parameters such as area replacement ratio, spacing, diameter and length of piles have been investigated in detail. The qualitative and quantitative improvements in the ultimate bearing capacity of improved ground and failure mode observed have been discussed. It has been found that even for settlements up to 20 mm, the failure does not take place and load-settlement plots are almost linear. The bearing capacity factor increases with increments in the pile length and diameter. No significant bulging was noticed, and floating granular piles were found to penetrate into clay bed up to some depth.

Keywords Stone columns · Clayey soil · Plaxis 3D

1 Introduction

Recent spurt in infrastructural growth has forced the reclamation of sites, otherwise unsuitable for construction, using various ground improvement techniques. Out of several techniques available, granular piles have been widely adopted for improving the load-carrying capacity, reducing the total settlement, increasing the unit weight and obtaining higher shear strength of treated ground [3, 4, 6]. The lateral confinement pressure from surrounding soils plays a dominant role to achieve

M. Hasan (✉)
Government Engineering College Bharatpur, Rajasthan 321001, India

N. K. Samadhiya
Indian Institute of Technology Roorkee, Roorkee, Uttarakhand 247667, India

T. G. Sitharam et al. (eds.), *Earthquake Geotechnics*, Lecture Notes in Civil Engineering 187, https://doi.org/10.1007/978-981-16-5669-9_38

load-carrying capacity of granular piles. The granular piles are most effective in clay soils having shear strength with range in 7–50 kPa, for supporting liquid storage tanks, embankments, raft foundations, granular piles are commonly used. Various researchers have carried out large and small size laboratory tests, field studies and numerical analyses to study the influence of granular piles installed in soft clays [2, 7, 11, 12, 14, 16, 18,]. Most of the laboratory studies on granular piles installed in soft clay are associated with end-bearing piles; however, limited information is available on the performance of floating granular piles. Therefore, the aim of present investigation is to study of series of variable parameters on the efficiency of floating granular piles installed in soft clay.

2 Experimental Programme

Details of the programme of experimental investigations are given in Table 1. Unit cell idealization was adopted by assuming piles in a triangular pattern. In their experimental investigations on granular piles, several researchers [2, 3, 5, 9] have used the unit cell concept. All laboratory model experiments were performed out on granular piles in clay with diameters of 60 mm, 75 mm and 90 mm. In the case of floating granular piles, the length of granular piles was maintained as 4d, 5d and 6d, whereas 7d for end-bearing piles (d = pile diameter). For 60 mm, 75 mm and 90 mm diameter granular piles, the area replacement ratio A_r was found to be 9, 14 and 20% with corresponding spacing (s) of 3.1d, 2.5d and 2d, respectively. In a 200 mm cylindrical tank diameter, the granular piles were placed in the clay bed and during the experiments, the height of the clay bed was held 7d. Throughout the experimental work, the undrained shear strength (c_u) of the clay bed was maintained near 20 kPa.

2.1 Materials Used

The clay is classified as CI as per IS: 1498:2000 which was transformed into powder by grinding and stored in air-dried space. Table 2 presents the physical properties

Table 1 Details of experimental programme

Test description	d (mm)	Length of granular pile			
		4d	5 d	6d	7d
Clay bed	–	–	–	–	–
Granular piles	60, Ar = 9%, s/d = 2				
	75, Ar = 14%, s/d = 2.5				
	90, Ar = 20%, s/d = 3.1				

Table 2 Results of laboratory model tests and FEM analysis

Length of pile	Granular pile diameter								
	60 mm			75 mm			90 mm		
	c_u (kPa)	Ultimate bearing capacity (kPa)		c_u (kPa)	Ultimate bearing capacity (kPa)		c_u (kPa)	Ultimate bearing capacity (kPa)	
		Exp	FEM		Exp	FEM		Exp	FEM
Clay bed	20.15	190.49	186.31	19.88	190.49	186.31	20.12	190.49	186.31
4d	20.19	249.20	240.83	20.14	286.03	281.88	19.94	355.98	343.48
5d	20.69	276.38	259.41	20.58	312.20	300.06	20.54	381.63	372.71
6d	19.95	290.97	268.78	20.11	334.51	320.16	20.26	400.77	380.91
7d	20.28	303.43	275.21	20.36	351.68	338.86	20.65	429.60	404.43

of clay. To construct granular piles, crushed granite aggregates were kept in sizes of 2–6.3 mm. It has been found that the maximum and minimum dry unit weights of stone aggregate were found to be 15.04 kN/m^3 and 13.41 kN/m^3, respectively. 70% relative density of aggregates during granular piles construction was kept to minimize bulging. The strong quality aggregates were used to avoid crushing during the test. The dry unit weight and internal friction angle of stone aggregates were found to be 14.51 kN/m^3 and 42° at 70% relative density, respectively.

2.2 *Clay Bed Preparation and Granular Pile Construction*

In the 200 mm diameter cylindrical tank, clay beds with undrained shear strength about 20 kPa were formed in all experiments. The clay bed heights were maintained as 420 mm, 525 mm and 630 mm for granular pile diameters of 60 mm, 75 mm and 90 mm, respectively, in order to maintain the clay bed height as 7d. To find the water content corresponding to the shear strength of 20 kPa, the hit and test method was used. The dry unit weight and water content corresponding to 20 kPa shear strength were calculated as 15.12 kN/m^3 and 27%, respectively. The calculated amount of clay was uniformly mixed with water in a large size container. To create frictionless boundaries between inner wall and clay, a thin layer of grease was added to the interior surface of the cylindrical tank. The tank was filled into layers having a height of 30 mm and properly compacted. The wet jute bag was used to cover the clay bed and left for two days after completion of each clay bed to gain uniformity. On an undisturbed sample, collected from the centre of the clay bed, unconfined compressive strength tests were carried out to verify the undrained shear strength before construction of granular piles.

The replacement technique was used in the construction of granular piles. The detailed procedure is explained in the research papers [8, 9].

2.3 Experimental Set-Up and Testing Procedure

The load has been applied with help of circular steel plate having diameter 10 mm less than the inside diameter of test tank up to a settlement of 20 mm. For laboratory model experiments, the detailed procedure of set-up and procedure is explained in the research papers by Hasan and Samadhiya [8, 9].

3 Finite Element Analysis

The numerical analyses were carried out using professional software Plaxis 3D. The model was generated and analysed using Mohr–Coulomb failure criterion in the present research. The results obtained through Plaxis 3D was in good agreement with experimental results through model tests. The detailed explanations are given in the research papers by Hasan and Samadhiya [8, 9].

4 Results and Discussion

The entire area loading condition represents similar behaviour of an interior granular pile in the field when granular piles in large group are loaded simultaneously. The model tests in laboratory and FEM examinations were carried out to evaluate the stiffness of granular piles treated ground. In the present research paper, the laboratory model tests were conducted on 60 mm, 75 mm and 90 mm diameters single granular piles. The length of granular pile has also been varied from 4 to 7d. The undrained shear strength (c_u) all through the experimental work of the clay bed was kept close to 20 kPa. Samadhiya et al. [15] suggested critical length to be 4–5d required to develop the full limiting axial stress on the granular pile. In case of floating granular piles, failure stress on granular piles has been found to increase up to pile length, 9d as reported by Ali et al. [1]. The experimental and FEM analysis results for untreated ground as well as treated ground with 75 mm diameter floating (pile length = 4d) and end-bearing granular piles were compared and have been presented in Fig. 1.

The results of laboratory model tests, as can be seen from the figure, are in close agreement with numerical analyses outcomes. The results of the experimental and FEM analysis for untreated and treated soil with granular piles having diameters of 60 mm in the form of load-settlement plots are shown in Fig. 2.

The vertical load for granular piles treated ground was found to increase with the increase in the settlement and did not show failure up to 20 mm settlement. It may be noted that curves for granular piles treated ground have been linear in the behaviour, up to a 20 mm settlement. For comparison purpose, the ultimate bearing capacity of granular piles reinforced ground has been evaluated through the linear part of load-settlement plot, up to 20 mm settlement. This settlement (20 mm) represents

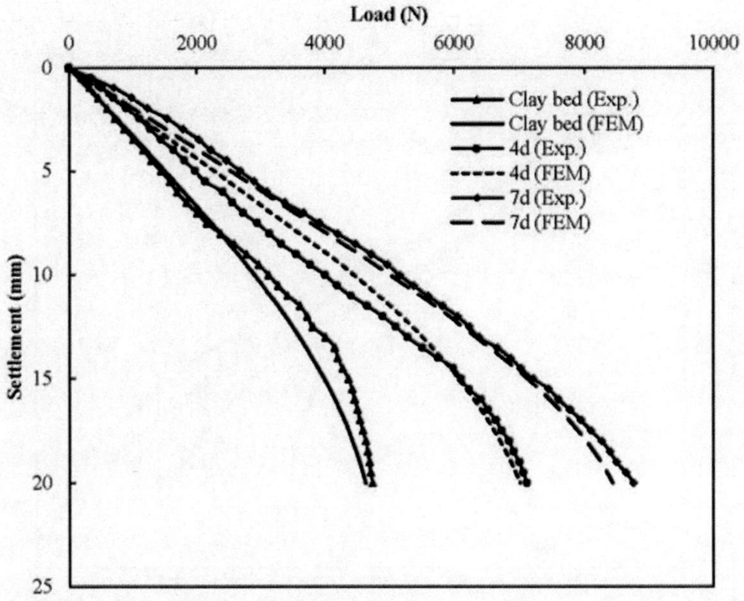

Fig. 1 Experimental and FEM analysis results comparison for 75 mm diameter granular pile

about 11% of loading plate diameter. Laboratory model test results in FEM analysis for granular piles reinforced ground as shown in Table 2.

In terms of the bearing capacity factor (β), the increase in the ultimate bearing capacity of the treated ground due to the presence of the granular pile was measured. The bearing capacity factor (β) has been expressed as the ratio of ultimate bearing capacity of treated ground to that of untreated ground corresponding to 20 mm settlement.

4.1 Effect of the Length of Granular Pile

With the increase in the length of granular piles, the ultimate bearing capacity of treated ground was found to increase, which may be attributed to the replacement of higher percentage of soft soil by granular material. The variation of ultimate bearing capacity of granular piles treated ground with length of pile is shown in Fig. 3.

In comparison with untreated ground, the ultimate bearing capacity of treated ground with 60 mm diameter granular piles of length 4d, 5d, 6d and 7d was found to increase by 30%, 45%, 53% and 59%, respectively. The ultimate bearing capacity of treated ground with 75 mm diameter granular piles of length 4d, 5d, 6d and 7d has been found to improve by 50%, 64%, 75% and 84%, respectively, with respect of untreated ground. In comparison with untreated ground, the treated ground with

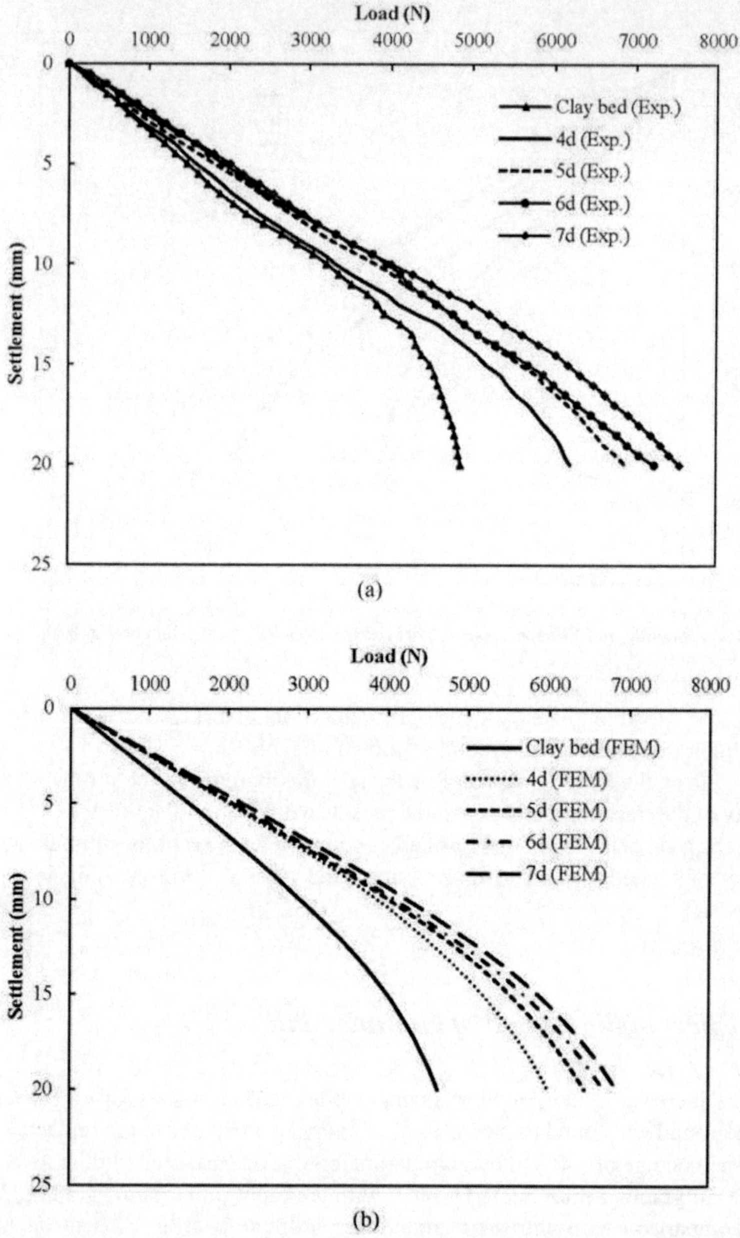

Fig. 2 Load-settlement behaviour of 60 mm diameter granular pile, **a** experimental results and **b** FEM analyses

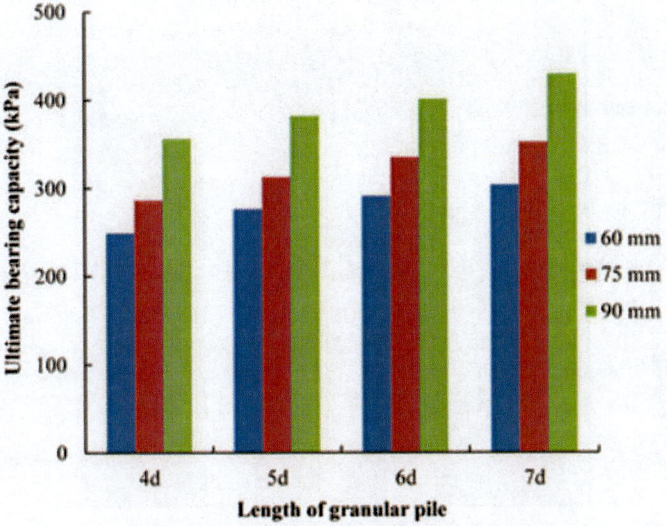

Fig. 3 Variation of ultimate bearing capacity with length of pile

90 mm diameter granular piles of length 4d, 5d, 6d and 7d; it was increased by 87%, 100%, 110% and 125%, respectively. The increment in ultimate bearing capacity of treated ground for 60 mm, 75 mm and 90 mm diameters end-bearing pile of length 4d was found to be almost 22% in comparison with floating granular pile (length of pile = 4d) having the same diameter. The bearing capacity factor (β) for 60 mm diameter granular pile varied in the range of 1.31–1.59 (exp.) andc1.29–1.48 (FEM). For 75 mm diameter piles, it was varied from 1.50–1.85 (exp.) and 1.51–1.82 (FEM). In the range of 1.87–2.26 (exp.) and 1.84–2.17 (FEM) for 90 mm pile diameter, the bearing capacity factor (β) of treated soil was found.

4.2 *Effect of the Diameter of Granular Pile*

With the increase in pile diameter, the ultimate bearing capacity of granular piles treated soil has been found to increase. The improvement in the ultimate bearing capacity of granular piles treated ground in the variation of pile diameter has been shown in Fig. 4. The ultimate bearing capacity of treated ground with 60 mm, 75 mm and 90 mm diameters floating granular piles (pile length = 4d) increased by 30%, 50% and 87%, respectively, over untreated ground, whereas it increased by 15% and 43% for 75 mm and 90 mm diameters floating granular piles treated ground, respectively, over the 60 mm diameter floating granular piles treated ground.

The ultimate bearing capacity of treated ground with 60 mm, 75 mm and 90 mm diameters end-bearing granular piles increased by 59%, 84% and 125%, respectively, as compared to untreated ground. The range of bearing capacity factor of granular

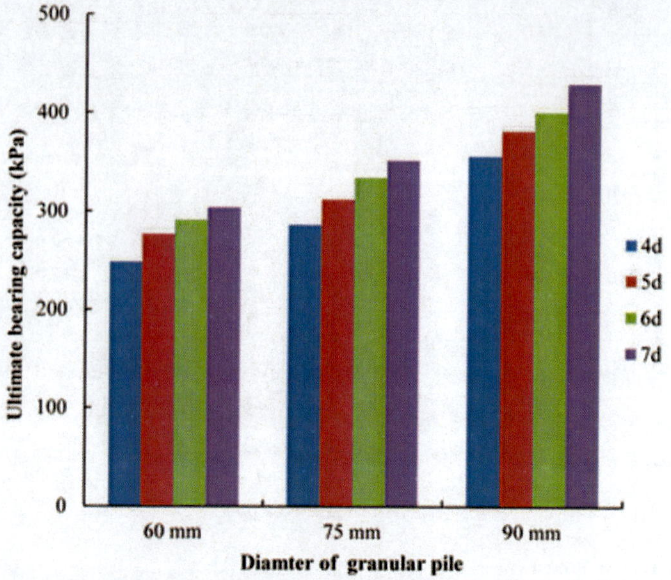

Fig. 4 Variation of ultimate bearing capacity with diameter of pile

piles treated ground with 60 mm, 75 mm and 90 mm diameter granular piles has been found as 1.31–1.87 (exp.) and (1.29–1.84) (FEM); 1.45–2.0 (exp.) and 1.39–2.0 (FEM); 1.523–2.10 (exp.) and 1.44–2.04 (FEM); 1.59–2.26 (exp.) and 1.48–2.17 (FEM) corresponding to the length of pile 4d, 5d, 6d and 7d, respectively.

4.3 Failure Modes of Granular Piles

The deformed forms obtained from laboratory model experiments and FEM analyses for granular piles with a diameter of 60 mm are shown in Fig. 5. No major bulging was observed, and all the floating granular piles were found to penetrate up to some extent into the clay bed. It should be noted that the vertical load was also shared by the surrounding soil, which simulates the surcharge loading on the surrounding soil, since the passive resistance provided against the lateral bulging of granular piles is higher compared to the case of loading of the pile alone, which can be due to less bulging. These findings are similar to that observed by Ambily and Gandhi [2], Shivashankar et al. [17] and Mohanty and Samanta [13]. Same deformation patterns were also observed for 75 mm and 90 mm diameter granular piles.

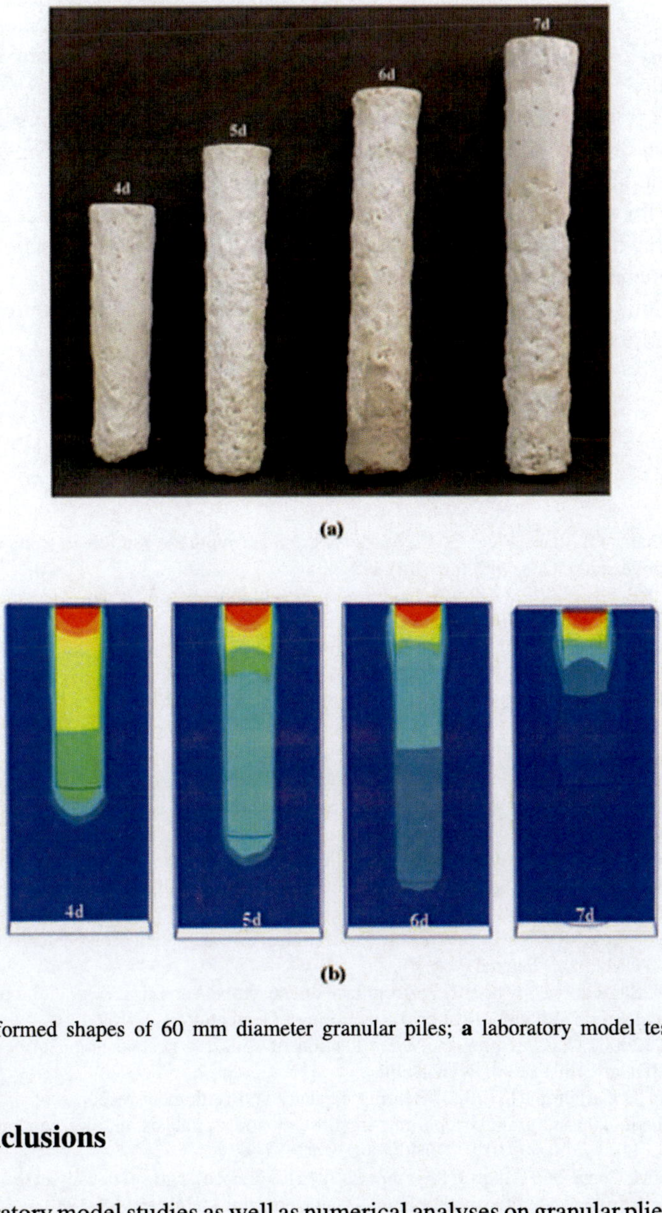

Fig. 5 Deformed shapes of 60 mm diameter granular piles; **a** laboratory model tests, **b** FEM analyses

5 Conclusions

The laboratory model studies as well as numerical analyses on granular plies (floating and end bearing) of 60 mm, 75 mm and 90 mm diameter were carried out in the present investigation. The effects of various parameters like area replacement ratio, spacing, diameter and length of granular piles on the performance of granular piles were studied in details. From this research, the following conclusions can be drawn.

1. With increment in pile length as well as diameter, the ultimate bearing capacity of granular piles improved ground also increased.

2. The ultimate bearing capacity of 90 mm diameter granular piles increases from 1.87 to 2.26 times of untreated ground with length increment from 4 to 7d of the pile.
3. It was found that ultimate bearing capacity for 75 mm and 90 mm diameters end bearing granular piles increased by 64% and 123%, respectively, over the 60 mm diameter end bearing granular piles.
4. Even for settlements up to 20 mm, the failure does not take place and load-settlement plots are almost linear. The bearing capacity factor increases with increments in the pile length and diameter.
5. No significant bulging was noticed, and floating granular piles were found to penetrate into clay bed up to some depth.

References

1. Ali K, Shahu JT, Sharma KG (2012) Model tests on geosynthetic-reinforced stone columns: a comparative study. Geosynth Int 19(4):292–305
2. Ambily AP, Gandhi SR (2007) Behavior of stone columns based on experimental and FEM analysis. J Geotech Geoenvironmental Eng ASCE 133(4):405–415
3. Barksdale RD, Bachus RC (1983) Design and construction of stone columns. Federal highway administration, RD-83/026
4. Castro J, Sagaseta C, Costa AD (2013) Consolidation and deformation around stone columns: Comparison of theoretical and laboratory results. Comput Geotech 49(1):326–337
5. Gniel J, Bouazza A (2009) Improvement of soft soils using geogrid encased stone columns. Geotext Geomembr 27(3):167–175
6. Greenwood DA (1970) Mechanical improvement of soils below ground surfaces. In: Proceedings of ground engineering conference, pp 11–22. Institution of civil engineers, London
7. Greenwood DA (1975) Vibroflotation: rationale for design and practice. In: Bell FG (ed) Methods of treatment of unstable ground. Newness-Buttersworth, London, pp 189–209
8. Hasan M, Samadhiya NK (2016) 3D numerical analysis of granular piles with internal horizontal geogrid strips in layers. In: Proceedings of the Indian geotechnical conference, pp 1–4. IIT Madras, Chennai
9. Hasan M, Samadhiya NK (2017) Performance of geosynthetic-reinforced granular piles in soft clays: model tests and numerical analysis. Comput Geotech 87:178–187
10. IS 1498 (2000) Classification and identification of soils for general engineering purposes. Indian Standards Institution, New Delhi
11. Jayarajan J, Karpurapu R (2020) Bearing capacity and settlement response of ordinary and geosynthetic encased granular columns in soft clay soils: analysis and design charts. Indian Geotech J https://doi.org/10.1007/s40098-020-00457-9
12. Miranda M, Costa DC, Castro J, Sagaseta C (2015) Influence of gravel density in the behaviour of soft soils improved with stone columns. Can Geotechn J 52(12):1968–1980
13. Mohanty P, Samanta M (2015) Experimental and numerical studies on response of the stone column in layered soil. Int J Geosynthetics Ground Eng 1:27
14. Rao NS, Prasad YVS, Rao HV (1992) Use of stone columns in soft marine clays. In: Proceedings of the 45th Canadian geotechnical conference, Toronto, (1992)
15. Samadhiya NK, Maheswari P, Basu P, Kumar MB (2008) Load settlement characteristics of granular piles with randomly mixed fibres. Indian Geotechnical Journal 38(3):345–354
16. Shahu JT, Reddy YR (2014) Estimating long-term settlement of floating stone column groups. Can Geotech J 51(7):770–781

17. Shivashankar R, Babu MRD, Nayak S, Rajathkumar V (2011) Experimental studies on behaviour of stone columns in layered soils. Geotechn Geol Eng 29:749–757
18. Tai P, Indraratna B, Rujikiatkamjorn C (2018) (2017) Experimental simulation and mathematical modelling of clogging in stone column. Can Geotech J 55(3):427–436

Finite Element Simulation of Vertical Pile Load Tests for Piles Casted Partly in Soil Cement and Natural Alluvium Soil Deposits

Soumalya Das, Shrikant D. Mishra, Apurba Mondal, R. N. Sarangi, Raghupati Roy, and Arvind Shrivastava

Abstract One of the proposed site in Northern India for four unit 700 MWe nuclear power plant (NPP) is being founded on alluvial soil. Soil medium at the site mainly consists of silty sand/sandy silt. Detailed analysis shows that top few meters of soil may liquefy under extreme earthquake scenario. To mitigate the liquefaction potential, ground improvement by means of using compacted soil–cement mix is placed for a certain liquefiable depth of the soil sub-strata. Safety-related structures of NPPs typically configured with thick shear walls for shielding as well as structural purposes that make the structures quite stiff and heavy. The combined pile-raft foundation (CPRF) system has been adopted for some of the safety-related structures of NPP to control both total as well as differential settlement. Performance of CPRF system is assessed through detailed soil structure interaction analysis where different pile-soil-raft interactions are appropriately simulated. As per 'ISSMGE Combined Pile Raft Foundation Guideline', for the design of CPRF system, detailed knowledge of single standalone pile behavior is required. Pile-soil interface parameters are calibrated based on site-specific pile load tests data. In the present work, an effort is made to simulate the field pile test results through finite element analysis. Single pile behavior along with the supporting anchor/reaction piles is modeled in Abaqus software to perform finite element analysis with due consideration of pile-soil interaction aspect. The interface parameters are suitably calibrated such that simulated

S. Das (✉) · S. D. Mishra · A. Mondal · R. N. Sarangi · R. Roy · A. Shrivastava
NPCIL, Mumbai, India
e-mail: soumalyadas@npcil.co.in

S. D. Mishra
e-mail: mishrasd@npcil.co.in

A. Mondal
e-mail: apurba@npcil.co.in

R. N. Sarangi
e-mail: rnsarangi@npcil.co.in

R. Roy
e-mail: rroy@npcil.co.in

A. Shrivastava
e-mail: ashrivastava@npcil.co.in

T. G. Sitharam et al. (eds.), *Earthquake Geotechnics*, Lecture Notes in Civil Engineering 187, https://doi.org/10.1007/978-981-16-5669-9_39

pile behavior closely matches the observed pile behavior established in pile load tests. The final calibrated soil-pile interface parameters, thus, obtained are to be used to perform detailed soil-structure interaction analysis and subsequent qualification of CPRF system of safety-related structures.

Keywords Combined piled-raft foundation (CPRF) · Soil–cement · Soil-pile interaction · Pile-soil interface parameters · Load transfer behavior · Mohr–Coulomb material model · Cohesive behavior

1 Introduction

Accurate prediction of behavior of structures, systems, and components (SCCs) of a nuclear power plant (NPP) is extremely important for ensuring adequate safety of public, occupational workers, and protection of environment. Safety-related structures of a NPP are designed for a very low probability of failure. Since, alluvium soil deposits are prone to higher settlements, combined pile-raft foundation (CPRF) system has been adopted for controlling the total and differential settlements of various safety-related structures. To mitigate the liquefaction potential of the site, compacted soil cement mixing has been placed up to a certain liquefiable depth of soil for the entire site. The piles of the CPRF system are, thus, embedded in soil cement layer up to certain depth and in existing alluvium soil deposit thereafter.

In the present study, an effort is made to appropriately simulate the pile behavior embedded partly in compacted soil cement layer (top portion) and partly in natural soil. Results of four initial pile load tests (Tests 01, 02, 03, and 04) have been considered in the present work. Each pile has a diameter of 1200 mm and a depth of 36 m.

Behavior of single pile or group of pile is generally analyzed using specialized software like Plaxis [2]; in the study presented in this paper, an attempt has been made to simulate the accurate behavior of pile using a general purpose finite element software Abaqus.

1.1 Test Pile Details and Layout

The tests were carried out on bored-cast in situ piles for vertical load up to the failure that is up to a settlement corresponding to 10% of pile diameter (i.e., 120 mm). All the tests were conducted in a comparable soil condition. Piles of 1200 mm diameter and 36 m length were cast for vertical load tests. The details of the typical arrangement of test piles are given in Fig. 1. Test piles are loaded by hydraulic jacks, which take support from structural steel reaction beams. Reaction beams transfer the load to four anchor/reaction piles. Load cells and pressure gages were used to measure applied load. Hydraulic jacks are calibrated at different loadings step up to the capacity

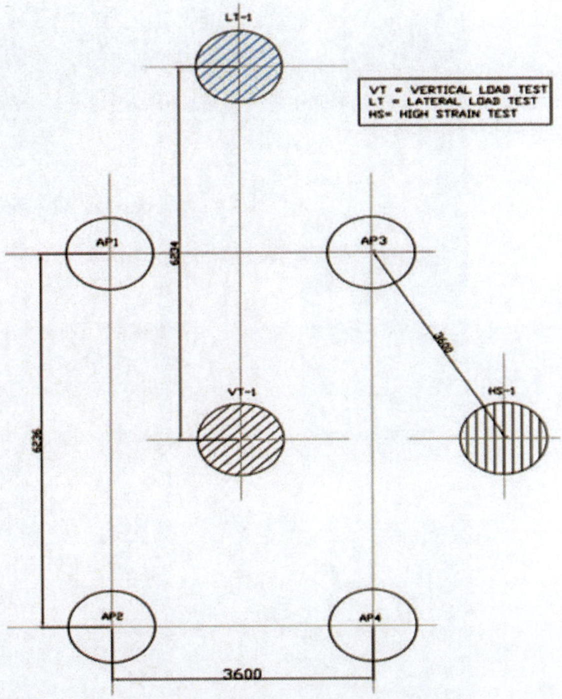

Fig. 1 Arrangement of test piles and reaction piles (plan)

of respective jacks, and jack efficiency factor is established. Load measured though pressure gauges/jacks systems is corrected using the applicable jack efficiency factor at respective load levels. Dial gauges and linear variable differential transducers (LVDTs) were used for measurement settlement. In addition, test pile number 4 was instrumented with vibrating wire sister bar (VWSB) strain gages placed at different depths along the pile shaft to establish the load transfer curve along the pile shaft. The mobilization of end bearing resistance with respect to settlement of the pile is measured from the strain gage data located near pile toe. Fig. 2 shows the site pile load test setup.

1.2 *Field Pile Test Results*

Vertical load settlement behavior Load settlement curves obtained for the four pile field tests are shown in Fig. 3. Average of four measured load settlement curves is considered as backbone curve for the simulation purposes. Ultimate capacity of pile has been considered as the load at settlement level corresponding to 10% of pile diameter (i.e., 120 mm). The mobilization of end bearing resistance with settlement, as obtained from the instrumented pile load test data (Test 04), is presented in Fig. 4.

Fig. 2 Test setup at site

Fig. 3 Load settlement behavior of pile field tests

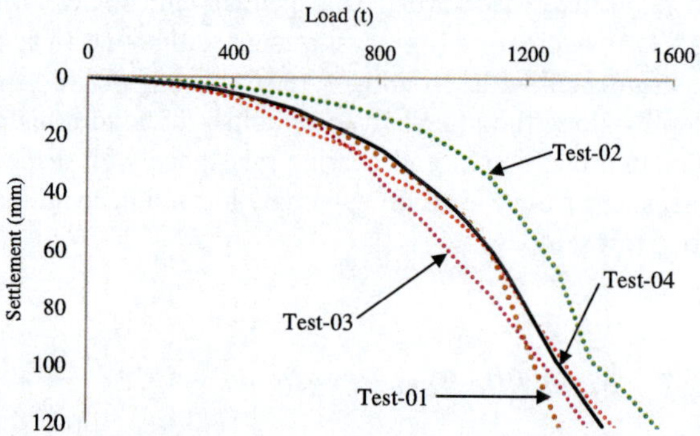

The load transfer curves along the depth of pile shaft obtained from strain gages embedded at different depths for different loading steps are shown in Fig. 5.

It appears from the load transfer curve (estimated from strain gage data) that very little friction is mobilized along the pile shaft, which seems to be unrealistic. Possible reason may be due to malfunction of embedded strain gages. This aspect is presently under investigation.

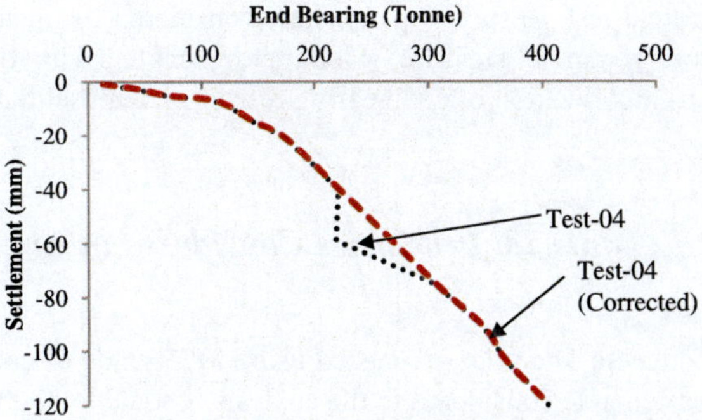

Fig. 4 Mobilization of end bearing resistance with settlement

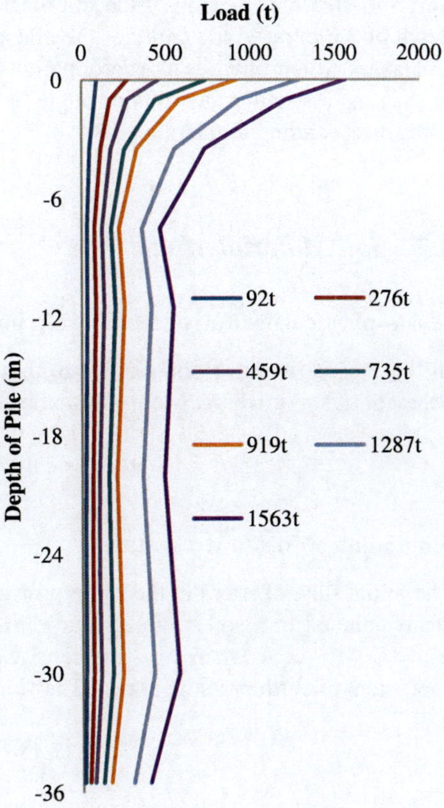

Fig. 5 Load transfer curves for various load increments along the depth of the pile shaft

2 Analytical Simulation of Field Test Results

In the present work, the entire pile test setup (test pile along with reaction piles) as shown in Fig. 1 has been modeled in Abaqus software [3]. Pile-soil interface parameters are calibrated to obtain better match with pile test behavior observed in field test. Response parameters considered for matching with pile test behavior observed in field test are (i) load settlement behavior, (ii) load transfer behavior, and (iii) mobilization of end bearing resistance with settlement.

2.1 Material Properties Considered in Analytical Simulation

Concrete The piles are casted using M35 grade of concrete. Modulus of elasticity for concrete considered in the analysis is equal to 5000 $\sqrt{f_{ck}}$ (in MPa) [4]. Poison's ratio is considered as 0.2.

Soil Sub-strata Top 3.6 m of the soil block consists of soil cement layer. Below, a depth of 3.6 m up to the depth of the pile and beyond consists of natural alluvium soil layers. Soil properties like deformation modulus of soil (E_s), cohesion (c), angle of shearing resistance (ϕ), dilation angle (ψ) are obtained from the nearest borehole data corresponding to the pile tests.

2.2 Soil Material Model

Elasto-plastic constitutive relationship for the soil medium

In the present work, Mohr–Coulomb plasticity model is used for simulating the behavior of soil medium. Mohr–Coulomb criterion is expressed as:

$$\tau = c + \sigma \tan \varphi \qquad (1)$$

Simulation of in situ stress state

The initial state of stress in the soil is considered to be geo-static in nature. That is, at any point of time, soil is subjected to vertical stress, $\sigma_z = \gamma \times h$, and lateral stress, $\sigma_L = k_0 \times \gamma \times h$, where γ is the density of soil, h is depth, and k_0 is the lateral coefficient of earth pressure at rest. The value of k_0 is specified as follows:

$$k_0 = 1 - \sin \varphi \qquad (2)$$

2.3 Modeling of Semi-infinite Soil Medium

The unbounded or infinite soil medium is approximated by extending the finite element mesh to a far distance, where the influence of the surrounding medium on the region of interest is considered small enough to be neglected. In the present work, finite element model of soil medium of 50 m × 50 m and 100 m depth has been considered based on the iterative analysis wherein effect of the boundary condition is not observed.

2.4 *Modeling of Contact Zones (Pile-soil Interface)*

The soil-structure interaction analysis requires considerations of the interaction or coupling between the structure (pile) and the founding media.

Pile-soil interaction

Tangential Behavior: The interface behavior of the raft and soil [5] (frictional force vs. relative displacement) is pictorially shown in Fig. 6. The interaction behavior is divided into two parts (a) *sticking phase* and (b) *slipping phase.* Initially, during the sticking phase, pile-soil interface allows some relative motion (elastic slip) between the pile and the soil surface. Once the induced shear stress between the pile and soil reaches the critical shear stress, the interface enters in the slipping phase.

The above sticking and slipping phase of the raft-soil interface is simulated using elastic Coulomb friction model is shown in Fig. 5. In this model, the critical shear stress (τ_{crit}) is defined as:

$$\tau_{\text{crit}} = \min(\mu p', \tau_{\max}) \tag{3}$$

Normal Behavior: Normal contact pressure develops, when pile moves normally toward the soil. Under such situation, surface area of the pile, which is in contact with soil, will develop normal stress. This phenomenon is simulated in the analysis using relationship as shown in Fig. 7 (normal hard contact). Normal hard contact assumes very high contact stiffness between master (pile) and slave (soil media) surfaces. In the present analysis, the end bearing formulation has been changed in the various approaches presented to arrive at the desired results.

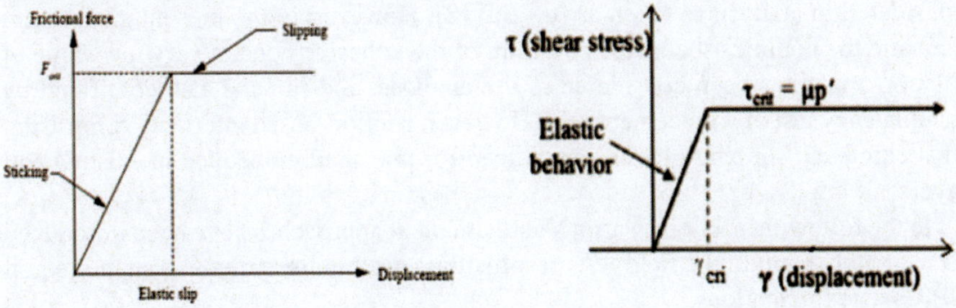

Fig. 6 Raft-soil interface behavior (tangential behavior)

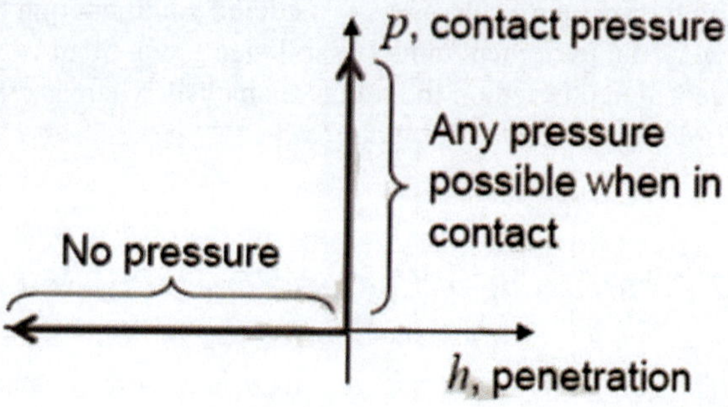

Fig. 7 Behavior of pile-soil interface in normal direction

2.5 *Finite Element Model of Pile Test Set Up*

A detailed 3D finite element model of pile test setup (i.e., test pile and reaction pile) along with the soil has been modeled in Abaqus to simulate the pile load test results. Finite element model of soil medium of 50 m × 50 m and 100 m depth has been modeled. The geometric and finite element details of the pile setup are shown in Figs. 8 and 9.

At pile bottom region, up to 5 m, below the pile tip, soil element size is restricted between 0.9 and 1.25 m in the finite element model. Further, element size along periphery of pile is kept fine as shown in Fig. 8c. As a boundary condition, *x* (north–south), *y* (east–west), and *z* (vertical) translations at the bottom of the soil are restrained. Similarly, *x*-translation is restrained in *yz* plane of soil, and *y*-translation is restrained on *xz* plane of soil.

2.6 *Interaction Between Pile and Soil Cement Surface*

It is observed that same set of contact parameters cannot be assigned to the interactive surfaces between pile shaft and compacted layer of soil cement and that between pile shaft and existing soil layer below. Cohesive-based interface is assumed in for the pile shaft embedded compacted layer of soil cement. Theoretically, dilation angle considered in analysis is taken as ($\varphi-30$) [2]. However, in the present analysis, to simulate the nonlinear behavior or failure of the cohesive bond, a very low value of dilation angle has been considered in the analysis, and same is also confirmed by the dilatency test of soil–cement mix. However, friction-based interface, comprising of sticking and slipping phases, is assumed for pile shaft embedded in existing soil layers.

In the following sections, various mathematical approaches have been introduced to accurately capture the field tests results to be used in the structures supported on CPRF-type foundations.

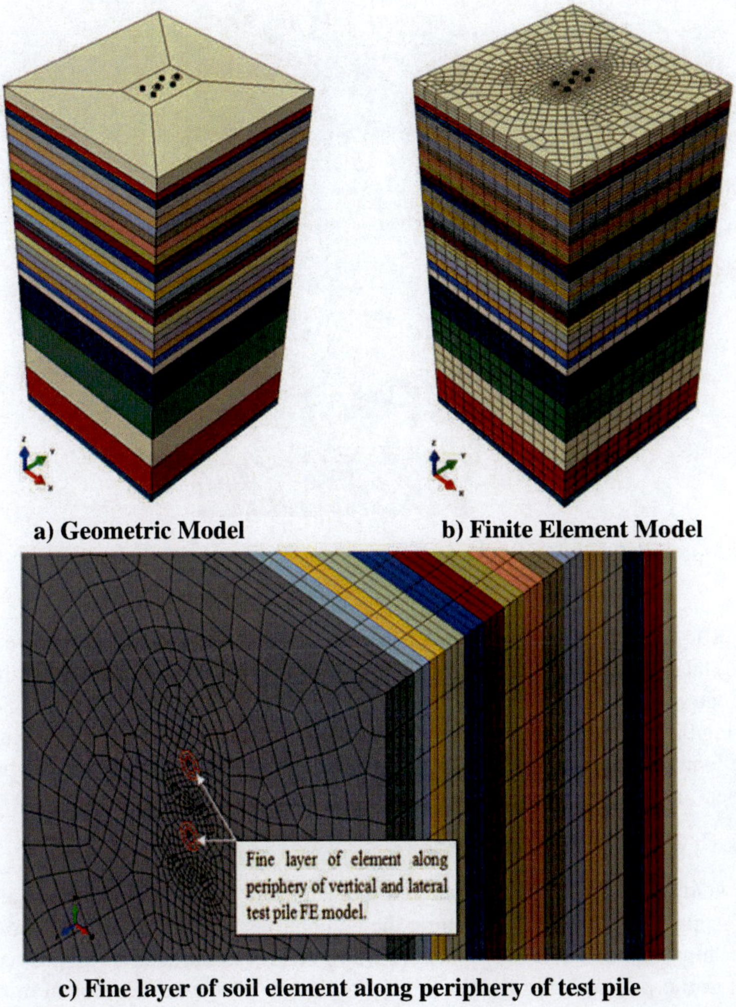

Fig. 8 Geometric and FE model of pile test setup

2.7 *Finite Element (FE) Approaches to Simulate Field Test Results*

Approach 1

In this approach, cohesive behavior between pile and soil–cement surface has been considered with a very low dilation angle of 2.5°. A very low friction factor (0.05) is between pile and soil surface for the soil layer below the depth of 7.5 m. Stress in the pile bottom tip has been limited to 3.47 MPa to limit the end bearing to 400 ton, as observed from pile load test, using pressure overclosure relationship as given

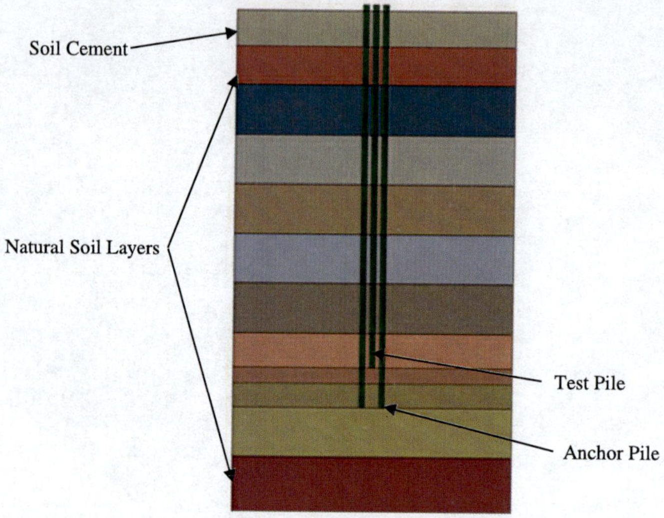

Fig. 9 Cross-section view showing soil layering with test and anchor pile arrangement

in Fig. 10. In this approach, normal hard contact behavior with very high contact stiffness has been replaced with a pressure overclosure relationship having finite and predefined contact stiffness. Soil properties of natural soil layers are obtained by averaging the parameters obtained from the nearest boreholes for each test locations.

The load deformation curve, progression of end bearing with increasing pile tip settlement, and load transfer curves are shown in Figs. 11, 12, and 13.

Approach 2

In this study, the behavior between soil–cement and pile surface is kept same as that of Approach 1. Instead of capping the end bearing using a pressure overclosure relationship considered in Approach 1, alternately, soil parameters (E_s, ϕ, and ψ values) in the pile-soil interface at the toe region in particular have been modified to simulate the progression of end bearing to ~400 ton at a settlement of 120 mm. Along with the modified soil properties in the pile-soil interface at the toe region,

Fig. 10 Description of interaction properties considered

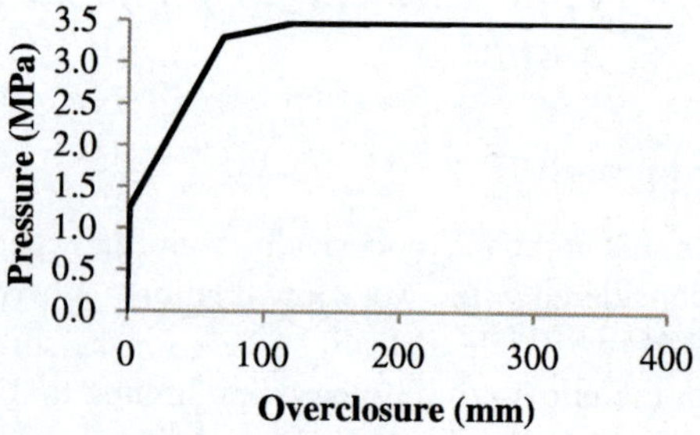

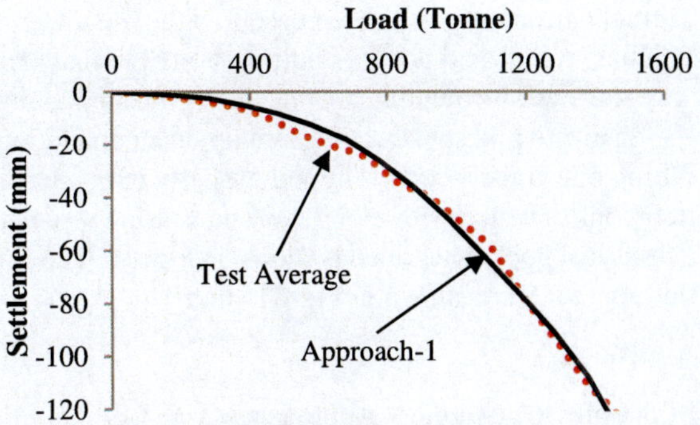

Fig. 11 Comparison of load settlement behavior obtained from FE simulation (Approach 1) and field test results of Test 04

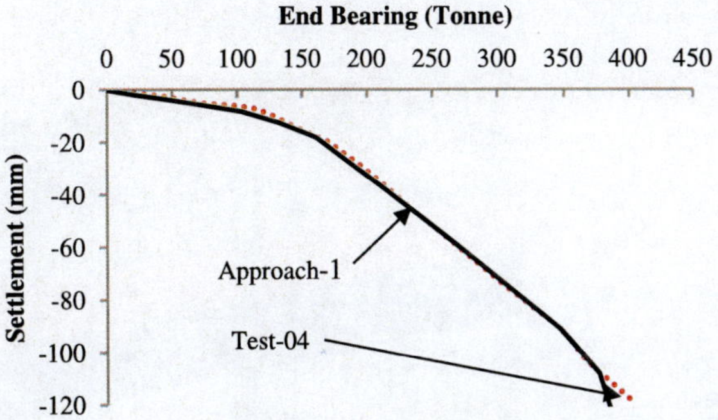

Fig. 12 Comparison of end bearing progression of FE simulation (Approach 1) and field test results of Test 04

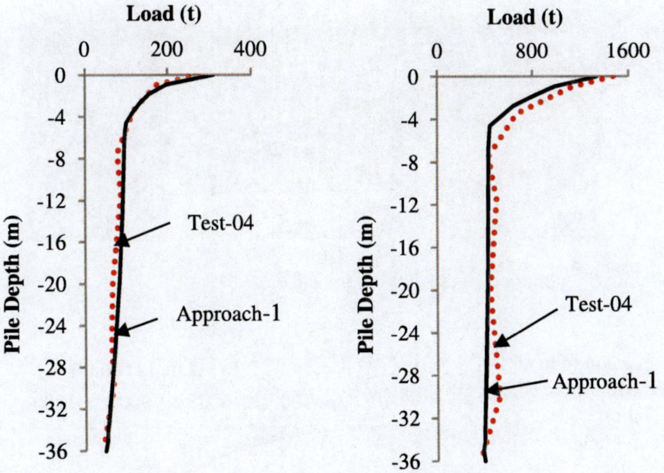

Fig. 13 Comparison of load transfer curves between FE simulation (Approach 1) of test pile and actual field test results (Test 04) for pile tip settlement ~4 and 125 mm

normal hard contact has been used to simulate the end bearing behavior. A separate volume (cylindrical pocket of dimension: 3 m height and 2.4 m diameter) in the soil was modeled surrounding the pile toe region simulating the possible disturbed pocket region due to pile construction activity or accumulation of muck in the pile toe region during pile construction. The soil property in this localized region is arrived through iterations to match with mobilized end bearing resistance obtained from the test. The cylindrical pocket modeled is shown in Fig. 14. The simulation results obtained from this approach are shown in Figs. 15 and 16.

Approach 3

In the previous studies, Approaches 1 and 2, the simulation was carried out with respect to the average field load settlement curve considering average soil properties

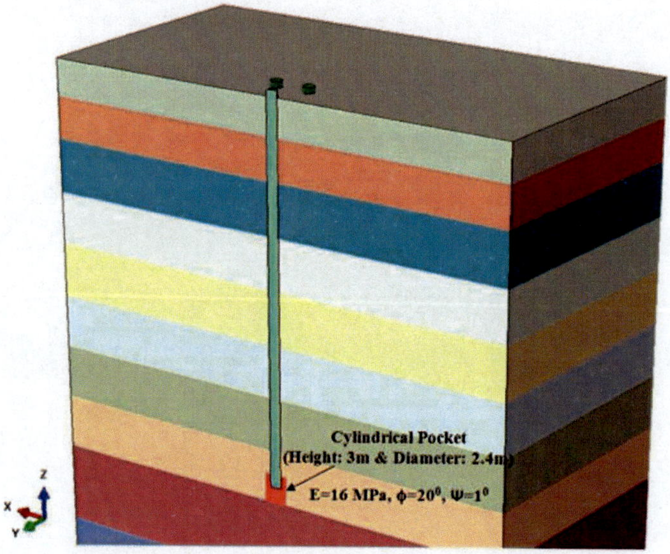

Fig. 14 Modeling of cylindrical pocket for Approach 2

Fig. 15 Comparison of load settlement behavior obtained from FE simulation (Approach 2) and field average test results

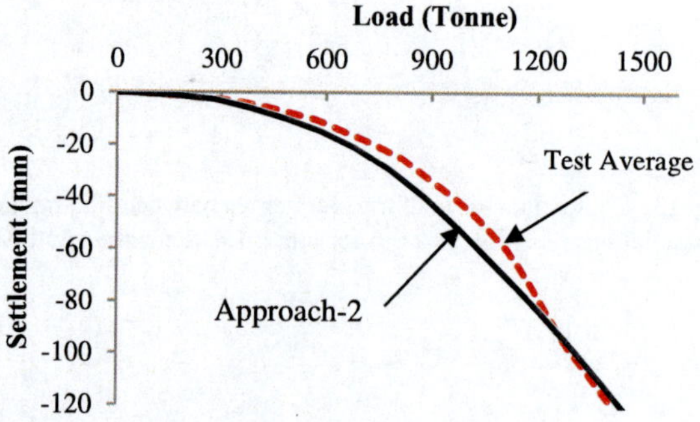

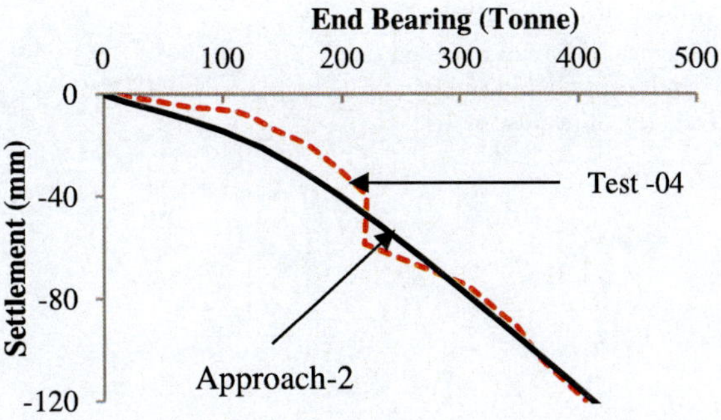

Fig. 16 Comparison of variation of end bearing with settlement obtained from post-test FE simulation and field test result

Table 1 Test pile and borehole locations

Test pile	Borehole location
Test 01	BH 01
Test 02	BH 02
Test 03	BH 03
Test 04	BH 01

for all the boreholes. In this Approach 3, simulation has been performed for individual test results considering the nearest borehole properties of the test pile location. Out of the four pile tests conducted, Tests 01 and 03 were located close to Borehole 01, Test 02 was near to Borehole 02, and Test 03 was located near to Borehole 03. The summary of test and the respective borehole locations are shown in Table 1. In this approach, the interface properties have been arrived by modifying the dilation angle of soil–cement layer and entire end layer of natural soil where the pile rests.

The load settlement curves obtained for each test simulation and its comparison with field test results are shown in Figs. 16, 17, and 18. Since, field progression of end bearing is available only from Test 04, comparison of FE simulation results for

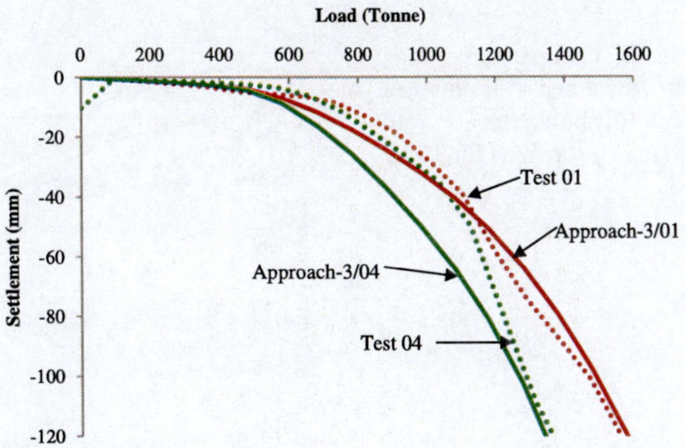

Fig. 17 Comparison of load settlement behavior obtained from FE simulation and field average test results for Tests 01 and 04

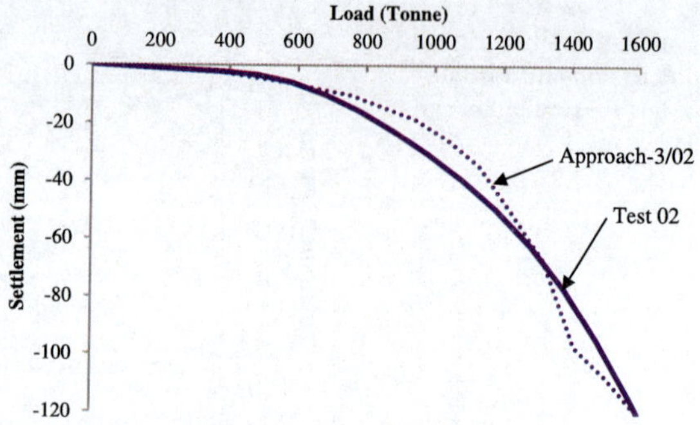

Fig. 18 Comparison of load settlement behavior obtained from FE simulation and field average test results for Test 02

all the test piles are compared with field end bearing progression results of Test 04 as shown in Fig. 19.

The dilation angle values used for FE simulation of all the four tests are averaged, and same is proposed to be used at the interface of pile and soil–cement layer for buildings supported on CPRF type of foundation (Fig. 20).

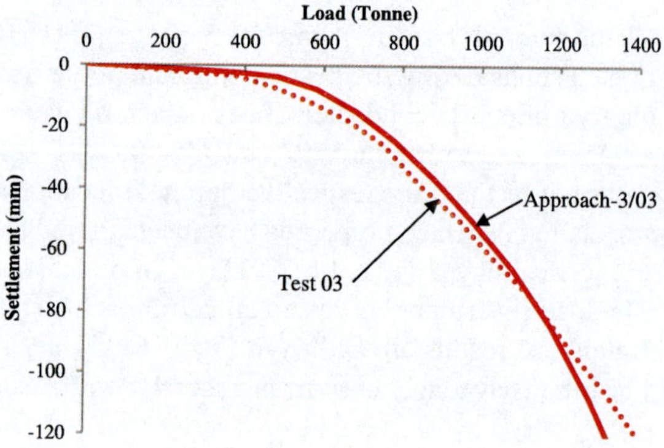

Fig. 19 Comparison of load settlement behavior obtained from FE simulation and field average test results for Test 03

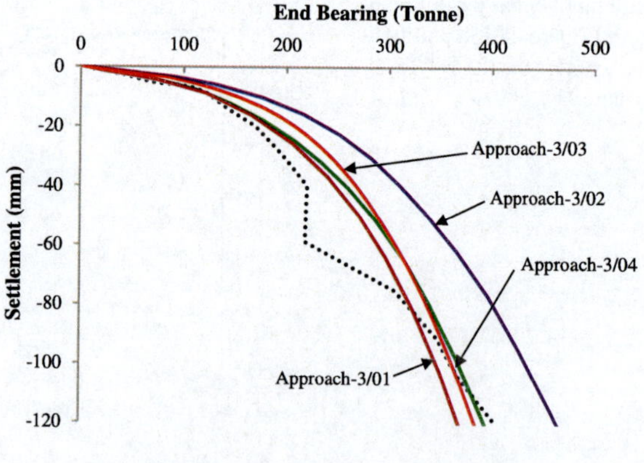

Fig. 20 Comparison of load settlement behavior obtained from FE simulation and field average test results for Test 03

3 Conclusion

For appropriate simulation of the soil-pile interaction for structure supported on CPRF, calibration of soil-pile interface parameters needs to be finalized considering the various attributes of the pile tests, namely load-deformation behavior, load-transfer pattern along the depth of pile shaft, and deformation-based mobilization of end bearing resistance. All the three methods presented in this study give reasonable approximation of the pile behavior under applied vertical loading.

In Approach 1, the soil property surrounding the pile was kept intact as obtained from detailed geotechnical investigations. Only, soil-pile interface properties were modified to obtain the desired results. In the actual scenario also, it is the interface parameters which will undergo changes due to various construction activities like concrete placement and use of bentonite. The soil properties of the surrounding media will largely remain unaffected. Hence, Approach 1 appropriately captures this phenomenon.

In Approach 2, soil properties were largely kept intact as obtained from geotechnical investigations barring the pocket region near the pile bottom tip where soil properties were degraded to obtain the desired progression of end bearing. This method though suitable for standalone pile analysis may lead to complicated geometric modeling for large NPP structures supported on CPRF.

In Approach 3, simulation was carried out for each of the pile load tests separately, and the interface parameters, thus, obtained were averaged. In this study, the entire bottom layer of the soil supporting the pile was modified to obtain the desired results. However, modification of entire soil layer though applicable for standalone pile simulation may lead to erroneous results when applied to actual structures supported on CPRF type of foundations.

From all the three studied, it may be concluded that Approach 1 gives a better approximation of single pile behavior which can be used to simulate the soil-pile interaction for structures supported on CPRF.

Acknowledgements The authors are grateful to Executive Director (Engineering) and Director (Technical), NPCIL, for their constant support and encouragement throughout the present work.

References

1. Sundaram R, Gupta S (2019) Foundations for tall buildings on alluvium deposits-geotechnical aspects. In: Madhvi Latha G (Ed), Frontiers in Geotechnical Engineering, Springer
2. Kowalska M (2014) Numerical study of the influence of dilatancy angle on bearing capacity and rotation of a gravity retaining wall. In: European conference on geotechnical engineering (DECGE 2014), Vienna, Austria. Paper No. 186
3. Manual A (2019) Dassault Systèmes (2019), Documentation of Abaqus Software, Dassault Systèmes Simulia Corp., Providence, RI, USA
4. IS 456-2000: Indian Standard, Plain and Reinforced Concrete-Code of practice

5. Patil G, Mondal TK (September, 2016) Analysis of a combined piled-raft system (CPRF) of a typical NPP safety related structure founded on alluvium soil. In: International Geotechnical engineering conference on sustainability in geotechnical engineering practices and related urban issues, Abstract ID 159, ISSMGE, IGS Mumbai chapter

Reliable Estimation of Shear Wave Velocity Using Various V_S—N Correlations

Jithin P. Zachariah and Ravi S. Jakka

Abstract Shear wave velocity (V_S) plays a key role in seismic hazard analysis. However, the direct measurement of shear wave velocity is impractical in some field conditions. The use of a viable field evaluation method such as standard penetration test (SPT) is conducted to obtain the field data at these situations. These SPT N values are correlated to the shear wave velocity using various V_S-N correlations. Yet, the direct application of SPT data from different boreholes at a particular site involves multiple uncertainties. The borehole data of multiple boreholes and the availability of numerous correlations are the primary reasons for these uncertainties. This paper discusses a statistical approach to account for the uncertainties mentioned above involved in V_S estimation using V_S-N correlations. All the available borehole data from the site and multiple correlations were incorporated in this study to develop the V_S profile.

Keywords VS-N correlations · Shear wave velocity · Reliable evaluations

1 Introduction

The seismic waves generated during an earthquake are transferred into the ground surface by wave propagation. This propagated wave is subjected to amplification in the transmission process which leads to various ground and structural failures at the ground surface. Shear wave velocity (V_S) is the velocity of the propagating S waves generated in any seismic activity that is directly influenced by the mechanical properties of the medium of propagation, i.e., the soil mass. V_S is a key tool in the seismic hazard analysis and seismic microzonation of any region. Several studies have been conducted all over the world on seismic microzonation and hazard analysis using V_S. A V_S30 map has been developed of Dinar region in Turkey and predicted

J. P. Zachariah · R. S. Jakka (✉)
Department of Earthquake Engineering, Indian Institute of Technology Roorkee, Roorkee, India
e-mail: ravi.jakka@eq.iitr.ac.in

J. P. Zachariah
e-mail: jpzachariah@eq.iitr.ac.in

T. G. Sitharam et al. (eds.), *Earthquake Geotechnics*, Lecture Notes in Civil Engineering 187, https://doi.org/10.1007/978-981-16-5669-9_40

the earthquake failure susceptibilities of that region [1]. It is evident from the study that the poor soil conditions lead to the failure susceptibilities that occurred in the region. Parallel studies were also conducted by different researchers for different regions and drawn similar conclusions [2–8]. A few researchers have performed the seismic hazard analysis of different parts of India [9–11]. The susceptibility of the Indian subcontinent toward moderate to severe earthquakes indicates the need for hazard estimation before any construction process.

The ground motion amplification and ground response are directly influenced by the shear wave velocity at a particular site. The evaluation of V_S can be conducted in both fields and in the laboratory. Piezometric bender element test, ultrasonic pulse test, resonant column test were the commonly used laboratory experiments while seismic reflection tests, seismic refraction tests, spectral analysis of surface wave tests (SASW), seismic downhole test were conducted at the field to measure the V_S. Nevertheless, the expenses involved in the tests limit the application of these tests. Standard penetration resistance (SPT N) of a soil, obtained from standard penetration test (SPT), is a commonly adopted field measurement worldwide which has a direct influence on the soil characteristics [12]. The SPT N values are correlated to the shear wave velocity using a power law, as shown in Eq. 1.

$$V_S = A.N^B \tag{1}$$

where A and B are coefficients depending on the type and properties of the soil [13]. Various modified versions of this form of the equation have been developed and applied in different regions of the world [12, 14–29]. Many V_S-N correlations for Indian cities were also developed in the timeline [9, 14, 30–33].

Even though V_S-N correlations provide reliable results, various uncertainties and ambiguities were raised in the effective application of these correlations and data set of SPT N values from different boreholes. This paper discusses a reliable method of estimation of V_S from a set of available borehole data and a number of V_S-N correlations. A statistical method of evaluation is approached to override the various uncertainties. The V_S profile is plotted in terms of mean (μ), mean ± standard ($\mu \pm \sigma$) deviation to obtain a range of possible V_S limits.

2 Uncertainties and Statistical Methodology

Estimation of V_S from SPT N values is widely practiced in seismic hazard analysis. However, the estimated V_S values are subjected to different uncertainties. For a set of borehole data available, selection of a particular borehole to evaluate the V_S is perplexing because the borehole data varies from borehole to borehole. Soil heterogeneity and measurement errors are the primary reasons for this uncertainty. Moreover, a single borehole cannot represent the entire site precisely as it generates a biased V_S profile of the site. This problem is solved by converging every borehole data available to a set of statistical parameters, μ, $\mu + \sigma$ and $\mu - \sigma$. The new set of SPT N values reflects the behavior of every available borehole data in the site, and also, the range of possible variation is observed.

To account for the variations in the estimation of V_S by different correlations, two sets of correlations were selected. The first set contains correlations developed generally for all soil, and the last set has specific correlations for sand and silt. The statistical borehole data and the virgin borehole data are passed through these set of correlations for general soil and specific soils, respectively. Further statistical analysis has been carried out to generate a second degree of μ values of V_S profile. Second-order $\mu + \sigma$ and $\mu - \sigma$ are ignored to reduce the width of the range of values. This helps in incorporating the behavior of all correlations into a set of values of V_S. The procedure can be further extended using more number of correlations which helps in improvising the quality of the results. The flow of procedure of the proposed methodology is shown in Fig. 1.

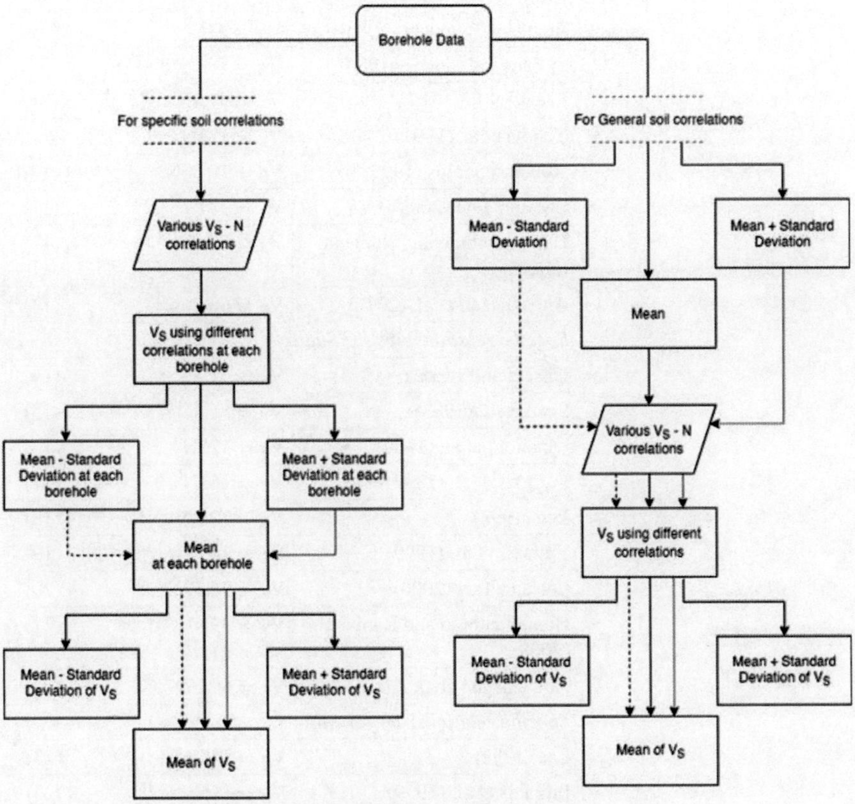

Fig. 1 Methodology adopted for estimating Vs accounting uncertainties

3 Correlations Used in This Study

Various correlations were available to evaluate V_S from the measured SPT N values. The correlations used in this particular study are listed in Table 1 (Eq. 2–25). The development period and the region of study are the two criteria for selection of correlations.

Table 1 Various correlations selected in the study

Correlation Name (Year)	Correlation	
General relationships for all types of soil:		
Ohsaki and Iwasaki (1973)	$V_S = 82N^{0.39}$	(2)
Imai and Yoshimura (1975)	$V_S = 92N^{0.329}$	(3)
Ohta and Goto (1978)	$V_S = 85.35N^{0.348}$	(4)
Seed and Idriss (1981)	$V_S = 61N^{0.5}$	(5)
Imai and Tonouchi (1982)	$V_S = 97N^{0.31}$	(6)
Sykora and Stokoe (1983)	$V_S = 100.5N^{0.329}$	(7)
Lee (1990)	$V_S = 57.4N^{0.49}$	(8)
Yokota et al. (1991)	$V_S = 121N^{0.27}$	(9)
Athanasopoulos (1995)	$V_S = 107.6N^{0.36}$	(10)
Lee and Tsai (2008)	$V_S = 137.153N^{0.229}$	(11)
Hanumanthrao and Ramana (2008)	$V_S = 82.6N^{0.43}$	(12)
Anbazhagan et al. (2012)	$V_S = 68.96N^{0.51}$	(13)
Specific relationships for Sand:		
Ohsaki and Iwasaki (1973)	$V_S = 59.4N^{0.51}$	(14)
Seed et al. (1983)	$V_S = 56.4N^{0.9}$	(15)
Okamoto et al. (1989)	$V_S = 125N^{0.7}$	(16)
Pitilakis et al. (1992)	$V_S = 162N^{0.21}$	(17)
Lee (1992)	$V_S = 104(N + 1)^{0.338}$	(18)
Pitilakis et al. (1999)	$V_S = 145.6N^{0.182}$	(19)
Lee and Tsai (2008)	$V_S = 98.07N^{0.309}$	(20)
Hanumanthrao and Ramana (2008)	$V_S = 79N^{0.438}$	(21)
Anbazhagan et al. (2012)	$V_S = 60.17N^{0.60}$	(22)
Specific relationships for Silt:		
Lee (1992)	$V_S = 104(N + 1)^{0.338}$	(23)
Pitilakis et al. (1999)	$V_S = 145.6N^{0.182}$	(24)
Hanumanthrao and Ramana (2008)	$V_S = 86N^{0.46}$	(25)

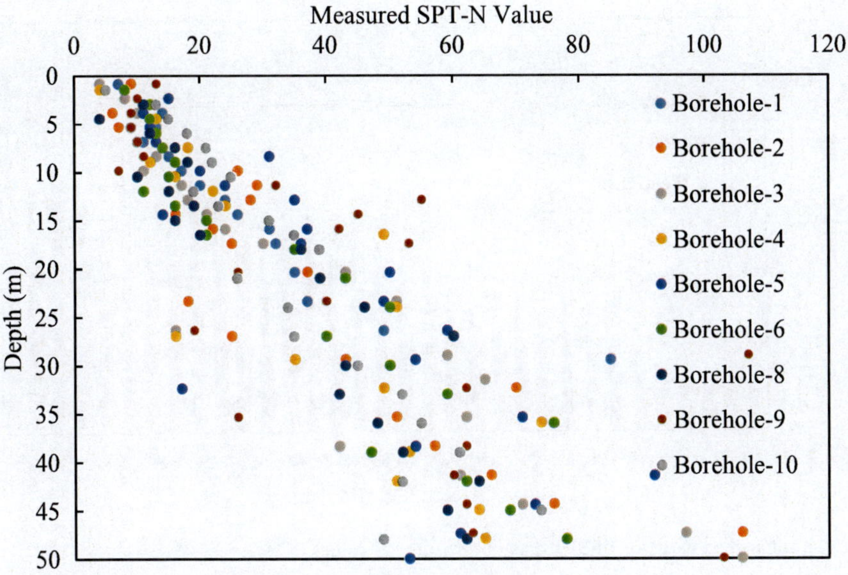

Fig. 2 Variation of measured SPT *N* value with depth

4 Site Study and Geotechnical Investigation

A project site at Delhi, India, is considered to conduct this particular study. 6.5–7 magnitude earthquakes are expected to occur in these regions. SPT has been conducted in 9 boreholes of 100 mm diameter at different locations of the site at different depths of 1.5 m interval. Most of the depth of the soil strata is observed to have sand with traces of silt or clay. The SPT *N* values observed at different depths of each borehole are plotted in Fig. 2.

5 Statistical Evaluation

To eliminate various uncertainties involved, statistical analysis has been incorporated in the V_S estimation procedure. From the ground conditions, it is seen that the available measured SPT values of different boreholes are at different depths. To take care of this aspect, all the SPT measurements and corresponding depths were converted in terms of actual depth elevation (ADE) from a common reference point at road level. Calculations are done to obtain the μ and σ of SPT data at different depth elevations. The statistical variation of the SPT N values is represented with depth elevations in Fig. 3.

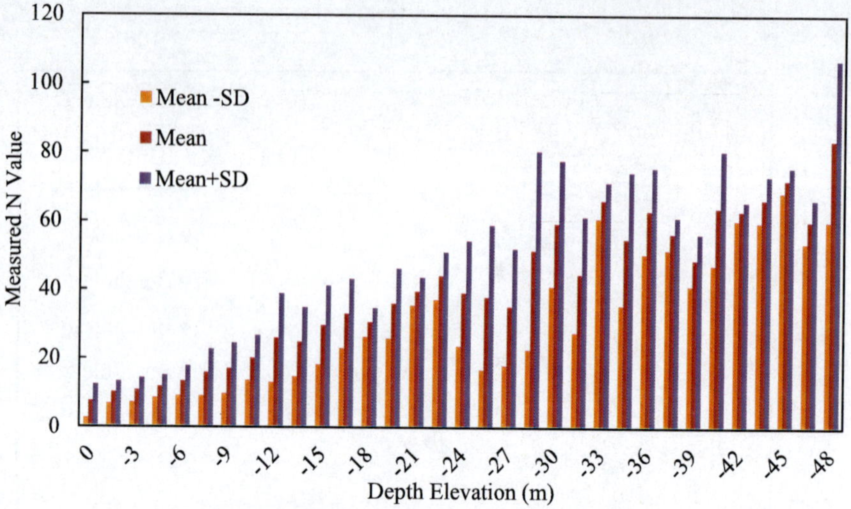

Fig. 3 Statistical variation of SPT data of measured *N* values

6 Results and Discussions

Shear wave velocity profile has been developed based on the statistical SPT N values using the above correlations. The general soil correlations are applied in the first stage of estimation. The V_S profile undergoes a broader range of values in μ, $\mu + \sigma$ and μ $\mu - \sigma$ σ sets when calculated using various correlations. This ambiguity is solved by converging the values into μ and $\mu \pm \sigma$ V_S profiles. Figure 4 shows the band of V_S values along the depth after the statistical evaluation of mean of V_S profiles developed using different correlations.

The statistical evaluation is further extended to $\mu \pm \sigma$ V_S profiles. In the case of $\mu + \sigma$ and $\mu - \sigma$ of $\mu + \sigma$ and $\mu - \sigma$ of V_S profiles, the range of values is too broad, such that the reliability of the results is questioned. In order to take care of this problem, the second-order $\mu + \sigma$ and $\mu - \sigma$ are ignored. Mean shear wave velocity obtained from various relationships based on $\mu + \sigma$ and $\mu - \sigma$ SPT N values is shown in Fig. 5.

The V_S calculation based on specific soil type correlation has also been carried out in the study. All the borehole data were subjected to different V_S-N correlations, and corresponding, V_S is calculated. In line with the previous case, the width of the band of values seems to be very large for V_S calculated using different correlations. The incorporation of statistical parameters takes this care. Variation of μ, $\mu + \sigma$ and $\mu - \sigma$ of shear wave velocity using mean shear wave velocity profiles of individual boreholes is represented in Fig. 6. Further, Fig. 7 shows the mean variation in shear wave velocity profiles estimated based on $\mu + \sigma$ and $\mu - \sigma$ shear wave velocity profiles of all boreholes.

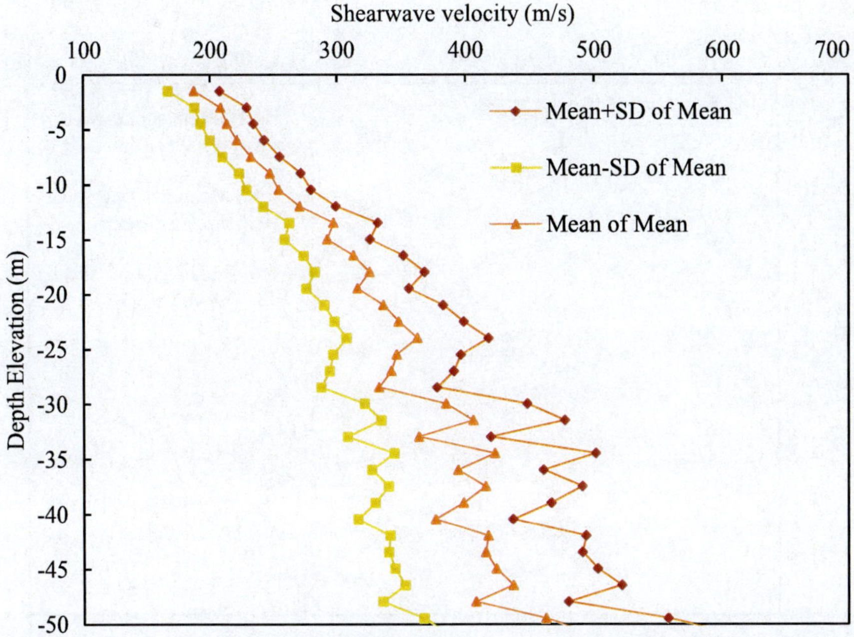

Fig. 4 Statistical variation of the estimated shear wave velocities based on mean SPT values

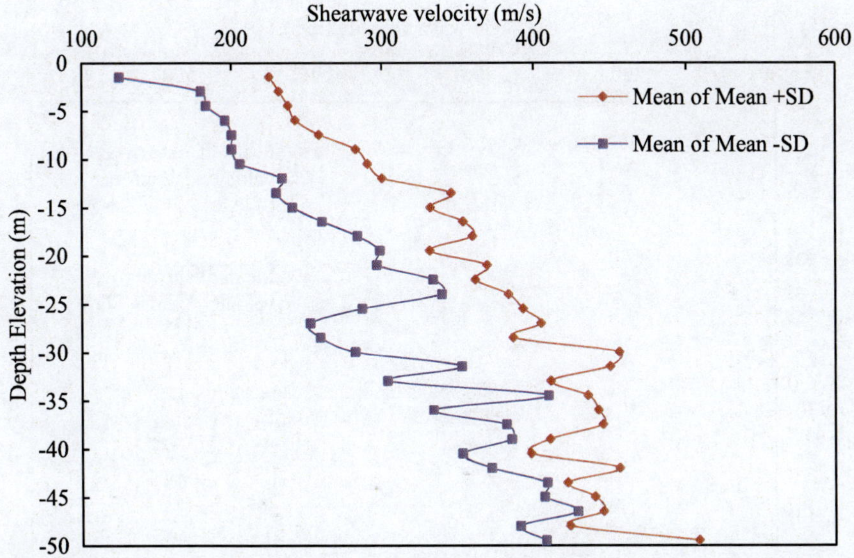

Fig. 5 Mean shear wave velocity obtained from various relationships based on $\mu + \sigma$ and $\mu - \sigma$ SPT N values

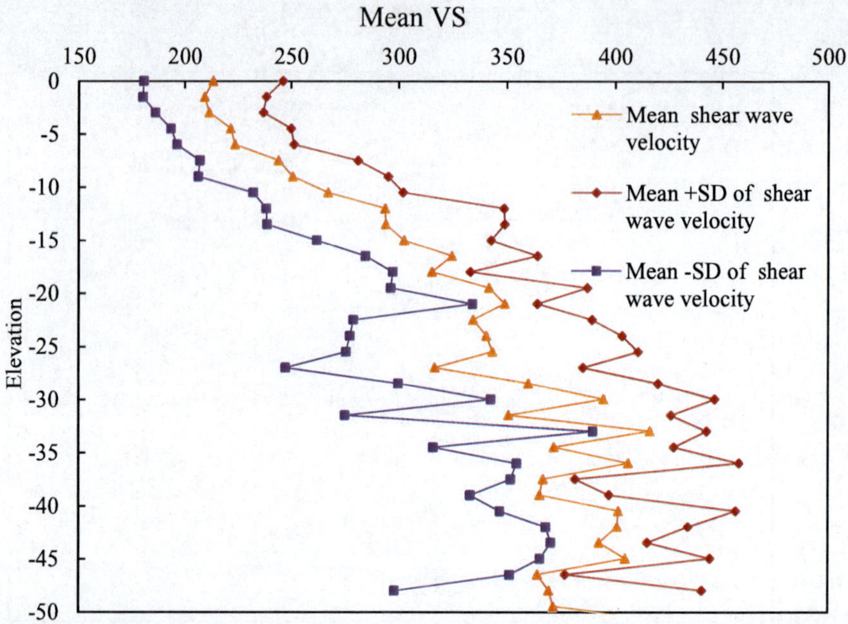

Fig. 6 Variation of mean, mean+SD, mean −SD of shear wave velocity using mean shear wave velocity profiles of all boreholes and correlations

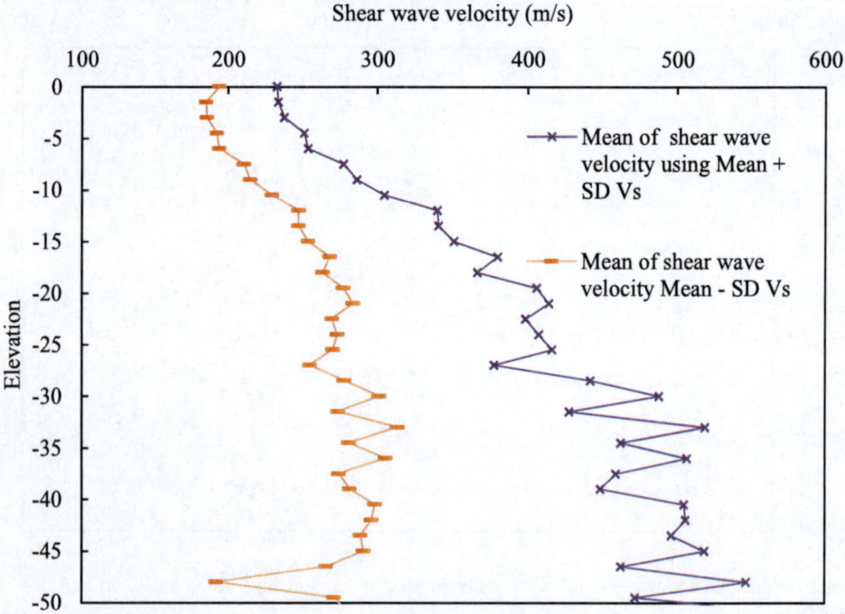

Fig. 7 Variation of mean of mean+SD, mean −SD of shear wave velocity

7 Summary and Conclusions

A statistical approach to develop the V_S profile of a particular site using different V_S-N correlations is detailed here. The method eliminates various uncertainties involved in V_S calculations. The range of values of SP N and VS was converged into μ, $\mu + \sigma$ and $\mu - \sigma$.

The SPT data shows a scattering nature throughout the site. The values of SPT N vary from 4 to 15 at the top of the boreholes and 53–106 at deeper points. This lead to a high value of standard deviation from borehole to borehole. These variations are directly reflected in the estimated V_S values. The proposed approach eliminates the uncertainties involved in the previous procedures using statistical comparisons. The range of values in which the shear wave velocity is represented using mean + SD and mean −SD band. The range of V_S values lies in between 500 and 600 m/s at the bottom levels. A similar width in the range is observed throughout the entire site, such that the new set of values are close to each other. This represents a set of reliable V_S values to further proceed to any seismic hazard analysis.

The elimination of various uncertainties involves several mathematical calculations, which can be swiftly performed using any computational application. The methodology provides a new reliable way of analysis of field data for practicing engineers as well as researchers.

References

1. Ismet KA, Tildy P, Prónay Z et al (2006) Vs30 mapping and soil classification for seismic site effect evaluation in Dinar region, SW Turkey. Geophys J Int 165:223–235. https://doi.org/10.1111/j.1365-246X.2006.02882.x
2. Ansal A, Kurtuluş A, Tönük G (2010) Seismic microzonation and earthquake damage scenarios for urban areas. Soil Dyn Earthq Eng 30:1319–1328. https://doi.org/10.1016/j.soildyn.2010.06.004
3. Lee VW, Trifunac MD, Bulajić B et al (2017) Seismic microzoning of Belgrade. Soil Dyn Earthq Eng 97:395–412. https://doi.org/10.1016/j.soildyn.2017.02.002
4. Rosa-Cintas S, Clavero D, Delgado J et al (2017) Characterization of the shear wave velocity in the metropolitan area of Málaga (S Spain) using the H/V technique. Soil Dyn Earthq Eng 92:433–442. https://doi.org/10.1016/j.soildyn.2016.10.016
5. Scott JB, Rasmussen T, Luke B et al (2006) Shallow shear velocity and seismic microzonation of the urban Las Vegas, Nevada, basin. Bull Seismol Soc Am 96:1068–1077. https://doi.org/10.1785/0120050044
6. Thitimakorn T (2019) Seismic microzonation maps of Phrae city, Northern Thailand. Geomatics Nat Hazards Risk 10:2276–2290. https://doi.org/10.1080/19475705.2019.1693705
7. Tuladhar R, Yamazaki F, Warnitchai P, Saita J (2004) Seismic microzonation of the greater Bangkok area using microtremor observations. Earthq Eng Struct Dyn 33:211–225. https://doi.org/10.1002/eqe.345
8. Wang Z (2008) A technical note on seismic microzonation in the central United States. J Earth Syst Sci 117:749–756. https://doi.org/10.1007/s12040-008-0060-8
9. Anbazhagan P, Sitharam TG (2008) Mapping of average shear wave velocity for Bangalore region: a case study. J Environ Eng Geophys 13:69–84. https://doi.org/10.2113/JEEG13.2.69

10. Bashir A, Basu D (2018) Revisiting probabilistic seismic hazard analysis of Gujarat: an assessment of Indian design spectra. Nat Hazards 91:1127–1164. https://doi.org/10.1007/s11069-018-3171-9
11. Kumar SS, Krishna AM, Dey A (2014) Nonlinear site-specific ground response analysis : case study of amingaon. Guwahati 15See:1–12
12. Seed HB, Idriss IM, Arango I (1983) Evaluation of liquefaction potential using field performance data. J Geotech Eng 109:458–482. https://doi.org/10.1061/(ASCE)0733-9410(1983)109:3(458)
13. Singh M, Duggal SK, Pallav K, Sharma KK (2020) Correlation of shear wave velocity with standard penetration resistance value for Allahabad City
14. Anbazhagan P, Kumar A, Sitharam TG (2013) Seismic site classification and correlation between standard penetration test N Value and shear wave velocity for Lucknow City in Indo-Gangetic Basin. Pure Appl Geophys 170:299–318. https://doi.org/10.1007/s00024-012-0525-1
15. Ohta Y, Goto N (1978) Empirical shear wave velocity equations in terms of characteristic soil indexes. Earthq Eng Struct Dyn 6:167–187. https://doi.org/10.1002/eqe.4290060205
16. Okamoto T, Kokusho T, Yoshida Y, Kusuonoki K (1989) Comparison of surface versus subsurface wave source for P–S logging in sand layer. In: Proceedings of 44th annual conference JSCE, vol 3, pp 996–997
17. Pitilakis K, Raptakis D, Lontzetidis K et al (1999) Geotechnical and geophysical description of euro-seistest, using field, laboratory tests and moderate strong motion recordings. J Earthq Eng 3:381–409. https://doi.org/10.1080/13632469909350352
18. Pitilakis KD, Anastassiadis D, Raptakis D (1992) Field and laboratory determination of dynamic properties of natural soil deposits. Tenth World Conf Earthq Eng 1275–1280
19. Seed HB, Idriss IM (1981) Evaluation of liquefaction potential sand deposits based on observation of performance in previous earthquakes. ASCE Natl Conv 81–544
20. Sykora DW, Stokoe K (1983) Correlations of in situ measurements in sands of shear wave velocity, soil characteristics, and site conditions. Soil Dyn Earthq Eng 20:125–136
21. Yokota K, Imai T, Konno M (1981) Dynamic deformation characteristics of soils determined by laboratory tests. OYO Tec Rep 3:13–37
22. Athanasopoulos GA (1995) Empirical correlations Vso-NSPT for soils of Greece: A comparative study of reliability. Proc 7th Int Conf Soil Dyn Earthq Engg 19–25
23. Hanumantharao C, Ramana GV (2008) Dynamic soil properties for microzonation of Delhi, India. J Earth Syst Sci 117:719–730
24. Hsien-Heng Lee S (1992) Analysis of the multicollinearity of regression equations of shear wave velocities. Soils Found 32:205–214
25. Imai T, Tonouchi K (1982) Correlation of N value with S-wave velocity and shear modulus. In: Penetration testing proceedings 2nd European Symposium Amsterdam, pp 67–72
26. Imai T, Yoshimura M (1976) the relation of mechanical properties of soils to P and S wave velocities for soil ground in Japan. Urana Res Institue, OYO Corp
27. Lee CT, Tsai BR (2008) Mapping Vs30 in Taiwan. Terr Atmos Ocean Sci 19:671–682. https://doi.org/10.3319/TAO.2008.19.6.671(PT)
28. Lee SHH (1990) Regression models of shear wave velocities in taipei basin. J Chinese Inst Eng Trans Chinese Inst Eng A/Chung-kuo K Ch'eng Hsuch K'an 13:519–532. https://doi.org/10.1080/02533839.1990.9677284
29. Ohsaki Y, Iwasaki R (1973) On dynamic shear moduli and poisson's ratios of soil deposits. Soils Found 13:61–73. https://doi.org/10.3208/sandf1972.13.4_61
30. Kirar B, Maheshwari BK, Muley P (2016) Correlation between shear wave velocity (Vs) and SPT Resistance (N) for Roorkee Region. Int J Geosynth Gr Eng 2:1–11. https://doi.org/10.1007/s40891-016-0047-5
31. Maheswari RU, Boominathan A, Dodagoudar GR (2010) Use of surface waves in statistical correlations of shear wave velocity and penetration resistance of Chennai soils. Geotech Geol Eng 28:119–137. https://doi.org/10.1007/s10706-009-9285-9

32. Roy N, Shiuly A, Sahu RB, Jakka RS (2018) Effect of uncertainty in V_S—N correlations on seismic site response analysis. J Earth Syst Sci 127https://doi.org/10.1007/s12040-018-1007-3
33. Sil A, Sitharam TG (2014) Dynamic site characterization and correlation of shear wave velocity with standard penetration Test 'N' values for the City of Agartala, Tripura State, India. Pure Appl Geophys 171:1859–1876. https://doi.org/10.1007/s00024-013-0754-y

Utilization of Bagasse Fibre and Ash—An Open Door for Sustainable Development: Review and Future Insights

Jithin P. Zachariah and Ravi S. Jakka

Abstract Bagasse is the lignocellulosic biomass left behind after the process of extraction of juice from sugarcane. The disposal of this agricultural waste material is a severe challenge faced in various parts of the world. India is the second-largest cultivator of sugarcane, which generates a substantial quantity of bagasse annually. This paper attempts to review and summarize the extensive studies that have explored the potential applications of bagasse in different fields. The efficient utilization of sugarcane bagasse in the form of fibre, as well as ash, has been surveyed. Apart from the environmental benefits, the application of bagasse also improves the mechanical properties of soil and other construction materials. The physical properties and chemical composition of bagasse reveal its capability to be converted into valuable products or a versatile ingredient in biocomposites. Furthermore, the paper conducts a life cycle study of the sugarcane plant. An insight into some future applications of bagasse is also discussed in the article.

Keywords Bagasse fibre · Bagasse ash · Sustainable practices · Sugarcane

1 Introduction

The lignocellulosic biomass fibre residue left after the extraction of juice from sugarcane, which consists almost 12.3% of the mass, is called bagasse. The cultivation of large quantities of sugarcane generates a proportional amount of bagasse globally. This points towards a severe challenges in the efficient disposal of bagasse wastes. Application of bagasse in various fields is the only sustainable method to overcome this challenge. Currently, most of the generated bagasse is burnt for energy applications, which adversely affects the environment. However, the scope of environmental friendly utilization of bagasse is also being explored around the globe. India, with a

J. P. Zachariah · R. S. Jakka (✉)
Department of Earthquake Engineering, Indian Institute of Technology Roorkee, Roorkee, India
e-mail: ravi.jakka@eq.iitr.ac.in

J. P. Zachariah
e-mail: jpzachariah@eq.iitr.ac.in

T. G. Sitharam et al. (eds.), *Earthquake Geotechnics*, Lecture Notes in Civil Engineering 187, https://doi.org/10.1007/978-981-16-5669-9_41

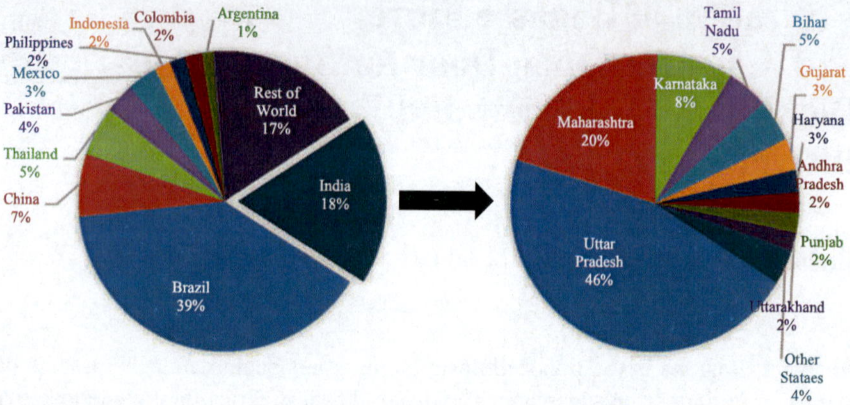

Fig. 1 Contribution of major sugarcane producing countries and states of India

large area of agricultural land, generates a considerable quantity of agricultural wastes annually. India produced 376.90 million tonnes of sugarcane in the year 2017–18 and sent it to the sugar mills for processing [1]. Figure 1 shows the contribution of major countries and Indian states in the production of sugarcane [2]. India holds the second-largest share in sugarcane production in the world.

18% of total sugarcane in the world is produced in India. Maharashtra, Uttar Pradesh, Uttarakhand, Tamil Nadu, etc. are the primary sugarcane cultivators in India. The production rate of sugarcane in India is also increasing annually. 70.84 million tonnes of additional sugarcane is produced in India in the last year. The growth in the production of sugarcane is directly reflected in the generation of bagasse fibre.

The deposition of the produced bagasse is a severe challenge to the industry as well as the government. In many industries, the bagasse biomass is burnt to meet the energy needs of the mill. The bagasse is transported to the burning chambers and set to fire. The heat thus produced is used to generate electricity [3, 4]. However, the combustion process results in various environmental problems like smoke emissions, ash emissions, ash depositions, climate changes, etc. Another method of removal of bagasse is the deposition of bagasse in land or pits. This also pollutes ground soil land and nearby water bodies. The water absorption nature of the biomass affects the natural flow of groundwater.

The need for sustainable utilization of bagasse is necessary for environmental safety. However, the decomposing nature reflects on the short life span of bagasse. This paper reviews the reported applications of bagasse fibre and bagasse ash in various fields. The works of literature discuss the bulk utilization of the bagasse. However, local, small-scale unreported utilization of bagasse is also practised worldwide. The paper also gives insight into the future scope of studies for the effective utilization of bagasse fibre and ash in bulk.

2 Process of Generation

Sugarcane bagasse is an agricultural waste produced from the sugar mills. The process of making sugar from sugarcane leaves the biomass waste material in fibrous form. The fibre mainly consists of cellulose, hemicellulose and lignin. The process of generation of bagasse starts with sugarcane harvesting. The harvested sugarcane, after cleaning, is transported to the factories and moved into the juice extractors. These extractors crush the sugarcane and press the canes at high pressure to extract the juice. This extracted juice is converted into commercial sugar and other products. The leftover obtained from the extractor, the bagasse, is moved from the machines for disposal or other value processes. In many industries, these bagasse fibres are burnt to meet energy needs. The burning of bagasse also helps in reducing the volume of waste to be disposed of. The burnt or incinerated form of bagasse fibre is known as bagasse ash. The bagasse ash is generally deposited into pits or lands because many studies have reported that the bagasse ash is non-toxic. Utilization of bagasse ash-based bricks, composites and soil stabilizers is studied widely to perform the sustainable disposal of bagasse ashes. It is also reported that the bagasse ash has high crop productivity nature. In most application cases, the bagasse ash and fibre are taken back by nature itself that completes the entire life cycle of sugarcane bagasse. A complete representation of the life cycle of sugarcane is shown in Fig. 2.

3 Properties and Composition of Bagasse

3.1 Physical Properties of Bagasse

The physical properties of bagasse change from one to another, depending on different factors, though the physical composition is uniform. The field conditions, moisture content, sugar extraction processes, etc. are the major factors that influence these fluctuations. Therefore, a series of values of specific gravity and tensile strength of bagasse fibre is tabulated in Table 1 to obtain a range of physical properties. The mean values and standard deviation are also reported.

3.2 Physical Composition of Bagasse Fibre

Bagasse extracted from sugarcane is composed of various materials. A detailed analysis of the composition of bagasse and bagasse ash is reviewed in this section. The composition of bagasse, bagasse fibre and pith explored by [18] is represented in Fig. 3. It is observed that more than 80% of raw bagasse is composed of carbon-rich cellulose and homo cellulose.

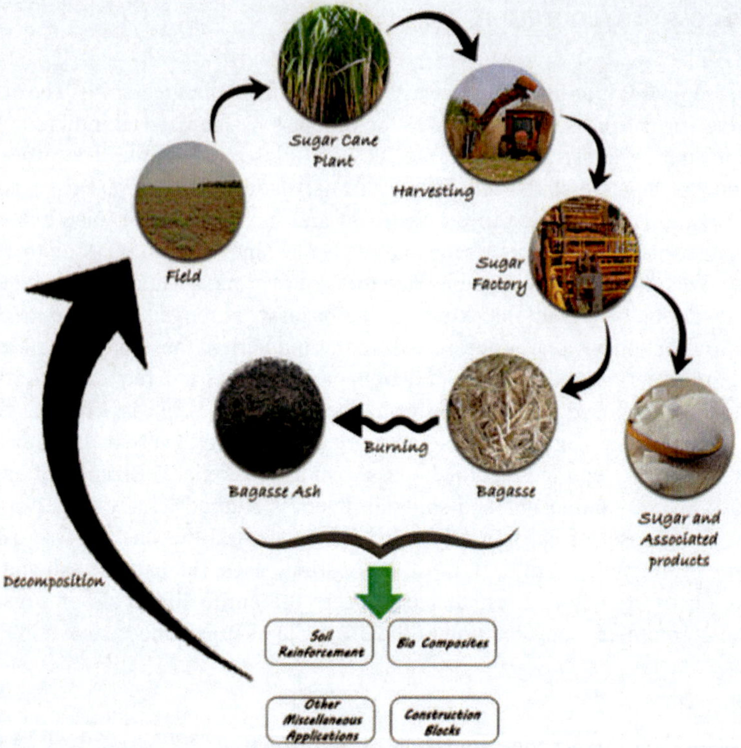

Fig. 2 Life cycle of sugarcane

Table 1 The physical properties of bagasse

References	Specific gravity	References	Tensile strength (N/mm^2)
[5]	2.38	[6]	132
[7]	1.91	[8]	62
[9]	2.03	[10]	42
[11]	2.04	[12]	70.9
[13]	1.92	[14]	96
[15]	2.22		
[16]	2.2		
[17]	2.08		
Mean	2.10		80.58
Standard deviation	0.16		34.67

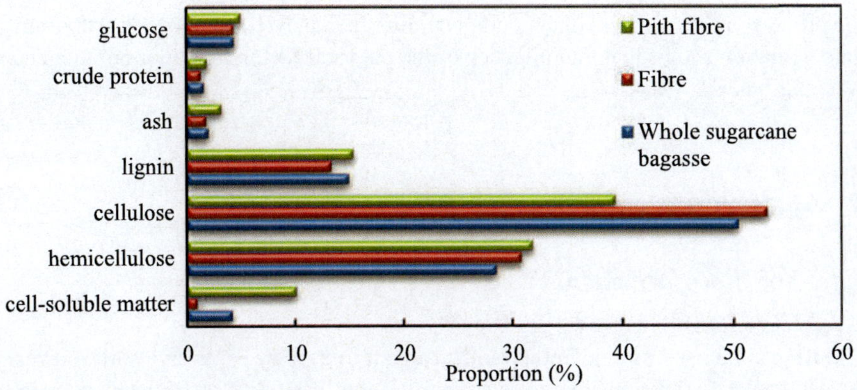

Fig. 3 Composition of bagasse [18]

3.3 Chemical Composition of Bagasse Ash

Similar to the physical properties, the chemical composition of bagasse ash also varies from sample to sample. The chemical composition of bagasse ash reported by various researchers is tabulated in Table 2. Many physical factors influence the variations in

Table 2 Chemical composition of bagasse

References	SiO_2(%)	Al_2O_3(%)	Fe_2O_3(%)	CaO (%)	MgO (%)	K_2O (%)	Na_2O (%)
[19]	57.95	8.23	3.96	4.52	4.47	2.41	–
[20]	64.38	11.67	4.56	10.26	0.85	3.57	1.05
[21]	67.33	3.45	3.39	12.51	2.04	4.11	0.05
[22]	66	6	5	2	2	6.5	1
[23]	71.63	9.37	5.1	5.9	1.67	2.5	–
[24]	54.9	7.8	10.0	4.9	2.5	3.0	0.2
[25]	85.5	5.3	1.3	2.1	1.1	3.5	–
[26]	50.36	–	18.78	8.81	–	19.34	–
[27]	85.58	5.25	1.31	2.08	1.09	3.46	–
[28]	43.32	2.67	1.17	10.01	1.85	17.41	0.04
[29]	51.66	9.92	2.32	2.59	1.44	2.1	1.23
[30]	72.95	1.68	1.89	7.77	1.98	9.28	0.02
[31]	55.05	3.87	0.42	5.09	4.82	4.00	0.94
[32]	87.40	3.60	4.90	2.56	0.69	0.47	0.15
[7]	72.95	1.68	1.89	7.77	1.98	9.28	0.02
Mean	65.80	5.75	4.40	5.92	2.03	6.06	0.47
Standard deviation	13.70	3.42	4.65	3.43	1.29	5.58	0.47

the proportions of each chemical component in the bagasse ash. However, the standard deviation shows that the influence of the physical factors in major components is negligible.

4 Applications of Bagasse Fibre (BF)

4.1 Soil Reinforcement

The BF can be used as a potential reinforcement to modify expansive soils. Nevertheless, the availability of literature supporting the feasibility of BF as a soil reinforcement is less. The initial stiffness and shear strength properties of the soil are significantly improved with the addition of BF reinforcement. A further strength improvement can be established with an additive like hydrated lime [33, 34]. Besides, the consolidation and CBR tests were also conducted on similar soil samples. The compression characteristics of 0.5% BF–2.5% lime–soil mix are reduced with the inclusion of fibre content, making the soil a non-swelling reinforced soil [35].

4.2 Cementitious Materials

In case of concrete mixes, a lower slump value is observed for concrete mixed with bagasse fibre. The addition of BF into the concrete mix shows a fluctuating trend in various properties of the mix. The ductility and flexural strength of the reinforced concrete show an enhancing nature, while there is a slight reduction in the compressive strength of the concrete mix. In addition, a reduced heat of hydration is observed for concrete mixed with BF [6]. The bagasse fibre can be effectively utilized as a reinforcement to cement mortar also. In line with the observations made for BF reinforced concrete, varying trends in the strength parameters are developed in mortar also [36].

4.3 Bio Composites

Bio composites are the future of many sustainable industries. Bio composites play a key role not only in construction fields, but also in automotive, aerospace and bio medical fields [37]. Natural fibre-reinforced composites show a lesser strength performance when compared with synthetic fibre. However, these bio composites have significant advantages like the environmental impacts, cost effectiveness, design flexibility etc. A mix of natural and synthetic fibre was also used in some industries like automotive industries. The chemical modification on the surface of these fibre-reinforced composites are used in furniture and electronic materials. Compared to some manufactured fibre like glass fibre, natural fibre shows better performance [38]. The modification of polyethylene using treated bagasse shows a significant improvement in the mechanical and structural properties of the composites. The chemical treatments modifies the tensile, thermal and morphological properties of the bio composites [39]. In addition, torsional modulus and water absorption capacity of the bio composites are also improved in the process [14, 15].

4.4 Soil Blocks

The BF reinforcement has the potential ability to improve the strength properties of soil blocks. The addition of the fibre reduces the density of the blocks, which can be used in lightweight structures. An optimum of 0.5% of BF by weight mixed with the soil mass shows a better results than any other mixes. However, an increased percentage of use of BF increases the water absorption of these blocks, which negatively affects the strength properties. An increase in the compressive and tensile strength of soil bricks will happen when reinforced with BF. The reduction in rate of erosion of soil blocks can also be achieved with the application of BF reinforcements [8, 10].

4.5 Miscellaneous Applications

Ethanol Production. The sugarcane bagasse from the mills can be converted into ethanol after conducting integrated biorefinery processing. Pre-hydrolysis, pre-treatment, enzymatic hydrolysis, fermentation, product recovery and separation, boiler/turbogenerator, and vanillin and syringaldehyde production are the basic processes which convert bagasse from mills to industrial ethanol. The use of this manufactured ethanol contributes towards the environmental as well as economic advantages [40].

Sound Absorber. [41] explored the feasibility of BF as a sustainable acoustic absorber that can be used as sound absorbers in lecture rooms, meeting room, theatre halls, music studios etc. A 6 cm-thick BF absorber layer shows an average absorption coefficient of 0.65 at frequency 1.2–4.5 kHz.

Paper Industry. The bagasse, in processed state, can be converted to pulp and paper. The process is practiced in various parts of South America that consumes nearly 10% of bagasse production in the world. The rapid growth of sugarcane plant and its widespread cultivation are the major advantages of using bagasse in industrial purposes [42].

Combustion. Non-renewable fuel sources like fossil fuels dominated the fuel word for a large period and then left its path to sustainable energy sources. The combustion of BF in large chamber produces a considerable amount of heat energy that can be converted into other forms of energy. However, the application BF in the combustion is not well appreciated because the burning of BF emits a large amount of ash residue and smoke into the atmosphere [43].

Apart from these, the treated BFs were also used as feeds to animals, cattle bedding, manufacturing of disposable utensils etc.

5 Application of Bagasse Ash (BA)

5.1 Soil Stabilization

Jakhar et al. [44] explored the application of bagasse ash as a property enhancer for expansive soil. A series of compaction, compression, CBR and direct shear tests were performed to evaluate the performance of BA added expansive soil. An increase in the mechanical properties of the BA–soil mix and BA–lime–soil mix is observed up to 7.5% addition of BA. Any further addition of the BA showed a decrease in the properties of the soil. It is also noticed that the presence of lime showed good performance in the properties of the BA–soil mix due to the formation of cementitious compounds from lime–BA reactions. Similar studies on the performance of BA has been reported by Hasan et al. [45]. An 80% increase in the compressive strength was observed by adding 25% of BA–lime to the expansive soil. Manikandan and Moganraj [46] also used lime with BA to study the consolidation and rebound characteristics of expansive soil, while [20] used virgin BA in the expansive black cotton soil from Maharashtra state of India. Kiran and Kiran [47] added both cement and lime into the BA–black cotton soil mix to observe a significant rise in strength characteristics of the mix.

5.2 A Supplementary Cementitious Material

A detailed evaluation of possibility of BA as a supplementary cementitious material was performed by Ganesan et al. [48]. Several tests were conducted on the BA-OPC mortar and BA-OPC concrete to study different characteristics of the composite mix. Well-burnt bagasse was added to the OPC in different percentages from 0 to 30%. Physical and chemical properties of the mix, consistency and setting time, compressive strength of the blended mix and strength of the concrete were studied using various experimental investigations. The BA particles are very finer than OPC. 20% of OPC cement can be efficiently replaced from the concrete using BA without any fall in strength characteristics. The pozzolanic activity of the BA can be increased using calcination process, which helps in enhancing the mechanical properties of the BA mixes [49].

5.3 *Geomaterial Bricks*

Sustainable bricks and construction materials always have a reasonable demand in the industry. Several natural and synthetic waste materials like fly ash [50, 51], PET [52], rice husk ash [53, 54] and natural fibre straws [55–57] were converted into useful construction bricks. Various authors have also reported an inline application of bagasse as a construction material and supplementary material in building blocks. However, the advantage of application of BA is sometimes limited to the waste disposal problem because of the poor performance of BA as a replacement for any geomaterial [30, 58] An improvement of BA-based bricks can be achieved by adding supplementary materials like quarry dust and lime. A compressive strength of 6.5 MPa has been reported for BA–quarry dust–lime (50%–30%–20%) bricks, which is significantly higher than that of common clay bricks [9]. Moreover, the BA mixed unburnt bricks show better performance in durability, water permeability and water sorptivity tests [30].

5.4 *Miscellaneous Applications*

Silica Production. Silica has widely been used in various industries as an adsorbent, catalyst host, in pigments, in cosmetics etc. With more than 50% of silica particles (SiO_2), bagasse ash is utilized as an alternative for silica extraction [24, 26].

Ceramic Materials production. Souza et al. [25] reused the bagasse ash as an additive to cast roof tiles. 60% of the clay is replaced with bagasse ash to produce tiles by burning above 1000 °C.

Waste Oil Treatment. The potential of bagasse ash in treating waste cooking oil was explored by Bonassa et al. [59]. The porous structure of the bagasse ash improves the adsorption capacity of BA. Moreover, when compared with other adsorbing agents like activated carbon, BA has high efficiency of treatment in low temperature in lesser time.

Quartz Replacement in Glass. The chemical composition of BA mainly consists of silicon oxides (quartz) and aluminium oxides along with iron, calcium and magnesium oxides. The abundant presence of quartz can be used in making glass and glass ceramics [27].

Crop Productivity. Direct application of BA into the agricultural fields improves the crop yield by a significant level. Thind et al. [60] performed a three-year field study on rice and wheat fields, which successfully utilize waste BA in the fields without any health issues.

6 Conclusion and Future Insights

A wide literature study on sugarcane bagasse has been carried out in this paper. The study reveals the various applications of bagasse in both fibre and ash forms. The study also reports the physical and chemical composition of the sugarcane bagasse. It can be concluded that the sugarcane bagasse can be converted into value products if it is processed properly. The bagasse fibre enhances the flexural and tensile strength of blocks and different composites. A high variation in the tensile strength of bagasse fibre is observed as evident from standard deviation values given in Table 1. Though some variability in the strength of the fibre is expected on account of variable testing conditions adopted by different researchers, the large degree of variability clearly indicating the variation in the physical properties of the bagasse fibre source. This is clearly evident from the variation in the chemical proportions in Table 2. The bagasse-reinforced materials show good water and sound absorption capability. The literatures reported that the use of bagasse ash in cementitious materials improve the mechanical properties of these materials. Thus, it can be effectively used as a replacement in concrete and mortar. The addition of bagasse ash also improves the properties of soil and soil-based blocks. Other miscellaneous applications of bagasse ash have also been reported in the paper.

From the above available literatures, it is observed that the application of bagasse as a soil reinforcement is scarcely evaluated. With good tensile and flexural strength, the bagasse fibre can resist the liquefaction susceptibility of soil. The fibre bonds between the soil particles induce a friction that helps in holding the particles together. The waste bagasse can be deposited in liquefaction susceptible areas and improve the utilization scope of the land. Even if the reinforced land could not hold much strength, it can be utilized for low-load structures like parking lots, recreation places etc. The bagasse ash can also act as a stabilizer for fine soils. The stabilized soil can be used to fill huge construction structures like road sub-base, embankments or making lightly loaded paver bricks.

The effective utilization of agricultural waste can contribute to the environment and economy simultaneously. The bagasse is a valuable waste that can be utilized in various unexplored areas because of its good properties. India, being the second-largest producer of sugarcane in the world stores a substantial quantity of these value products for future scopes.

References

1. Ministry of Agriculture & Farmers Welfare Government of India (2019) Agriculture Annual Report 2018–19, New Delhi
2. Directorate of Sugarcane Development. https://sugarcane.dac.gov.in/
3. Alves M, Ponce GHSF, Silva MA, Ensinas AV (2015) Surplus electricity production in sugarcane mills using residual bagasse and straw as fuel. Energy 91:751–757. https://doi.org/10.1016/j.energy.2015.08.101

4. Sampaio ILM, Cardoso TF, Souza NRD et al (2019) Electricity production from sugarcane straw recovered through bale system: assessment of retrofit projects. Bioenergy Res 12:865–877. https://doi.org/10.1007/s12155-019-10014-9
5. Anupam AK, Kumar P, Ransinchung RNGD (2013) Use of various agricultural and industrial waste materials in road construction. Procedia - Soc Behav Sci 104:264–273. https://doi.org/10.1016/j.sbspro.2013.11.119
6. Ribeiro B, Yamamoto T, Yamashiki Y (2020) A study on the reduction in hydration heat and thermal strain of concrete with addition of sugarcane bagasse fiber. Materials (Basel) 13https://doi.org/10.3390/ma13133005
7. Bahurudeen A, Wani K, Basit MA, Santhanam M (2016) Assesment of pozzolanic performance of sugarcane bagasse ash. J Mater Civ Eng 28:1–11. https://doi.org/10.1061/(ASCE)MT.1943-5533.0001361
8. Danso H, Martinson DB, Ali M, Williams JB (2015) Effect of sugarcane bagasse fibre on the strength properties of soil blocks. In: First international conference on bio-based building materials
9. Madurwar MV, Mandavgane SA, Ralegaonkar RV (2015) Development and feasibility analysis of bagasse ash bricks. J Energy Eng 141:1–9. https://doi.org/10.1061/(ASCE)EY.1943-7897.0000200
10. Danso H (2017) Properties of coconut, oil palm and bagasse fibres: as potential building materials. Procedia Eng 200:1–9. https://doi.org/10.1016/j.proeng.2017.07.002
11. Rajasekar A, Arunachalam K, Kottaisamy M, Saraswathy V (2018) Durability characteristics of ultra high strength concrete with treated sugarcane bagasse ash. Constr Build Mater 171:350–356. https://doi.org/10.1016/j.conbuildmat.2018.03.140
12. Cao Y, Shibata S, Fukumoto I (2006) Mechanical properties of biodegradable composites reinforced with bagasse fibre before and after alkali treatments. Compos Part A Appl Sci Manuf 37:423–429
13. Rerkpiboon A, Tangchirapat W, Jaturapitakkul C (2015) Strength, chloride resistance, and expansion of concretes containing ground bagasse ash. Constr Build Mater 101:983–989. https://doi.org/10.1016/j.conbuildmat.2015.10.140
14. Arrakhiz FZ, Malha M, Bouhfid R et al (2013) Tensile, flexural and torsional properties of chemically treated alfa, coir and bagasse reinforced polypropylene. Compos Part B Eng 47:35–41. https://doi.org/10.1016/j.compositesb.2012.10.046
15. Montakarntiwong K, Chusilp N, Tangchirapat W, Jaturapitakkul C (2013) Strength and heat evolution of concretes containing bagasse ash from thermal power plants in sugar industry. Mater Des 49:414–420. https://doi.org/10.1016/j.matdes.2013.01.031
16. Joshaghani A, Moeini MA (2017) Evaluating the effects of sugar cane bagasse ash (SCBA) and nanosilica on the mechanical and durability properties of mortar. Constr Build Mater 152:818–831. https://doi.org/10.1016/j.conbuildmat.2017.07.041
17. Chusilp N, Jaturapitakkul C, Kiattikomol K (2009) Effects of LOI of ground bagasse ash on the compressive strength and sulfate resistance of mortars. Constr Build Mater 23:3523–3531. https://doi.org/10.1016/j.conbuildmat.2009.06.046
18. Han YW, Catalano EA, Ciegler A (1983) Chemical and physical properties of sugarcane bagasse irradiated with 7 Rays. J Agric Food Chem 31:34–38. https://doi.org/10.1021/jf00115a009
19. Sadeeq JA, Ochepo J, Salahudeen AB, Tijjani ST (2015) Effect of bagasse ash on lime stabilized lateritic soil. Jordan J Civ Eng 9:203–213
20. Kharade AS, Suryavanshi V V., Gujar BS, Deshmukh RR (2014) Waste product 'bagasse ash' from sugar industry can be used as stabilizing material for expansive soils. Int J Res Eng Technol 03:506–512. https://doi.org/10.15623/ijret.2014.0303094
21. Surjandari NS, Djarwanti N, Ukoi NU (2017) Enhancing the engineering properties of expansive soil using bagasse ash. J Phys Conf Ser 909https://doi.org/10.1088/1742-6596/909/1/012068
22. Lal D, Kumar MJ, Kumar KN et al (2020) Soil stabilization using bagasse ash. Lect Notes Civ Eng 71:21–28. https://doi.org/10.1007/978-981-15-3662-5_3

23. Jamsawang P, Poorahong H, Yoobanpot N et al (2017) Improvement of soft clay with cement and bagasse ash waste. Constr Build Mater 154:61–71. https://doi.org/10.1016/j.conbuildmat.2017.07.188
24. Chindaprasirt P, Rattanasak U (2020) Eco-production of silica from sugarcane bagasse ash for use as a photochromic pigment filler. Sci Rep 10:1–8. https://doi.org/10.1038/s41598-020-66885-y
25. Souza AE, Teixeira SR, Santos GTA et al (2011) Reuse of sugarcane bagasse ash (SCBA) to produce ceramic materials. J Environ Manage 92:2774–2780. https://doi.org/10.1016/j.jenvman.2011.06.020
26. Affandi S, Setyawan H, Winardi S et al (2009) A facile method for production of high-purity silica xerogels from bagasse ash. Adv Powder Technol 20:468–472. https://doi.org/10.1016/j.apt.2009.03.008
27. Teixeira SR, De Souza AE, De Almeida Santos GT et al (2008) Sugarcane bagasse ash as a potential quartz replacement in red ceramic. J Am Ceram Soc 91:1883–1887. https://doi.org/10.1111/j.1551-2916.2007.02212.x
28. Eberemu AO (2013) Evaluation of bagasse ash treated lateritic soil as a potential barrier material in waste containment application. Acta Geotech 8:407–421. https://doi.org/10.1007/s11440-012-0204-5
29. Alavéz-Ramírez R, Montes-García P, Martínez-Reyes J et al (2012) The use of sugarcane bagasse ash and lime to improve the durability and mechanical properties of compacted soil blocks. Constr Build Mater 34:296–305. https://doi.org/10.1016/j.conbuildmat.2012.02.072
30. Deepika S, Anand G, Bahurudeen A, Santhanam M (2017) Construction products with sugarcane bagasse ash binder. J Mater Civ Eng 29:1–10. https://doi.org/10.1061/(ASCE)MT.1943-5533.0001999
31. Jagadesh P, Ramachandramurthy A, Murugesan R (2018) Evaluation of mechanical properties of sugar cane bagasse ash concrete. Constr Build Mater 176:608–617. https://doi.org/10.1016/j.conbuildmat.2018.05.037
32. Amin NU (2011) Use of bagasse ash in concrete and its impact on the strength and chloride resistivity. J Mater Civ Eng 23:717–720. https://doi.org/10.1061/(ASCE)MT.1943-5533.0000227
33. Dang LC, Fatahi B, Khabbaz H (2016) Behaviour of expansive soils stabilized with hydrated lime and bagasse fibres. Procedia Eng 143:658–665. https://doi.org/10.1016/j.proeng.2016.06.093
34. Dang LC, Hasan H, Fatahi B et al (2016) Enhancing the engineering properties of expansive soil using bagasse ash and hydrated lime. Int J GEOMATE 11:2447–2454. https://doi.org/10.21660/2016.25.5160
35. Dang LC, Khabbaz H, Fatahi B (2018) An experimental study on engineering behaviour of lime and bagasse fibre reinforced expansive soils Une étude expérimentale sur le comportement d ' ingénierie des sols expansifs renforcés de fibre de chaux et de bagasse. 2497–2500
36. Ribeiro B, Yamashiki Y, Yamamoto T (2020) A study on mechanical properties of mortar with sugarcane bagasse fiber and bagasse ash. J Mater Cycles Waste Manag. https://doi.org/10.1007/s10163-020-01071-w
37. Mulinari DR, Cipriano J de P, Capri MR, Brandão AT (2018) Influence of surgarcane bagasse fibers with modified surface on polypropylene composites. J Nat Fibers 15:174–182. https://doi.org/10.1080/15440478.2016.1266294
38. Chand N, Rohatgi PK (2008) Natural fibers and their composites. Tribol Nat Fiber Compos. https://doi.org/10.1201/9781439832592.ch1
39. Hossain MK, Karim MR, Chowdhury MR et al (2014) Comparative mechanical and thermal study of chemically treated and untreated single sugarcane fiber bundle. Ind Crops Prod 58:78–90. https://doi.org/10.1016/j.indcrop.2014.04.002
40. Vikash PV, Mandade PV, Shastri Y (2018) Assessment of bagasse and trash utilization for ethanol production: a case study in india. Environ Prog Sustain Energy 37:2165–2174. https://doi.org/10.1002/ep.12900

41. Putra A, Abdullah Y, Efendy H et al (2013) Utilizing sugarcane wasted fibers as a sustainable acoustic absorber. Procedia Eng 53:632–638. https://doi.org/10.1016/j.proeng.2013.02.081
42. Rainey TJ, Covey G (2016) Pulp and paper production from sugarcane bagasse. Sugarcane-based Biofuels Bioprod 259–280https://doi.org/10.1002/9781118719862.ch10
43. Cundy VA, Maples D, Tauzin C (1983) Combustion of bagasse. Fuel 62:775–780. https://doi.org/10.1016/0016-2361(83)90027-3
44. Jakhar VS, Singhvi BS, Shiv Kumar S, Sharma MK (2020) Effect of bagasse ash on the properties of the expansive soil. Int J Res Eng Sci Manag 3:545–550. https://doi.org/10.1061/(ASCE)1090-0241(2004)130:7(764)
45. Hasan H, Dang L, Khabbaz H et al (2016) Remediation of expansive soils using agricultural waste bagasse ash. Procedia Eng 143:1368–1375. https://doi.org/10.1016/j.proeng.2016.06.161
46. Manikandan A, Moganraj M (2014) Consolidation and rebound characteristics of expansive soil by using lime and bagasse ash. Int J Res Eng Technol 03:403–411. https://doi.org/10.15623/ijret.2014.0304073
47. Kiran R, Kiran L (2013) Analysis of Strength characteristics of black cotton soil using bagasse ash and additives as stabilizer. Int J Eng Res Technol 2:2240–2246
48. Ganesan K, Rajagopal K, Thangavel K (2007) Evaluation of bagasse ash as supplementary cementitious material. Cem Concr Compos 29:515–524. https://doi.org/10.1016/j.cemconcomp.2007.03.001
49. Yadav AL, Sairam V, Muruganandam L, Srinivasan K (2020) An overview of the influences of mechanical and chemical processing on sugarcane bagasse ash characterization as a supplementary cementitious material. J Clean Prod 245:118854https://doi.org/10.1016/j.jclepro.2019.118854
50. Çiçek T, Çinçin Y (2015) Use of fly ash in production of light-weight building bricks. Constr Build Mater 94:521–527. https://doi.org/10.1016/j.conbuildmat.2015.07.029
51. Cicek T, Tanrıverdi M (2007) Lime based steam autoclaved fly ash bricks. 21:1295–1300. https://doi.org/10.1016/j.conbuildmat.2006.01.005
52. Alex K, Zachariah JP, Joseph M et al (2020) Compressive strength characteristics of stabilized rammed earthen paver blocks. Indian J Environ Prot 40:642–646
53. Rahman MA (1987) Properties of clay-sand-rice husk ash mixed bricks. Int J Cem Compos Light Concr 9:105–108. https://doi.org/10.1016/0262-5075(87)90026-1
54. Hegazy BE-DE, Fouad HA, Hassanain AM (2012) Incorporation of water sludge, silica fume, and rice husk ash in brick making. Adv Environ Res 1:83–96. https://doi.org/10.12989/aer.2012.1.1.083
55. Allam ME (2015) Recycled chopped rice straw- cement bricks: mechanical, fire resistance & economical assessment recycled chopped rice straw- cement bricks: mechanical. Fire Resistance & Economical Assessment
56. Chan CM (2011) Effect of natural fibres inclusion in clay bricks: Physico-mechanical properties. World Acad Sci Eng Technol 73:51–57. https://doi.org/10.5281/zenodo.1331783
57. Ag Mumin AM, Tahir MM, Ngian SP et al (2020) Flexural behaviour of interlocking brick system with grout cement mixed with various fibre. IOP Conf Ser Mater Sci Eng 849 https://doi.org/10.1088/1757-899X/849/1/012064
58. Kulkarni A, Raje S, Rajgor M (2013) Bagasse ash as an effective replacement in flyash bricks. Int J Eng Trends Technol 4:4484–4489
59. Bonassa G, Schneider LT, Alves HJ et al (2016) Sugarcane bagasse ash for waste cooking oil treatment applications. J Environ Chem Eng 4:4091–4099. https://doi.org/10.1016/j.jece.2016.09.017
60. Thind HS, Yadvinder-Singh B-S et al (2012) Land application of rice husk ash, bagasse ash and coal fly ash: effects on crop productivity and nutrient uptake in rice-wheat system on an alkaline loamy sand. F Crop Res 135:137–144. https://doi.org/10.1016/j.fcr.2012.07.012

Seismic Stability Evaluation of an Indian Himalayan Slope: A Case Study

Koushik Pandit, Mahendra Singh, and Jagdish Prasad Sahoo

Abstract The seismic stability of any vulnerable landslide slope is of great importance, especially for those slopes stretching along the national highways connecting important places in the Himalayas. The present study area is located in the Garhwal Himalayas near Tehri town in Tehri Garhwal district of Uttarakhand. It is situated about 5–7 km south of the surface trace of the Tons thrust (TT) or the North Almora Thrust (NAT). The region falls in Zone IV of the seismic zoning map of India. Hilly regions of Uttarakhand have past record of several major earthquakes, the most well-known ones being the Uttarkashi earthquake (1991) of magnitude 6.6 and the Chamoli earthquake (1999) of magnitude 6.3. For the present study, a debris slope has been selected which is located along the highway connecting Tehri and Koteshwar townships. The connecting road is also of importance because it connects the Tehri Dam (the tallest dam in India) built on the Bhagirathi River and Koteshwar Dam built in the downstream of the Tehri Dam. The area is also prone to high seismic activities due to the reasons mentioned before. Overall, a careful and detailed stability analysis of the potential slopes is required for this region. To perform a generalized seismic stability analysis, the study is divided into three parts: Firstly, a stability analysis of the study slope is carried out for static and pseudo-static loading conditions as per the relevant standard guidelines. Then, the critical seismic coefficient has been evaluated for the critical target factor of safety of 1 for only horizontal vibration. In the final part, Newmark displacement of the slope along its most probable failure surface has been determined for two major earthquake signatures experienced by the region in the past. For the stability analysis and Newmark displacement evaluation, the two-dimensional limit equilibrium methods developed by several researchers have been utilized.

Keywords Garhwal Himalayas · Seismic stability · Pseudo-static analysis · Critical seismic coefficient · Newmark displacement

K. Pandit (✉) · M. Singh
Indian Institute of Technology Roorkee, Roorkee 247667, India
e-mail: kpandit@ce.iitr.ac.in

J. P. Sahoo
Indian Institute of Technology Kanpur, Kanpur 208016, India

T. G. Sitharam et al. (eds.), *Earthquake Geotechnics*, Lecture Notes in Civil Engineering 187, https://doi.org/10.1007/978-981-16-5669-9_42

1 Introduction

The seismic stability of any susceptible landslide slope is of substantial significance, especially for those hillslopes encompassing the major national and state highways connecting important pilgrimages, townships, and strategic border locations in the Himalayas. Uttarakhand is a hilly state in India which is subdivided into two zones as per the regional geological features, namely the Garhwal and Kumaun Himalayas. The present study area is located in the Garhwal Himalayas near Tehri town in Tehri Garhwal district of Uttarakhand. It is situated about 5–7 km South of the surface trace of the Tons thrust (TT) or the North Almora Thrust (NAT) [1] (Fig. 1). The region falls in Zone IV of the seismic zoning map of India [2]. Hilly regions of Uttarakhand have past record of several major earthquakes, the most prominent ones being the Uttarkashi earthquake (1991) of magnitude 6.6 and the Chamoli earthquake (1999) of magnitude 6.3. The recurrence interval for great earthquakes in the Dehradun region (nearly 27 km from Tehri) has been estimated as 290–980 years [3]. No super earthquake (magnitude $\geq$8 on Richter scale) has occurred in the past two centuries in this part of the Himalayas which makes the area more susceptible to a long impending mega earthquake in the near or remote future.

For the present study, a debris slope has been selected which is located along the highway connecting Tehri and Koteshwar townships. The connecting road is also of importance because it connects the Tehri Dam (the tallest dam in India) built on the Bhagirathi River and Koteshwar Dam built in the downstream of the Tehri Dam. These two dams are part of the Tehri hydropower complex and serve hydroelectricity of 1000 MW from Tehri Dam and 400 MW from Koteshwar Dam facilities. Geographically, the study area is occupied by highly undulating terrain, represented by high ridges or spurs, deep valleys, and sharp slopes. The rocks exposed in the study area belong to the inner Lesser Himalaya (represented by the rocks of

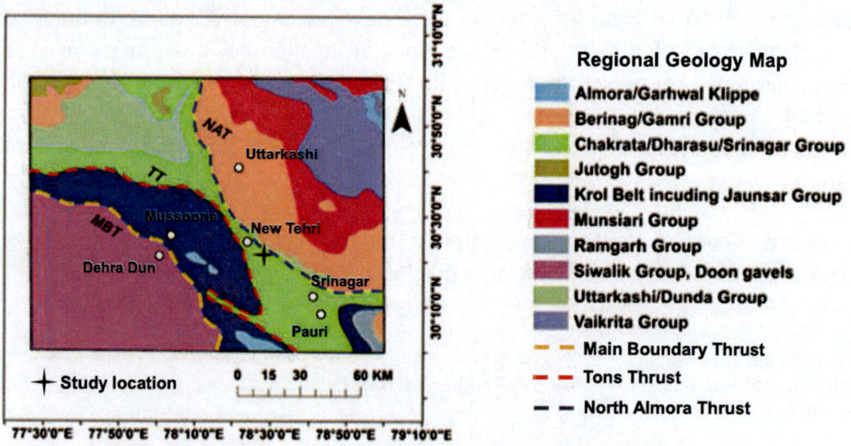

Fig. 1 Regional geology map of the study area showing the near thrust lines

Rautgara Formation of Damtha Group, Deoban and Mandhali Formation of Tejam Group, and Berinag Formation of Jaunsar Group) as well as the outer Lesser Himalaya (represented by the rocks of Chandpur and Nagthat Formation of Jaunsar Group and Blaini Formation of Mussoorie Group) [4]. The studied slope has bedrock of metamorphosed phyllites with moderate to high weathering which is overlain by a debris stratum. During the monsoons, the locality receives huge rainfall, thus making a large number of slopes vulnerable for sliding. The area is also prone to high seismic activities due to the reasons mentioned before. Overall, a careful and detailed stability analysis of the potential slopes is required for this region.

To perform a generalized seismic stability analysis, the present study is divided into three parts: Firstly, a stability analysis of the study slope is carried out for static and pseudo-static loading conditions as per the relevant standard guidelines as depicted in [5]. Secondly, the critical seismic coefficient has been evaluated for the critical target factor of safety of 1 for horizontal vibration. In the final part, Newmark displacement of the slope along its most probable failure surface has been determined for two major earthquake signatures experienced by the region in the past. For the stability analysis and Newmark displacement evaluation [6], the two-dimensional limit equilibrium methods developed by several researchers [7–13] have been utilized.

2 Data Collection and Material Properties

Before commencing the seismic stability analysis of any slope, certain parameters are required to be determined and evaluated. For this purpose, a field survey was carried out. Geotechnical investigations have been carried out for the 1.63 km chainage site. The cross section of the slope has been shown in Fig. 2. The slope height is approximately 70 m from the existing road level, and the average slope face angle is being about 42°. Since, borehole drilling was not performed, there is no information about the natural groundwater table depth, although from the locality, it was known that the domestic water supply is collected from neighboring springs. For geotechnical investigations, at first, representative debris soil material and rock samples from various locations on the slope were collected. The core cutter method was used in the field to obtain the in-situ density of the slope material. The laboratory testing included different index tests through which properties like moisture content, grain size distribution, and Atterberg's limits were evaluated. Direct shear tests were also performed to obtain the shear strength properties of the debris soil samples. The rock mass properties were determined from the rock mass quality assessed using the geological strength index (GSI) at field. The obtained properties used in the stability analysis have been summarized in a tabular form and are shown in Fig. 2. As it can be seen, there is a debris overburden over the rock mass which is basically weathered phyllite.

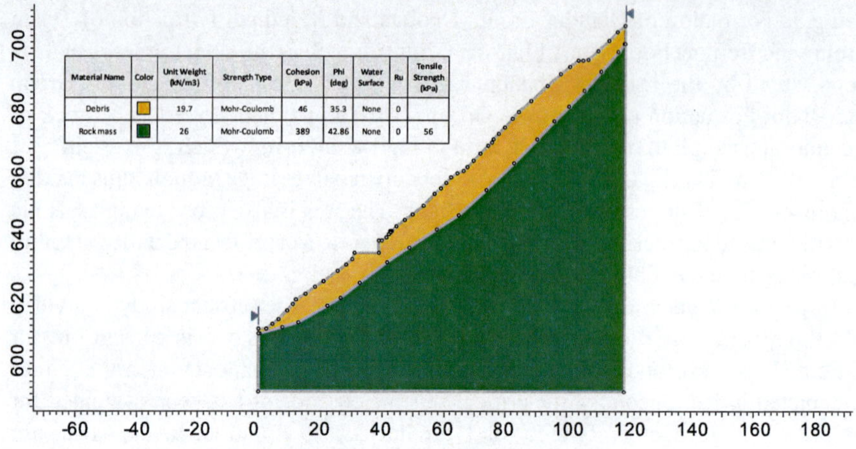

Fig. 2 Slope geometry and evaluated slope mass properties for Ch. 1.63 km

3 Static and Pseudo-Static Seismic Stability Analysis

After determining the various physical and engineering properties of the slope, the next step is to perform the stability analysis of it. For this, Slide2 program from Rocscience was used which incorporates various well-established and popular limit equilibrium methods (LEMs) for two-dimensional stability assessment. For the present study, the ordinary or Fellenius method [7], Bishop's method [8], Janbu's method [9], Spencer's method [10], Morgenstern and price method [11], and Sarma's method [12, 13] were used to determine the global safety factor of the slope under the static, i.e., gravity loading and pseudo-static loading. In the pseudo-static method, the seismic loading is modeled as a statically applied inertial force, the magnitude of which is a product of a seismic coefficient (k) and the weight of the potential sliding mass. Generally, the k-values are much smaller than a_{mh}/g, where a_{mh} is the maximum horizontal acceleration expected at the site, and g is the acceleration of gravity. In the present study, the horizontal and vertical coefficient values required for carrying out the pseudo-static stability analysis were evaluated as per [5]. As per the guidelines, in the absence of site-specific estimates of design peak ground, the design seismic inertia forces for equivalent-static slope stability assessment shall be taken as

$$F_H = \frac{1}{3} ZISW \tag{1}$$

where F_H is the horizontal inertial force, Z is the zone factor given in IS: 1893—Part 1 (2016) [2], I is the importance factor, S is an empirical coefficient to account for the amplification of ground motion between bedrock and the elevation of the toe of the dam or embankment, and W is the weight of the sliding mass.

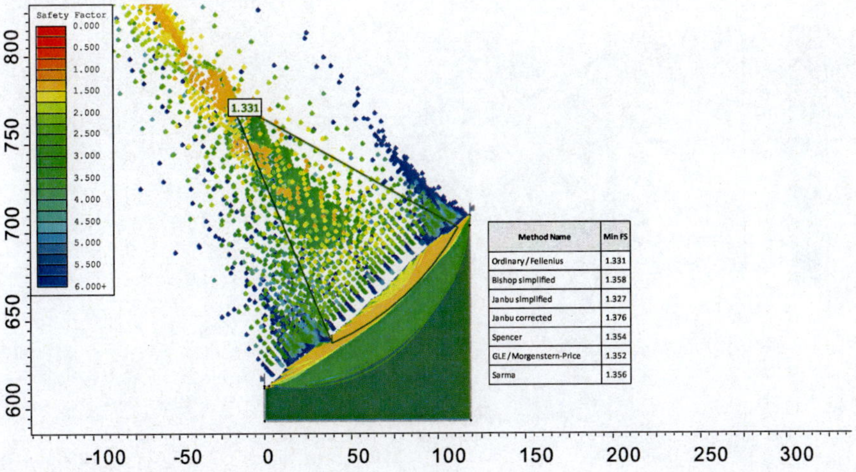

Fig. 3 Global minimum safety factors for the study slope obtained through various LEMs under the static loading condition

The present study area is located in seismic Zone IV, which makes the value of Z (zone factor) = 0.24; importance factor, I, value is taken as 1.5 (for major highways); the site amplification factor, S, is taken as 1.0 for Zone IV and S1 type soil (hard rock, soft rock, and hard soil). The computed horizontal (α_h) and vertical (α_v) seismic coefficients are as follows:

$$\alpha_{\mathrm{h}} = (1/3) * 0.24 * 1.5 * 1.0 = 0.12 \tag{2}$$

and

$$\alpha_v = (2/3) * 0.12 = 0.08 \tag{3}$$

The obtained safety factor values for the static and pseudo-static loadings from various LEMs are shown in the Figs. 3 and 4. The several points in these figures represent slip circle's center or point of rotation of each slip surface generated during the analysis and passing through the slope mass section.

4 Critical Seismic Coefficient (k_c) Analysis

In the next phase, an advanced seismic analysis was performed to determine the critical horizontal seismic coefficient (k_c) that results in a destabilized slope with a global minimum factor of safety (FoS) as 1. In this analysis, only the horizontal seismic coefficient was considered, and vertical seismic coefficient was ignored. From the analysis, different critical seismic coefficient ($\mathrm{k_c}$) values were obtained for

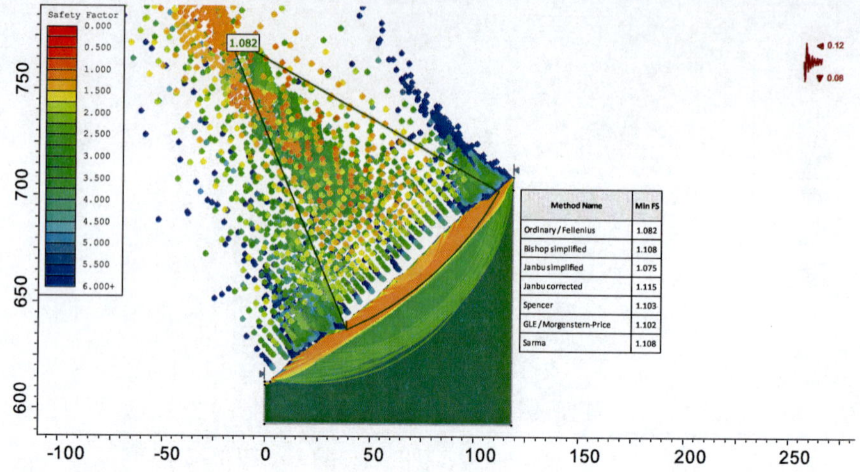

Fig. 4 Global minimum safety factors for the study slope obtained through various LEMs under the pseudo-static loading condition

the corresponding most critical slip surface depending upon the limit equilibrium method used to perform the stability analysis (Fig. 5).

It can be understood from this analysis that there are a number of unstable surfaces for this model, wherein a seismic coefficient more than or equal to 0.174 (obtained from Janbu's simplified method) would result in a destabilized slope. This makes sense since the global minimum factor of safety for the seismic = 0.12 scenario in the pseudo-static analysis where the horizontal (α_h) seismic coefficient was obtained

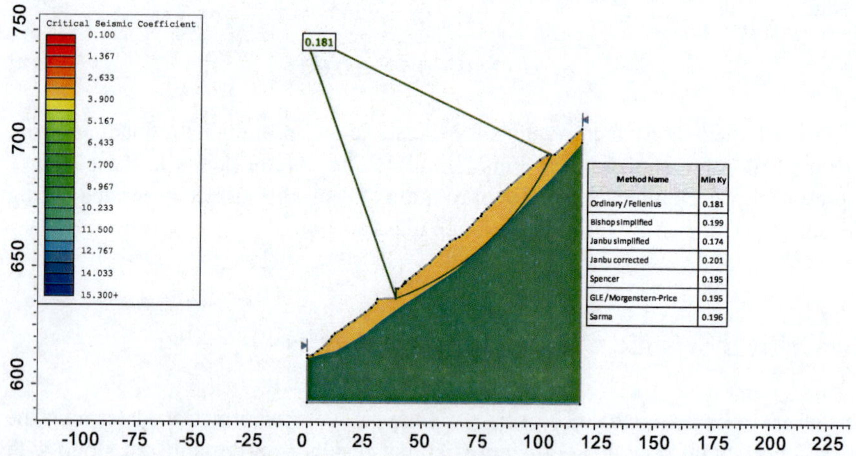

Fig. 5 Critical seismic coefficient (k_c) values obtained for the corresponding most critical slip surface using various LEMs

from [5], and the safety factor value was obtained as 1.075 from Janbu's simplified method which is just above the critical FoS = 1.0).

5 Newmark Displacement Analysis

To estimate the permanent slope displacements that may occur due to the inertial forces associated with an earthquake, Newmark displacement analysis [6] is popularly used. The principle of a Newmark displacement analysis is that during an earthquake event, there will be short moments in time when the inertial forces (mass times acceleration) plus the initial static forces will exceed the available shear resistance, and during these times, the temporary loss of stability will lead to unrecoverable deformations. The accumulation of the unrecoverable deformations will manifest itself as permanent deformation after the shaking has stopped.

In the final stage of analysis, Newmark displacement analysis was performed to determine the critical Newmark displacement that resulted from seismic loading generated during past two major and well-known earthquakes in the study region of Garhwal Himalayas, namely the Uttarkashi earthquake (1991) of magnitude 6.6 and the Chamoli earthquake (1999) of magnitude 6.3. These two earthquakes caused a large number of landslides, new and previously known in the public domain, thus initiating a chain of road blockages for several days, huge damages to residential and commercial buildings and bridges, and loss of human lives and animal stocks.

For the present study, the Newmark analysis was performed in Slide2 program from Rocscience which is based on the program SLAMMER, developed by the U.S. Geological Survey [14]. For the present case, the time and acceleration data points for the Uttarkashi (1991) and Chamoli (1999) earthquakes were provided as input seismic motions in the Slide2 program (Fig. 6). Uttarkashi and Chamoli are at about 113 km and 183 km away from the study area, respectively.

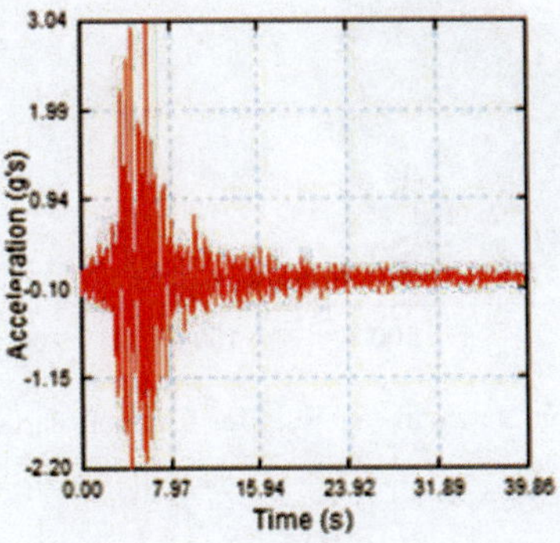

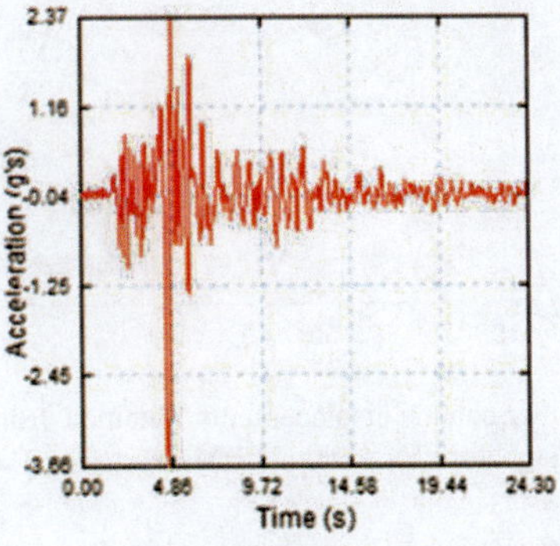

Fig. 6 Time vs acceleration history of **a** Uttarkashi (1991) and **b** Chamoli (1999) earthquakes

The permanent displacement values obtained from the Newmark analysis for the Uttarkashi and Chamoli earthquake signatures are shown in Figs. 7 and 8, where the displacements are measured along the most probable critical slip surfaces according to the corresponding limit equilibrium methods used for the stability analysis.

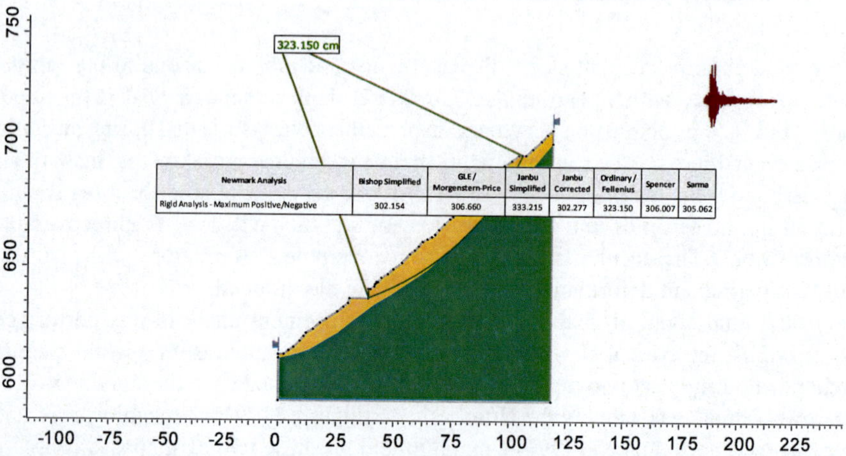

Fig. 7 Permanent displacements obtained from the Newmark analysis for Uttarkashi earthquake (1991) signature for various LEMs

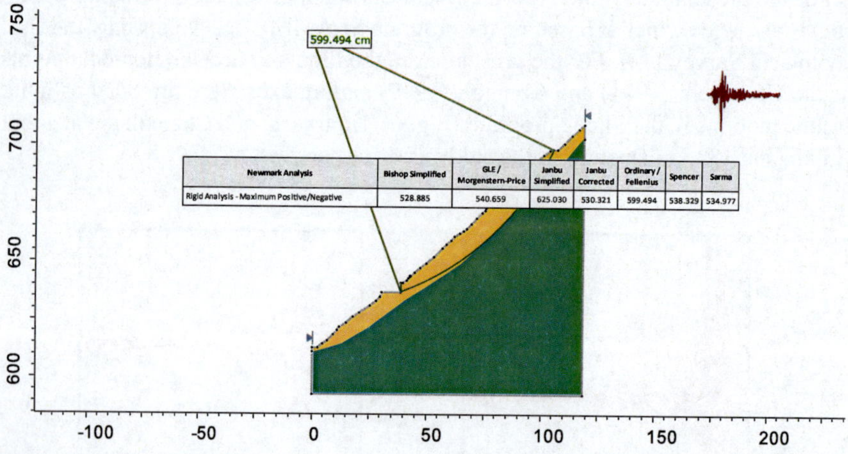

Fig. 8 Permanent displacements obtained from the Newmark analysis for Chamoli earthquake (1999) signature for various LEMs

6 Concluding Remarks

The present analysis demonstrated the various modes of seismic analysis for slope stability assessment, from the static (no seismic) and pseudo-static factor of safety analysis to determine the critical seismic coefficient (k_c) to finally evaluate the permanent displacements along the most critical slip surface, expected from any future earthquake events having seismic signatures similar to past two major earthquakes in the region. This study will encourage the research community to have an overall understanding of seismic slope stability evaluation methods for better slope design and choosing landslide mitigation measures accordingly. The presented methodology herein may further be improved once the sub-strata profiling of the studied slope is done and borelog data is made available. From boreholes, the underlying rock mass cores may be collected and tested for various critical strength parameters. Once back-integrated into the numerical models, this will further enhance the accuracy in prediction of the slope stability factor. The slope can further be monitored through instrumentations like inclinometers, pore water pressure gauges, etc., for pre-monsoon and post-monsoon slope displacement and induced effective pore pressure seasonal variation records. Thus, the overall methodology can be utilized for developing a more robust early warning framework which will further provide more safety and risk assessment options for the investigators.

References

1. Gupta S, Mahesh P, Sivaram K, Rai SS (2012) Active fault beneath the Tehri dam, Garhwal Himalaya—seismological evidence. Curr Sci 103(11):1343–1347
2. IS 1893 (Part 1): 2016: Criteria for earthquake resistant design of structures—Part 1: General Provisions and Buildings. Bureau of Indian Standards, Manek Bhawan, New Delhi
3. Bilham R, Gaur VK, Molnar P (2001) Himalayan seismic hazard. Science 293:1442–1444
4. Valdiya KS (1980) Geology of Kumaun Lesser Himalaya. In: Interim Report 291 of Wadia Institute of Himalayan Geology, Dehradun (1980)
5. IITK-GSDMA Guidelines for Seismic design of earth dams and embankments. http://www.iitk.ac.in/nicee/IITK-GSDMA/EQ09.pdf. Last accessed 01 Oct 2020
6. Newmark NM (1965) Effects of earthquakes on dams and embankments. Géotechnique 15(2):139–160
7. Fellenius W (1936) Calculation of the stability of earth dams. In: Proceedings of the second congress on large Dams, vol 4, pp 445–463
8. Bishop AW (1955) The use of the slip circle in the stability analysis of earth slopes. Geotechnique 5(1):7–17
9. Janbu N (1973) Slope stability computations. In: Hirschfeld RC, Poulos SJ (eds), Embankment-dam Engineering, Jon Wiley and Sons Inc., NY
10. Spencer E (1967) A method of analysis of the stability of embankments assuming parallel inter-slice forces. Géotechnique 17:11–26. https://doi.org/10.1680/geot.1967.17.1.11
11. Morgenstern NR, Price VE (1965) The analysis of the stability of general slip surfaces. Geotechnique 15(1):79–93. https://doi.org/10.1680/geot.1965.15.1.79
12. Sarma SK (1973) Stability analysis of embankments and slopes. Géotechnique 23(3):423–433
13. Sarma SK (1979) Stability analysis of embankments and slopes. J Geotech Eng Div ASCE 105(12):1511–1524

14. Jobson RW, Rathje EM, Jibson MW, Lee YW (2013) SLAMMER—seismic landslide movement modeled using earthquake records (ver.1.1, November 2014): U.S. Geological Survey Techniques and Methods, book 12, chap. B1, unpaged

Site-Specific Seismic Ground Response Analysis for Typical Soil Sites in Central Khartoum, Sudan

Mohammed Al-Ajamee, Mohamedelamin M. M. Mahmoud, and Ahmed M. El Sharief

Abstract The effect of shallow soil conditions on ground motion parameters is well-recognized; thus, it is important to characterize the site-specific response of soils, especially in the most critical locations in Khartoum like Central Khartoum, the capital state of Sudan. In this paper, site-specific seismic ground response analysis was conducted. The equivalent-linear (EQ-L) method of analysis was adopted using the computer program DEEPSOIL. Two different soil models from Central Khartoum were chosen, along with the average soil profile used in previous study about the liquefaction assessment for comparison purposes. The sites are predominantly silty clay soils in the top 8 m, grading into sandy and silty sand with depth. The correlation between the standard penetration test SPT N-value and shear wave velocity was used. It has been shown that the ground response significantly influences local soils. A substantial difference between the two considered sites was remarked, and the average profile for all ground response parameters was noticed. The peak ground acceleration was found to range from 1.7 to 2.5, and the response spectrum at the surface for all the sites was also illustrated. Fourier amplitude ratios were found to vary from 4.3 to 8.35.

Keywords Equivalent-linear analysis · Site response · Response spectrum · Fourier amplitude ratio

1 Introduction

In geotechnical earthquake engineering, site effects or "site-specific response" of local soil layers exposed to dynamic earthquake loading have a profound impact on the earthquake motions that the site is experiencing. Besides the topography, the composition of the bedrock, the nature, and geometry of the deposits are the key factors that influence the local modification of the earthquake motions [1, 2]. Wave

M. Al-Ajamee (✉) · M. M. M. Mahmoud · A. M. El Sharief
Department of Geotechnical Engineering, Building and Road Research Institute, University of Khartoum, Khartoum, Sudan
e-mail: mohammedalajamee@gmail.com

T. G. Sitharam et al. (eds.), *Earthquake Geotechnics*, Lecture Notes in Civil Engineering 187, https://doi.org/10.1007/978-981-16-5669-9_43

propagation theory has been utilized in the ground response analysis in order to estimate the local soil effects on the propagating shear wave oozing from the bedrock at the time of an earthquake. The energy released from the earthquake source is in the form of seismic waves propagating in all directions. It is commonly assumed that for the 1D ground response analysis, the waves propagate only in the vertical direction (e.g., from the bedrock to the ground surface vertically), which is the same presumption used in the work reported in this paper [3]. Investigated the applicability of the 1D ground response analysis using 34 downhole array sites and found that 69% of the sites they considered were suitable for the 1D analysis. The wave characteristics have changed significantly during the transmission of seismic waves based on the local soil properties and may result in serious repercussions (e.g., structures without sufficient ground motion resistance or earthquakes loading considerations may experience drastic damages). The model of the soil column consisting of all soil layers must be specified with the corresponding dynamic properties of each layer. Soil properties such as the unit weight of the soil and the soil layer depth must be defined as the inputs in the soil model. Based on the bore-log profile of the site representing different soil layers with thickness, the model of the soil column could be described.

Dynamic soil behavior is quite complex and requires models that describe the important elements of cyclic behavior, which could be evaluated experimentally using a cyclic triaxial test, resonant column test, or field measurements using existing correlations (transformation models) with field tests (e.g., cone penetration test (CPT) and standard penetration test (SPT)), with SPT N-value being the most widely used correlation. Theoretical simulation of 1D site response will be achieved eventually using equivalent-linear (EL) or nonlinear (NL) analyses. Equivalent-linear (EL) ground response analysis is commonly used in practice to replicate nonlinear soil behavior. The equivalent-linear (EL) approach belongs to the total stress type of analysis, in which the nonlinear stress–strain relationship is approximately represented by linear secant shear modulus, which is a function of the amount of shear strain. The equivalent-linear (EL) approach is more effective, requires minor computational efforts, and few input parameters compared to the nonlinear (NL) analysis, which captures the actual nature of nonlinearity. Useful guidelines for performing hazard-consistent 1D ground response analysis (GRA) could be found in [4] and commonly used equivalent-linear computer programs are DEEPSOIL, SHAKE, and EERO.

Equivalent-linear (EL) analysis for site response follows an iterative process in which estimates of the initialization of the shear modulus and the damping of each soil layer are given. For each layer, peak shear strains are examined, and effective shear strains are used as a fraction of the peak strains. This effective shear strain is used to determine the related equivalent shear modulus (G) and the damping ratio (ζ). The steps are replicated until the strain-compatible properties are consistent with the properties used to conduct a dynamic response. In this study, an attempt has been made to investigate the impacts of local soil conditions on modifying earthquake ground motions using one-dimensional (1D) equivalent-linear (EL) ground response analysis for important sites in Khartoum, namely Almogran sites.

The computer program DEEPSOIL was used in the analysis [5]. The response of the local soil at the presence of earthquakes in terms of surface accelerations (SA), amplitude ratio (AR), and response spectrum (RS) is a prerequisite, especially for densely populated regions with lots of high-rise buildings, like the Almogran area in Khartoum, which is the center of the business district (CBD), as well as many governmental agencies. The results deducted from this study will be useful, particularly for structural designers, as there is neither guidance for the professionals nor domestic design codes for buildings in Sudan.

2 Study Areas of Khartoum

Khartoum is the National capital of Sudan. It consists of three major cities: Khartoum, Khartoum North (Bahri), and Omdurman (see Fig. 1). The state is located in the northeastern part of the center of the country at the confluence of the White and Blue Niles (the main tributaries of the great River Nile). It lies between a longitude of (34–5.31) east and latitudes of (16–15) north approximately and is 410 m above sea level. The total area of the state is estimated to be 22,736 Km^2. The colonial rule left the country unplanned, except for the old city of Khartoum. Later on, the Sudanese people came and resided in these areas, which eventually led to the concentration of development in specific places. The state became a center for urbanization, and the people kept flowing from various parts of the country, which led to a dramatic rise in the population that was estimated in 2018 to be around 8 million people [6], which represents more than 25% of the gross population of the whole country. This rapid

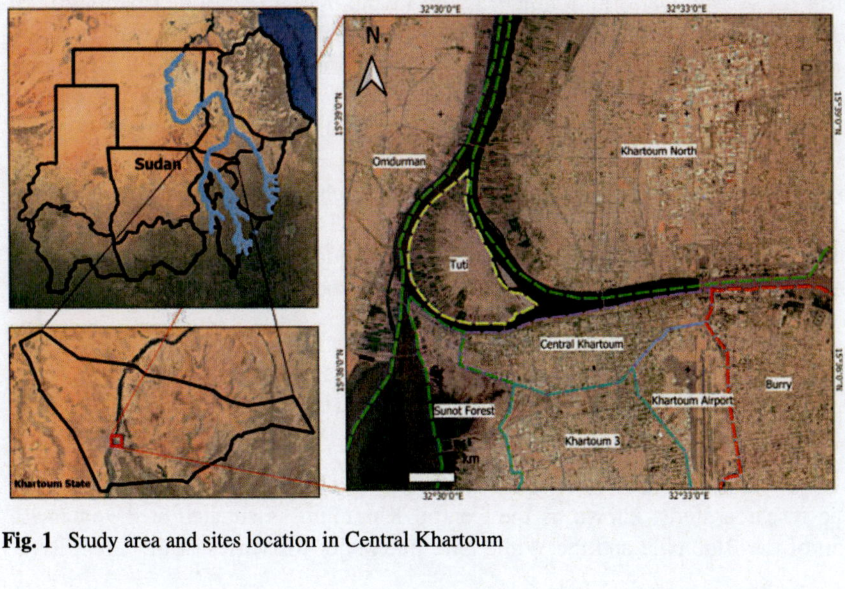

Fig. 1 Study area and sites location in Central Khartoum

urbanization, which is often synonymous with urban sprawl, results in exceptionally high population density, resulting in a steady reduction in urban infrastructure, including disaster risk management.

The following subsections report a quick review of the seismicity setting of Sudan, which was thought to be a stable intraplate region. However, the earthquake which occurred on 20 May 1990 with MS = 7.4—the highest earthquake in Africa at that time—has led to the reconsideration of these beliefs. Also, we briefly described the geological features at Central Khartoum accompanied by the geotechnical description of the area under study.

3 Seismicity and Seismotectonics Features of the Region

There are three primary rifting systems in Sudan, namely The Red Sea rifts, Central African rift, and Southern Sudan rifts. The latter is known as Abu Gabra rifts, and its branches include Blue Nile, White Nile, and Atbara River rifts. Earlier studies by [7] and [8] indicated that Sudan is seismically stable with infrequent earthquakes of low-to-moderate magnitudes that can give rise to damaging intensities. They noted that the Southern parts of Sudan were frequently subjected to moderate to high earthquake intensities capable of producing damage (e.g., 1990 earthquake with MS = 7.4 was one of the strongest and most devastating earthquakes in Africa [9]).

In Central Khartoum, earthquakes have been triggered from several sources (e.g., rifts and faults in the Western State (about 260 km west of Khartoum), rifts and faults in the Southern States (about 800 km south of Khartoum), faults in the Red Sea State (about 500 km northeast of Khartoum), and earthquakes in Lake Nassir (near the Aswan Dam) in Southern Egypt (about 900 km north of Khartoum). Of all these outlets, Abu Dulayq's source in western Khartoum (Latitude 15.9° and Longitude 33.84°) is the one that mainly affects Central Khartoum [10], which has been involved in a 5.3 magnitude earthquake on a Richter scale that occurred in Khartoum in November 2003. The duration of the shaking lasted for about 40 s and caused minor injuries, but no major damages were reported. Abu Dulayq was also the cause of another 3.5 magnitude shock on the Richter scale, which was also felt in Khartoum in July 2010, as reported by Warage [11]. This could be validated by the elastic rebound hypothesis, which notes that earthquakes are more likely to occur in any area where minimal or no seismic activity has been recently detected [12].

4 Geological and Geotechnical Description of the Study Area

The research study, known as the Central Khartoum, is situated at the intersection of the Blue Nile and the White Nile (the major tributaries of the Great River

Nile), creating a triangular shape. This triangular section consists of deposits ranging from the Quaternary to the Pre-Tertiary Sediments [13]. The superficial soils, locally known as "Gezira formation" include windblown sands, the White Nile alluvium, and Wadi-deposits (streamer channel), which consists of unconsolidated sand, silt sand, gravel, loosely compacted silt, sandy silt, and clay. The pre-Tertiary sediments known as Nubian sandstone were named by Whiteman [13], which was then used as a description to the Nubian Desert in North Africa and made up of conglomerates, grits, sandstone, mudstone, sandy-mudstone, and usually flat-lying sedimentary rock.

In the context of seismic hazard assessment for Khartoum, Seismic micro-zonation has been carried out by [14, 15]. Mohamedzein et al. [15] have assessed the soil liquefaction at Central Khartoum using extensive SPT boring logs. According to their data, the Gezira formation is comprised of clayey and silt–clay soils found in the upper 7 m with various types of sands found underneath from 8 to 25 m. Below, the Nubian Sandstone is found, with the fluctuation of subsurface water from 4 to 8 m near and away from the Nile, respectively. Table 1 shows the reported average soil profile used by [16]. They concluded by saying that saturated liquefiable sand layers are available at Central Khartoum. Abdalla et al. [8] carried out a probabilistic seismic hazard assessment of Sudan and the surrounding regions. They showed that the PGA is around 0.02–0.62 g for low and high seismically active regions, respectively. They reported that in Khartoum area, the PGA ranged from 0.1 to 0.2 for 50 and 250 years expose time, respectively. Warage [10] studied the tectonic structures in Central Sudan and considered the local site effects in Western Khartoum. Very limited research had been done in the context of seismic hazard assessment of Khartoum as a whole, and particularly in Central Khartoum, the area of businesses and government activities. To our knowledge, seismic loading is not accounted-for in the design practice in Sudan, as neither codes nor guidance for earthquake consideration are available.

5 Soil Models Used in the Present Study

The purpose of this study is to quantify local soil effects on the seismic risk at two represented sites at Central Khartoum. The soil models used in the present study are shown in Tables 2 and 3. The soils seem to be similar, comprising of silty clay soil in the upper 8 m with some of the soil layers as sandy soils. The boring-logs data was provided by a reputable geotechnical investigation institute, in which the SPT N-value and other geotechnical properties along the depth were acquired from the laboratory. For comparison, the average soil profile reported in a recent study by [16] for liquefaction assessment was also considered for the response analysis (see Table 1). All of the soil models reached 25 m at which bedrock was found, except for site (2) with a 24-m depth. The average unit weight for clayey layers, loose to medium dense sand, and dense layers was taken as 18.5 kN/m3, 19 kN/m3, and 18.5 kN/m3,

Table 1 Average soil profile in Central Khartoum [16]

Layer No.	Depth (m)	Soil profile	Description	Shear wave velocity (m/sec)	SPT N-value
1	1.0		Stiff brown, Silty Clay of low to medium plasticity with calcareous material.	150	
	2.0				
	3.0				
	4.0				
	5.0				
	6.0				
	7.0				
	8.0				
2	9.0		Loose to medium dense Sand and Silty Sand.	197	11
	10.0				
	11.0				
	12.0				
	13.0				
	14.0				
	15.0				
	16.0				
3	17.0		Medium dense, yellow, graded Sand.	237	19
	18.0				
	19.0				
	20.0				
	21.0				
	22.0				
	23.0				
4	24.0		Dense coarse grained Sand with some gravel.	285	32
	25.0				

respectively. The water table is in 10 m depth for site (2) and 8 m for the average and site (1) models, respectively.

6 Dynamic Properties of Soils

Determination of the dynamic soil properties is the next step in the ground response analysis. As no information was available about the dynamic soil properties in the considered soil sites, the correlation between SPT N-value and the shear wave velocity proposed in [17] for all soil types was used as follows:

$$V_s = 85.35\,\mathrm{N}^{0.348} \tag{1}$$

Table 2 Soil model for site (1)

Layer No.	Depth (m)	Soil profile	Description	Shear wave velocity (m/sec)	SPT N-value
1	1.0 2.0 3.0 4.0 5.0		Hard, dark brown, dry, Silty Clay of low Plasticity with calcareous material.	270	30
2	6.0 7.0 8.0		Hard, light brown, Silty Clay of low Plasticity with calcareous material.	330	49
3	9.0 10.0 11.0 12.0 13.0 14.0 15.0 16.0		Medium dense, yellow, graded Sand.	237	19
4	17.0 18.0 19.0 20.0 21.0 22.0 23.0		Hard, light brown, Silty Clay of low Plasticity with some gravel.	340	50
5	24.0 25.0		Very dense, light brown, Silty Sand.	424	> 100

In which V_s is the shear wave velocity in (m/s), and N represents the SPT recorded value.

The calculated shear wave velocity for each soil model using Eq. (1) is shown in Tables 1, 2, and 3. The estimated average shear wave velocities are about 195 m/s, 295 m/s, and 317 m/s for the average profile, site (1), and site (2), respectively. According to the average shear wave velocity [18] classified sites, the average profile is then classified as *DE* while site (1) is classified as *D* and site (2) as *CD*.

A variety of modulus reduction and damping ratio curves for various soil types have been assigned in the software DEEPSOIL 7.0 [5]. These modules are built in DEEPSOIL. The model propped by Seed and Idriss [19] for sandy soils was used, whereas Vucetic Dobry [20] model was used for silty and clayey soil layers. Plasticity index PI had to be defined in Vucetic Dobry's [20] model to calculate the modulus reduction and damping ratio curves. The PI values varied from 0 at non-plastic layers to 27 for silty layers.

Table 3 Soil model for site (2)

Layer No.	Depth (m)	Soil profile	Description	Shear wave velocity (m/sec)	SPT N-value
1	1.0 2.0 3.0 4.0		Dense, light brown, wet, Silty Clay of High plasticity with calcareous material.	280	37
2	5.0		Dense, light brown, wet, Silty Sand.	290	42
3	6.0 7.0 8.0 9.0		Hard, dark brown, Silty Clay of low Plasticity with calcareous material.	320	49
4	10.0 11.0 12.0 13.0 14.0 15.0 16.0		Medium dense, light brown, poorly graded Sand.	272	28
5	17.0 18.0 19.0		Very dense, light brown, Clayey Gravel.	330	> 50
6	20.0 21.0 22.0 23.0 24.0		Very dense, light brown, Clayey Sand.	430	> 100

7 Frequency and Period of Soil Layers

The frequency or period of soil layers plays a vital role in determining the site response of the soil during an earthquake. When shear waves propagate through soil layers, earthquake motions can be amplified and/or de-amplified; according to the fundamental frequency and damping characteristics, the fundamental period of soil column model could be calculated using the relation in [12] as:

$$T = \sum_{i=1}^{n} \frac{4(H)_i}{(V_s)_i} \quad (2)$$

where $H =$ thickness of soil layer "i", V_s is the shear wave velocity of layer "", $T =$ fundamental time period of the soil column, and $f =$ fundamental frequency of soil column $= 1/T$. For the current soil used in this study, the fundamental frequency and period of soils are given in Table 4.

Table 4 Fundamental natural period and frequency of soils at various sits

Site location	Profile Depth (m)	Profile natural frequency (Hz) (DEEPSOIL)	Profile natural period (Sec) (DEEPSOIL)	F (Hz) = 1/T	$T(sec) = \sum_{i=1}^{n} \frac{4(H)_i}{(V_s)_i}$	Frequency at max. amplification (Hz)	Diff
Average	25	2.12	0.476	1.91	0.524	1.7	9.91
Site (1)	25	2.88	0.3467	2.88	0.3467	2.5	0
Site (2)	24	3.064	0.3263	3.26	0.3068	2.8	6.01

8 Input Ground Motions Used in the Present Study

As Sudan is considered a region of low seismic activity, no seismic recording stations have been set up, and hence, no records of ground motion were reported. Strong ground motion data published in the literature was used in this research. The motions considered are 1989 Loma Prieta, 1979 El Centro, 1995 Kobe, and 1994 Northridge earthquakes. The same ground motions were considered in [15] for developing a design response spectrum curve for Central Khartoum using an average represented soil model similar to that in Table 1. Much consideration should be accounted for when selecting such strong ground motion records as similarity in motion characteristics (e.g., magnitude and peak ground acceleration) and soil conditions [12] occurs. The considered earthquake characteristics like date of occurrence, recording station, and moment magnitudes (Mw) are represented in Table 5. These ground motions are obtained from PEER strong motion database and scaled using the computer program "Seismosignal" (see www.SeismoSoft.com). Each of these records was scaled to the target peak ground acceleration PGA of 0.1 g to match the calculated one obtained from the seismic hazard analysis performed by [8], which corresponds to a 10% probability of exceedance in 50 years. The earthquake characteristics of these motions like predominant period (Tp), bracketed duration, significant duration, and mean period are shown in Table 5, with mean time period varying from 0.481 to 0.604 s. The scaled acceleration time histories for Loma Prieta (1989), El Centro (1979), Kobe (1995), and Northridge (1994) earthquakes are shown in Fig. 2.

9 Method of Analysis

Ground response analysis (GRA) can be performed using equivalent linear (EQ-L), linear, or nonlinear (Non-L) types of analysis according to a chosen stress–strain representation of soil. Based on the shear modulus, $G_{\tan}$ is associated with nonlinear (Non-L) GRA, $G_{\max}$ is for linear, and $G_{\sec}$ is for equivalent linear (EQ-L). In this study, the equivalent-linear (EQ-L) method of analysis was adopted using the computer program DEEPSOIL V 7.0 [5]. This approach uses a frequency domain solution by a transfer function in which the shear modulus and damping for each soil layer

Table 5 Earthquake parameters of scaled motions used in the present study. [*Source* PEER Strong Motion Database-Seismosignal software]

Earthquake parameters	Loma Prieta 1989	Imperial Valley, El Centro 1979	Kobe 1995	Northridge 1994
Date of occurrence	October 18, 1989	October 15, 1979	January 16, 1995	January 17, 1994
Recording station	090 CDMG STATION 47,381	USGS STATION 5115	KAKOGAWA (CUE90)	090 CDMG STATION 24,278
Moment magnitude (Mw)	6.9	7.0	6.9	6.7
PGA (g) applied at the bedrock	0.1	0.1	0.1	0.1
Predominant period (Tp) (sec)	0.220	0.140	0.160	0.260
Bracketed duration (sec)	34.770	34.310	35.840	27.50
Significant duration (sec)	11.410	8.930	12.860	9.070
Mean period (Tm) (sec)	0.604	0.481	0.529	0.540

are approximated based on the effective shear strain, which is generally taken as 65% of the maximum shear strain at each layer. The bedrock was considered rigid; therefore, energy dissipation due to the reflection of seismic waves in the boundaries is not considered. The model of soil column is modeled such that each stratum is composed of several layers, each having a uniform SPT N-value at a respective depth.

10 Results and Discussions

Equivalent-linear (EQ-L) seismic ground response analysis has been carried out using the software DEEPSOIL for two soil models from Central Khartoum subjected to the following scaled strong seismic motions: Loma Prieta (1989), El Centro (1940), Kobe (1995), and Northridge (1994) Earthquakes. Then, the soil models have been analyzed using the above motions as input motions applied at the bedrock. Results were presented in terms of surface acceleration time history, spectral acceleration versus period, and Fourier amplitude ratio versus frequency for surface layers, respectively. Calculated frequencies from DEEPSOIL were compared to those obtained by the closed-form expression in [12]. Further, explanation of these findings and their impact on local soils is illustrated in the subsequent sections.

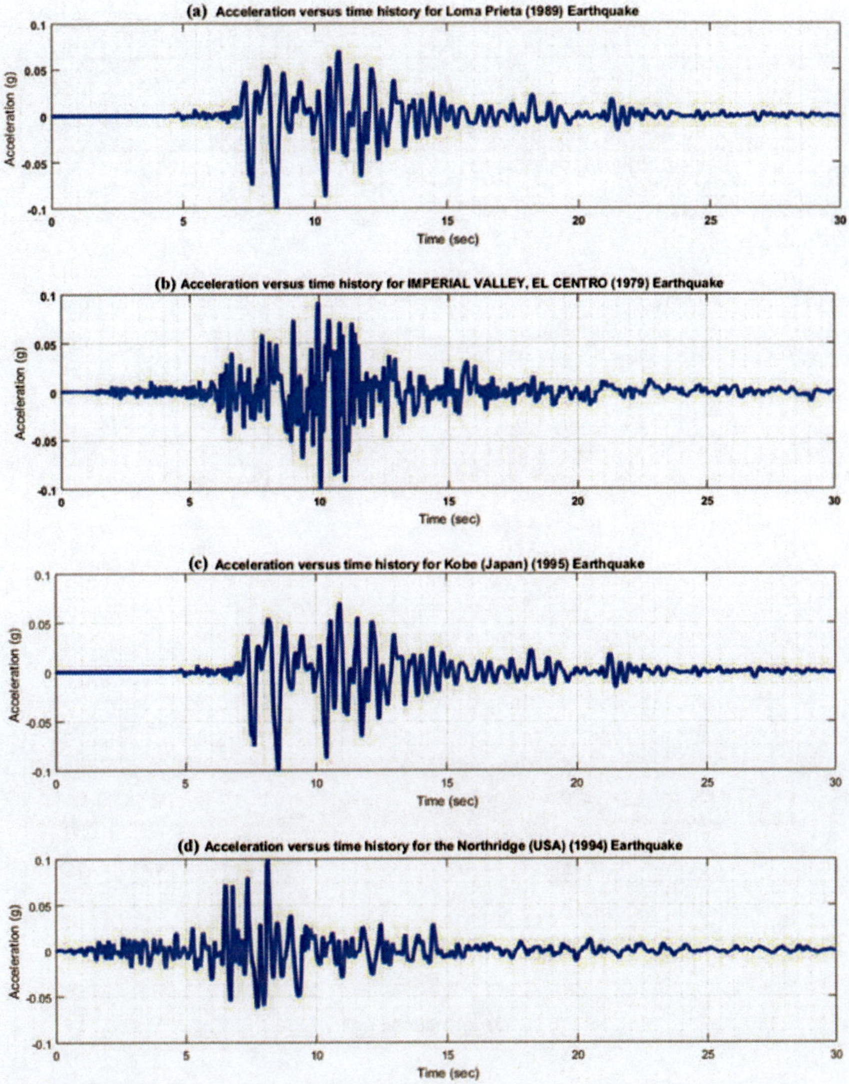

Fig. 2 Scaled input time histories for Loma Prieta (1989), El Centro (1979), Kobe (1995), and Northridge (1994) earthquakes, respectively

10.1 Surface Acceleration Time Histories

The surface acceleration time histories obtained at all sites in response to Loma Prieta (1989), El Centro (1979), Kobe (1995), and Northridge (1994) Earthquakes applied at the bedrock are shown in Fig. 3. For site (1), the maximum accelerations were found to be 0.17 g, 0.21 g, 0.22 g, and 0.25 g, respectively for the above-mentioned

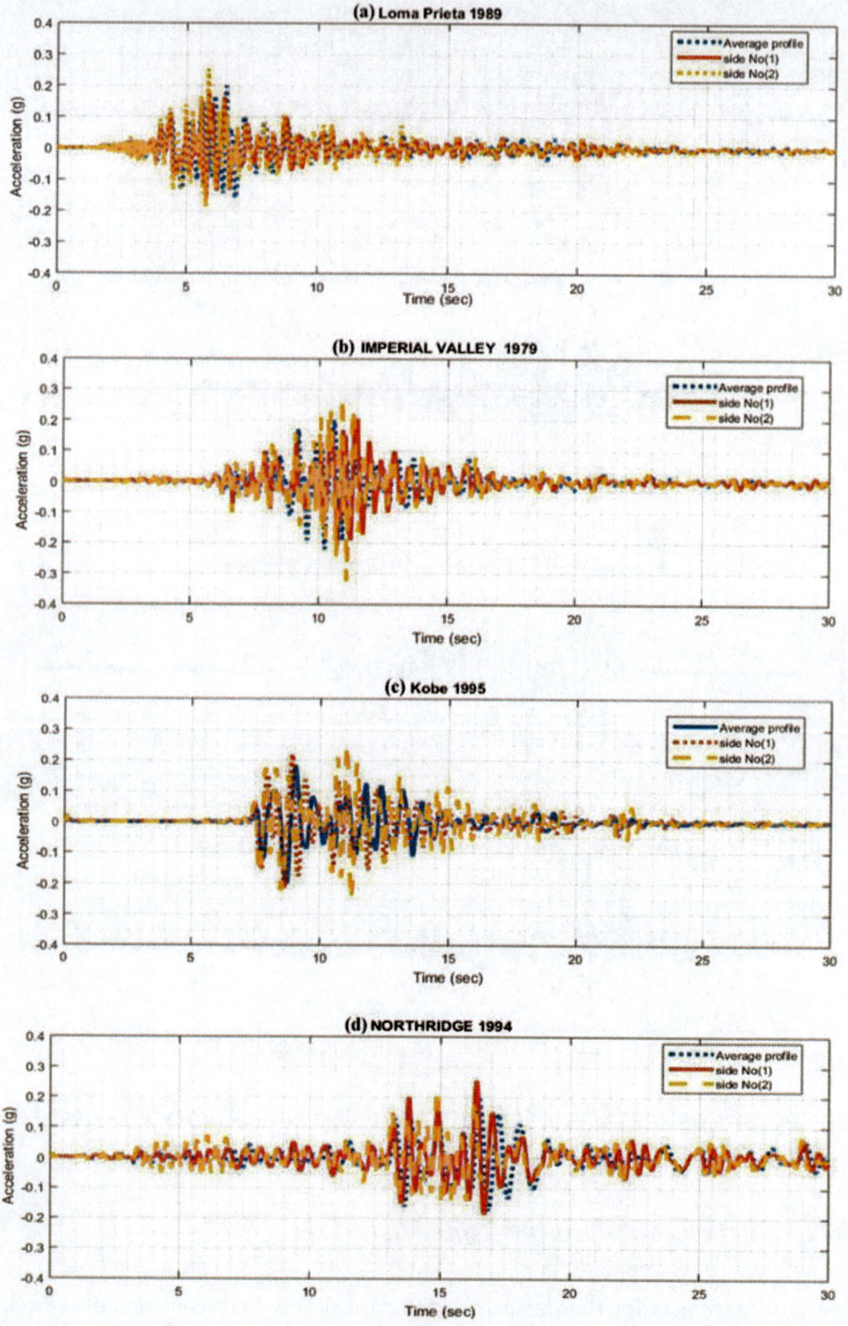

Fig. 3 Surface acceleration time histories at all sites in response to Loma Prieta (1989), El Centro (1979), Kobe (1995), and Northridge (1994) earthquakes applied at the bedrock, respectively

earthquake motions. The amplification ratios compared to the scaled ones applied at the bedrock (0.1 g) were observed to be 1.7, 2.1, 2.2, and 2.5 respectively, with the maximum amplification corresponding to The Northridge motion. Also, Fig. 3 shows the maximum acceleration for site (2), which was found to be 0.195 g, 0.226 g, 0.25 g, and 0.22 g for Loma Prieta (1989), El Centro (1979), Kobe (1995), and Northridge (1994) Earthquakes, respectively. The amplification ratios were found to be 1.95, 2.26, 2.5, and 2.2, respectively, for the same motions, with maximum amplification of 2.5 in response to Kobe's motion. Relating to the average profile, this was adopted for comparison purposes. Maximum acceleration was found to be 0.2 g for Loma Prieta (1989) and Kobe (1995) motions, whereas it was about 0.22 g for El Centro (1940) and Northridge (1994), given a maximum amplitude of 2.2.

10.2 Response Spectrum

The variation of spectral acceleration versus period for site (2) in response to deferent seismic ground motions is shown in Fig. 4c. The damping ratio for generating these plots is taken as 5%. The maximum acceleration for El Centro input motion was found to be 0.94 g at 0.45 s, while that of Loma Prieta, Kobe, and Northridge were found to be 1.02 g, 1.34 g, and 0.37 g at periods of 0.3 s, 0.39 s, and 0.37 s, respectively, which fall around the fundamental period of the soils considered in site (2). However, these factors must be taken into account when designing a structure in Central Khartoum. The response spectrum versus the period at the ground surface for site (1), in response to the deferent input motions considered, is shown in Fig. 4b.

The maximum acceleration of Northridge motion was found to be 1.16 g at a time period of 0.74 s. For Loma Prieta motions, the maximum acceleration was 0.68 g corresponding to 0.35 s, which is also around the fundamental period of the soils in site (1). Maximum acceleration was found to be 1.13 g at a time period of 0.7 s, and 1.12 g at 0.68 s for Kobe and El Centro motions, respectively. It was observed that the maximum acceleration for the average profile in response to Loma Prieta motion was 0.9 g occurred at a time period of 0.4 s (see Fig. 4a). It was also shown that Loma Prieta motion is affecting more the shorter periods for soils. For Northridge input motion, the maximum peak acceleration was observed to be 0.94 g at 0.85 s, whereas, for El Centro and Kobe motions, it was 0.808 g at 0.8 s and 0.725 g at 0.75 s, respectively. It can be shown that the Loma Prieta earthquake had more impact on the average profile, even though the peak spectral acceleration was obtained by referring to the Northridge subordinated motion, this is due to the looser soils in the average profile compared to soils in site (1 and 2).

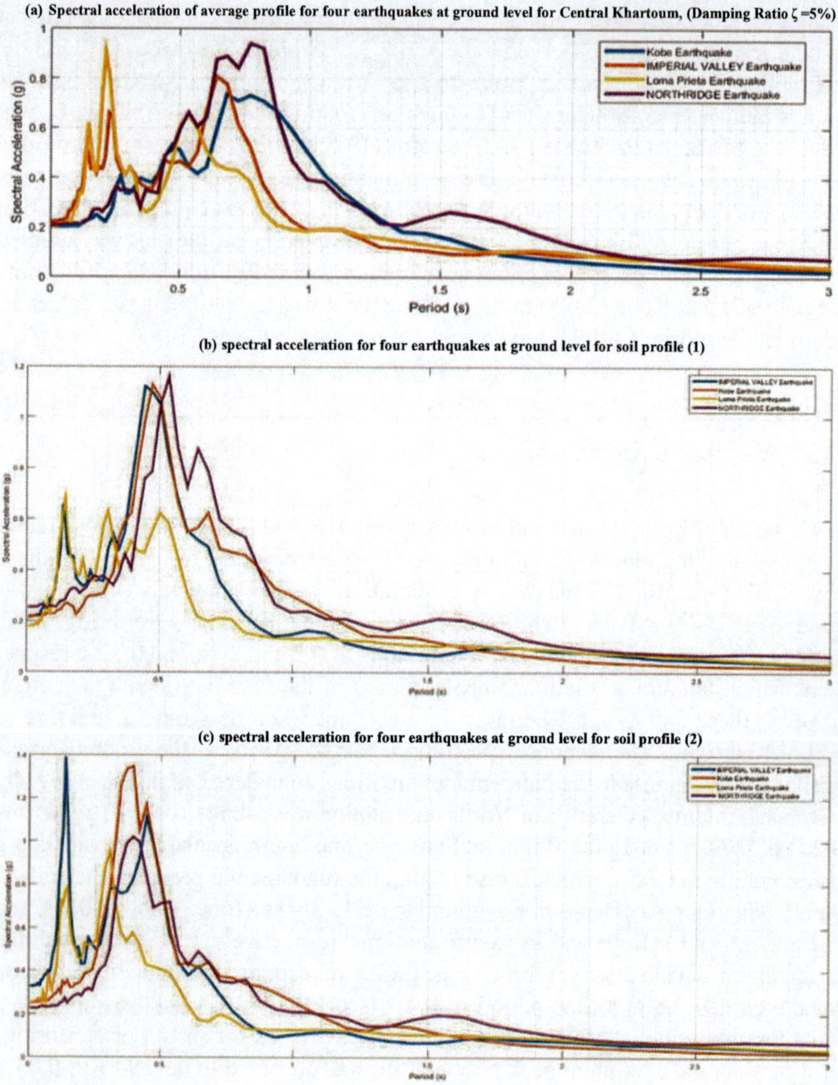

Fig. 4 Spectral acceleration for average profile, site (1) and site (2), respectively

(a) Amplitude ratio for different earthquake for average profile.

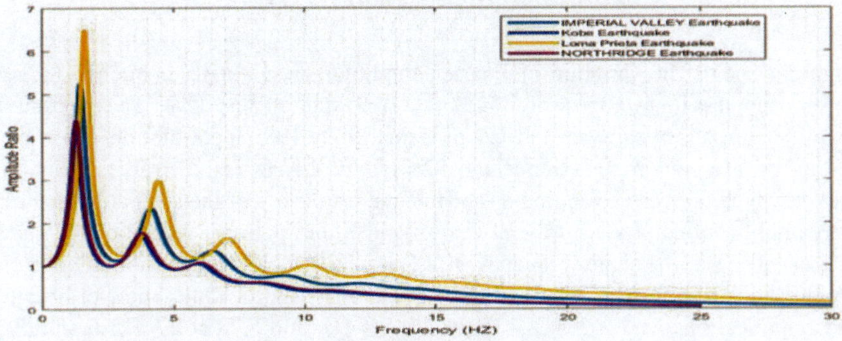

(b) Amplitude ratio for different earthquake for soil profile (1).

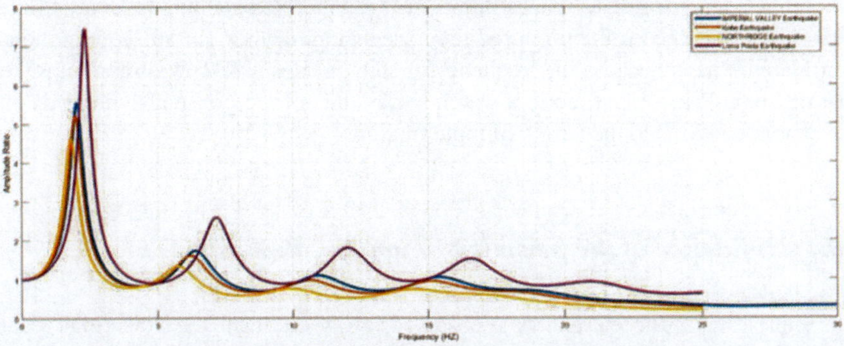

(c) Amplitude ratio for different earthquake for soil profile (2).

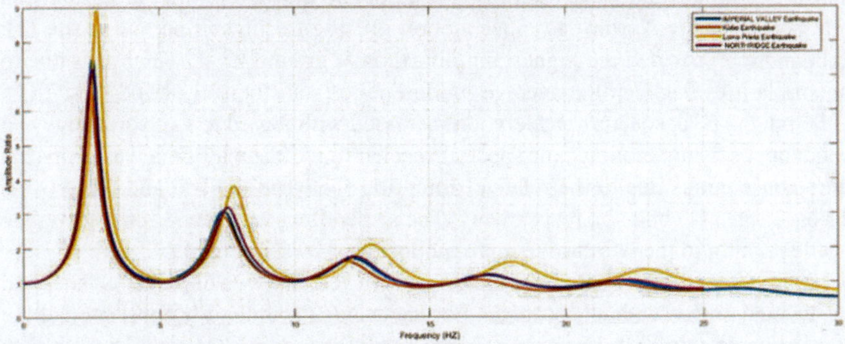

Fig. 5 Variation of Fourier amplitude ratio versus frequency for all seismic motions for average profile and sites (1&2) at the ground level, respectively

10.3 Fourier Amplitude Ratio Spectrum

Figure 5 shows the variation of Fourier amplitude ratio versus frequency for all seismic motions for average profile and sites (1&2) at the ground level. The maximum amplification observed at sites (1&2) of Loma Prieta 1989 input motion was about 7.4 and 8.3 at a predominant frequency of 2.5 Hz and 2.8 Hz, respectively (see Fig. 5b and c), which are very close to the fundamental frequency of the soils at both sites. The amplifications produced by El Centro and Kobe motions were about 5.5 and 5.2, occurring at the same frequency of 2 Hz, whereas Northridge motion gave an amplification of 4.6 at 1.9 Hz at site (1). Also, For El Centro, Kobe, and Northridge, the amplifications were 7.3 at 2.7 Hz, 6.2 at 2.5 Hz, and 6.78 at 2.6 Hz at site (2), respectively, in comparison with the average profile with a maximum amplitude ratio of 6.5 occurring at 1.7 Hz as shown in Fig. 5a. Corresponding to Loma Prieta motion, the average profile produced the lowest amplification for all input motions considered; this is because the frequency of soils in sites (1& 2) was more close to that of Loma Prieta input motion, which shows that site response also depends on the characteristics and the period of input motions.

10.4 Influence of the Seismic Motion on Site Response

The highest surface acceleration was observed corresponding to the Northridge input motion with respect to that at the bedrock, where the frequency content of the Northridge input motion was similar to that of the average profile, and the mean time period of the Northridge input motion is close to the soil period at site (1). Kobe motion provided the highest amplification at ground level, which was due to the similarities between the average motion period and the soil period at site (2). Although the peak spectral acceleration increased with regard to the Northridge and Kobe input seismic motions, the spectral acceleration of Loma Prieta was primarily more conspicuous due to the relative similarities between the soil and the seismic period at sites (1) and (2), respectively. The same effect was found for the average profile relating to the Northridge input motion, as shown in Fig. 4a.

Figure 6 shows the variations of the shear wave velocity with depth for all sites and the average profile with all locations. For the average profile, it is apparent that the shear wave velocity falls in a narrow range, compared to sites (1) and (2). Both sites have a higher shear wave velocity at the lower layers and also at the bedrock level. The highest Fourier amplitude acquired corresponded to Loma Prieta input motion, as shown in Fig. 5c. Natural frequencies of soils were those of peak amplitude values in Fourier amplitude ratio vs frequency plots (see Table 4). High amplifications illustrate the crucial role of the soil type in the specific site in modifying the ground response.

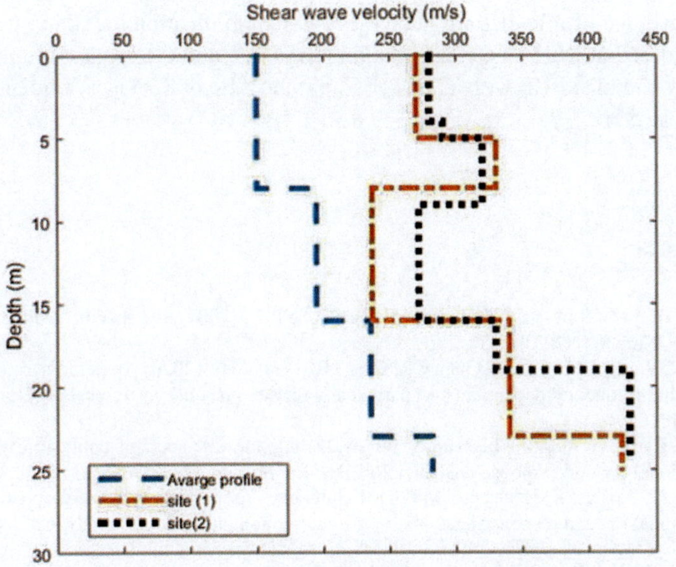

Fig. 6 Variations of the shear wave velocity with depth for all sites and the average profile with all locations

11 Conclusions

The effects of local soil sites on modifying seismic ground motions are studied at two different locations, and the average profile from Central Khartoum using the equivalent-linear (EQ-L) seismic ground response analysis was obtained. The input motions considered are Loma Prieta (1989), El Centro (1940), Kobe (1995), and Northridge (1994) earthquakes using the software DEEPSOIL, and the various output parameters were shown.

The conclusions obtained from this study are as follows:

1. The role of ground response in influencing local soil sites is site-specific, where the type of soil underlying the site of interest greatly affects the site response. A substantial difference between the two considered sites was noticed, and the average profile was observed for all output parameters reported in this study.
2. The natural periods of the two sites were compared to those obtained from the DEEPSOIL program and used a closed formula. The results were in strong alignment, which effectively demonstrated the reliability of DEEPSOIL.
3. The peak ground acceleration has been fluctuating from site to site, even for the same earthquake input motion. The amplification factors ranged from 1.7 to 2.5.
4. Response spectral curves have been obtained for all of the soil sites considered in this study with a 5% damping ratio. The maximum spectral acceleration at these locations varied from 0.68 to 1.34 g.

5. The average profile that provided the lowest amplification for all input motions considered at the Fourier amplitude ratio is 4.3, and the highest amplification was found in site (2) with a Fourier amplitude ratio of 8.35 in a frequency range of up to 5 Hz.

References

1. GovindaRaju L, Ramana GV, HanumantaRao C, S. T. G. (2004) Site specific ground response analysis. Curr Sci 87(10):1354–1362
2. Meite R, Wotherspoon L, McGann CR, Green RA, Hayde C, (2020) An iterative linear procedure using frequency-dependent soil parameters for site response analyses. Soil Dyn. Earthq. Eng 130:105973
3. Tao Y, Rathje E (2020) Taxonomy for evaluating the site-specific applicability of one-dimensional ground response analysis. Soil Dyn Earthq Eng 128:105865
4. Stewart JP, Afshari K, Hashash YM (2014) Guidelines for performing hazard-consistent one-dimensional ground response analysis for ground motion prediction. Berkely, California
5. Hashash YMA et al. (2020) DEEPSOIL 7.0, User Manual. Urbana, IL
6. Elnimeiri MKM, Satti SSM, Ibrahim MKM (2020) Barriers of access and utilization of reproductive health services by adolescents-Khartoum state-Sudan-2020: study protocol. Reprod Health 17(1):121
7. Ambraseys NN, Adams RD (1986) Seismicity of the Sudan. Bull. Seism Soc Am 76:483–493
8. Abdalla JA, Mohamedzein YE, Wahab AA (2001) Probabilistic seismic hazard assessment of Sudan and its vicinity. Earthq. Spectra 17(3):399–415
9. Person WJ, Jacobs JM (1992) Significant earthquakes of the world 1990
10. Warage A (2007) Seismotectonics in Central Sudan and local site effect in Western Khartoum. University of Bergen, Norway
11. Ibrahim KY, Babiker NM, Ali MH, Warage AA, Humeda KA (2015) A unified mw-based earthquake catalog of sudan and adjacent regions for the period 1408–2014. Am J Earth Sci 2(5):150–156
12. Kramer SL (1996) In: Geotechnical earthquake engineering. Upper Saddle River, New Jersey, Prentice-Hall, Inc.
13. Whiteman JA (1971) In: The geology of the sudan republic. Oxford
14. Mohamedzein YE, Abdalla JA, Abdelwahab, OO E (2001) Seismic microzonation of Central Khartoum. In: Proceedings of the forth international conference on recent advances in geotechnical earthquake engineering and soil dynamics, San Diego, CA, USA, March 26–31
15. Mohamedzein YE-A, Abdalla JA, Wahab A (2006) Site response and earthquake design spectra for Central Khartoum, Sudan. Bull Earthq Eng 4:277–293
16. Al-Ajamee M, Mustafa MM, Ali AM (2021) Khartoum Geohazard: an assessment and a future warning. In: Adhikari BR, Kolathayar S (eds.), Geohazard Mitigation, Lecture Notes in Civil Engineering 192. https://doi.org/10.1007/978-981-16-6140-2_8
17. Ohta Y, Goto N (1978) Empirical shear wave velocity equations in terms of characteristic soil indexes. Earthq Eng Struct Dyn 6(2):167–187
18. NEHRP (2020) Recommended seismic provisions for new buildings and other structures—Volume I: Part 1 Provisions. Washington, D.C.
19. Seed HB, Idriss IM (1970) Soil moduli and damping factors for dynamic response analysis. California, Berkeley
20. Vucetic M, Dobry R (1991) Effect of soil plasticity on cyclic response. J Geotech Eng ASCE 117(1):89–107

Assessment of Earthquake Recurrence for North India and Surrounding Region Using Probabilistic Models

Monalisha Nayak and T. G. Sitharam

Abstract The northern part of India is one of the seismically active regions of the world, which lies near the plate boundary region of Indian plate and Eurasian plate. The continuous tectonic movement between Indian plate and Eurasian plate accumulates lots of strain energy in the plate boundary region and this energy may come out in near or far future generating big earthquakes in this region. The 25thApril 2015 Nepal earthquake (Mw 7.8) is one of the outcomes of such huge stored strain energy. So, in the attempt of minimizing loss of lives and properties, this paper presents probabilistic models for forecasting of earthquakes of magnitude $(M_w) \geq 7.0$ in North India (NI) and surrounding region. In this present study, three probabilistic models namely Weibull, Gamma and Lognormal probabilistic models were used to estimate the probability of occurrence of earthquake $M_w \geq 7.0$ using an earthquake catalogue 1900–2016. The goodness of models was assessed by the logarithmic likelihood function (ln L). As per the results, the Weibull models shows highest probability of occurrence whereas the lognormal model shows the lowest probability of occurrence. The estimated conditional probability reaches 0.999 after a time interval of 11, 13 and 15 years for Weibull, Gamma and lognormal Distribution respectively from the last earthquake of 2015. This means by the year 2026, 2028 and 2030 by Weibull, Gamma and lognormal model respectively will achieve a conditional probability of 99.9% in and around the present study area.

Keywords Conditional probability · Earthquake recurrence · Maximum likelihood estimation (MLE) · Probability models · Recurrence interval

---

M. Nayak (✉)
IISc, Bangalore, India
e-mail: monalishan@iisc.ac.in

T. G. Sitharam
IIT, Guwahati, Assam, India

T. G. Sitharam et al. (eds.), *Earthquake Geotechnics*, Lecture Notes in Civil Engineering 187, https://doi.org/10.1007/978-981-16-5669-9_44

1 Introduction

Earthquakes are natural phenomena that can happen at any time in any location in earth crust. The danger of loss of lives and property due to earthquake is much more than any other natural hazard as it can destroy the locality within very short time and can trigger other kinds of disastrous hazards like tsunami, landslide, liquefaction, lateral spreading etc. As the natural phenomena can't be stopped, the alternative is to minimize the loss due to the natural hazards. In this context, every country has building codes for earthquake resistance construction method. But in a country like India which has a huge population (2nd highest populated country in world) and due to many differences and problems in different regions, it became impossible to strictly follow the codal provisions for all kind of structures. In this regard another alternative could be prediction of occurrence of future earthquake, as earthquakes occur repeatedly along certain tectonic regime in world and in India. But the real time prediction of occurrence of a future potentially damaging earthquake is very difficult. But considering the past earthquake events of any region, many researchers had attempted to predict this complex natural phenomenon of earthquake by probabilistic approaches. By the use probability tool, the earthquakes are mathematically modeled considering the time interval of successive earthquake events.

Many researchers have come up with several probabilistic models for forecasting of earthquakes in any region [6, 19, 9, 25, 15, 2, 11, 16, 10, 7]. Among these stated previous studies, Lindh [9] and Sykes and Nishenko [25] used Normal probability distribution, Nishenko and Buland [25] and McCalpin and Slemmons [11] used Lognormal probability distribution, Hagiwara [6] and Rikitake [19] used Weibull probability distribution. Ellsworth et al. (1995) along with the above three probability distributions, introduced one more probability distribution, Gamma distribution whose properties are intermediate between Weibull and Lognormal distribution. Ogata [16] used Bayesian estimation of optimum model for four different distribution models (Lognormal, Gamma, Weibull and double Exponential) and one stationary Poisson Model using paleo-earthquakes. Mathews et al. [10] constructed a model where time interval between events follows Brownian passage time (BPT) distribution. Hasumi et al. [7] found superposition of Weibull and log-Weibull probability distribution can justify the time interval between successive earthquakes using JMA (Japanese Meteorological Agency) earthquake catalog. The time dependent study of probability of occurrence of large earthquakes based on recurrence interval for Indian sub-continent had carried out by Parvez and Ram [17], Shanker and Papadimitriou [22], Sharma and Kumar [23], Tripathi [26, 29, 30], Sil et al. [24], Chingtham et al. [1] using different probabilistic models. Parvez and Ram [17] had carried of time dependent seismicity analysis of for two areas of Himalayan region, i.e. NE India and Hindukush region. The probability models used were Exponential, Gamma, Lognormal and Weibull. Parvez and Ram [18] had performed similar kind of analysis using the same probability models as Parvez and Ram [17] for the Indian subcontinent. Shanker and Papadimitriou [22] had carried out the time dependent assessment of probabilities of occurrence future large earthquakes ($M_S \geq 6.0$) in Hindukush–Pamir–Himalayas region. Tripathi [26] and Yadav et al. [29], both had estimated time dependent earthquake recurrence for Gujarat, India using three models named Gamma, Lognormal and Weibull distribution. Similarly, Yadav et al. [30] and Sil et al. [24] worked on forecasting of earthquakes in North East India and surrounding region using the same above three probabilistic models. The random variable for both these studies were the time interval between successive earthquake (recurrence interval) of respective catalogue. This time dependent seismic studies for NE India suggested that, the best fit model is Gamma model as per Yadav et al. [30] whereas the same by Sil et al. [24] was Weibull model in terms of logarithmic of likelihood (ln L) function. The time dependent seismic study was done by Chingtham et al. [1] for NW Himalaya and adjoining region using earthquake catalogue from 1900–2010, using Lognormal and Weibull model, along with time independent Poisson model. But there has been no such probability model fitting is done for the northern part of India.

In the present study an area is selected in northern part of India that includes seismically more active western Himalaya, central Himalaya and moderately active Indo-Gangetic plain and its surround region for probability model fitting to earthquake recurrence. An attempt was made for forecasting of probability occurrence of major earthquake ($M_W \geq 7.0$) in the study area using thee different probabilistic distribution models. The probabilistic models considered in this present study are Lognormal, Gamma and Weibull model. An earthquake event catalog of magnitude greater than equal 7.0 was prepared for a region of 500 km around the study area. The time interval between successive earthquake events of the catalog was model through the above stated probabilistic distribution functions. For best fit model, the negative Logarithmic likelihood function (ln L) was estimated for all the three models. The probability of occurrence of next big event was assessed using conditional probability.

2 Study Area and Its Seismotectonics

The present study area lies between latitude of 19°N-38°N and longitude of 72°E-91°E (Fig. 1). It includes mostly northern parts of India and some parts of central and eastern India. This selected study region primarily falls to two seismotectonic regimes of India, the Himalayan region and Indo-Gangetic plain, out of three seismotectonic regions, named the Himalayan region, Indo-Gangetic plain and the peninsula shield region [8]. The seismic activity of the present study area is mainly controlled by the tectonic setting of Himalayas.

A lot of important tectonic features are present in and around the present study area. They are mainly spread along the length and width of Himalaya. As the Himalaya extend about 2400 km from west to east and width varies 250–300 km, so these tectonic features extend from few meters to few kilometers. The primary tectonic features in the Himalayan region from north to south are; the main central thrust (MCT), the main boundary thrust (MBT) and the Himalayan frontal thrust

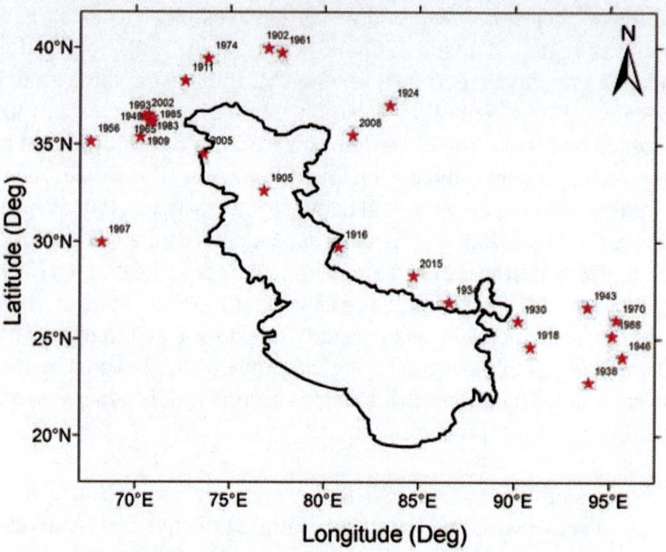

Fig. 1 Earthquake data of *Mw* ≥ 7.0 for probabilistic forecasting of earthquake in seismic study area as given in Table 1

(HFT) [12, 28 14, 21]. The thrust action starts in north from MCT and progressed southward giving rise to MBT and the youngest one HFT.

The moderately seismically active Indo-Gangetic plain separates the Himalaya from peninsula shield region. Being one of the most fertile river basins, the population is more than 500 million (population density as per 2011 census is 520 persons per km^2) and this makes it most populated river basin of the world. The tectonic movement in Indo Gangetic plain leads to formation of Aravalli ranges. The floor of Gangetic basin is uneven and interrupted by faults [20, 3]. Several hidden faults and ridges are also present in Ganga basin [4, 28]. The seismic activity of this basin is mainly associated with strike slip faulting [5]. The seismic activity of this region is less as compare to Himalayan belt, but seismic risk could be more in moderate earthquakes due to very high population density. The nature of propagating seismic waves from focus could be changed dramatically and became more hazardous due to local site effects of soft alluvial deposits of the basin.

In this study, the tectonic features of the study area were collected from SEISAT (Seismotectonic Atlas of India) 2000 published by Geological Survey of India (GSI). The present study has considered the earthquake catalogue prepared by Nayak and Sitharam [13]. The required earthquake events were selected for the analysis. All the earthquakes of magnitude $M_W \geq 7.0$ are selected for the seismic study area. Considering one big event per year, the final dataset for probabilistic forecasting of earthquakes in seismic study area are 28 in number. The details of these 28 earthquakes are given in Table 1 and corresponding figure showing the distribution of these earthquakes in seismic study area is shown in Fig. 1.

Table 1 Earthquake data of $M_W \geq 7.0$ for probabilistic forecasting of earthquake

Sl. No	Long	Lat	Year	Month	Date	Hour	Min	Depth (km)	Mw
1	77.00	40.00	1902	8	22	3	0	0.0	7.7
2	76.79	32.64	1905	4	4	12	49	20.0	7.9
3	70.25	35.39	1909	7	7	9	37	200.0	7.7
4	72.63	38.33	1911	2	18	6	41	15.0	7.3
5	80.75	29.73	1916	8	28	6	39	20.0	7.0
6	90.85	24.60	1918	7	8	10	22	15.0	7.2
7	70.71	36.20	1921	11	15	8	36	240.0	7.8
8	83.45	37.06	1924	7	11	7	44	10.0	7.2
9	90.18	25.93	1930	7	2	9	3	15.0	7.1
10	86.59	26.89	1934	1	15	8	43	15.0	8.0
11	93.92	22.75	1938	8	16	4	28	75.0	7.0
12	93.85	26.64	1943	10	23	5	23	15.0	7.2
13	95.67	24.05	1946	9	12	3	17	15.0	8.0
14	70.70	36.56	1949	3	4	10	19	228.7	7.5
15	67.61	35.16	1956	6	9	23	13	25.0	7.3
16	77.71	39.76	1961	4	13	4	34	35.0	7.0
17	70.72	36.41	1965	3	14	3	53	207.8	7.4
18	95.34	25.98	1970	7	29	10	16	76.1	7.0
19	73.83	39.46	1974	8	11	1	13	9.0	7.3
20	70.74	36.37	1983	12	30	11	52	214.5	7.4
21	70.90	36.19	1985	7	29	7	54	98.7	7.4
22	95.13	25.15	1988	8	6	12	36	90.5	7.3
23	70.87	36.38	1993	8	9	12	42	214.5	7.0
24	68.21	29.98	1997	2	27	21	8	33.0	7.1
25	70.48	36.50	2002	3	3	12	8	225.6	7.4
26	73.59	34.54	2005	10	8	3	50	26.0	7.6
27	81.47	35.49	2008	3	20	10	32	10.0	7.2
28	84.73	28.23	2015	4	25	6	11	8.2	7.8

3 Probabilistic Models

The question of 'When' and 'Where' an earthquake will take place can't be predicted exactly. Probability gives a tool for mathematical modelling of any physical phenomenon with respect to some random variables. Earthquake is a phenomenon of building up of strain energy in earth crust and release of that same stored strain energy, when it exceeds the strength of the rock mass. The theory of strain energy built up is well explained by elastic rebound theory. For seismically active regions whose tectonics has a great potential for future earthquake, may be mathematically modelled by suitable probabilistic models, considering recurrence interval as the random variable for the probabilistic models. Recurrence interval is the time interval between successive events. So, the time interval between successive past earthquakes may give the future recurrence interval from the last event using the conditional probability concept. In this present study the forecasting of future earthquakes of magnitude *Mw* ≥ 7.0 was carried out by three probabilistic models named, Lognormal, Gamma and Weibull probability Distribution. Utsu [27] has given a detail of these three models and a brief description of these model are presented below.

3.1 Lognormal Probability Distribution

Lognormal distribution is a continuous distribution where logarithm of random variable is distributed normally. The probability density function, (PDF, $f(T)$) for two parameter lognormal model is given as.

$$f(T) = \frac{1}{\sigma T\sqrt{2\pi}} \exp\left[-\frac{(\ln T - \mu)^2}{2\sigma^2}\right], \mu > 0, \sigma > \tag{1}$$

Where, T is the time interval between successive earthquakes, μ and σ are model parameters. The cumulative distribution function (CDF, $F(T)$) is given as

$$F(T) = \frac{1}{2} erfc\left[-\frac{\ln T - \mu}{\sigma\sqrt{2}}\right] = \Phi\left(\frac{\ln T - \mu}{\sigma}\right) \tag{2}$$

where 'erfc' is the complementary error function and Φ is the CDF of standard normal distribution (i.e., normal distribution with $\mu = 0$ and $\sigma = 1$) which is given by

$$\Phi(x) = \frac{1}{\sqrt{2\pi}} \int_{-\infty}^{x} \exp\left[-\frac{x^2}{2}\right] dx \tag{3}$$

If 't' is the time elapsed from the last earthquake in a region, the probability of occurrence of next earthquake can be calculated with the help of cumulative probability $\varphi(t)$, which is 1–CDF.

$$\varphi(t) = 1 - \Phi\left(\frac{\ln t - \mu}{\sigma}\right) \tag{4}$$

The conditional probability (probability of future age with reference to current age), $p(\tau/t)$ gives the probability of occurrence of next earthquake within a time interval of t ant $(t + \tau)$ which can be given as

$$p(\tau/t) = 1 - \frac{1 - \Phi\left(\frac{\ln(t+\tau)-\mu}{\sigma}\right)}{1 - \Phi\left(\frac{\ln t-\mu}{\sigma}\right)} \tag{5}$$

3.2 Gamma Probability Distribution

It is also a two-parameter continuous distribution with parameters such as shape parameter (r) and scale parameter (θ). Gamma distribution also given in terms of shape parameter (r) and inverse scale parameter, called rate parameter (c). In this present study, the parameters considered for Gamma distribution are parameter (r) and rate parameter (c). The PDF $f(T)$, cumulative probability $\varphi(t)$ and the conditional probability $p(\tau/t)$ are given as follows:

$$f(T) = [c/\Gamma(r)] * [(cT)(r-1)] * \exp[-cT]; \quad r > 0,\ c > 0 \tag{6}$$

$$f(T) = [c^\wedge r/\Gamma(r)] * [T^\wedge(r-1)] * \exp[-cT]; \quad r > 0,\ c > 0 \tag{7}$$

$$\varphi(t) = \Gamma(r,\ ct)/\Gamma(r) \tag{8}$$

$$p(\tau/T) = 1 - [\Gamma\{r,\ c(t+\tau)\}/\Gamma(r,\ ct)] \tag{9}$$

where r and c are model parameters and other symbols are same as in Lognormal distribution. The term $\Gamma(k, x)$ is the incomplete gamma function.

3.3 Weibull Probability Distribution

Please Weibull distribution is one of the most widely used probability model due to its versatility. The two parameter Weibull distribution with model parameters, scale parameter (α) and shape parameter (β) is used in this present study. The PDF $f(T)$ with model parameter α and β is given as

$$f(T) = (\beta/\alpha) * (T/\alpha)^\wedge(\beta - 1) * \exp -[(T/\alpha)^\wedge\beta] \tag{10}$$

Normalizing the scale parameter with respect to shape parameter, the normalized scale parameter (λ) is given as

$$\lambda = 1/\alpha^\wedge\beta \tag{11}$$

So, the PDF in terms of normalized scale parameter (λ) and shape parameter (β) is given as

$$f(T) = \{\lambda\beta T\hat{}(\beta - 1)\} * \exp -(\lambda T\hat{}\beta) \quad (12)$$

The cumulative probability $\varphi(t)$ and the conditional probability $p(\tau/t)$ are given as follows

$$\varphi(t) = \exp[-(t/\alpha)\hat{}\beta] \quad (13)$$

$$p(\tau/t) = 1 - \exp[(t/\alpha)\hat{}\beta - ((t + \tau)/\alpha)\hat{}\beta] \quad (14)$$

In the above equations α and β are model parameters and other notations are same as in Lognormal distribution.

4 Results and Discussions

The data set contains 28 earthquake events from 1900 to 2015, with 27 recurrence intervals. The probability of occurrence of any major events with magnitude $M_W \geq 7.0$ was estimated using three probabilistic models named Lognormal, Gamma and Weibull. The mean of the sample was estimated to be 4.176. The model parameters of these three distribution models were evaluated using maximum likelihood estimation (MLE) and the values are listed in the Table 2. The estimated expected mean values of Lognormal, Gamma and Weibull models are 4.1892, 4.1760 and 4.1857 respectively. Among these expected mean values, the mean for Gamma model matches to the mean of sample space exactly. To select best fitting model for the data set, the negative log-likelihood function (ln L) for all three models were estimated. Depending on the ln L value, highest value ln L indicates best fit model and lowest value of ln L indicated worst fit model. According to this consideration, the Gamma model is best fit having highest value of ln L (−52.2598) and Weibull is the worst fit among the

Table 2 Estimated model parameters for the three probability distributions

	Lognormal	Gamma	Weibull
Model parameters	$\mu = 1.3350$	$r = 5.4614$	$\alpha = 5.332$
	$\sigma = 0.4416$	$c = 1.3078$	$\beta = 2.205$
Expected mean	4.1892	4.1760	4.1857
ln L	−52.2863	−52.2598	−53.2115

Sample mean time interval between successive earthquakes = 4.1760

Standard deviation of the sample = 1.8503

three models having lowest value of ln L (−53.2115). Lognormal model with ln L value −52.2863 is intermediate between Lognormal and Weibull.

Table 3 gives the conditional probability of occurrence of major earthquake of magnitude $M_W \geq 7.0$ from last earthquake of 2015 and also 10 years elapsed time from last earthquake of 2015 (i.e. 2025) for all the three models. The plots of conditional probability values listed in the Table 3 are shown in Figs. 2 and 3 for $t = 0$ and $t = 10$, respectively. As per ln L value, Gamma model was found to be most suitable model for seismic study area. The conditional probability reaches 99.9% by the year 2027 after a time interval of 12 years from 2015 by the best fit model, Gamma model. The other two models, Weibull and Lognormal model estimates 99.9% probability of occurrence in time intervals of 11 years and 15 years from base period of 2015

Table 3 Conditional probability of major earthquake of magnitude $M_W \geq 7.0$

τ	For $t = 0$ (i.e. year 2015)			For $t = 10$ (i.e. year 2025)		
	Lognormal	Gamma	Weibull	Lognormal	Gamma	Weibull
1	0.0013	0.0054	0.0222	0.4345	0.6010	0.8075
2	0.0730	0.0834	0.1152	0.6761	0.8456	0.9706
3	0.2962	0.2790	0.2811	0.8120	0.9418	0.9965
4	0.5462	0.5175	0.4868	0.8894	0.9786	0.9997
5	0.7329	0.7173	0.6838	0.9341	0.9923	1.0000
6	0.8495	0.8508	0.8345	0.9602	0.9973	1.0000
7	0.9167	0.9275	0.9274	0.9757	0.9990	1.0000
8	0.9541	0.9670	0.9736	0.9850	0.9997	1.0000
9	0.9746	0.9857	0.9922	0.9906	0.9999	1.0000
10	0.9858	0.9941	0.9981	0.9940	1.0000	1.0000
11	0.9920	0.9976	0.9996	0.9962	1.0000	1.0000
12	0.9954	0.9991	0.9999	0.9975	1.0000	1.0000
13	0.9973	0.9997	1.0000	0.9984	1.0000	1.0000
14	0.9984	0.9999	1.0000	0.9989	1.0000	1.0000
15	0.9991	1.0000	1.0000	0.9993	1.0000	1.0000
16	0.9994	1.0000	1.0000	0.9995	1.0000	1.0000
17	0.9997	1.0000	1.0000	0.9997	1.0000	1.0000
18	0.9998	1.0000	1.0000	0.9998	1.0000	1.0000
19	0.9999	1.0000	1.0000	0.9999	1.0000	1.0000
20	0.9999	1.0000	1.0000	0.9999	1.0000	1.0000
21	0.9999	1.0000	1.0000	0.9999	1.0000	1.0000
22	1.0000	1.0000	1.0000	1.0000	1.0000	1.0000
23	1.0000	1.0000	1.0000	1.0000	1.0000	1.0000
24	1.0000	1.0000	1.0000	1.0000	1.0000	1.0000
25	1.0000	1.0000	1.0000	1.0000	1.0000	1.0000

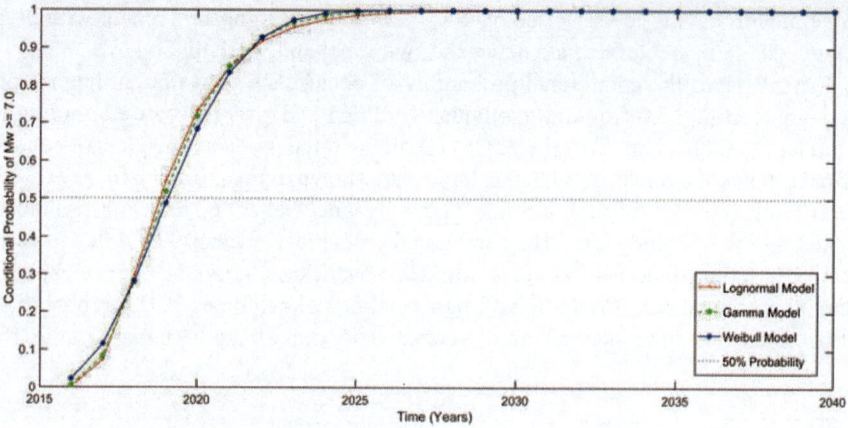

Fig. 2 Conditional probability for occurrence of an earthquake $M_W \geq 7.0$ for an elapsed time $t = 0$, from last strong event in seismic study area in next τ year

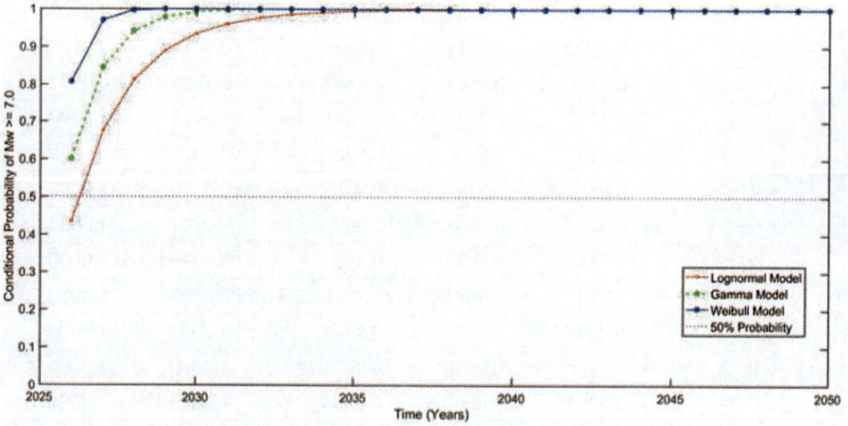

Fig. 3 Conditional probability for occurrence of an earthquake $M_W \geq 7.0$ for an elapsed time $t = 10$, from last strong event in seismic study area in next τ year

respectively. This means by the year 2026 and 2030 respectively by Weibull and Lognormal model, there is 99.9% probability of occurrence of an earthquake of $M_W \geq 7.0$ in the seismic study area.

The present study has modeled the earthquake phenomenon mathematically by the probabilistic tool using time interval between successive big earthquakes as the random variable. Though elastic rebound theory explains the occurrence of major earthquake considering the built up of crustal stress and its release, consideration of the time interval between successive earthquakes may not be just sufficient parameter for prediction of future earthquakes. The catalog used for prediction of future earthquake in this article has considered only event $Mw \geq 7$. But along with these selected big events, a lot of small to moderate earthquakes occur very frequently in the present study area and seismic energy is being released regularly. In recent years with advanced instrumentation and recording, it has been reported that the active tectonics of the region involves much slower form of aseismic release of stored strain energy in this selected study region. This aseismic slow slip may continuously relieve the built-up tectonic stress and may avoid the future big earthquakes in the region. On the other hand, it may also happen that the low strength of rock mass (due to younger rocks of geological origin) may not sustain such high amount of tectonic stress and release it much more than the expected time in the form of great earthquakes. So, the results obtained in this work only give a percentage of probability for a major earthquake to occur in future from the previous earthquake in the used catalog. But there is pretty good and absolute chance that a major earthquake may occur in future or it may occur much earlier than the expected time.

5 Conclusion

In this present study an attempt was made to forecast the occurrence of a major earthquake with magnitude $M_W \geq 7$ in the study area and surrounding region using probability models. A total of 28 earthquakes of magnitude $M_W \geq 7.0$ were listed out for analysis. The forecasting was carried out using three probabilistic models namely Lognormal, Gamma and Weibull model. Two parameter distribution models were considered and the model parameters were estimated by maximum likelihood estimation (MLE) tool. For best fitting of model, the negative log-likelihood (ln L) value was estimated for the three models using the dataset and model parameters. As Gamma model has highest ln L (−52.2598) value, so it is best fit for earthquake occurrence and that of Weibull and Lognormal model. The conditional probability gave the probability of occurrence of any big earthquake in future with respect to the last major event (present time) with different time intervals. Among the three models, overly Weibull model gave highest probability followed by Gamma model and least values are for the Lognormal model. The estimated conditional probability reaches 0.999 after a time interval of 11 years (by year 2026), 12 years (by year 2027) and 15 years (by year 2030) for Weibull, Gamma and lognormal Distribution respectively from base last earthquake of 2015. This means by Weibull model there is 99.9% chance an earthquake will occur in study area by the year 2026 and that of for Gamma and Lognormal model will occur by year 2027 and 2030 respectively. The conditional probability for 10 years elapsed time from last earthquake of 2015 (i.e. from 2025) will reach a value 0.999 by Weibull, Gamma and lognormal Distribution respectively by the year 2029, 2032 and 2040.

References

1. Chingtham P, Yadav RBS, Chopra S, Yadav AK, Gupta AK, Roy PNS (2016) Time-dependent seismicity analysis in the Northwest Himalaya and its adjoining regions. Nat Hazards 80(3):1783–1800
2. Ellsworth WL, Matthews MV, Nadeau RM, Nishenko SP, Reasenberg PA, Simpson RW (1999) A physically-based earthquake recurrence model for estimation of long-term earthquake probabilities. US Geol Surv Open-File Rept 99(522):23
3. Fuloria RC (1996) Geology and hydrocarbon prospects of the Vindhyan sediments in Ganga valley. Memoirs-Geol Soc India. pp 235–256
4. Gansser A (1964) Geology of the Himalayas. Inter-science, New York, p 289
5. Gupta ID (2006) Delineation of probable seismic sources in India and neighbourhood by a comprehensive analysis of seismotectonic characteristics of the region. Soil Dyn Earthq Eng 26(8):766–790
6. Hagiwara Y (1974) Probability of earthquake occurrence as obtained from a Weibull distribution analysis of crustal strain. Tectonophysics 23(3):313–318
7. Hasumi T, Akimoto T, Aizawa Y (2009) The Weibull–log Weibull distribution for interoccurrence times of earthquakes. Physica A 388(4):491–498
8. Kayal JR (2008) Microearthquake seismology and seismotectonics of South Asia. Springer Science and Business Media
9. Lindh A (1983) In: Preliminary assessment of long-term probabilities for large earthquakes along selected fault segments of the san andreas fault system, California. vol 83. No. 63. US Geological Survey
10. Matthews MV, Ellsworth WL, Reasenberg PA (2002) A Brownian model for recurrent earthquakes. Bull Seismol Soc Am 92(6):2233–2250
11. McCalpin JP, Slemmons DB (1998) Statistics of paleoseismic data: final technical report prepared for the US geological survey. contract
12. Nakata T (1972) Geomorphic history and crustal movement of the foot-hills of the Himalayas. Science Report Tohoku Univ 7th series (Geography) 22:39–177
13. Nayak M, Sitharam TG (2019) Estimation and spatial mapping of seismicity parameters in western Himalaya, central Himalaya and Indo-Gangetic plain. J Earth Syst Sci 128(3):45
14. Ni JF, Guzman-Speziale M, Bevis M, Holt WE, Wallace TC, Seager WR (1989) Accretionary tectonics of Burma and the three-dimensional geometry of the Burma subduction zone. Geology 17(1):68–71
15. Nishenko SP, Buland R (1987) A generic recurrence interval distribution for earthquake forecasting. Bull Seismol Soc Am 77(4):1382–1399
16. Ogata Y (1999) Estimating the hazard of rupture using uncertain occurrence times of paleoearthquakes. J Geophys Res: Solid Earth 104(B8):17995–18014
17. Parvez IA, Ram AVADH (1997) Probabilistic assessment of earthquake hazards in the north-east Indian peninsula and Hindukush regions. Pure Appl Geophys 149(4):731–746
18. Parvez IA, Ram AVADH (1999) Probabilistic assessment of earthquake hazards in the Indian subcontinent. Pure Appl Geophys 154(1):23–40
19. Rikitake T (1976) Recurrence of great earthquakes at subduction zones. Tectonophysics 35(4):335–362
20. Sastri VV, Bhandari LL, Raju ATR, Datta AK (1971) Tectonic framework and subsurface stratigraphy of the Ganga basin. Geol Soc India 12(3):222–233
21. Schelling D (1992) The tectonostratigraphy and structure of the eastern Nepal Himalaya. Tectonics 11(5):925–943
22. Shanker D, Papadimitriou EE (2004) Regional time-predictable modeling in the Hindukush–Pamir–Himalayas region. Tectonophysics 390(1–4):129–140
23. Sharma ML, Kumar R (2008) Conditional probabilities of occurrence of moderate earthquakes in Indian region. In: 14th World conference on earthquake engineering, Beijing, China
24. Sil A, Sitharam TG, Haider ST (2015) Probabilistic models for forecasting earthquakes in the northeast region of India. Bull Seismol Soc Am 105(6):2910–2927

25. Sykes LR, Nishenko SP (1984) Probabilities of occurrence of large plate rupturing earthquakes for the San Andreas, San Jacinto, and Imperial faults, California, 1983–2003. J Geophys Res: Solid Earth 89(B7):5905–5927
26. Tripathi JN (2006) Probabilistic assessment of earthquake recurrence in the January 26, 2001 earthquake region of Gujrat India. J Seismol 10(1):119–130
27. Utsu T (1984) Estimation of parameters for recurrence models of earthquakes. Bullet Earthquake Res Instit Univ of Tokyo 59:53–66
28. Valdiya KS (1976) Himalayan transverse faults and folds and their parallelism with subsurface structures of north Indian plains. Tectonophysics 32(3–4):353–386
29. Yadav RBS, Tripathi JN, Rastogi BK, Chopra S (2008) Probabilistic assessment of earthquake hazard in Gujarat and adjoining region of India. Pure Appl Geophys 165(9–10):1813–1833
30. Yadav RBS, Tripathi JN, Rastogi BK, Das MC, Chopra S (2010) Probabilistic assessment of earthquake recurrence in northeast India and adjoining regions. Pure Appl Geophys 167(11):1331–1342